D0429317

Fundamentals of
Spatial Information
Systems

The A.P.I.C. SERIES
General Editors: M. J. R. SHAVE and I. C. WAND

* Out of Print
† Now published in the Computer Science Classics Series

The A.P.I.C. Series
Number 37

Fundamentals of Spatial Information Systems

Robert Laurini
Université Claude Bernard Lyon 1, Lyon, France
and
Derek Thompson
University of Maryland at College Park, College Park, USA

ACADEMIC PRESS
Harcourt Brace & Company, Publishers
London   San Diego   New York
Boston   Sydney   Tokyo   Toronto

ACADEMIC PRESS LIMITED
24–28 Oval Road
London NW1 7DX

United States Edition published by
ACADEMIC PRESS INC.
San Diego, CA 92101

A catalogue record for this book is available from the British Library
ISBN 0-12-438380-7

*To Christiane and Judith
and our families*

Typeset by Columns Design and Production Services Ltd, Reading
and printed in Great Britain by St Edmundsbury Press, Bury St. Edmunds, Suffolk

Contents

PART THREE   CONCEPTUAL MODELLING FOR SPATIAL DATA

PART FOUR   SPATIAL DATA RETRIEVAL AND REASONING

Chapter 13   Algebras: Relational and Peano tuple   479

Preface

This book came about because the physical separation of the two authors by the Atlantic Ocean was a surmountable space in the pursuit of a blending of different ideas, approaches and teachings about spatial information systems. Something of a coincidence in our time–space activity paths led us to co-operate in offering an introductory course on geographic information systems for graduate geography students who knew little about information from a computer science perspective. This kind of audience, representative, we believe, of a much larger group of students and practitioners interested in computer processing of spatial data, has an intuitive feel for space, yet an ignorance of relevant principles or formalisms from mathematics and data processing.

Our combination of perspectives from informatics and geography, in the interest of furthering the understanding of spatial information systems, has led to the writing of a book that emphasizes spatial **semantics** as an organization principle. Our concern is with concepts, principles and ways of organizing; not directly with bits and bytes or the latest algorithm. So we deal with the nature of spatial data and spatial problems. We treat a variety of geometries that are necessary for positioning features in space and representing how they might be related. We present methods of conceptual modelling developed in computer science that provide valuable intellectual aids for sorting out different spatial problems.

Consequently, this book is in the context of **geomatics**, the fusion of ideas from geosciences and informatics. It is about spatial information systems, the assembly of computer hardware and software, data and intellectual capital that, after only a twenty-five year lifetime, have come to be seen by some people as the combined microscope and telescope for better solving a variety of problems in the world.

We see spatial information systems as toolboxes, as resources, and as a sign of a frame of mind. The practitioner, whether in a government agency for global natural resources monitoring, or working in geographic data processing for a public utility company, or drafting in a house basement with a personal computer to create the latest plan for political districts, can look at a software system as a set of tools for undertaking varied tasks. We see a spatial information system as more than this, as a substantial resource, the tools and the data, for problem solving or for academic research or educational purposes. In time it may be that such a resource will change the way in which some academic disciplines are taught in schools and colleges. In any event, we see the tools and resources, in their current form, as valuable aids leading to a better understanding of a dynamic spatial world.

Yet spatial information systems represent, too, a perspective, a way of looking at problems or thinking about the world. It is interesting to note that the commercial geographic information systems software industry has discovered the value of 'seeing relationships based on geography'. We believe that the fields of geography and cartography, even with at least a two thousand year history in describing and measuring properties of the earth, and an appreciation already for the spatial relationships point of view, are being enriched by the recent developments in computer software science and engineering.

We also believe that the development of an understanding of how to use computers for spatial data is improved by a knowledge of disciplines that deal with space, not only geography and cartography, but also others like architecture, cognitive science, geology, geometry, geotechnics, meteorology and surveying. So this book is also oriented to the informaticians interested in developing spatial information system products. Indeed, spatial information system engineers often under-estimate the complexity of spatial data, the semantics of which are much richer than those found usually in conventional computer use domains like office automation, accounting, student records, library archiving, management, and so on. For this group of people, with this document we can introduce not only some tools and models which can be useful in solving practical problems in geomatics, but which can also serve as foundations for the development of a general data theory.

Serving a mixed audience with one book has meant we have had to bridge several worlds. Not only have we had to blend computing and spatial problem solving, but we have aimed to reach both academician and practitioner, both system designer and database manager, and the intuitive and the formal learner. We have had to deal with different, and at times conflicting, uses of language and particular terminology. Not only are there different word uses for essentially the same phenomena depending on the field or discipline, but there are also differences from the point of linguistics. Absent any standard terminology, establishing a common ground for a French informatician and an originally British geographer living in the USA was part of the effort in putting the book together.

Our approach is conceptual, having a strong pedagogic orientation with many examples presented for edification and assimilation. Moreover, the variety of examples also provides a framework for evaluation, and a realization that there can be a solid systematic conceptual organization for many concepts, tools and perspectives. We have concentrated on showing how to organize and use spatial information rather than on how to process or manage spatial data. Just as an understanding of art is facilitated by looking at many pictures, so we think spatial data organization can be better understood by seeing many examples. The variety is inherent in the cultural and natural realms of this Earth; it is also present in the many viewpoints and needs that users of spatial data have. We see this book as helping others to appreciate the varied world of spatial information systems.

We do not know exactly whom 'others' might be; therefore we have not oriented the book to just one group. For example, it is likely that informaticians and software engineers with no background in a spatially oriented science can benefit from the first two parts of the book; for geographers and cartographers much of the first two parts may be a review. For some spatial scientists, Part Three could be the starting point. We have tried to accommodate the reader who may wish to delve only into certain topics by having some repetition of material. Perhaps the practitioners who have different backgrounds and contexts for working with spatial information systems may need to consult only selected chapters. The book has not been written to accompany any particular course (indeed, it has been oriented to people outside as well as in teaching institutions), and we believe it offers an alternative to other introductory books in the field by reason of its conceptual semantic approach, combining information handling and spatial concepts.

In the beginning, we briefly identify components of a spatial information system (Chapter 1, Geomatics), including the reasons for their existence (Chapter 2, Purposes and Types of Spatial Problem). Then we review the nature of spatial data (Chapter 3, Semantics: Objects, Surfaces, Data). In several chapters in Part Two we present details of the foundations for treating and organizing spatial data: different geometries (Chapter 4, Geometries: Position, Representation, Dimensions, and Chapter 5, Topology: Graphs, Areas, Ordering), and structures for organizing and representing (Chapter 6, Tessellations: Regular and Irregular Cells, Hierarchies), and manipulating and processing spatial entities (Chapter 7, on the manipulation and transformation of spatial data and objects, and Chapter 8, on spatial analysis operations for spatial data).

The synthesis in Chapter 8 leads into, in Part Three, Conceptual Modelling for Spatial Data, the presentation of the techniques for conceptual modelling of spatial situations, chiefly the entity-relational modelling approach (Chapter 9, on design methodologies for information systems). Many examples for point and line entities (Chapter 10, Conceptual Modelling of Line-oriented Objects), and for areas and solids (Chapter 11, Conceptual Modelling for Areas and Volumes) are then provided as tastes of spaghetti and pizza, respectively. Chapter 12, Spatial Object Modelling: Views, Integration, Complexities, concludes the third part by a review of the process used in synthesizing different models, and then raises some issues to be dealt with in the fourth part of this book.

Part Four emphasizes techniques and principles for accessing, retrieving and using information from databases containing a variety of spatial entities. The basis of relational algebra is presented (Chapter 13, Algebras: Relational and Peano Tuple), followed by examples of queries (Chapter 14, Spatial Queries). After a discussion in Chapter 15 of the challenging topics of spatial indexing and integrity constraints, the book then moves on to discussions of currently developing fields, at first looking at multimedia and hypermedia concepts (Chapter 16, Multimedia Spatial Information Systems and Hypermaps), and then presents some ideas about intelligent spatial information systems (Chapter 17). These last two chapters (perhaps these are the dessert!) establish how the stage is changing, pointing the way to the anticipated developments of the next few years.

While we have presented this book in traditional form, we wish we could have had the time and resources to prepare it as a hyperdocument. We should be embarrassed about this – Chapter 16 talks about hypermaps, and there already are several 'Hypermap' products in the marketplace. But addressing the medium as part of the message will have to wait until next (?) year.

Robert Laurini
Villeurbanne, France

Derek Thompson
Greenbelt, Maryland, USA

March 1991

Acknowledgements

Many people have helped with the preparation of this book. Large in number, students and professionals in several places have sat through lectures and less formal presentations of much of the material contained in the book, albeit under a different label. We appreciate their interest, their suggestions, and their enthusiasm about spatial information systems.

We especially wish to acknowledge the intellectual contributions made by faculty colleagues, partly as collaborators on projects, as co-authors of papers or presentations, or generally via discussions: in Lyon, Françoise Milleret-Raffort and Jean Marie Pinon both with the Computing Department of Institut National des Sciences Appliquées (INSA-Lyon); in College Park, PohChin Lai, currently associated with the Department of Geodetic Science at the Ohio State University, was an enthusiastic computer-oriented colleague in the Department of Geography at the University of Maryland. In particular, we thank these three colleagues for being willing to spend hours reading earlier drafts of the manuscript.

A few other colleagues have helped in particular ways. Fred Broome of the US Census Bureau, Washington, DC, and John Townshend, Chairman of the Geography Department of the University of Maryland, commented on an earlier draft of this work; Lisa Wolfisch Nyman created a large number of figures (with electronic aid, of course), and Guinn Cooper produced photographic versions for some figures, and executed the cover illustration. The latter was created by the cartographer, Joseph Wredel, a professor of geography at the University of Maryland, College Park, Maryland, USA. We appreciate his assistance. We are indebted to the City of Padova, Italy for the air photograph used in that illustration.

Derek Thompson also wishes to acknowledge the extensive hospitality provided by various institutions which aided his efforts to learn directly about spatial information systems teaching and practice in Western Europe: The Department of Geography, University of Edinburgh, Scotland, UK; the Laboratory for Urban Studies of the Université de Paris–X at Nanterre, and L'Ecole des Hautes Etudes en Sciences Sociales, both in Paris, France; and the Instituto de Economía y

Geografía Aplicadas, Madrid, Spain. He is especially grateful for the financial support from the Centre National de la Recherche Scientifique for his three-month visit to France, and to the Consejo Superior de Investigaciones Científicas of the Ministerio de Educación y Ciencia for his three-month sojourn in Spain.

While most figures have been created by the authors specially for this book, some illustrations appear by the grace of others. Therefore we acknowledge the following individuals and organizations for granting permission to use their material: the Association of American Geographers, Washington, DC, USA for Figure 4.31; the Cartographic Laboratory of the University of Wisconsin at Madison, USA for Figure 3.24; the Bureau of the Census of the United States Department of Commerce (and to Fred Broome, Robert Marx and Alan Saalfeld) for Figures 5.8, 5.16, 8.14 and 12.14; the Washington Post Company, Washington, DC, USA for Figure 1.7; the Environmental Systems Research Institute, Inc., Redlands, California, USA for Figures 1.9, 1.10, 2.14, 8.15 and 9.24; The Maryland Office of Planning, Baltimore, Maryland, USA, for Figure 6.6; The National Center for Geographic Information and Analysis, Santa Barbara, California, USA, for Figure 4.10; the Maryland Department of Natural Resources, Annapolis, Maryland, USA, and Dr. PohChin Lai, Ohio State University, Columbus, Ohio, USA, for unpublished material used as Figures 1.6, and 2.15; The Cambridge University Press and Dr. Mahes Visvalingham, The University of Hull, Hull, England, UK, for Figures 5.20 and 5.24; to Mr. Jan Van Est of IRIS International, Voorburg, The Netherlands, for Figure 4.6; Dr. Marc Armstrong, the University of Iowa, Iowa City, USA, and Taylor and Francis, Ltd., London, UK for Figure 12.7; Dr. John Herring, The Intergraph Corporation, Huntsville, Alabama, USA, and the American Society for Photogrammetry and Remote Sensing, Bethesda, Maryland, USA, for Figures 8.16 and 12.12; Dr. Ramez Elmasri and the Bejamin/Cummings Publishing Company Inc., Redwood City, California, USA, for Figure 13.2, and Dr. Peter Taylor, Newcastle University, Newcastle-upon-Tyne, UK, for Figure 7.20.

Several illustrations are loosely based on the work of others; in this case we rely on source citations within figure captions as our acknowledgement. We also point out that numerous figures in this book have appeared earlier in the collection of materials prepared for the Seminar on Multimedia Urban Information Systems presented at Lyon, France, November 13–17, 1989, and distributed as A Primer on Multimedia Urban Information Systems Concepts (edited by Robert Laurini and Françoise Milleret-Raffort), through the Urban and Regional

Spatial Analysis Network for Education and Training (URSA-NET), Computers in Planning Series, Athens, Greece: Figures 1.13, 9.6, 10.6, 10.8, 10.16, 10.17, 10.20, 10.22, 11.3, 11.4, 11.11b, 12.9, 12.10, 13.15, 14.2, 14.3, 14.7, 14.11, 14.12, 15.13, 15.14, 16.7, 16.8, 16.11, 16.12, 16.13, 16.15, 16.17, 16.19, and 16.20. Additional figures appeared in the monograph for a later seminar held at Patras, Greece, June, 1990: Figures 17.2, 17.3, 17.4, 17.5, 17.6, 17.7, 17.8, 17.9, 17.25, 17.26, and 17.27. These seminars were delivered under the auspices of the European Community Programme on Cooperation between Universities and Industry (COMETT), and organized by Nicos Polydorides of the University of Patras, Greece.

Throughout this book we refer to several commercial products and official government products. Our mention does not constitute an endorsement, or, indeed, the opposite, a refutation; we refer to the items as illustrations of different approaches or concepts. Even though some details mentioned may soon be out of date, we are not concerned with implementation specifics of the software; we offer the examples so that readers can relate somewhat more easily to the practical world. We believe that the progress made by many software, and other, companies in the past decade is one of the exciting aspects of spatial information systems.

APIC is a trademark for the software of the company APIC Systèmes, a subsidiary of Lyonnaise des Eaux, Paris, France.

ARC/INFO is a registered trademark, and *ARC/INFO NETWORK* and *ARC/INFO TIN* are trademarks of the Environmental Systems Research Institute, Inc., Redlands, California, USA. ESRI is the registered company name.

DLG is the abbreviation for the official name of the Digital Line Graph data of the US Geological Survey, Reston, Virginia, USA.

Domesday Disks are a resource produced by the British Broadcasting Corporation, London, UK.

GBF/DIME is the acronym for the official name of the Geographic Base File/Dual Independent Map Encoding computer system and database of the US Census Bureau, Washington, DC, USA.

GEMSTONE is the name of software marketed by Serviologic, Oregon, USA.

GEO/SQL is a trademark for software of Generation 5 Technology, Inc., Westminster, Colorado, USA.

GEOVIEW is the name of a database design created at the Department of Geography, University of Edinburgh, UK.

GIMMS is a trademark for software of GIMMS Ltd., Edinburgh, Scotland, UK.

GIRAS is the official name of the Geographic Information Resources Analysis System of the US Geological Survey, Reston, Virginia.

GRASS is the official name for the Geographic Resources Analysis Support System created by the US Army Construction Engineering Research Laboratory, Champaign, Illinois, USA.

GTV is a trademark for the videodisk resource produced by the National Geographic Society, Washington, DC, USA.

HBDS is a data structure devised by François Bouillé, Université de Paris, France.

HYPERTIES is a trademark of Cognetics Corporation, Princeton Junction, New Jersey, USA.

HYPERCARD is a registered trademark of Apple Computer Inc., Cupertino, California, USA.

IDRISI is the official name of software developed by Ron Eastman, at the Department of Geography, Clark University, Worcester, Massachusetts, USA.

INGRES is the trademark of a database system product line of the Ingres Corporation (now a division of ASK Computer Systems, Inc., Alameda, California, USA.

MAGI is the official name of the Maryland Automated Geographic Information System, of the Office of Planning, Baltimore, Maryland, USA.

MAP is the name for the software and map algebra, Map Analysis System, developed at Harvard University, Cambridge, Massachusetts, USA by Dana Tomlin.

O_2 is a trademark of O_2 Technologies, Paris, France.

ODYSSEY is the name of software developed by the Laboratory for Computer Graphics, Harvard University, Cambridge, Massachusetts, USA.

ORACLE is a trademark of a database system product line of the Oracle Corporation, Belmont, California, USA.

ORION is a product marketed by MMC, Austin, Texas, USA.

POSTGRES is the name for a relational database base extension concept developed by Dr Michael Stonebraker, University of California, Berkeley, California, USA.

QUILT is the name for an experimental quadtree geographic information system developed at the University of Maryland, College Park, Maryland, USA by Hanan Samet.

SALADIN is a trademark of the Institute for Urban and Regional Information Systems, Inc., Voorburg, The Netherlands.

SICAD is a trademark of Siemens AG, Munich, Germany.

SMALLWORLD is a trademark of the Smallworld Systems Limited, Cambridge, England, UK.

SPANS is a trademark of Tydac Technologies, Inc., Ottawa, Ontario, Canada.

SPOT is a trademark for the SPOT Image Société, Toulouse, France.

SQL is the name for a standard computer query language adapted by the National Bureau of Standards, USA.

SYSTEM 9 is a trademark of Prime Wild GIS, Inc., Natick, Massachusetts, USA.

SYMAP is a software product created by the Laboratory for Computer Graphics, Harvard University, Cambridge, Massachusetts, USA.

TIGER is an acronym for the Topologically Integrated Geographic Encoding and Referencing System of the US Bureau of the Census, Washington, DC, USA.

TIGRIS is a trademark of the Intergraph Corporation, Huntsville, Alabama, USA. Intergraph is a registered company name.

TRANSCAD is a trademark of Caliper Corporation, Newton, Massachusetts, USA.

TRIPTIK is a trademark of the American Automobile Association, Washington, DC.

xGEM2 is an official name of a software system developed at the CSIRO, Canberra, Australia.

World Data Bank is the official name for two cartographic databases developed by the Central Intelligence Agency, Washington, DC.

Part One

Introduction to the Spatial Context

1
Geomatics

Introduction to spatial information systems

Today in Redlands, California, a group of people is building a digital data equivalent of the world land map represented on almost three hundred sheets of aeronautical charts at a scale of approximately one-to-one million, showing cultural and physical features of inhabited continents. In Bracknell, England, the Meteorological Office uses the world's fastest computer to solve each minute thousands of equations applied to data for 345,000 sample points in the atmosphere to provide global macroscale six-day weather forecasts. In Beijing, China, the National Geography Institute uses data from satellites in space to map the distribution of land resources. In Lyon, France, informaticians have designed computer programs to simulate landscapes with buildings yet to be built. And in England a few years ago, schoolchildren, after walking around their neighbourhoods marking paper maps to show the location of grocery stores, land use, and other features of the landscape, sent their data to the British Broadcasting Corporation for incorporation in a computer-based visual portrait of Britain.

Our conceptual view of spatial information systems begins with three chapters emphasizing the comprehensive nature of spatial data use and organization. We start with an introduction to the general context of the needs for and nature of automated information systems for aiding the solution of practical and research problems in many fields. We move from the stage-setting first chapter to a more detailed discussion of functions and component parts in Chapter 2. In the third chapter we provide some details of the concepts and practicalities involved in working in the domain of spatial concepts.

1.1 SPATIAL DATA ORGANIZATION

As we move towards the end of the twentieth century, we are positioned at a time of exciting developments in the ability to obtain data about the physical and cultural worlds, and to use those data to do research or to solve practical problems. Digital and analog electronic devices facilitate the inventory of resources, the display of futuristic scenarios and the rapid execution of many arithmetic or logical operations.

Contemporary computer-based tools for working with data for phenomena on, above or below the earth's surface (known as **spatial information systems**) are numerous; they can utilize data of many forms and are still undergoing much improvement. In a period of increasing uses of digital computers, we still use photographic film cameras, filing cabinets, paper maps and mathematical tools developed centuries ago. But we are now able much more rapidly to create, manipulate, store and use spatial data by means of information systems. **Spatial** is a term used here to refer to located data, for objects positioned in any space, not just geographical, a term we will use for the world space.

An **information system**, a collective of data and tools for working with those data, contains data in analog form, for example, handwritten notes or photographic slides; or digital form, for example, by computer binary encoding, about the phenomena in the real world. Our perceptions of the world, involving cognitive processes like selection, generalization and synthesis, give us information which can be shared among and used by people other than the creators (Figure 1.1). The physical representations of this information, that is, the data, constitute a model of those phenomena. So the collection of data, the databank or **database**, is a physical repository of varied views of the real world representing our knowledge at one point in time.

In the database context the terms data, information and knowledge are differentiated. **Information** is derived from the individual data elements in a database – information that is not directly apparent. In a sense, the information is produced from data by use of our thought processes, intuition or whatever, based on our knowledge. For example, a bicycle can be represented by circles for the two wheels, various lines for the frame, wheel spokes, and so on. The geometric entity of the bicycle is the assembly of the data elements; human **knowledge** is needed to deduce the reality of a bicycle as a means of transportation and the steps necessary for the physical object to function.

In a less complex case, knowledge about world latitude and longitude can allow us to make sense of some numbers in a database. An x or y coordinate is a data element. Extra data are needed to make sense of the magnitude of those numbers. We know that high positive values for x must refer to a place near the North Pole, provided that we have information about the scaling of the axes and how they fit the real world. Having the 'extra data' allows us to produce the information from the data. Interpretation of discrete, apparently unrelated, pieces of data is akin to having 'added value' as we progress from data, to information, to knowledge.

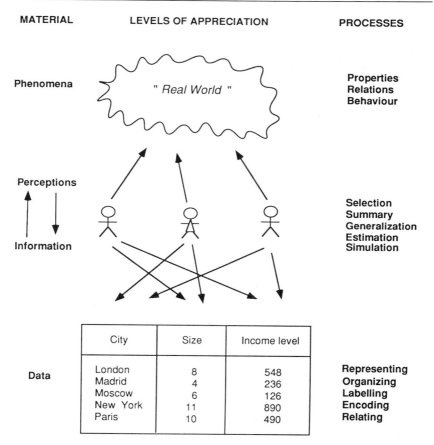

Figure 1.1 Different interests about the 'Real World'.

The material in this mélange of 'data' has many conceptual origins and forms (Figure 1.2). The data may be any of the following:

1. Real, for example the terrain conditions or buildings.
2. Captured, that is, recorded by physical devices like electronic sensors, and film cameras; for example, seismic signals or landscape pictures.
3. Interpreted, that is, involving some human intervention as in field sketches of landscapes, a questionnaire, or writing in books.
4. Encoded, as in paper maps, digital data for depths of oceans or summary statistics for median income levels.
5. Structured or organized in some way, such as tables in census reports or data in geographic information systems.

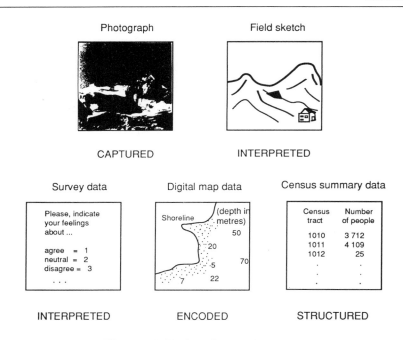

Figure 1.2 Various forms of data.

It is the structuring for this variety of information in an environment of digital computers that is the subject of this book. As you might expect, just as there are many ways to organize furniture in a house, so there are many ways systematically to arrange information about the world. Various commercial software packages are available to aid this structuring; there are some dominant forms or models for working with spatial data, and there is a growing body of fundamental principles and formalisms to guide us in the process of information and data organization. By principles we mean a conceptual framework for information organization; and by formalisms we refer to mathematically based representations and organization of spatial data.

One common organization (Figure 1.3) is that of layers of sets of maps. Each layer, representing a thematic approach to a particular purpose or set of needs, may contain one or several different kinds of information. For example, for studies of natural resources, bedrock geology, subsurface conditions, land use, soil types, drainage, elevation above sea level, slope and aspect, and transportation features, may constitute the data themes. For urban planning, an information set may contain data for streets, utility lines, transportation features, tax and zoning conditions, landowners, and improvements to property. This layered data set approach is associated with photogrammetry, computer-assisted design, and map overlay modelling using paper maps. The layer view supposes that we are observing something everywhere. However, the layers may not necessarily be, or only be, themes; they can be different elevations, as for the floorspace use type for different stories of buildings (Figure 1.3c); or they may be different points or intervals of time for a given theme, for example, data from different censuses (Figure 1.3b).

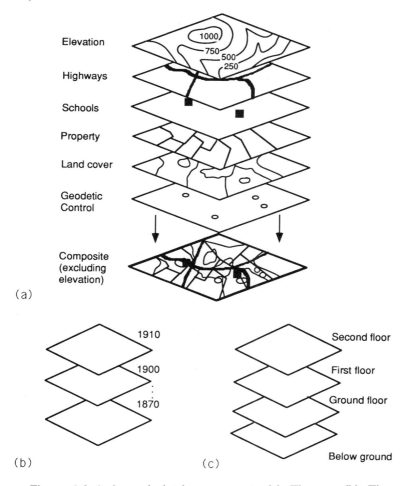

Figure 1.3 A layered database concept. (a) Themes. (b) Time periods. (c) Vertical slices.

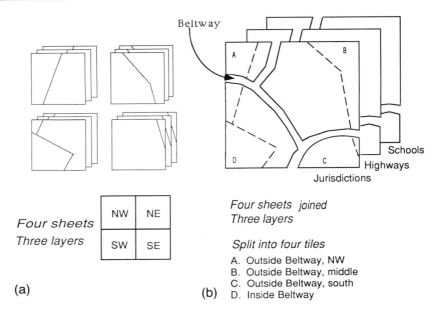

Beltway

Schools
Highways
Jurisdictions

NW | NE
SW | SE

Four sheets
Three layers

Four sheets joined
Three layers

Split into four tiles

A. Outside Beltway, NW
B. Outside Beltway, middle
C. Outside Beltway, south
D. Inside Beltway

(a)

(b)

Figure 1.4 Layers and tiles for map data. (a) Map sheets and thematic layers. (b) Tiles and thematic layers.

A structuring into themes and pieces of space (Figure 1.4), perhaps counties or map sheets or special units called tiles, is very common in analog information systems as represented by photographic film positive plates for different categories of information gathered by the British Ordnance Survey, or the plates for the 55,000 topographic quadrangles at a scale of 1:24,000 of the United States Geological Survey. This organization is also common for automated spatial information systems, reflecting the fact that these commercial products were first developed as digital equivalents to the paper map collections.

Another arrangement (Figure 1.5) works with a single geographic area containing a variety of objects, some, or all, of which may be of interest at any one time. These entities are imagined to be in one layer only, thereby allowing the third space dimension to represent the world vertical variations as well as providing a latitude/longitude reference frame. This **object orientation** is associated with human cognition and reasoning, but is not so good for dealing with surfaces of real terrain, or gradients of the human condition like spatial variations in personal income.

For all of these general types of organization, there is a more detailed

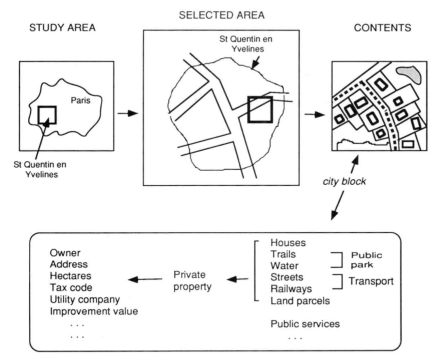

Figure 1.5 The object-based form of organization.

level concerning the **physical representation** of the real world phenomena. For example, we can imagine tables of numbers for all the roads in a province, or we can think of the roads expressed as a set of lines with associated characteristics represented symbolically (for width, number of lanes, amount of traffic, etc.). In addition, the physical structuring may not match with a conceptual layering, a concept very useful for user comfort. The details of forms of systematic representation are as much a part of this book as the general conceptual structuring into layers or entities. We treat the foundations for this topic extensively in Part Two.

Independent of the world of automated mapping, users of spatial data have long worked in a particular mode: the world of the paper map. Such a method of recording and using data for the earth, which we will call the **map model**, is one of the steps along the way to the modern-day digital environment, and is seen in the replication of manual procedures, use of similar terminology, and frame of mind associated with some computer-based spatial data processing, as is demonstrated in subsequent chapters.

1.2 HETEROGENEITY OF USES OF SPATIAL INFORMATION SYSTEMS

Because the different domains of use of automated spatial information systems have particular needs, it appears at a general level that there are many, perhaps too many, commercial software systems or theoretical models to establish principles or order. The language confusion that can, and does, arise from the use of similar tools in different fields, in part a reflection of the development of the contemporary information systems from different roots over a twenty-five year history, adds to the difficulties in understanding, evaluating, and discussing these automated resources.

1.2.1 Uses of spatial information systems

Today, looking at the worlds of business and government, we see spatial information systems applications, activities and tools for dealing with spatial data with names such as:

> Land information systems
> Automated mapping and facilities management systems
> Computer-aided design
> Thematic mapping software
> Marine cartography
> Remote sensing systems
> Surface modelling
> Environmental modelling
> Resource management
> Transportation planning
> Emergency response
> Geomarketing
> Geotechnics
> Archaeology
> Military exercises

Written in no particular order, this deliberately mixed bag of terms conveys descriptive characteristics of: type of subject matter (e.g. land), purpose or activity (e.g. city planning), and instrumentation (e.g. remote sensing).

In the large realm of practical uses of spatial information systems, the general purposes include: inventory, a data and measurement orientation, as for land resources (Figure 1.6); communication or transportation

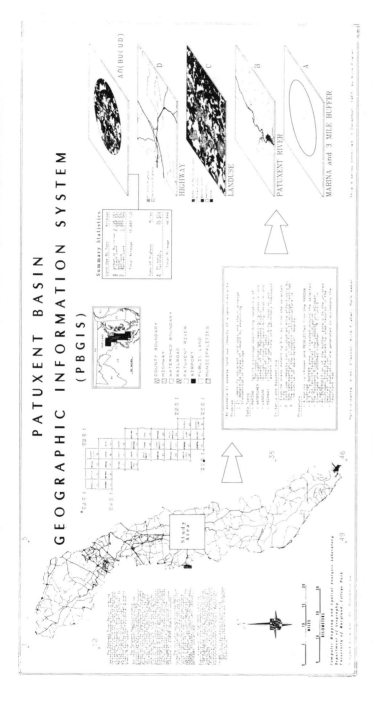

Figure 1.6 Land resources management. This figure, a much reduced photograph of a large coloured poster, illustrates several major components of geographic information systems in widespread use for resources inventory, management, and analysis. Created via the ARC/INFO software, it shows the concepts of map sheets as tiles, a single study area, thematic layers, map layer combinations using Boolean algebra, map display capabilities like insets and north arrows, legends and annotations, and the creation of reports for statistical data. (The display was created by, and used with the permission of Professor PohChin Lai, Ohio State University, Columbus, Ohio, and with the permission of the Maryland Department of Natural Resources, Annapolis, Maryland, USA.)

utilities infrastructures (Figure 1.7); and administrative record-keeping, as for property-based taxation. In these domains, typically large quantities of real world objects are mapped at large scales. Whether or not analytical or querying procedures are used, spatial information systems are often employed as support resources for decision making, as in bus route planning or determination of market areas. There is also an increasing use of such systems for educational purposes, either in schools or at the individual's home.

Knowledge of particular requirements is important for appropriate and effective design of the entire system of tools for processing data for earth phenomena. For example, the layered style, also well demonstrated by Figure 1.6, reflects the manual procedures used in composite mapping and map modelling of the late 1960s and early 1970s. This design type may not be the most efficient for all needs. We illustrate this topic of information use tasks at this time by only two examples, simply to establish the general content and orientation for this book. More extensive treatment follows in Chapters 2 and 3.

1.2.2 Examples of data requirements

Way-finding is a basic human need. Knowing how to navigate through a city street system, how to find the best route from home to the park, or giving directions to strangers who show up lost in one's neighbourhood, are a few varieties on this basic theme, assuming that persons are involved (for they do not have to be, as seen in the current developments of autonomous land vehicles). If we imagine, though, way-finding in a car, we can imagine that we need the following:

1. To know the current location of the vehicle.
2. A process to find a destination.
3. The retrieval of a map to help in the location and way-finding task.
4. The presentation of information to the navigator.

Figure 1.7 Utilities networks—an artist's view of subterranean Washington. Public utility companies worldwide are beginning to invest huge sums of money to create computer databases for pipes, poles, and pathways, cables, and conduits, sewers, valves, and tracks, etc., below, on, and above ground. (This caricature appeared in The Washington Post newspaper, November 11, 1988, under the caption 'What is under the streets of Washington?' The picture is used with the permission of The Washington Post, Inc., Washington, DC.)

1. Electrical system cables, transformers, vaults

2. Cables of telephone wires

3. Fibre optic communications cables in conduits

4. Natural gas pipes and access manholes

5. Water supply pipes and manholes

6. Hot and chilled water pipes

7. Sewer pipes

8. Storm drains

9. Conveyor belt for books for the US Congress

10. Pedestrian walkways and subway for the US Congress

11. Tunnels carrying the water of Tiber Creek

12. The METRO rail subway system, stations, escalators, etc.

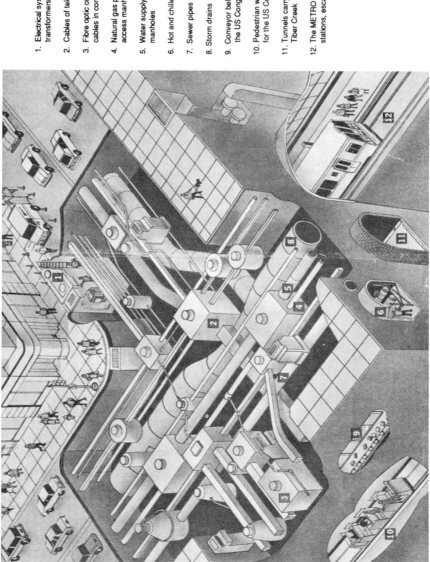

Assuming that national legislation allows vehicles to carry visual display devices to aid in this task, then imagine a small television display screen at the front of the vehicle. The monitor shows a detailed map of part of the city, with a blinking cursor identifying the vehicle location (Figure 1.8). The screen also has a small-scale map inset for a much larger area. A destination address or even place name is entered by some means, and the on-board computer determines a good route to get to that destination. Assuming that national laws allow drivers to activate video displays while vehicles move, as the vehicle moves along streets, the map changes, showing only features necessary to help in the driving task, for example by signalling when a turn should be made, and identifying temporary barriers like road works.

Rather than consider the electronic devices needed – and they all exist today – let us focus on the data requirements (White, 1987). We need coordinate information to draw detailed maps revealing the shapes of the roads along which a journey might be made. We need to know street addresses, names of locations, possibly important features that can be used in a landmark oriented guidance system, and street names (recognizing that in some countries one street can have many names). Some ancillary data, like the existence of traffic works, might be helpful in choosing a route, or knowing the average travel times. Also, it will be important to know if and how particular streets are connected, in order to indicate the sequence from origin to destination or for determining how

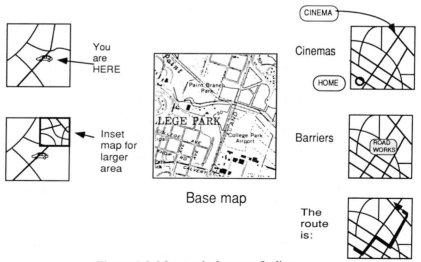

Figure 1.8 Map tools for way-finding.

and where to make a turn. Some filtering tools will be necessary for production of the right kinds of map for assisting the navigator, for not all details are needed for following a particular route. Many data items, appropriately structured, are thus needed for effective and efficient (timely responses to the user) vehicular based way-finding.

Consider, next, a research problem: analysing the impact of the pattern of land uses of a landscape on the amount of chemical nutrients that are discharged into streams (Figure 1.9). Single event or long-term patterns of precipitation create conditions of oversurface wash, channelization, seepage and underground flow. The amount and speed of exodus into the river or stream channels is affected by many factors including slope angle, soil permeability, underground hydrology and type of vegetation (for example, forested or not). It is therefore important to have information about relative elevations, surface gradients and aspects, and the sequence of land uses along the river courses. It may also be necessary to know what is under the surface, in case some underground drainage causes flows to bypass the study area. To deal effectively with the analysis may necessitate certain kinds of data structuring; not simply the acquisition of information.

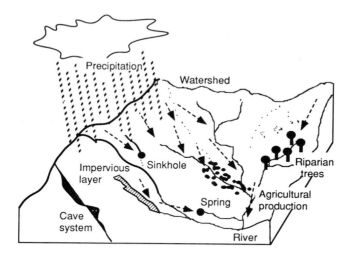

Figure 1.9 A view of overland flows. A simplified diagram showing surface and subsurface conditions affecting overland flow of water and other matter. (Based on a diagram originally in Haber and Schaller, cited in Chapter 2, and published in ARC News, Volume 11, Number 1, 1989, by the Environmental Systems Research Institute, Inc., and used, with modifications, with the permission of that company.)

The common elements in the many domains of the use of spatial information systems, as demonstrated by the two more detailed examples are, however, not extensive. Whether we are navigating through city streets or following chemical laden water over a landscape; or whether we are inventorying city telephone poles, or counting forest fires; or designing the layout of EuroDisneyworld near Paris or forecasting the weather for the Commonwealth Games in Auckland, New Zealand; or planning the extension of a sewer system underneath the streets of Moscow, Russia, we deal in a general sense with just the following few features like:

1. Phenomena that vary in character from place to place.
2. Natural features with unclear boundaries or no boundaries at all.
3. Person-made phenomena with clear limits.
4. Phenomena located in space, either geographic (earth) or arbitrary.
5. Entities that are related or unrelated to each other by location.

In other words, we deal with properties of spatial entities positioned in a spatial reference system.

1.3 SOME COMPONENTS OF SPATIAL INFORMATION SYSTEMS

An automated spatial information system is a toolbox for representing views of the real world via data about locations (Figure 1.10). Spatial information systems are a technology for processing spatial data. The tools, which may be activated by pushing a button or typing a command, represent processing functions or operations for example, drawing a map, or measuring the distance from Paris to Madrid. The tools work on some or all of the information stored in some systematic way in a database. The information system requires data, software, hardware, 'brainware' and other resources, and exists within some institutional setting as a **resource** to solve problems.

1.3.1 The toolbox view

The support for decision making, practical or academic, may consist of virtually instantaneous answers to questions, a **query** type of system. For example, the best route for crossing a city is displayed in map form for the vehicle navigator, or the urban geographer can quickly find which cities in Europe are less than 100 kilometres from any national frontier.

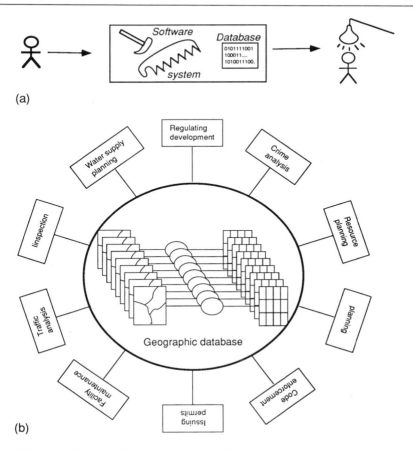

Figure 1.10 A toolbox approach. (a) The toolbox concept. (b)
Different user needs for data in a shared database. (Figure 1.10b is a
slightly modified version of an original diagram of the Environmental
Systems Research Institute, Inc., Redlands, California. The diagram
is used with the permission of that company.)

Or the information system may be oriented to producing information,
perhaps in the form of maps or tables, for subsequent study. **Products**
may be detailed maps for public briefing purposes in a re-zoning case, or
may be the complete set of possible routes through a city street system.
Whatever the medium, the output is designed for further study, is
generally extensive and is not designed for immediate consumption. In
contrast, the query mode provides quick responses to highly focused
questions usually in the form of small amounts of information.

In providing these capabilities to the user, the spatial information systems must, at a general level of detail, fulfil the following:

1. Provide tools for the creation of digital representations of the spatial phenomena, that is, data acquisition and encoding.
2. Handle and secure these encodings efficiently, by providing tools for editing, updating, managing and storing; for reorganization or conversion of data from one form to another, and for verifying and validating those data.
3. Foster the easy development of additional insight into theoretical or applied problems, by providing tools for information browsing, querying, summarizing and the like: that is to say, facilities for analysis, simulation and synthesis.
4. Assist the task of spatial reasoning, by providing for efficient retrieval of data for complex queries.
5. Create people compatible output in varied forms of printed table, plotted map, picture, scientific graph and the like.

In a sense, the spatial information system lies within a larger framework of information requirements of the human world (Figure 1.11). Just as the plumbing system in a house has separate but functionally related water, odour and waste subsystems monitored by the householder with the aid of a set of plans (or the map in the plumber's head), so there are physical, informational and guidance systems for phenomena at large. In a spatial information system context, there is the component of observable phenomena, the component of 'knowledge' about those phenomena, and the component of guidance (the planner,

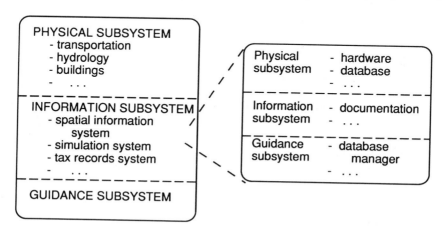

Figure 1.11 A conceptual structure for spatial information systems.

forester, politician). Then the information subsystem itself has a physical component (the hardware and data encodings), the documentation component describing the characteristics of the spatial information system, and the guidance component (the person who looks after the database and tools, and solves malfunctions).

Consequently, beyond a general level of the domain of the use of spatial information systems, we can begin to appreciate differences in terms of

General purposes
Themes of information
Types of data
Particular tools
Specific processing requirements
Specific forms of data organization

It is the goal of this book to provide the foundation necessary to understand these components.

1.3.2 The physical components

A spatial information system is a software product which has several components and makes connections with other devices in its environment (Figure 1.12). In structure it includes a database management system (DBMS) for storing and managing data, linked with a graphics management system for cartographic or other visual displays. These two software subsystems are connected in one direction to the computer operating system, and in the other direction, through graphic work-stations, to users by means of an interface and a command language interpreter.

The main devices used in a spatial information system are disks for storing various data (alphanumerical, graphic and image), digitizing tablets and scanners to enter graphic data, and plotters and printers to present the results. Similarly, as for all computer systems, it is valuable to have the spatial information system computer linked to a communication network allowing the exchange of data with other people or companies providing or working with geographic or other spatial data.

1.4 THE ROLE OF AUTOMATION: GEOMATICS

'Seeing relationships based on geography' is a selling point today for

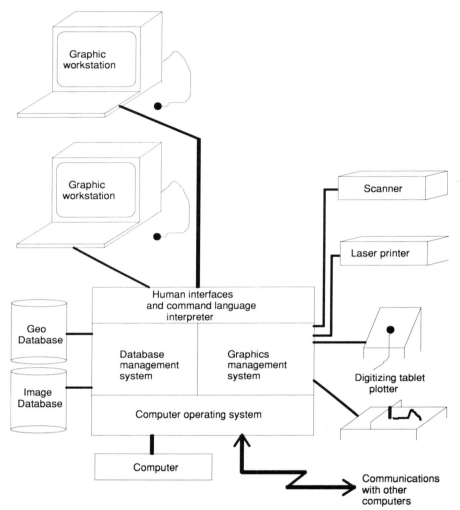

Figure 1.12 The physical elements of a spatial information system.

many practically oriented fields. Perhaps 80 per cent of decisions by state and local governments involve a spatial component either directly or by implication. Geographers are used to thinking spatially, and geography has the same common intellectual root as geometry. Architects, either building or landscape, work with spaces. So our search for organization of material about spatial information systems has to look to many disciplines and fields of study. These are:

1. Disciplines that have developed concepts for dealing with space: cognitive science, geography, linguistics, psychology.
2. Fields that develop practical tools and instruments for obtaining or working with spatial data: cartography, geodetic science, photogrammetry, remote sensing, surveying engineering.
3. Disciplines that provide formalisms and theories fundamental to our working with space and automation: computer science, geometry, informatics, artificial intelligence, semiology, statistics.
4. Fields making substantial use of automated spatial information systems: archaeology, civil engineering, forestry, geotechnics, landscape architecture, urban and regional planning.
5. Fields that provide guidance about information: law, economics.

The strength and intellectual interest of the spatial information systems subject derives from the cross-fertilization among these many fields. While some writers use the term geographic information systems to refer to the composite field of study, we prefer the term **geomatics** as an umbrella covering all the fields listed above that are today important for understanding and further developing spatial information systems.

Ideas and tools from many fields are, then, necessary for a full understanding of spatial information systems. As user requirements and expectations increase for such resources and toolboxes, so there are intellectual pressures on the theoreticians and practitioners of geomatics to cooperate in interdisciplinary research studies and development projects to further improve our automation tools. For example, in the USA the National Science Foundation is providing about $6 million over a five year period for a group of scholars to undertake a thirteen-item research plan. This plan covers much theoretical research into topics like spatial statistics, methods of spatial analysis, database structures, visualization of spatial data and results of empirical research involving spatial entities, the use and value of geographic information in decision making, modelling time alongside space, the integration of remotely sensed data into geographic information systems, and understanding human spatial cognition. A nationally funded research programme is also active in Regional Research Laboratories in several British universities, in a major university in The Netherlands, and elsewhere in the world.

Notwithstanding the impressive and comprehensive statement of research tasks indicating, according to many scholars, that we know little about spatial information systems, we have attempted to organize much of what we do know in a systematic fashion. Our focus is on organizing principles and instruments that help us to understand, evaluate and design spatial information systems – in a sense, a course in appreciation based on many examples, much like an understanding of fine art comes about by looking at and studying many pictures.

We may accept automation as the way of the last decade of the twentieth century, but we must also have a view of non-automated spatial information systems, for most of the information about and guidance of the world's physical system are not yet within a computerized environment. While the concepts of data structuring, concepts of spatial relationships, and matching tools of drawing and tabulating existing in the non-automated world have been taken to the automated world, it has not been without cost. So we presume, on the one hand, that the spatial information system in a computerized environment allows some problems to be handled more efficiently and cost effectively. On the other hand, we expect to be able to undertake problems otherwise considered insoluble. Automation allows faster creation and use of large quantities of information, speeds up mathematical or logical operations on data, and makes possible tasks that are otherwise unmanageable.

The **spatial information system**, then, is a computerized environment whereby utility programs performing specific functions are used in an integrated environment, in which the user is shielded from the details of computer processing, to achieve some goal of research, education or decision making. The inherent form of spatial data representation and organization must be designed to support effectively and efficiently the kinds of query and analysis required by many users. The performance of computer systems is a reflection of hardware technology and software engineering, but also reflects the data structures and the quality of algorithms.

We approach the question of design of spatial information systems at the conceptual level. Elaborating on Figure 1.1 as Figure 1.13, we present a process (taken up further in Chapter 9) demarcated for practical convenience into four stages, proceeding from users' views of their needs to the computer's physical encoding at the level of hardware (known in computer jargon as the byte level). Using the terminology from the field of informatics, the real world is seen from an external view, representing the needs, objectives and process of extraction on the part of different people or organizations. Each such subset of the real world can, referring to the information system toolbox, be thought of as an application.

Conceptual modelling, which we treat extensively in Part Three, covers the synthesis of the external models. It is here that we can think of common ground among users, in part the universe of discourse, in part the universe of modelled entities. Using tools and formalisms discussed later in this book, the phenomena of interest are identified, their pertinent characteristics described, and how objects relate to each other is

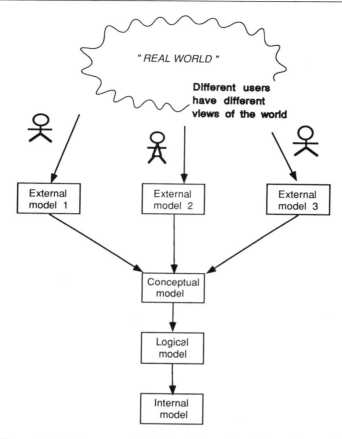

Figure 1.13 A framework for the design of an information system.

mapped out as well as possible. Different views are integrated via a common language and structure. The **logical modelling** stage translates the conceptual organization into something more practical, perhaps simply thought of as putting numerical values into tables of data, but avoiding the details of storage of data on physical media. It is at this level that phenomena are organized into a database as tables of data records and connections to other tables. The internal modelling, which we do not address, treats the organization of data on hardware storage media.

As automation for spatial data handling has proceeded from making graphs and then maps, to automated mapping and thematic cartography, to tools for inventory and management, and then to database resources for decision making or education, so it has become more important to think about the value of the information in the database. The contents, certainly reasonably viewed as a commodity, are presumably valuable for one or more purposes, yet such data alone are not enough. Users provide the added value by interpretation and analysis, and by understanding the nature and quality of the data. It is important to focus on the message, the contents of the envelope, not the messenger, the postal system. Consequently, our discussion of data structuring in an automated system is put in the context of the semantics of the data.

Automated spatial information systems, then, are many things to many people. By way of summary, they are:

A data depository
A toolbox
A resource
A technology
Frames of mind, one or several

As a set of software processing routines in a hardware setting they are a new kind of toolbox for practical problem solving. As a new resource compared with paper based map making, they represent a new technology; and, through their emphasis on spatial data, they stand, for many people, for a different approach to thinking about problems and knowledge.

Computers provide compact forms of data storage, fast retrieval and mathematical equation solving; can store vast quantities of information, but especially allow linkages to be made among separate and apparently unrelated pieces of data. But computers are not enough. The users of information systems must have a good sense of human error and be capable of making wise judgements. The path from the real world to people's views to digital encoding must be trodden with care; otherwise we will not be able to make the return journey, from data to information to knowledge.

1.5 BIBLIOGRAPHY

The references given at the end of each chapter as Bibliography are the principal works used in the preparation of each chapter. Many other books may have been utilized peripherally, or represent writings on recent technical developments. In some places important references for historical purposes are likewise not indicated by name. We do not incorporate long detailed bibliographic lists or in-text citations. The recent major works cited can be consulted for additional literature. In-text citations, here, as in introductory books in general, used sparingly, indicate sources for particular illustrations or techniques.

For the first chapter, we provide a selection of items broad enough in scope to provide material on the general nature of spatial information systems. This includes introductory texts on geographic information systems, geomatics, and closely related topics that readers may consult for alternative approaches to the same topic of spatial information systems. In subsequent chapters the bibliography consists of particular works, often very recent, that may be consulted for details on specific topics, although not all items are mentioned in the text. Each chapter may have a few introductory remarks, providing some guidance to general-purpose works or comprehensive treatments on major topics.

Abler, Ronald F. 1987. The National Science Foundation National Center for Geographic Information and Analysis. *International Journal of Geographical Information Systems* 1(4): 303–326.

Aronoff, Stanley. 1989. *Geographic Information Systems: A Management Perspective.* Ottawa, Ontario, Canada: WDL Publications.

Burrough, Peter A. 1986. *Principles of Geographical Information Systems for Land Resources Assessment.* Oxford University Monographs on Soil and Resources Survey, no. 12. Oxford, UK: Clarendon Press.

Chrisman, Nicholas R. *et al.* 1989. Geographic information systems. In Gary L. Gaile and Cort J. Willmott (eds), *Geography in America.* Columbus, Ohio, USA: Merrill, pp. 776–796.

Dangermond, Jack. 1988. Trends in GIS and comments. *Computers, Environment, and Urban Systems* 12(1): 137–159.

Department of the Environment. 1987. *Handling Geographic Information.* Report of the Committee of Enquiry Chaired by Lord Chorley. London, UK: Her Majesty's Stationery Office.

Lai, PohChin. 1988. Maryland Department of Natural Resources Patuxent Basin geographic information system. Report prepared for the Maryland Department of Natural Resources, University of Maryland, College Park, Maryland, USA.

Laurini, Robert and Françoise Milleret-Raffort. 1989. A primer of multimedia database concepts. In Robert Laurini (ed.), *Multi-media Urban Information Systems.* Urban and Regional Spatial Analysis Network for Education and Training, Computers in Planning Series, vol. 1, N.D. Polydorides (ed.) Athens, Greece, pp. 7–75.

Laurini, Robert and Françoise Milleret-Raffort. 1989. *L'ingénierie des Connaissances Spatiales.* Paris, France: Hermès.

McHarg, Ian L. 1971. *Design with Nature.* New York: Natural History Press.

Ministère des Affaires Municipales, Gouvernement du Québec. 1984. *Introduction au Système d'Information Urbain à Référence Spatiale.* Québec, Canada: Ministère des Affaires Municipales, Gouvernement de Québec.

Muehrcke, Phillip C. 1986. *Map Use, Reading, Analysis, and Interpretation*, 2nd edn. Madison, Wisconsin, USA: JP Publications.

Muehrcke, Phillip C. 1990. Cartography and geographic information systems. *Cartography and Geographic Information Systems* 17(1): 7–15.

Nagy, George and Sharad Wagle. 1979. Geographic data processing. *Association for Computing Machinery Computing Surveys* 11(2): 139–181.

Raper, Jonathan (ed.). 1989. *Three-dimensional Applications in Geographical Information Systems.* London, UK: Taylor and Francis.

Robinove, Charles J. 1986. *Principles of Logic and the Use of Digital Geographic Information Systems.* U.S. Geological Survey Circular 977. Washington, DC, USA: US Government Printing Office.

Samet, Hanan. 1990. *Applications of Spatial Data Structures: Computer Graphics, Image Processing and GIS.* Reading, Massachussetts, USA: Addison-Wesley.

Smith, Terrence R., Sudhakar Menon, Jeffrey L. Star and John E. Estes. 1987. Requirements and principles for the implementation and construction of large-scale geographic information systems. *International Journal of Geographical Information Systems* 1(1): 13–31.

Star, Jeffrey L. and John E. Estes. 1990. *Geographic Information Systems: An Introduction.* Englewood Cliffs, New Jersey, USA: Prentice Hall.

Teicholz, Eric and Brian J. L. Berry (eds). 1983. *Computer Graphics and Environmental Planning.* Englewood Cliffs, New Jersey, USA: Prentice Hall.

Tomlin, C. Dana. 1990. *Geographic Information Systems and Cartographic Modelling.* Englewood Cliffs, New Jersey, USA: Prentice Hall.

Tomlinson, Roger F. 1988. The impact of the transition from analogue to digital cartographic representation. *The American Cartographer* 15(3): 249–262.

Unwin, David. 1981. *Introductory Spatial Analysis.* London, UK: Methuen.

White, Marvin. 1987. Digital map requirements of vehicle navigation. *Proceedings of the Auto Carto 8 Conference, Baltimore, Maryland, USA,* pp. 552–561.

2
Needs
Purposes and types of spatial problem

Jane Forester wants to know how much timber there is in the landscape she can see from a lookout tower. Jean Paul Pompier needs to determine the fastest route to send a fire truck on a Saturday evening to deal with a fire at a downtown restaurant. Chuck Plants intends to demonstrate that the average distance of the location of manufacturing premises from the port of Baltimore has not changed in forty years. María Proprieteria hopes to learn whether a proposal to build a chemical plant near her house will cause her property value to decline. Olga Mole has been called in to track down a noxious odour in a city sewer system.

In this chapter we treat the specific spatially oriented tasks that can arise within and across different domains of study or analysis or decision making. We present several approaches, each with a crude classification of specific spatial questions. In time we expect there will be a more refined taxonomy of spatial problems at a high level of detail, possibly created as a result of co-operative work combining several disciplines, but the young field of geomatics is not yet so developed.

2.1 PROBLEMS TO BE SOLVED; TASKS TO BE PERFORMED

A sensible approach to designing, evaluating, buying and using spatial information systems is to know what particular, preferably well-demarcated problems, scientific or practical, have to be solved. A study of general categories of use, or application domains, provides only a broad framework, leading to a conclusion that these information systems are virtually universal tools. A study of specific questions needing answers allows us to begin to appreciate the data and processing requirements of many users, and especially to reveal the tools required in the automated toolbox.

The approach represented by Figure 1.10 serves to demonstrate the many kinds of task in which a particular agency might be engaged. The

WHAT IS THE HEIGHT OF WATCH TOWERS ?

WHERE ARE THE WATCH TOWERS ?

WHICH TOWERS ARE CONNECTED BY TRAILS ?

WHICH PATHS WILL SHOW A JOURNEY TO EACH TOWER ?

WHICH PATH WILL INVOLVE THE LEAST TRAVEL ?

HOW MUCH TERRAIN CAN BE SEEN FROM A
PARTICULAR WATCH TOWER ?

WHAT KIND OF VEGETATION IS THERE IN THE
FOREST MANAGEMENT REGION ?

WHAT IS THE AREA OF EACH TYPE OF VEGETATION ?

WHAT IS THE AREA OF EACH TREE TYPE FOR
DIFFERENT ZONES OF ELEVATION ?

WHAT IS THE AREA OF DIFFERENT TREE TYPES
IN 500 FEET BANDS ON EACH SIDE OF THE TRAILS ?

Figure 2.1 Some specific spatial questions.

identification of application subdomains, the boxes, reveals the typical situation in large government organizations of many separate uses of different pieces of data contained within the organizations' files, whether automated or not. This approach highlights the concept of user views of a database resource.

In contrast, Figure 2.1 focuses on the specific questions being asked within a general area of forest management. The questions may be easily transferred to other domains of study. For example, we may wish to know the height of telephone poles or of buildings. We may wish to measure distances along footpaths within a complex of buildings in a city or ascertain the solar shadow reach on the ground. So our approach to structuring an information system in the area of spatial analysis is to identify types of problem based on parameters of space and location. We will imagine the questions via the mind of a nine year old in school. At about this age, the young mind is able to handle concepts like the qualities of space, such as, neighbours and neighbourliness; and the more rigid characteristics of space measurements incorporated in Euclidean geometry and exemplified, at least in some parts of the world, by precise boundaries used for demarcating private property.

Schoolchildren are often asked to look at maps and to find a particular feature or to identify some particular attribute of an object at a given location, for example, what kind of forest is found near the Equator. That is:

WHAT IS IN A PARTICULAR PLACE?

FIND THE POSITION OF SOME OBJECT.

Thus, City M in Figure 2.2 is positioned at the intersection of Freeways D and L; or alternatively, at 25 units going east from the left side of the map, and 12 units up. The question 'What is in a particular place?' may be answered by finding all things within the circle. Alternatively, it may result in a specific kind of phenomenon like the transportation networks, as might be necessary, in the spirit of Figure 1.7, in cases of maintenance or repairs.

The determination of both categories of information, akin to the informatician's notion of database queries, may include **bounds** placed on specific questions. These constraints may be of several types:

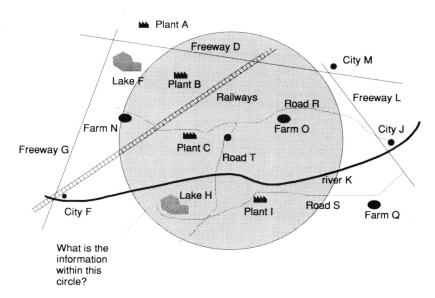

Figure 2.2 A part of the world.

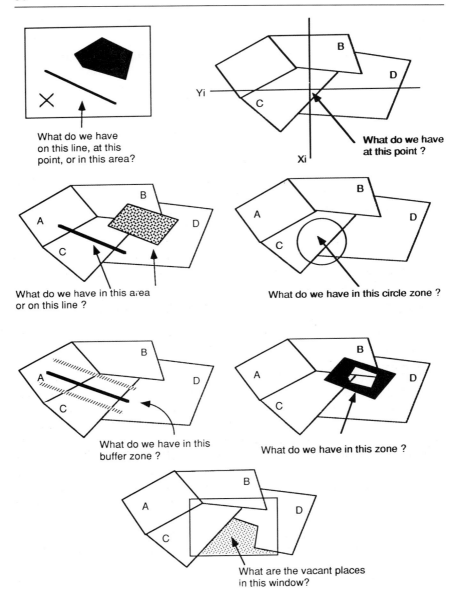

Figure 2.3 Some types of spatial queries.

1. Based on **character**; that is, by a range of specified properties (values, qualities, combinations), for example, the kind of forest.
2. By **count**: that is, looking for only a certain number of objects, as in finding three plants in Figure 2.2.
3. By **distance** from a specific object; for example, what exists at 100 miles north from Darwin, Australia?
4. **Included** within zones defined arbitrarily by geometric figures, such as the circle in Figure 2.2, or existing as real phenomena, like a continent or country.

Some of the many varieties of queries that are spatially bound, such as the distance and inclusion categories, are demonstrated in Figure 2.3. The variety exists for both the nature of the **spatial unit** that is found by the query, and the object that defines or constrains it, as shown in the top-left box.

Students may be asked to make measurements, perhaps on the ground or from maps, for example, the number of paces in walking from home to school, or for the airline distances between cities. That is:

MAKE MEASUREMENTS FOR OBJECTS IN SPACE

or, indicate how places relate to each other; for instance, who are their neighbours at home or in the classroom? In other words:

IDENTIFY SPATIAL RELATIONSHIPS

Spatial measurements involving one or more objects on the earth may relate to many properties: the distance apart of administrative centres, the pattern of cities as seen at night from a spaceship, the size of lakes, the total area of grandfather's several scattered pieces of land, or the distance to the nearest gas pipe or bus route (Figure 2.4a). The category of determining **spatial relationships** also has varied content, such as drawing cognitive maps, or demonstrating a knowledge of world issues by telling friends the names of the countries which border China (Figure 2.4b).

Children like maze games; they may be asked to solve problems like where on a beach they could set up an ice cream stand to make the biggest profit. A tough earth science instructor may even ask the pupils to guess what is under the ground of their state or county. That is, we have topics falling into two categories of **spatial problem solving**:

CHOOSE A LOCATION TO MEET SOME REQUIREMENTS

INFER THE FORM OF OBJECTS NOT DIRECTLY OBSERVABLE

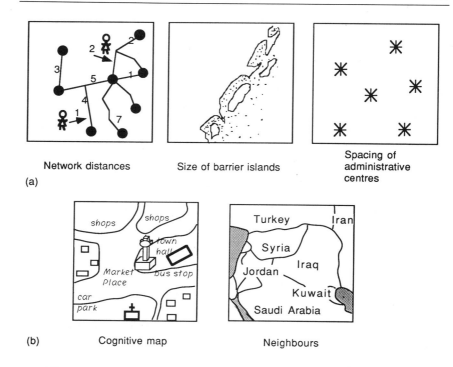

Network distances   Size of barrier islands   Spacing of administrative centres

(a)

(b)   Cognitive map   Neighbours

Figure 2.4 Some types of measurements and spatial relations. (a) Examples of measurements for different kinds of phenomena. (b) Examples of spatial relationships.

The first of these, representative also of **spatial decision making** problems, may include a variety of phenomena with different spatial form – determining the best location for a hospital or restaurant, finding a path through a maze, creating a path over mountainous terrain, deciding which children go to which school, laying out the attractions in an amusement

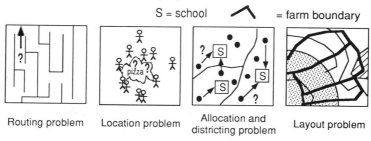

Routing problem   Location problem   Allocation and districting problem   Layout problem

Figure 2.5 Some types of spatial problem.

park, or allocating different types of land to particular farmers (Figure 2.5).

Spatial inference may be necessary for domains like mineral prospecting where it is impossible to have many direct observations, or for interpreting digital data transmitted from satellites. Figure 1.9 suggests the conditions that might need to be assessed to be able to determine the existence of an aquifer based on limited information for surface water hydrology, and a few boreholes or tunnels, perhaps also natural caves, providing subterranean data.

It is not inconceivable that children could also engage in more challenging activities like making maps that represent some view of the future, or maps that represent the combination of specified conditions. That is:

PREDICTIVE SPATIAL MODELS

This category includes activities like answering 'what if?' questions or undertaking **simulations** of known processes to produce different outcomes. For example, the best places for building more houses in a region in the urban periphery of Washington, DC (Figure 2.6) may be different according to the specific perspectives of an environmentalist, construction company or local government housing authority. This category of activity also includes scientific simulations to understand better the existing observed phenomena on the earth; for example, that of predicting patterns of rural land use or of modelling the development of rivers in a geologic time span.

Figure 2.7 provides one summary of the variety of spatial situations that arise. Determining position and character are perhaps the most basic geographic and spatial questions. They can, however, be very different depending on the restrictions or qualifications of the query. Measuring properties of spatial entities or relationships among several objects underlies much scientific endeavour. Choosing places or other spatial entities by comparison of alternatives reflecting different opportunities or constraints is a major type of public or private decision problem. Spatial prediction, perhaps used for inferring the location or existence of objects, or identifying and assessing different scenarios, can be part of many tasks.

2.2 LOCATION AND CHARACTER

The themes of location and character, fundamental to the other activities presented above, represent two different views, two contrasting thought processes. In a practical sense, they can be difficult to handle. For example, imagine a blindfolded person dropped into a forest from a

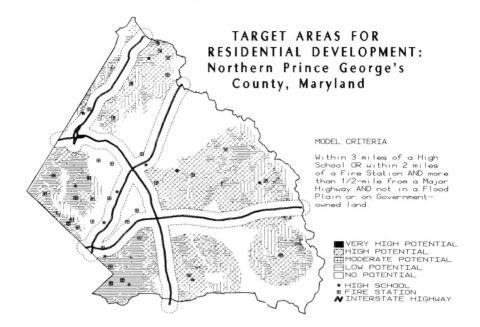

Figure 2.6 Predictive modelling. This map portrays land in five categories of potential value for residential development. A spatial information system was used to create digital versions of maps for many thematic variables: to remove some land areas from consideration, for example, too close to highways or owned by government; and to undertake mathematical modelling before converting scaled integer values into five categories for mapping purposes.

helicopter: 'Where am I?' What provides a frame of reference for this person? Equivalently, 'what is the nature of the Sahara Desert?' But what determines the boundary of the desert? Or what provides orientation to a non-resident coming to the surface out of the Paris rail Metro system into that city's Chinatown?

Frames of reference, products of at least two thousand years, may be personalized, such as distance and direction from where he landed for the man in the forest, or external, as in the worldwide systematic grid of latitude and longitude. Such references may be **polar**, also called azimuthal, based on distance and compass direction from a specific point; or **Cartesian**, that is, having two values representing distances in two perpendicular directions. They may have real origins, such as a railway station, or be virtually real by some internationally accepted standard like

NATURE OF ACTIVITY

GIVEN	FIND
Name of entity	Position Character
Character of entity	Name Location
Location	Characteristics Names of entities
CATEGORY	PARTICULARS
Make measurements	Area Distance Shape, etc.
Ascertain spatial relationships	Inclusion Direction Adjacency Connection
Make decision	Allocation Location Route or path Layout
Make prediction	Geometric inference Simulation Modelling

Figure 2.7 Categories of spatial activities.

the intersection of the Equator and Greenwich Meridian. Or they may be 'arbitrary', based on a local frame of reference not tied to any standard, such as for grids used in city street atlases. Also, they may vary in time, as in the ancient world's flat earth compared with the modern world's sphere.

Today, spatial information systems invariably use a Cartesian frame of reference, usually planar, and generally cross-referenced to the geographic coordinate system of the world (latitude and longitude). They use

Euclidean geometry as a basis for identification of absolute position and for distance measurements. Yet there are alternative methods that may need to be available for some purposes, and there are **cognitive** and **linguistic parameters** that are important in the development of information systems, partly for reasons of personalized frames of reference, and partly for designing the style of and tools for the transfer of instructions from person to machine.

While we do not yet know much about mental spatial data processing, there are some fundamental matters that can be included in this book. Firstly, there is not a universal acceptance of a Euclidean rectilinear frame of reference. People often use personalized frames based on their most familiar locations, like their home, or they may use also a focused territorial system, like that used by helicopter pilots. Or an azimuthal system may be employed, like the one Hawaiian Islanders use to determine positions based on distance out from or into the middle of an island or around the perimeter.

Secondly, landscape objects may be used for reference purposes, as is quite common in Europe, where the same street may have many different names, and where there are many landmarks. In contrast, in the USA, highway or street road numbers are more often used in an urban landscape with a smaller density of unusual objects.

Thirdly, the language of space varies considerably – words like up and down, in and on, inside or outside, have different interpretations. And common words for some spatial relationships, like north, have different interpretations depending on circumstances. For example, north may be identified as a line concept, a single line pointing to the compass point north; or as distance from a line running east and west, the basis for identifying latitude and the concept of north implemented in spatial information systems.

Customarily, earth phenomena are located in some kind of reference system. We use the term **position** to refer to where an existing object is located, as differing from **location**, which implies solving a problem like where to put something; although in practice these terms are often used as synonyms. Assuming for the moment that phenomena can be clearly demarcated, that is, we can recognize (bounded) entities; then some method of **access** is needed to get to a part of the entire world, either by marking on a paper map or doing its equivalent in an automated system. **Reference**, that is, a method of describing an absolute or relative placement of an object or its representation, can be provided by an address, key or index. Later we make much use of the term **locator**, rather than positioner, when talking about a means of getting at the objects in a spatial database.

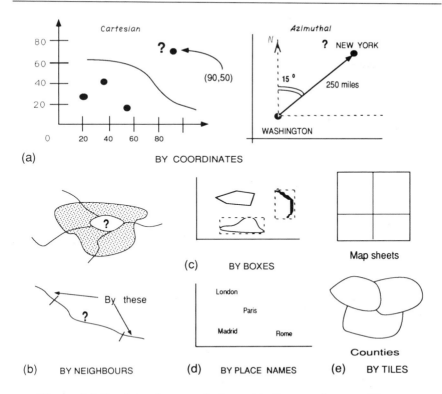

Figure 2.8 Spatial reference devices. (a) By coordinates. (b) By neighbours. (c) By boxes. (d) By place names. (e) By tiles.

The concept of a **spatial reference** takes on several practical forms (Figure 2.8):

1. Specific Cartesian or azimuthal coordinates.
2. Identification of neighbours; that is, the entities, either line or area type, that touch a specific entity.
3. A linear locator, such as a mile point on a road.
4. Enclosing rectangles or boxes of minimum size (based on the lowest and highest Cartesian coordinates in x and y metrics).
5. A place name or numerical code.
6. References to chunks, tiles, or other blocks of space, either regular or irregular in shape, such as map sheets or countries.

Leaving until later a discussion of refinements to the basic concepts of positioning and spatial access to a database, we conclude this section with a reminder that differentiation is needed between identifying some arbitrary portion of the world in order to learn its character, and retrieval of information for an entity. The third and fifth items in the list above serve to refer to space, but not to particular phenomena, whereas the second and fourth imply that spatially defined entities exist. This fourth implies that the geographic name, like Sahara Desert, or a numerical identifier like census tract 1012.02, has an associated spatial definition, or possibly several. And, at times, the same name is used for different landscape features; for example, Geneva for a lake and city, or New York for a city and state.

2.3 MEASUREMENTS AND SPATIAL RELATIONSHIPS

Taking a view that entities of the world can be clearly demarcated and positioned in some **coordinate** frame of reference, different tasks using spatial information may require the specification of **spatial properties**, such as shape, and tools to provide some concrete measure of those characteristics (Figure 2.9).

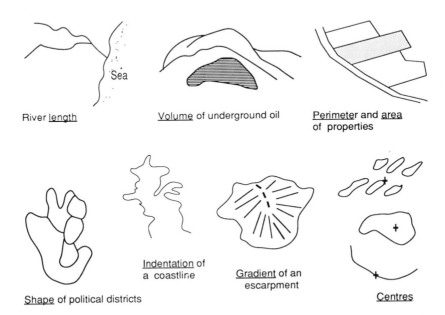

Figure 2.9 Some basic spatial properties.

2.3.1 Spatial properties

Taking individual entities first, we may have:

1. Length, as for a river or boundary of a lake.
2. Surface area, as in a lake, or a barrier island, or a piece of property.
3. Volume, for example, the amount of earth to be removed to build the English Channel Tunnel.
4. Shape, such as circular, or not, as for a political district, or a straight railway.
5. Irregularity in shape, as for an indented coastline.
6. Orientation, such as the principal axis of drumlins.
7. The middle of a line or area such as the centreline of roads or the middle of a city.
8. Slope as in the case of a mountainside.

Superficially, the measurement of such properties is straightforward, in the sense that if we already have a lake shown on a map or in a database (or on the ground for that matter) we can go around its perimeter. However, we will get different answers as to length depending on the precision of the measuring device, and on the scale at which the data were originally encoded. As we explore more fully later, the length of the coastline of England and Wales is much shorter if measured from a generalized map at a scale of 1 : 100,000 than it is if based on a walk (by a person, let alone an ant) along the entire shore. Other aspects of variations in measurement include the measuring instrument used, the source of data, approximations being satisfactory or acceptable, and the amount of error that can be tolerated for a given situation. For example, the association of form and function may simply utilize visual judgements or may require more precise measurements.

Moreover, some properties are **multi-dimensional**. Shape may be seen as the degree of conformance to a regular geometric figure like a straight line or circle, or it may involve extra properties like indentedness. Measurements may not be exact, or several may be required. If there are different definitions for objects, then several measures may be required, or a probability concept utilized. For example, outlying counties around the New York urban area may or may not be included in the metropolitan area definition. That is, there could be a measure of the extent to which they are attached, referred to as membership degree, usually indicated by a percentage of probability.

Entities of the same general thematic class, for example lakes or highways, then may be observed for their group characteristics. The same or varying shape of barrier islands may be indicative of how they were created. The specific statistics like area or shape can be defined for one instance, and summarized or generalized across a set of instances, including:

1. Minima, maxima, ranges of properties.
2. Averages and variabilities for those properties.

Moreover, some spatial properties can be defined only for sets of instances, not single units. While area and shape can be measured for one lake, the nature of the puncturedness of a forest requires measures for all holes, even if shape is the basic property. Examples of these more complex concepts (Figure 2.10) are:

1. The pattern of distribution of unconnected features like schools.
2. The layout of a housing development, or the separate land pieces of a farm.

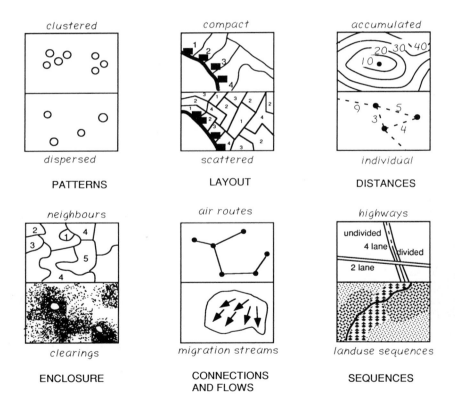

Figure 2.10 Spatial measurements for sets of instances.

3. The distance between traffic lights, or the aggregate travel time to reach a city centre from different places.
4. The number of neighbours for countries, or the puncturedness of forest land use.
5. The connections in an air traffic system.
6. The predominant flow of migrants across a country.
7. The sequences of phenomena, as for road character, or land use.

For our purposes in building up the structure of particular tools required in an information system, we concentrate on different kinds of property or measurement rather than treat the issue of precision and error.

Some properties may be defined only for the combination of an attribute of an entity with a locational parameter. For example, the degree to which nearby places are similar or different with regard to some attribute like temperature or population density, is measured by correlating the values of the attribute across space, either for the geometric case or topological variety. Or, as shown in Figure 2.10, the **sequences** of attributes across space or along linear features may be of interest. The index of localization used to measure the degree of spatial concentration of economic activities compares the proportionate share of one attribute in a given place out of a set of places with the proportionate share of some other attribute. Simply put, for Paris as an example, the ratio compares that city's share of one-fifth of the population of France, with the two-fifths portion of lawyers in the country.

2.3.2 Spatial relationships

Implicit in the examples given above for properties of single entities and sets of entities of the same class, are some other fundamental distinctions. First of all, we differentiate between properties that require measurement using coordinates, that is **(geo)metric** information, for example the Cartesian coordinate centre of a country, or the distance between boreholes, and properties based on **non-metric** information for the qualities of space, like connections between places. These **topological** properties are:

Connectivity
Orientation (to, from)
Adjacency
Containment

as illustrated indirectly in Figure 2.10. Connectivity is the property for air routes; orientation is illustrated in the direction of flow of rivers; adjacency is seen in the case of neighbouring countries and layout, and containment is involved in the bare places in forests.

Some spatial concepts may be measured in both geometric and topological domains. For example, the distance between cities in the global air routes system can be represented by the miles travelled, or the number of legs of a route, where leg is defined as each part of a journey between stops. Proximity may be thought of as adjacency or a Euclidean distance.

Secondly, identification or measurement of properties of spatial relationships like distance or connectedness requires that we define a special class of entity, the object pair or **dyad**. The chart of driving times between cities, with lists of the places in both the rows and columns of the table of information, has a city pair as the basic unit, not the individual city. The driving time is a property of the combination of two places. While the dyad is not necessarily a physical entity, it is nonetheless an important spatial unit in many fields of study, such as population migration, social contacts and weather prediction: and the physical representation may be very elusive as in the case of linking a water flow via a sinkhole and spring.

Thirdly, additional spatial properties are encountered if we extend to different attribute themes. The concept of **spatial colocation** or **covariation**, the extent to which different phenomena have similar occurrences in space, involves a comparison of the distribution of one phenomenon with another, giving rise to concepts like overlap (for example, of soil type A and wheat production, or sales territories), and intersection (such as railways with roads) or not intersections (as for sewers with gas pipes), by the addition of attributes to the spatial conditions (Figure 2.11). These concepts apply, too, to the three-dimensional world, although they are illustrated only for intersections, via the example of airspace. Other examples are the three dimensional covariation of water content and geological structures, or oceanic properties of temperature and fish density.

In this context it is important to recall the differentiation, first made in Chapter 1, of information systems with a layered architecture and those oriented to objects. In a layered system, where each theme is separately encoded, all spatial covariation tasks require bringing together different layers of information very much like the manual process of overlaying maps on a light table. In this process it is usually necessary to demarcate new objects, for example the new polygons containing particular combinations of soils and land use, the area of overlap of newspaper circulation territories, or the zones in which airspace can be used by particular categories of aircraft.

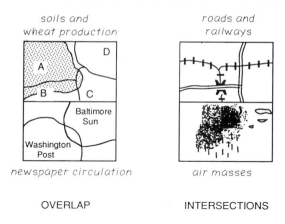

soils and
wheat production

roads and
railways

newspaper circulation

air masses

OVERLAP   INTERSECTIONS

Figure 2.11 Spatial covariation.

2.4 CATEGORIES OF SPATIAL PROBLEMS

Our structuring of particular questions in the space domain can now treat the categories defined earlier of where to locate something. Many public and private authorities make decisions on locating specific phenomena like traffic lights or railroads, or make determinations about sets of phenomena like the intercity rail services to be provided by a national railway company or where to locate a set of hospitals. Scientists grapple with questions of how human societies are organized and particular topics like the impact of an earthquake in Japan on the coastline of Peru.

2.4.1 Types of spatial problem

Geographers in particular have attempted to organize problems semantically from a spatial perspective. The spatial concepts in this discipline, and others, provide a valuable additional perspective on spatial information systems. Particular categories of **spatial organization** (Haggett, 1965) are (Figure 2.12):

1. Movements or flows, such as population migration or air masses movements.
2. Networks, such as highway systems or water pipes.
3. Nodes, the intersections of railways.
4. Hierarchies, or ordering of cities in terms of economic functions.

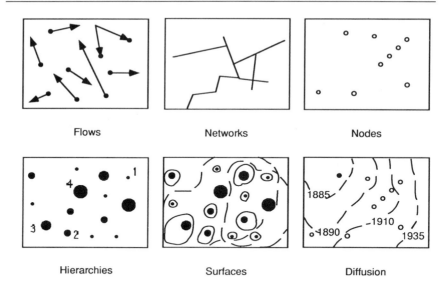

Figure 2.12 Categories of spatial organization.

5. Fields or surfaces, as in newspaper circulation territories.

With the addition of a specific time element, a sixth category would be:

6. Diffusion of disease, ideas or innovations in space.

This different approach to spatial relationships reflects processes of natural and social ordering and organization. Movements of goods match up places of supply and demand; networks are the channels of communication or flow; possibly, by reasons of greater connectivity, some places become more important than others or have to deal with greater volumes of water; and entities may have influences extending over shorter or greater distances. Some conditions may be thought of as spatial structure, like morphology and dimensionality; others as movement, including circulation and diffusion; and others may be processes like growth and organization.

While this conceptual framework is particularly valuable for dealing with human organization and artefacts, it is also applicable to physical and biological phenomena, for example, animal territoriality, bird migration paths, and overland flows of pollutants.

Generally, on the basis of this kind of framework, we can begin to identify narrowly defined categories of problem, involving:

1. The types of entity involved, and their combinations.
2. The spatial properties involved.
3. The elements that are given, and those that have to be found.
4. The objectives.

For example (Figure 2.5), finding a route through a maze is essentially a linear sequence problem – putting together different segments of the maze in an order to guarantee a path from entrance to exit. It also may have an overtone of finding the best path out of a set of alternatives. In the case of locating a 'fast-food' restaurant, the spatial elements may be a point (the restaurant building) in a suburban area, may be a point at a service station on a highway (line feature), or may be a piece of real property (area) next to other land plots.

The **location** problem may begin with a known location of customers, a known transportation system, or a known set of available plots on which to build, and require the determination of which point in space is good for maximizing sales. Or two unknowns may have to be determined, such as the location of hospitals and the hospitals to which people will go, in other words, a combined **location–allocation** problem. Otherwise, the locations of supply points and demand points for some mobile resource, for example, bottled mineral water, are known, and the element being addressed is the **allocation** of quantities among supply and demand points.

Specific types of location problem (Figure 2.13) are:

1. Location of some point entity, like retail commercial centres.
2. Allocation of resources from points to other points, as in petroleum from oilfields to refineries.
3. Route-finding, like getting from Heathrow Airport to Trafalgar Square in London, England.
4. Path-finding, as in avoiding swamps and mountain peaks.
5. Creation of area units, as in redistricting for school areas.
6. Layout, as in housing estates design.

To demonstrate potential variations on these themes, we take just one category. Path-finding situations may consist of the choice of one route out of several to get through city streets, or may be more complex, as demonstrated in later chapters, as in Figures 17.26–17.28. A path may have to be laid out through different terrain, recognizing elevation, barriers (penetrable or absolute), or utilizing different modes. Indeed, whole systems of routes may have to be designed, as for the provision of intercity rail passenger services by many national railway companies.

Any one type of location problem may be set up in different ways, although alternatives are not always intuitively obvious. Trading centres,

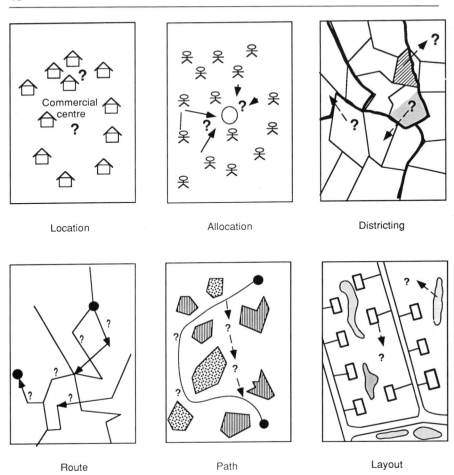

Figure 2.13 Some types of location problem in two space dimensions.

whether extensive modern supermalls or local convenience stores, may be seen as points in territory. There are models and tools to solve a retailing distribution of this kind by demarcating the market area in which a store or commercial centre dominates competitors, that is to say, an area entity unknown. However, it is much easier and just as effective to treat the problem as location–allocation by matching up consumers to retailing locations as flows of discretionary income. This obviates the need to undertake area definition, and, in an information system sense, does not require an area type unit to be recognized as a data type. At the same time, though, it requires that the dyad be a recognized spatial unit.

2.4.2 Other aspects

The particular problems are further differentiated by goals and by the spatial properties of the unknown. Generally some quantity is satisfied or optimized, for example minimizing the overall distance that people must travel to hospitals from their homes, or minimizing the costs in delivering pizza. Usually, a geometric spatial property is specified, but in some problems a topological element is also present, as in the creation of voting districts that do not contain enclaves or do not have exclaves. The condition(s) being optimized usually combine explicit spatial properties like miles of travel, with other attributes such as retail sales profit or absence of bias in voting districts. An interesting example from today's world of fast-food is the pizza delivery company in New York that uses a spatial information system to transfer a telephone order from the receiving store to another store whose territory covers the caller's address in order to have a more effective service.

Some spatial problems involve a **time** element. Perhaps this is only time of travel, which can be represented as an attribute of a pair of places, but it may be a framework for a process of spread of urban areas over landscape, or the diffusion of a disease. Or we may be interested in improving scientific knowledge about landscape-forming processes operating in geologic time. Space and time are to some extent complementary and substitutable, as in the measurement and depiction of accessibility by travel time isochrones, lines of equal travel time, or the use of spatial cross-sectional data to represent different points in time, as in observing the process of industrialization, based on contemporary conditions for different countries.

The implications of clearly identifying the spatial tasks in the context of spatial information systems are few, but very important. Some types of processing will be difficult, perhaps impossible practically, if the information system does not treat particular types of spatial unit. We have already mentioned the need for the spatial dyad. Another example is the ability to characterize nodes in networks as to vehicular movement restrictions, as occurs for variations in the width of a tunnel, or for left-turning traffic for highways on which vehicles use the right side. Redistricting is easier to implement if the database allows the fundamental property of adjacency to be explicitly coded: it is easy to combine the basic units of election precincts if we know they are adjacent.

2.5 SOME EXAMPLES OF MULTI-FACETED NEEDS

The complexities and uncertainties of spatial problem solving are best demonstrated by examples. A first example is a scientific activity to ascertain the effect of patterns of land use on nutrient discharges into a river system. A second task is a natural resources management activity to determine the land uses associated with wetland habitats. The third illustration is an activity to discover the existence of phenomena that are not directly observable, that is, where inference is necessary and a fourth is an example of simulation.

2.5.1 Example of flows over landscapes

Ecologists at the Smithsonian Institution in the USA are interested in the effects of landscape characteristics on the flows of chemicals, particularly nutrients, over a surface (Figure 2.14). It is clear that water containing chemicals will flow over a surface gravitationally, will sooner or later become channelled, and may also seep below the surface, possibly being channelled into subterranean streams. Not only does the terrain slope affect flows, but surface land use and land cover influence the location of

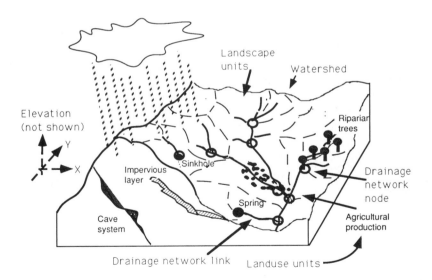

Figure 2.14 Data elements for the study of overland flows. (For source see Figure 1.9.)

nutrients that can be picked up by the water. It is important to know not simply how much agricultural land (receiving fertilizers, for example) there is, but especially if it is adjacent to rivers.

A model of the level of nutrients at different places must, therefore, recognize many spatial elements and types of spatial problem. The point sources for chemicals may be regularly or haphazardly distributed in space; the slope and aspect of the facets of terrain must be known; data are needed for conditions beneath the surface; water must be tracked through a network; some water bodies may have substantial surface area; and the pattern of land use must be known. Objects are seen in three-dimensional space, and consist of polygons (of land use type), line segments (of the river system), paths (of the flow of nutrients), elements (of terrain), and coordinates for establishing the position of specific features or attributes.

2.5.2 Resources inventory

Government planners have been charged with the responsibility of ascertaining the extent of different kinds of land use occurring within surface areas designated by legislation as environmentally sensitive wetlands (Figure 2.15). Specifically, a process was required to determine accurately the land use or land cover in zones of one thousand feet from boundaries of wetlands (Lai, 1988).

Starting with original maps and air photographs of terrain conditions, two types of wetland were identified: those that could be represented at a geographic scale as having areal extent, and those represented as linear features (Figure 2.15b). Critical zones, or buffer regions, drawn at the prescribed distance from the single demarcations of the wetlands were established, and combined (Figure 2.15f). The polygon zones were then used as 'pastry cutters' on the land use map, although in practice the particular software could not directly detect the interior holes (Figure 2.15h). After unwanted pieces were eliminated (Figure 2.15i), common boundaries dissolved, and true wetland areas omitted (Figure 2.15j), the area measures were obtained.

2.5.3 Predicting the location of mineral ore deposits

Geologists often need to infer the occurrence of phenomena that cannot be directly observed with instruments (Figure 2.16), relying possibly on limited information from just a few boreholes. A combination of

This set of maps illustrates the sequence of actions applied to a series of map layers to produce some statistics. These, given at the right, indicate the amount of land used, by major category, in acres, within critical zones created by legislation. These critical areas lie within one thousand feet from water bodies. In this case these are wetlands of class PF01A, the Palustrine forested broad-leaved deciduous temporary. Details of the processing are discussed in Chapter 8.

LAND USE CATEGORY	ACREAGE
Urban or built-up land	
residential	203
park, racetrack, golfcourse	775
Agricultural land	
cropland and pasture	2235
Forest land	
deciduous forest	524
evergreen forest	135
mixed forest	2449
Water	
bays and estuaries	21
Wetland	208
Barren land	
strip mines, quarries, pits	15
transitional areas	42

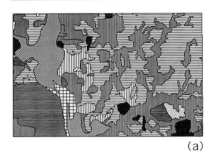

(a)

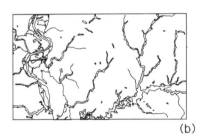

(b)

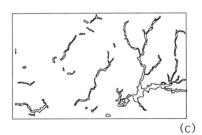

(c)

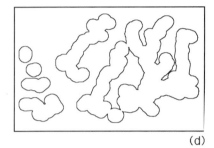

(d)

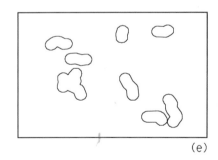

(e)

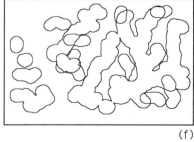

(f)

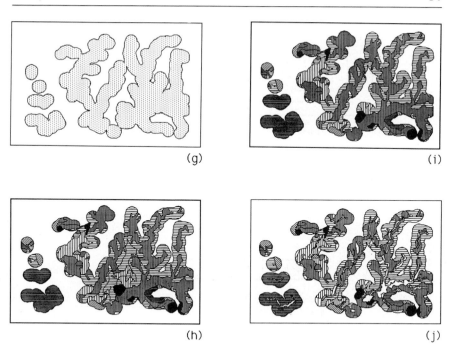

Figure 2.15 Land use inventory for critical areas. (a) Inventory of land use. (b) Map of surface hydrology. (c) Extraction of wetland polygons. (d) Creation of 1000 feet zones around areal hydrology features. (e) Creation of 1000 feet zones about linear hydrology features. (f) Combination of two sets of 1000 feet zones. (g) Removal of common boundaries. (h) Use of 1000 feet zones to cut out pieces of the land use map. (i) Elimination of land use for the undetected interior hole. (j) Removal of wetland areas from the land use map. (This example was created by, and is used with the permission of Professor Poh-Chin Lai, Ohio State University, Columbus, Ohio.)

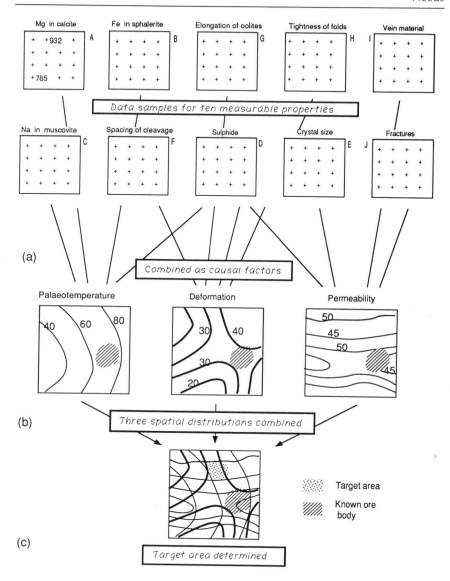

Figure 2.16 Spatial prediction. (a) Data samples for ten measurable properties (only a few pieces of data are shown). (b) Measurable properties combined statistically to get causal factor values. (c) Map based prediction. (This example is based on material in the paper by Klovan, 1968.)

statistical and spatial tools can help in such tasks. Not only is it necessary to guess boundaries of the subsurface strata containing the minerals or water, but it may be necessary to use some observable attributes like iron content or porosity to stand for the preferred conceptual criteria like susceptibility to deformation (Klovan, 1968). In three-dimensional space there is no explicit framework except that provided by boreholes and tunnels. Surfaces must therefore be extrapolated or interpolated from just a few triads of positional coordinates.

It is important to realize that analytical or inferential activites like these may have processing requirements beyond what spatial information systems can today undertake. The development of software has been segmented historically. Statistical data processing operations implied by the first stages of the mineral prospecting example (Figure 2.16b), or the mathematical modelling involved in the nutrient discharges example, do not necessarily have to be done within a spatial information system, and historically were accomplished by separate computer programs. Today, regarding software tools, there is debate as to how much processing there should be within the spatial information system itself, and how much should be done by separate special purpose programs.

2.5.4 Engineering network simulation

In the domain of utility network engineering, it is often important to have the facility for simulating the behaviour of infrastructure elements. For instance, let us examine (but possibly not too closely!!) a sewerage management system. Suppose we have a database integrating the sewerage pipe description, including features like size, location, gradient and flow capacity, together with the topology, that is, pipe connections. We also may need to determine the location of filtering and purification plants.

By means of a spatial information system, we can simulate the effect of a rainfall in order to see whether the sewerage system can absorb it and where the overflows might be located. The same model can be executed as if the storm centre is going upstream or downstream relative to the sewerage. When downstream, it leads to a sort of accumulation in which the pipes must absorb more and more water, perhaps suggesting pipes should get larger in that direction. For the daily routine use of the purification, we can also compare the actual and simulated behaviours in order to know where the muddy water is located in order to cleanse the sewerage. In addition, a sewerage-cleansing timetable can be established to schedule the work to coincide with the times of the best weather conditions. Moreover, when city developers are designing new allotments (property subdivisions), it is possible to check whether the actual level of sewerage can absorb new dwelling discharges; and, if not, determine where to enlarge or double some pipes. One can also test the necessity of installing more purification plants.

2.6 MAIN CATEGORIES OF TOOLS IN A SPATIAL INFORMATION SYSTEM

A toolbox in the context of physical, documentation and guidance subsystems in an automated spatial information system, much as for a non-automated system, must include resources and capabilities to keep track of data, bring information into the database, produce output, provide security and allow for a range of non-spatially oriented processing. Several classifications have been made at varying levels of detail, and there is no one universally accepted framework.

Some criteria for undertaking a classification are:

1. The type of data being processed: words, pictures, numbers, graphs and maps, sound.
2. The stage of processing: input, intermediate point, or output of data.
3. The management: control, guidance, documentation, security, book-keeping.
4. The type of operation on the information: spatial, non-spatial, combined.

These are not mutually exclusive: number data can be input or output, and many specific operations being performed on the encoded data will involve several types of processing. Some need may be met by a single command, by a single command that invokes a series of operations unknown to the user, or by a series of commands given by the user. How a task is undertaken depends on the level at which the user wishes to, or is forced to, work. The style of the person–machine interface is not as important for our purposes as a discussion of the functionality of software.

Data acquisition involves the direct capture of data by electronic imaging, either remote (for example, *SPOT* satellite imagery) or close at hand (for example, by still video camera used on a field trip). It may or may not cover attributes of phenomena, depending on the nature of the original materials. Thus, these may be maps, engineering drawings, field notes, or tables of data in a census, all of which have involved some kind of prior processing or human activity. Direct digital encoding via satellite is followed by a step of interpretation.

Information about the same phenomena may be held in a database in different ways, or may be put there via a different process. In any event, though, there will be a need for tools to **manage** the material, to provide security, to allow for verification and validation, to control access, to correct and update, to convert from one form of internal encoding to another, to organize efficiently, to help the user, to identify and describe the contents of the database, and to facilitate import and export of digital data.

Tools for **output** may be many, depending in part on the kind of information in the system, and in part on the mode of use. Analytically oriented systems will most likely provide capabilities for the creation of physical documents – maps, scientific graphs, presentation graphs or other images, tables of numbers or narrative text. Query oriented systems will focus more on the preparation of virtual output in the form of quickly produced maps or graphs, the contents of particular sections of the database, or derived statistics, all made available immediately in response to a user request. The same physical resources and capabilities of the information system may be used in both cases, such as basic drawing capabilities.

More important are the **processing** functions, as these are the tools by which the spatial analysis is undertaken. Again, depending on factors like type of data, which will be discussed in more detail later, we can identify functions involving or not involving spatial properties. Many require-ments have already been reviewed. There also are categories of processing for just the attributes of phenomena, such as arithmetic operations, logical operations and parsing. For example, census data in published tables may require aggregation. A query oriented system is more likely to emphasize functions of search and retrieval. A product oriented system is more likely to emphasize measurements. The one is oriented to bringing together disparate pieces of information via browsing, by cognitive associations or providing capabilities to proceed from general to more detailed information. The other emphasizes the computing as opposed to information handling capabilities of computers.

Generally speaking, we can distinguish among several major types of orientation (Figure 2.17):

1. Automated mapping, simply making computer based replicas of paper documents.
2. Some variety of thematic mapping, perhaps in conjunction with census data.

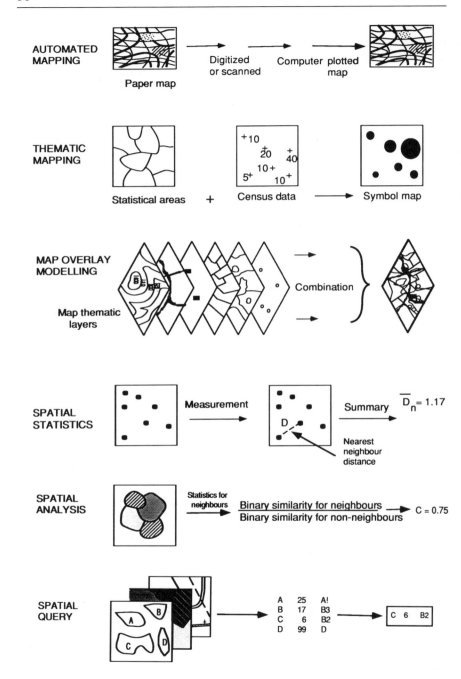

AUTOMATED
MAPPING

Paper map   Digitized   Computer plotted
                   or scanned   map

THEMATIC
MAPPING

Statistical areas   +   Census data   ⟶   Symbol map

MAP OVERLAY
MODELLING

Map thematic
layers   Combination

SPATIAL
STATISTICS   Measurement   Summary $\overline{D}_n = 1.17$

                                          D

                                    Nearest
                                    neighbour
                                    distance

SPATIAL
ANALYSIS   Statistics for   Binary similarity for neighbours   ⟶   C = 0.75
                neighbours   ─────────────────────────────────
                                  Binary similarity for non-neighbours

SPATIAL
QUERY   A   25   A!
                                  B   17   B3   C   6   B2
                                  C   6   B2
                                  D   99   D

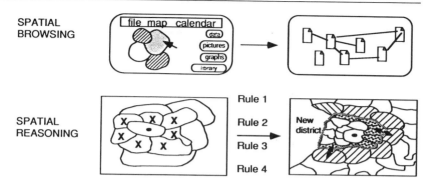

Figure 2.17 Orientations for spatial information systems. (This diagram is an extension of one devised by Patrice Boursier, Université de Paris Sud, France.)

3. The variety known as composite mapping, characterized by a map overlay concept.
4. Analysis of spatial data, encompassing tasks which deal with the attributes of entities, like the average size of cities or the degree to which crop yields are related to climate conditions.
5. Creating spatial statistics covering tasks that require measurements of spatial properties of phenomena, like the shape of drumlins or the volume of timber in a forest.
6. Analysis of spatial statistics, that is tasks which treat spatial properties as attributes, for example the correlation between highway network connectivity and levels of economic development in the world.
7. Spatial analysis, encompassing tasks, including simulation, which use a variety of tools of spatial statistics and location based problem solving.
8. Spatial browsing, or generally exploring the contents of a database containing spatial information of some kind.
9. Spatial querying, obtaining information from a database in response to identification of particular conditions.
10. Spatial problem solving, such as geometric reasoning, for example deducing inclusions of points in polygons, or for spatial decision making incorporating both spatial and logical deductive reasoning.

These different orientations reflect not application domains but general processing requirements pertaining to spatial data, spatial properties and measurements, or the types of information needed. We shall make reference throughout this book to these categories, and associated terminology, much more than we will to subject matter application domains.

2.7 SOME IMPLICATIONS FOR THE DESIGN OF SPATIAL INFORMATION SYSTEMS

The foregoing discussions of types of investigations, themes of location and character, measurements and spatial relationships, categories of spatial problems, and major categories of tools are intended to provide a perspective useful for both evaluating and designing spatial information systems. It might be regarded as trivial to say that information systems will be better if we know what functionality is required. However, there are some important practicalities that complicate what otherwise might be regarded as a simple process of matching demand and supply:

1. Users do not know clearly what they want to do.
2. Users may think specific software systems can solve their problems when, in fact, they cannot be dealt with via the data structures used.
3. Data and processing requirements are mutually dependent.
4. Some operations may be theoretically possible but practically very costly.
5. The means of implementing particular types of function may be confused with what is needed.
6. The commercially available systems must be judged from a market-place perspective in a historical context.
7. Choices are made because a neighbour or a sister agency used a particular type or brand for sharing costs or knowledge.

While there are many lists of requirements associated with software procurements on the part of businesses and government organizations, often following extensive formal needs assessments, there are few systematic academic studies that give a profile of the functions expected of spatial information systems. One study is described here as an example of the process to obtain good material.

A selection of users of spatial information in the field of natural resources management have provided some indication of their needs by indicating how important certain kinds of activity were for them in their ordinary work (Tsai, 1987). A questionnaire produced patterns of response to sixteen items, scaled in six categories from very frequent to not at all. The items covered geometric and topological properties, spatial and non-spatial properties, querying and browsing. All activities were present, but shape measurement and comparative characteristics on either side of a boundary were infrequent. Area measurement, and attribute and location specified searching were most common for this set of responses.

The study clearly showed, too, that there is confusion over terminology, in part because of differences among disciplines. Moreover, users of spatial information systems are often not oriented to spatial concepts, let alone knowledgeable about specific measurement tools like spatial autocorrelation.

From the perspective of automation, it appears that tools for efficiently handling spatial relationships and spatial problem solving are not well developed in the context of present comprehensive conventional information system toolboxes. Because of the huge number of spatial relationships that are possible, we may not be able to compute and store some attributes for future use, for example, all possible distances for a set of points, or to anticipate all possible combinations of spatial entities, recognizing their type. One implication is, of course, that we should not rely on general purpose information systems, but have a number of special purpose systems. After all, we have many specialists such as plumbers, lorry drivers, teachers, geographers, informaticians and creators of spatial information systems.

2.8 BIBLIOGRAPHY

General, comprehensive reviews of spatial and locational concepts are provided by Abler *et al.*, Cole and King, Haggett, and Unwin.

Abler, Ron, John Adams and Peter Gould. 1971. *Spatial Organization*. Englewood Cliffs, New Jersey, USA: Prentice Hall.

Berry, Joseph K. 1987. Fundamental operations in computer-assisted map analysis. *International Journal of Geographical Information Systems* 1(2): 119–136.

Charré, Joël and Pierre Dumolard. 1989. *Initiations aux Pratiques Informatiques en Géographie: Le Logiciel INFOGEO*. Paris, France: Masson.

Cole, John P. and Cuchlaine A. M. King. 1968. *Quantitative Geography*. London, UK: John Wiley.

Dangermond, Jack. 1983. A classification of software components used in geographic information systems. In Donna J. Peuquet and John O'Callaghan (eds), *Design and Implementation of Computer-based Geographic Information Systems*. Amherst, New York, New York, USA: International Geographical Union Commission on Geographical Data Sensing and Processing.

Goodchild, Michael F. 1987. A spatial analytical perspective on geographic information systems. *International Journal of Geographical Information Systems* 1(4): 327–334.

Goodchild, Michael F. 1987. Towards an enumeration and classification of GIS functions. *Proceedings of the International Geographic Information Systems Symposium: The Research Agenda*, Arlington, Virginia. Greenbelt, Maryland, USA: NASA, vol. 2, pp.67–77.

Haber, Wolfgang and Jörg Schaller. 1988. Ecosystem Research Berchtesgaden – Spatial relations among landscape elements quantified by ecological balance methods. Paper presented at the European ESRI Users Conference, Munich, Germany, 1988.

Haggett, Peter. 1965. *Locational Analysis in Human Geography*. London, UK: Edward Arnold.

Klovan, J. E. 1968. Selection of target areas by factor analysis. *The Western Miner* (February): 44–53.

Lai, PohChin. 1988. Maryland Department of Natural Resources Patuxent Basin geographic information system. Report prepared for the Maryland Department of Natural Resources, University of Maryland, College Park, Maryland, USA.

Tsai, Bor-Wen. 1987. A query approach for GIS database systems design: the case for national resources analysis and management. MA thesis, Department of Geography. University of Maryland, College Park, Maryland, USA.

Unwin, David. 1981. *Introductory Spatial Analysis*. London, UK: Methuen.

3
Semantics

Objects, surfaces, data

The manager of a geographic data vendor decided to try to sell an electronic atlas of countries of the world, but emphasized the tabular and graphing aspects rather than the map displays. The map librarian created a system to select map sheets that contained many features but could not allow for selection of the maps based on those features. The politician bought a statistics package to help in redistricting but was unable to ascertain if one election precinct was next to another. The geologist tried to infer the boundaries for petroleum deposits in salt domes that could not be seen. A Londoner says that Eskimos are to be found just 'after' Scotland. The visitor to a restaurant in 'The South' of the United States was intrigued by the curious paper placemat map depicting 'Old Dixie'.

The content of this chapter extends the discussion of semantics from needs to a more particular level, that of the nature of spatial data. Spatial problems may be evaluated differently if they deal with phenomena from special viewpoints represented by idiosyncratic treatments of data and space. We distinguish primarily between an entity- or object-oriented view and a continuous space perspective. The chapter also reviews the different kinds of data, not just spatial, that may be part of a spatial information system.

3.1 THE INFORMATION IN A SPATIAL INFORMATION SYSTEM

The discussion in Chapter 2 of the tasks that may need to be performed with the aid of a spatial information system indirectly introduced the topic of the nature of spatial information. Let us now treat the topic of the meaning associated with phenomena more directly. In simple terms, imagine that for some purpose we have gone through a process of definition, the conceptual description of phenomena, for example, specifying the properties that are necessary and sufficient for identifying a lake; recognition, the observation of the existence of those properties such that a claim can be made that something observed is a lake; and demarcation, the delimitation of the feature on the ground or a representation of it as in a photograph.

3.1.1 Spatial entities

We use the term **entity** to refer to a phenomenon that can not be subdivided into like units. A house is not divisible into houses, but can be split into rooms. An entity is referenced by a single identifier, perhaps a place name, or just a code number. Even though it may be composed of several pieces, it does have an overall identity, for example, a city or a watershed. The entity may be a mappable feature like the lake, but perhaps not readily mappable, like the salt dome, because it cannot be seen or because it is an arbitrary unit like a statistical unit. It may also be a conceived unit, necessary for argument or imagining, but not yet having an existence in the real world. The indivisibility is based on the properties used in the definition, and there is not a necessary implication of complete homogeneity within the spatial extent of the entity because some characteristics, such as depth of lake, are not included in the criteria used to define and demarcate the entity. In the spatial information systems field, there is no comparable standard structuring of phenomena as in the physical world's molecules, atoms and quarks, to help us in identifying the **scale** of phenomena as implied by divisibility.

While in everyday language the terms entity and object (or thing) are synonymous, we shall follow the terminology of the field of informatics to use entity when speaking of the conceptual organization of phenomena, and object when referring to the digital representation. However, as the choice of term is context dependent, we will occasionally use entity and object interchangeably. At times, other terms like geo-object or feature may be encountered. The former term implies phenomena pertaining to the earth. The latter, a more confusing term in the context of spatial information systems, has the connotation of geo-objects that are shown on maps. Because of cartographic conventions used to ameliorate the difficulties of showing the large curved earth on small flat pieces of paper, features on a map may not be positioned as they are on the earth, and some real entities may not appear at all as features on maps.

The phenomena exist as perceived by people (Figure 3.1), so there may be different definitions for lakes. But there may also be different demarcations on the ground or in the abstract as a result of 'nature' being at odds with the intellectual definitions. For instance, the territorial extent of a body of water can vary at different times as a result of tidal,

Figure 3.1 Geographic entities. This sketch demonstrates just a few of the semantically based elements in a geographic landscape. Boundaries may be imprecise, and data representation is not a simple one-to-one match.

earth crustal or other processes, only one extent being compatible with the definition. Lack of clarity may be represented by a concept of membership function. For example, determining the extent of low-income neighbourhoods in London requires a double measure: low income and neighbourhood. The spatial extent of the entire area as imagined in the stated requirement could be seen as a probability (percentage in practice) of association with a core.

3.1.2 Categories of information

Assuming that the earth's surface contains clearly definable and demarcatable phenomena, we can consider five categories of information for specific entities (Figure 3.2):

ENTITIES	IDENTIFIER	POSITION	SPATIAL PROPERTY	CHARACTER ATTRIBUTE	BEHAVIOUR ATTRIBUTE
School	Name	Coordinates for one point	Distance to another school	Number of students	Education
Borehole	ID number	Coordinates for location at surface	Depth	Size of hole	Purpose
Aquifer	Name	Coordinates for surfaces	Volume	Water quality	Change in water level
Cloud	??	Coordinates for one point in time	Volume	Type	Dissipation
Railways	Code number	Coordinates for endpoints	Connection of two cities	Traffic volume	Use
Barrier island	Number or name	Coordinates for perimeter	Total area	Composition	Change position
Country	Name	Coordinates for boundary	Neighbours	Gross national product	Wars with neighbours?

Figure 3.2 Example data for spatial entities.

1. An identifier (name, code number, or other device).
2. The position on the earth's surface (locator).
3. The character of the entity.
4. The role, behaviour or function of the entity.
5. Spatial properties of the entity.

An **identifier** provides a means to refer to different entities, although it is possible that the naming system used is not very precise, consistent or able to guarantee uniqueness. For example, there is more than one Derek Thompson in the world, and more than one Paris, although apparently only one Robert Laurini. Identifiers may be geographic names, either in local usage or officially recognized by national or international organizations. For example, the United States Geological Survey has an extensive list of official place names for the USA, compiled largely but not solely on the basis of nomenclature appearing on topographical maps, and approved by committees at different levels.

Entities may be given arbitrary names solely for the purposes of use by a particular person or organization. At times, numerical coding may be used instead of, or in conjunction with, place names. On a global scale it is likely that not all entities have place names, and there are types of entity in a spatial information system that do not need names, but are referred to by unique numbers. Some phenomena may have different names at different times, like the Volgograd and Stalingrad in the USSR, or the change from Cape Canaveral to Cape Kennedy and back again to Cape Canaveral for a promontory in Florida.

Information about **position** (discussed more fully in Chapter 4) is needed for locating or delimiting natural or man-made objects on the earth's surface. Usually numerical coordinates are used, but there are other forms of spatial referencing, such as a postal code or a narrative description. Positional information may be used in conjunction with names for identification purposes. For example, the centre of the city of Paris, France is located at (approximately) 48°50′ north latitude, and 2°20′ east longitude, but it is unwise to use such information as the sole identifier for that settlement. Very few place names give any indication of position, although there are some that do, and even then not always correctly.

An entity may not change in any regard except for where it is positioned, for example, a realignment in a highway. But if that occurs, then any identifier that is based on coordinates will have to be altered. Change may also occur via earth processes; for example, a point in three-dimensional space may no longer be there after an earthquake or other disturbance. Positional information is also subject to limitations in accuracy, reflecting principally the instruments used for recording purposes. Not only is accuracy limited if position is measured from moving vehicles; but it is also affected by the models used to establish global measurements, that is the shape of the geoid and the datum used for recording distances above and below sea level.

Because **spatial properties** are conditions that use positional information in some way, we treat them as a separate category, although the data for basic characteristics like perimeter of area can be, and sometimes are, customarily contained in the information system in the same manner (tabulated) as other attributes. Positional information is often used as a basis for spatial properties, like shape, areal extent, or volume; but some spatial characteristics like adjacency and containment do not require absolute positional data. However, sometimes shapes are not clearly defined over time. For instance, rivers can change their riverbed shape and position, creating, perhaps, new temporary islands or enlarging the river banks very much during flooding.

Entities are usually defined and demarcated on the basis of selected characteristics. An **attribute** is a defined characteristic of an entity, for example the composition of a bridge, the size of a lake, the wetness of soil. Entities may have conditions which are not used in their definition, but for which data are collected and assembled in an information system. Some descriptions of entities may refer to **behaviour** in time in contrast to static conditions, as with the property of erosion of a barrier island. Some characteristics may represent **functions** of an entity, for example the activities undertaken in a building. Descriptions of form, the spatial characteristics, may be insufficient but, most likely, necessary for many phenomena.

Whatever the kind of attribute, a particular **instance** of an entity category may be distinguishable from other instances by the particular condition. For example, particular occurrences of the entities labelled cities can be differentiated on the basis of size (number of inhabitants). Instances may be differentiated solely on the basis of a non-spatial attribute value or for a combination of characteristics.

The entities may also be characterized by spatial type, as discussed quite fully in later parts of this chapter. For now, we mention simply point, line or area type of expression for phenomena. Entities of similar nature may be conveniently grouped. For example, highways, railroads and canals can be grouped as transportation features. The grouping may be done on the basis of spatial form. For instance, the category of transportation features could be only those of linear form, the railways, etc., but not the stations, conceived as point elements; or the categorization could ignore spatial form entirely.

Figure 3.3 identifies the commonly encountered terminology, although we must point out that there is no standard nomenclature.

ENTITY	:	Primitive spatial (possibly non-spatial) unit as perceived (existing) or conceived (projected), for example a school
OBJECT	:	Physical representation of an entity
INSTANCE	:	A particular occurrence of a given type, for example Roosevelt School
SPATIAL TYPE	:	Point, line, area, etc.
ATTRIBUTE	:	Descriptive characteristics of an entity, for example, type of school
ATTRIBUTE CLASS	:	Category comprising similar attributes, for example grades of students
ATTRIBUTE VALUE	:	Particular measure of an attribute for an instance of an entity, for example, number of students, or type of school
ENTITY CLASS	:	Category of entities with some similarity, for example all educational establishments
FEATURE	:	A combination of several entities making up a complex unit

Figure 3.3 Definitions for categories of information.

3.1.3 Metainformation

In an information system, indeed in any collection of data, the entities carry a variety of properties and names, but there is also material about the criteria and procedures used for definitions and demarcations. Sometimes this documentary material is referred to as metadata or **meta-information**. Without going into details at this point, we see the ancillary and complementary information including such data as:

1. Definitions of entities.
2. Definitions of attributes.
3. Explanations for measurements of attributes or coding practices.
4. Explanations for false colour encodings used in maps based on remotely sensed image data.
5. Accounts of rules used for delimiting entities spatially.
6. Guidance as to data sources, quality and date.
7. Explanations for missing values or inappropriateness of a measure.
8. Any other information to provide clear explanations of the data.

Indeed, in a spatial database there may be spatial data for some ancillary information. Think of maps showing the variation from place to place of the accuracy of the representation of entities. Furthermore, some material can itself be recorded as attributes: for instance, a column of data could carry the name of source documents or of who encoded the information. Textual data could be stored separately, but associated with the numerical tabular information. We return to the topic of meta-information in section 12.6.1.

3.2 NONSPATIAL ATTRIBUTES

Properties of static conditions, function or behaviour may be numerous. Even though only a few are needed for definition or demarcation, a comprehensive inventory may produce hundreds of attributes for encoding in an information system. Accordingly, attributes can be characterized according to purpose or value, possibly by creating **attribute classes**: for example, those characteristics needed to define wetlands can be specially marked. Groupings of attributes or entities are done for convenience of working with data, but may reflect some particular uses of an information system, or an organizational structure that facilitates many functions of an agency.

By way of example, Figure 3.4 represents an organization of

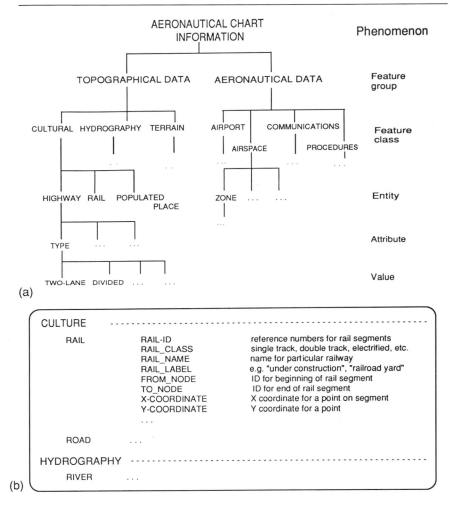

Figure 3.4 Example of classifying information on aeronautical charts. (a) Classification. (b) Some specific items.

phenomena shown on United States aeronautical charts. An initial distinction is made between topographical and aeronautical in accord with the data handling realities of the organization producing the maps, essentially representing two different flows of material. The topographical and aeronautical categories, here called feature groups, each contain major types of entity, here called feature classes. This division recognizes the data flow within the organization for the air travel information and the customary classification of topographical map data, partly itself reflecting map symbology and production processes. Within feature classes are the different entities that, for the purpose of map-making, are meaningful and are divisible no further, for example, highways.

So at the next level, highways are described by attributes like Road_class, Road_name, Road_ID, Road_type, Road_number, some of which are identifiers for record keeping, like ID; or official names, like Pennsylvania Turnpike. For each attribute there are **attribute values**, the specific quality or quantity assigned to an attribute, for example, National, for class of road; or 2,500 metres for road length.

Definitions are affected by cultural differences, for example, the physical entities of city and town are defined differently in many countries. Definitions may change in time, as in the USA with the concept of poverty threshold. A single term may be used for different attributes. In France, for instance, the word for landowner propriétaire, has two meanings (polysemes): the persons who have signed the property document to a notary (holders of property documents) or the persons paying property taxes.

The quantitative or qualitative properties may result from simple measurements, may be produced by many or complex mathematical operations, or may be narrative descriptions. Some characteristics may reflect a state of possession, as with the presence or absence of forest; or categories of land use, referred to, respectively, as binomial or multinomial attributes, or in a conventional classification of **levels of measurement** as nominal data. Some may be based on sequence, such as a ranking of cities (ordinal data); some carry a property of difference in magnitude, like temperature differences (interval data); other attributes are measured according to equality or difference in a ratio (ratio data), such as volumes of traffic on turnpikes. Characteristics which are inherently qualities or result from a classification process using other types of data are referred to as categorical data. For example, the designation of highways as concrete, gravel or bituminous surface, or the use of stages of small, moderate or large towns, are both categorical variables, albeit in the latter case a progression is also implied.

Recognition of such levels of measurement (Figure 3.5) is important for knowing what logical, mathematical or graphical operations are appropriate for specific attributes. For example, it is inappropriate to add or subtract for the nominal variables; here classification is the correct operation. Addition and subtraction operations are legitimate for interval and ratio data, but multiplication and division are legitimate only for ratio data. The diagram also demonstrates the matching of geographical display techniques with the basic nature of the attribute data.

Attributes may also be classified or characterized with respect to:

1. Specific values or ranges of values that are correct for a given attribute.
2. If an attribute is original or derived.
3. If it is necessary for the definition of the entity, such as an identifier.

Specification of different values that are possible provides a means of

	"CATEGORICAL"		"SCALAR"	
	NOMINAL	ORDINAL	INTERVAL	RATIO
Properties				
	Presence/absence	Sequence	Differences	Ratio
	Difference in degree or quality	Relative position	Arbitrary zero	Real zero
	Counting	Greater, lesser	Addition	Multiplication
	Equality of category	or equality	Subtraction	Division
Process	CLASSIFICATION	ORDERING	MEASURING	
Examples				
	Name of city	Large city	Bearing in degrees	Number of people
	Type of landuse	Wettest soil		Number of air passengers
	Name of highway	Primary highway		Distance between two places

Displays

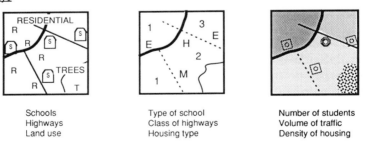

Schools	Type of school	Number of students
Highways	Class of highways	Volume of traffic
Land use	Housing type	Density of housing

Figure 3.5 Levels of measurement. Attributes are measured in several ways; mapping techniques match the numeric properties of the attributes.

checking and **validating** the encoded information. For example, only five types of land use may be recognized, or lakes must have an area greater than zero, or people may be no older than 120 years. In a database, one or more attributes may have no information recorded, or if they do, they may have been derived from other attributes contained there rather than being independent measures or qualities. For example, population density is derived from the number of people and area of census zones. At a given moment, or even permanently, the values for density may not actually exist. Good documentation about the attributes and entities may be undertaken by specially marking those properties used for definitions.

Many techniques are available for creating **derived measures**, and many different operations may be involved. Some types of operation are:

1. Value reclassification from real numbers, that is, interval or ratio scales, to ordered categories.
2. Value substitutions, such as averages, particular values like a maximum.
3. Value creation through combinations of values for different attributes via arithmetic operations, perhaps using advanced statistical techniques.
4. Value creation through Boolean logic combinations; via trigonometry or interpolation.
5. Value assignment, substituting values from one variable for another.
6. Those which involve a spatial property as well as non-spatial attributes.
7. Those which result from simulations, or produced by 'what if?' modelling.

Thus, we distinguish between measured properties and values which are produced for simplification, because some properties are inherently measurable, or because new scales are needed for modelling purposes. The example of Figure 2.16, showing the prediction of the location of a target area for mineral discovery, uses data for several measurable variables, such as iron content in rocks, to stand for unmeasurable qualities like rock deformation. The example of Figure 2.6 uses a procedure of combining different original attributes to estimate the potentiality for residential development by means of a new scale of values.

Because our primary purpose is to emphasize spatial data, we shall not systematically elaborate on the logical, mathematical or statistical operations possible for attribute data. However, in different contexts in the next few chapters, some of these items will be encountered in discussions of data representations or demonstrations of analytical procedures.

3.3 SPATIAL CHARACTERISTICS OF ENTITIES

Entities may also be classified according to spatial properties. Accordingly, the entities are often thought of as basic **spatial units**, the fundamental, practical, non-divisible element matching a perceived entity and holding the encoded data. These units may be seen from three perspectives:

1. From that of the number of spatial dimensions.
2. From that of the type of spatial property.
3. From that of how they may be combined for a particular purpose.

3.3.1 Dimensionality of entities

Suppose we see a part of the earth's surface on which are scattered phenomena; we can think of the entities as points, lines, areas, volumes or combinations of those. Assuming we can recognize and demarcate separate entities, we have different **spatial dimensions** (Figure 3.6). In principle, the point, line, area and volume descriptions refer to the zero, one, two and three dimensions of geometric figures, with respective properties of no possible measures, length, length and width, length, width, and depth. The idea of zero, one or two dimensions is person-made, for in reality everything is three-dimensional.

As a practical matter, a dot on a map has size. On the other hand, the dimensionless concept of a point for a city will, upon enlarging a scale of viewing, as for the map representation, become an area object, and at times a volume entity as in a physical three-dimensional model. The well-known adage of 'the further away you are the less you see' is as important for spatial information systems as for many other topics. This **scale** of observation or mapping will affect what might be represented in a database. For example, some entities may disappear altogether because a minimum size of unit is established for a particular scale. This concept of **resolution**, the minimum addressable spatial unit, for example a line no less than 0.005 of an inch, or a square of 2.5 hectares, is related to scale but is not the same.

Point entities, defined for a given scale of observation, will appear in a database as separate lines of information in a table, with two items of information for position, one, or possibly two, items for identification, and other attributes of interest. Each point is seen as independent of every other point, until such time as derived statistics like distances to neighbours might be involved, so that the ordering in the table by rows is

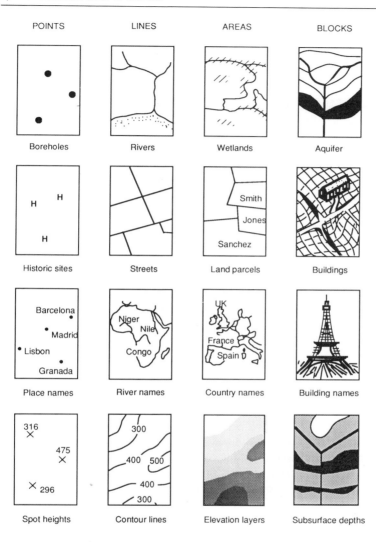

POINTS　　　LINES　　　AREAS　　　BLOCKS

Boreholes　　Rivers　　Wetlands　　Aquifer

Historic sites　　Streets　　Land parcels　　Buildings

Place names　　River names　　Country names　　Building names

Spot heights　　Contour lines　　Elevation layers　　Subsurface depths

Figure 3.6 Types of spatial units.

not important. As we shall discuss later, points may represent phenomena of higher dimensions, or may simply be locations for the placement of names on maps. Point spatial units are especially important in spatial sampling, for example where to drill boreholes. In utility and transportation systems inventory and analysis, such zero-dimensional units record the location of telephone poles or sewer manholes, or are used in recording data for individual houses with discrete addresses. In geology and surveying, point positions on the earth's surface, known as control points, are important zero-dimensional entities.

Line entities (also referred to as polylines, arcs or edges) in a simple view of phenomena like pipes and railroads, are separately occurring phenomena, unrelated to each other in any way. So we think of the Mississippi River, the M1 motorway, or an air link from Madrid to Miami. But these may be parts of a set of linear elements, all the rivers of the Mississippi system, all motorways in Britain, all air routes of the world. The entities may be directly observable or may be conceptual only, as is the case of air routes, and the most interesting combination of a sinkhole and spring in a karst (limestone) landscape. They are not necessarily only in a horizontal plane, for example a borehole or a water level in a mine. Linear features are important in transportation studies, hydrology, utilities management and geology, and are prominent features on many types of map.

Some spatial properties defined for linear features are:

1. Length of entity: for instance, length of canals.
2. Sinuosity, for example, the Lower Mississippi River bends.
3. Orientation, such as compass direction of mineral veins.

Area entities (sometimes referred to as polygons, regions or zones) may be identified for natural or man-made phenomena, assuming in most cases a world of the earth's surface without height. The area spatial units may be natural entities like lakes, islands, territory with a particular soil type, or the top of a building. The units may be artefacts used for statistical reporting like census zones or delivering mail like postal zones, or discretizations (the creation of pieces or segments) of continuous space like climate regions. Boundaries may be unclear, multi-attribute, changing in time, variable according to definition, and may not be directly observable. Area units are important in socio-economic studies, analysis of terrain conditions, land use and natural resources inventory, and recording of real estate.

Particular spatial properties associated with area entities are:

1. Areal extent: for example, the size of lakes.
2. Perimeter length: for example, the extent of a shoreline.

3. Being isolated, or connected to others: for example, the separate elements of the Hawaiian Islands, or the contiguity for countries of the Middle East, or the historical case of the West Berlin exclave of West Germany.
4. Being punctured or indented: for example, the District of Columbia 'hole' within the United States, or the small Spanish village, Llivia, completely surrounded by France, or, in North America, the Chesapeake Bay with islands and a very irregular shoreline.
5. Overlapping: for example areas of circulation of different newspapers, or non-overlapping, for example school districts.

Volume entities (solids or blocks or polyhedra in some contexts) bring the third dimension of height or depth into the picture, giving us recognition of buildings or earth features like volcanoes or chasms. The phenomena may have clearly marked boundaries, or they may not be directly observable with current measuring instruments, like aquifers or mineral zones below the surface. The volumes may be three-dimensional equivalents of statistical artefacts, for example, the faces of polyhedra used to model mountains. Volume entities are important in geology, architecture, oceanography, geotechnics, meteorology and climatology. The phenomena may be from the physical world, for example, ocean currents, or geological facies; or man-made or man-visualized, for example airspace or drifts and stopes in mining; or three-dimensional utility networks, or shops in the underground area of downtown Montreal.

Some particular properties pertaining to three-dimensional spatial units are:

1. Volume, as in engineering cut-and-fill operations.
2. Area for each two-dimensional component.
3. Length of perimeters.
4. Being punctured or indented.
5. Having isolated or connected pieces.
6. Profiles and cross-sections.

3.3.2 Geometric elements

From the viewpoint of fundamentals of geometry, the object point is created by every crossing or termination of lines (Figure 3.7a). Consequently, points are said to be without size. The notion of a line comes from our visualization of spatial separation between two distinct points, and the connection of those two points by a line growing from one

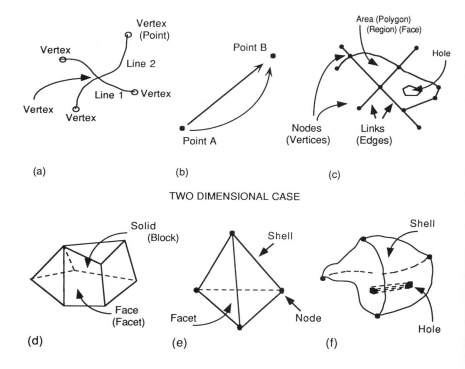

TWO DIMENSIONAL CASE

THREE DIMENSIONAL CASE

Figure 3.7 Geometrical units. (a) Linear units. (b) Concept of a line. (c) Units for complex objects. (d) Solid object. (e) Simplest solid. (f) Irregularly shaped solid with hole.

to the other (Figure 3.7b). The idea of growth in one direction leads us to say that lines are one-dimensional, and to distinguish between extent and dimensionality. The latter concept holds whether the lines between points are straight or curved.

Considering sets of fundamental geometric entities (Figure 3.7c), we recognize a **node** (usually called a vertex in geometry and engineering) as a point that terminates a line or is the point at which lines cross. Therefore it has a property of connectedness, being related to the lines. These lines are called **links**, sometimes **arcs**; and are referred to as **edges** in geometry, if both ends terminate in a node. Every link has two nodes. A set of three links is a minimum needed to bound a two-dimensional surface, or **area**. For the third dimension the **solid** is made up of a set of areas or faces (facets), which combine to produce an object with an inside and outside (Figure 3.7d). In this type of entity some faces touch and some have common vertices. For the same set of points, three-dimensional figures may be regular or irregular in appearance (Figures 3.7e and f). The outer boundary of the complex figure, a polyhedron, is known as the shell, and any set of line segments is a loop. Solids may be punctured, having one or more holes, and may be connected.

Governments, academicians, and others, besides geomaticians, have produced classifications of spatial objects which often use different terminology from that used in the fields of geometry or computer graphics, a field very important to the practical creation of spatial information systems. One recent effort in the USA (National Committee for Digital Cartographic Data Standards, 1988) has produced a scheme proposed as a standard for two-dimensional cartographic entities. In this scheme zero-, one-, and two-dimensional cartographic objects are classified according to three circumstances of spatial property:

1. Geometry-only operations (as generally involved in map drawing).
2. Topology-only operations (as might be required or sufficient for some kinds of spatial analysis, like accessibility of nodes in networks).
3. Geometry and topology.

In cartography, zero-dimensional entities may be perceived real phenomena, an **entity point**; points with text, a **label point;** points for carrying attributes of areas, an **area point**; and topological junctions or end points, a **node**, when dealing with networks of linear features (Figure 3.8). A set of coordinates specifies the location of any type of point.

A line or one-dimensional object may be a straight line connecting two points, a **line segment**; or non-straight, a **string** (called **polyline** in

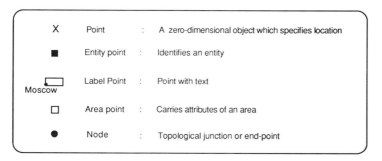

Figure 3.8 Point cartographic objects. (Based on material in the recommendations of the US National Committee for Digital Cartographic Data Standards, as later published as National Committee for Digital Cartographic Data Standards.)

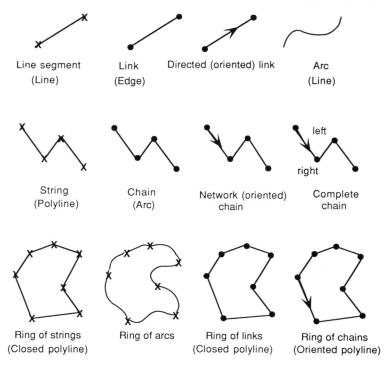

Figure 3.9 Line cartographic objects. There is no standard terminology for the names of cartographic entities; some synonyms are
indicated. (Based on material in the recommendations of the US
National Committee for Digital Cartographic Data Standards, as later
published as National Committee for Digital Cartographic Data
Standards, 1988.)

computer graphics), being defined as a locus of points or a set of line
segments (Figure 3.9). The locus of points, an **arc**, may be defined by a
spline curve or polynomial mathematical function, or otherwise. If the
topological property of connectedness is considered, the basic unit is
termed a **link** (part of a set in a network), which may be directed (usually
by specifying origin and destination nodes) or a **chain** if several line
segments are involved. If the property of neighbours for the line features
is involved, then we define an area chain; if both connection and
adjacency properties are involved then the set of links produces a
complete chain. To demonstrate the variations in terminology, note that
the word arc is used in several ways: as in this proposed standard, in some
software systems as a synonym for complete chain, or, in France, as an
oriented edge. If the line feature begins and ends in one point, then the

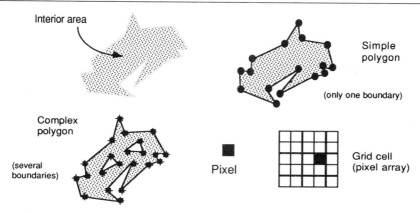

Figure 3.10 Area cartographic objects. The boundary is not of interest for the 'area' unit. A complex polygon may also have disconnected pieces. (Based on material in the recommendations of the US National Committee for Digital Cartographic Data Standards, as later published as National Committee for Digital Cartographic Data Standards, 1988.)

spatial unit is called a **ring**. Rings may be created from strings, arcs, links, or chains (Figure 3.9).

Two-dimensional entities (Figure 3.10) are bounded spatial units which may or may not include the boundary for carrying information – the boundary being necessary for demarcation purposes. If the boundary is excluded, the term **interior area** is used; if the boundary is included then we have a **polygon** consisting of the interior area and an outer ring. A polygon may be simple if it has no interior rings, or complex if it has at least one interior ring. (In some contexts the terms closed polyline and polygon are synonymous with ring.)

Three-dimensional objects, not covered by this proposed standard classification, with the addition of the vertical dimension can be treated similarly to two-dimensional objects. Thus we can recognize the primitive elements of line segments and area segments, or faces, and whether or not topology properties are involved.

Depending on circumstances, the same primitive elements may be needed for two different definitions: area entities in which the boundary line is the feature of interest, or area entities in which information is carried for the character of the enclosed territory. These differences, and the possible complex mix of entities, are demonstrated by the example of aeronautical charts (Figure 3.11), possibly the most complex and congested of any kind of thematic map. Civilian aviation needs to know of the existence (and, of course, the position) of airspace restricted to

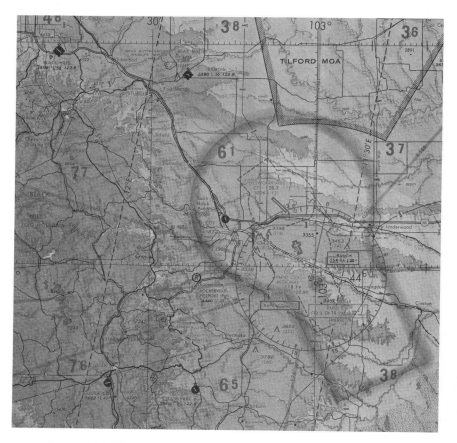

Figure 3.11 The example of aeronautical charts. This chart is included as an example of source documents with a high density of data, and many different types of spatial entity. This example is a portion of an original multicolour aeronautical chart of the area of Cheyenne, Wyoming, USA. The example shows the elevation zones, airport characteristics, airspace, communication facilities, urban areas, and other cultural and aeronautical information. (The chart is one of a series, Sectional Aeronautical Charts, at a 1:500,000 scale, published on a regular six month cycle by the National Ocean Service, US Department of Commerce, Washington, DC, USA.)

military aircraft – the boundary condition is important, not the character of the airspace itself. But at the same time, the civilian aircraft may need to recognize zones within which particular radio frequencies can be used. In some fields the three-dimensional objects are discrete, for example, separate architectural structures. In others, they are joined, as in the example of geological strata.

The character of entities and space is often expressed in databases by the use of discrete primitive recording units as demonstrated above. How well the data retrieval part of the information system works depends on the particular circumstances of the uses of the data. It is in this regard that we now look at combinations of spatial entities and alternative ways of recording spatial variations.

3.4 COMBINATIONS OF ENTITY TYPES

Spatially oriented processing as described in Chapter 2 can involve many types of entity at one time, or for one problem on many occasions. Cartographic representations can deal with many types at one time.

3.4.1 Combinations of spatial units

Spatial units may be combined in different ways:

1. Spatial problems can be described according to which spatial units are involved, and particular tasks may require several types of spatial unit.
2. Complex entities may be created by combining units of different spatial dimensions and type.
3. Spatial units of one type are combined to produce a new type, a compound entity.
4. One type of spatial entity may be transformed to another type.
5. Some spatial entities have duals, that is, a matching entity of different dimensionality.

Location problems or spatial queries may require many combinations of units. Taking only pairwise or **binary combinations** for point, line, or area entities, there are nine possibilities (Figure 3.12). It may appear at first sight that there are only six: point–point, point–line, point–area, line–line, line–area, and area–area, but it is important to distinguish between one entity as conceptually dependent on another. For example, we may be interested in the location of cases of AIDS within low income census tracts (point within area), or the areas containing talented and gifted students (areas containing points). A problem may be posed differently conceptually in terms of apparent causal dependence or, as a practical matter, the processing sequence for a spatial problem of points and areas can give different results.

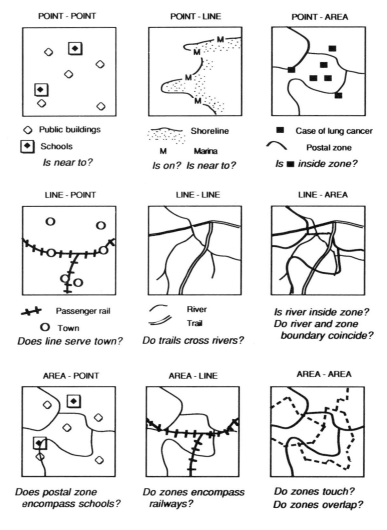

Figure 3.12 Binary combinations of spatial units.

Each combination has associated operations that may require specific tools in an information system, as illustrated. For example, point–point relations can involve distance measures or comparisons of coordinate values. Point–area relations are inclusion queries or problems like finding the extent of visibility from a helicopter. Line–line relations include such elements as coincidence of features, or determination of intersections.

Secondly, entities may be defined as combinations of more than one spatial unit (Figure 3.13a). If these combinations consist of different types of entity, we have **complex objects**. A parcel of land with a house, water utility pipe, frontage onto a road and a garden may for some purposes be treated as a spatial entity, called an improved lot. A collection of parking spaces and adjacent kerbs, a parking lot, is another example.

If the combination consists of different instances of one type, for example links in a transportation network, we have **compound objects**, also, of course, a type of complex object. In many spatial problems, the unit of interest is a combination of two places, or spatial **dyad** (Figure 3.13b). Most often this is a pairing of point objects, like cities at the beginning and end of air routes, or points representing area entities as in the study of streams of migration between countries. However, more generally, it is a linking in space of particular entities as implied in measuring distances, or finding direction from an origin to a destination. Linear and areal spatial entities may also be either continuous or discontinuous: that is, there may be breaks in a set of pipes, or gaps in a land parcel system.

In the transportation case we have the unit of the **net** or network, or graph, a set of links (Figure 3.13c). This unit has properties in addition to those of an individual link, for example the possibility of identifying paths, and the definition of network connectedness. Point objects may be combined to produce **lattices** of points (Figure 3.13d); these can be thought of as networks or arrangements of points in different patterns, like squares or triangles. Area entities are combined to produce **tessellations** (Figure 3.13e), sets of areal objects of either regular or irregular form that cover all space in a defined area. Facets are combined to produce **polyhedra** (Figure 3.13f). In summary, we produce new entities at a higher level than the elements from which they are constituted.

3.4.2 Substitutions of spatial units

Thirdly, particular types of unit may be substituted for others, involving either a reduction or increase in dimensionality. Points or lines may stand

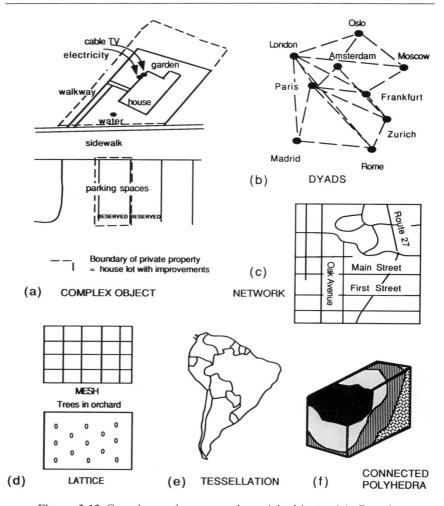

Figure 3.13 Complex and compound spatial objects. (a) Complex objects. (b) Dyads. (c) Network. (d) Lattice and mesh. (e) Tessellation. (f) Connected polyhedra.

for area units, although in practice it is usually a point like a mathematical **centroid** or an arbitrary location within or without the spatial extent of the area that represents the two-dimensional object, rather than the boundary line (Figure 3.14a). In any event, the attributes for the areas are carried by the units of lower dimensionality. Such changes may facilitate mapping or mathematical operations.

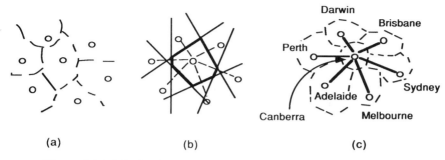

Figure 3.14 Substitutions and duals for spatial objects. (a) Centroids for polygons. (b) Proximal areas for points. (c) Points and polygons for a graph.

The reverse, an increase in dimensionality, sometimes occurs. There are ways to create **proximal areas** from data for only points (Figure 3.14b). That is, territory about specific points is demarcated so that all possible locations defined by the areas are associated with the nearest specific point. Such a technique has been used for demarcating statistical units representing the postal areas based on zip codes of the United States Postal Service, because few postal areas are delimited. The postal units are seen by the Postal Service more from the point of view of streets followed by walking or driving mail carriers, not for obtaining statistics for zones. Similarly, in the Netherlands, the postal zones are the set of streets making up a daily workload of a single postperson.

In some cases, point and area spatial units are **duals** for each other (Figure 3.14c), that is, pairs of matching entities. In other words, there exists a set of polygons for a set of points, as in the example of the postal zones created from points for post offices. For transportation networks, line elements, a set of polygons and matching points can be defined as duals. Imagine European countries with capital cities, mapped as points within polygons. The cities can be connected by links, and polygons produced for those points.

3.4.3 Mixed uses

Fourthly, spatial problem solving and querying may use information for several types of spatial entity at one time (Figure 3.15). Indeed, we can recognize categories of location problems in this way. Derived measures of spatial properties can involve combinations. For example, the regularity with which point features are distributed over a landscape or in

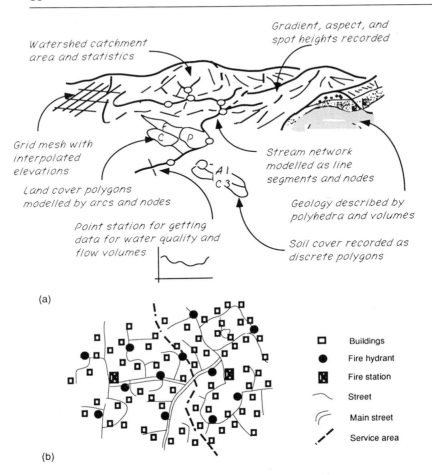

Watershed catchment
area and statistics

Gradient, aspect, and
spot heights recorded

Grid mesh with
interpolated
elevations

Stream network
modelled as line
segments and nodes

Land cover polygons
modelled by arcs and nodes

Geology described by
polyhedra and volumes

Point station for getting
data for water quality and
flow volumes

Soil cover recorded as
discrete polygons

(a)

(b)

	Buildings
	Fire hydrant
	Fire station
	Street
	Main street
	Service area

Figure 3.15 Examples of the use of many types of spatial unit.
(a) Terrain analysis. (b) Fire protection services.

a study area can be measured by a summary spatial statistic requiring
initial measurements between nearest neighbour points. The flow of
chemically laden water over a landscape involves point-to-point properties
of gradient and flow through channels (Figure 3.15a). The determination
of the area served by fire stations involves working with point, line and
area units (Figure 3.15b).

Dealing with spatial relationships is quite varied, often complicated,
and sometimes difficult to handle. The possible set of all relationships is
huge in number – for example, there are over nine million dyadic pairs

for the binary connections among counties of the United States. As discussed, the relationships may be metric, topological or ordered. They may include many instances of one class of entity, or mixtures of several types; and they may be represented in ways that are indirect and not obvious. Some may be encoded fully in a database as a complex object; some software systems may not allow this, recognizing only primitive elements of point, line, or area.

3.5 CONTINUOUS VARIATION OVER AND IN SPACE

A view of phenomena as entities is not the only conceptual model that can be applied to spatial information. Basic English expressions contrast phenomena that occur *on* the earth's surface or *in* the atmosphere from phenomena that are distributed *over* space. For example, Finland has many lakes, and coniferous forest covers much of its land. The continuity of space is also represented by concepts of gradients and mathematical integration over area. For our purposes, we will discuss first the distinction between an entity oriented view and a field oriented view of space; and, secondly, address the condition of isotropicity.

3.5.1 A field view of spatial variations

Discussions in sections 3.3 and 3.4 have dealt with the **entity-oriented** view. While that treatment of spatial entities did include reference to the possibility of continuity, as in the case of a network or lattice, and implied by a mention of contour maps, the discussion was focused on spatial units as separate entities. In the **field** view of space (Goodchild, 1987) we explicitly recognize **continuity** as the principal element. That is, any position in two- or three-dimensional coordinate space has a value of an attribute. Even though in practice there is a limit to the number of possible cases, in theory we have: $\{a, x, y, z, t\}$ where x, y, and z are the conventional symbols for position, t is the time element, and a is a single or set of attributes. The value of a is, in principle, empirically confirmable. That is, if a person is dropped from a plane to a location on the earth, the condition of one or more phenomena at that exact place is observable.

In contrast, the **entity-oriented view** has the condition that a is not everywhere defined, representing phenomena as $\{e, a, s, t\}$ where e is an entity, a the single or multiple attributes, and t is time. The entity also has associated positional information, s, being as complex as necessary for

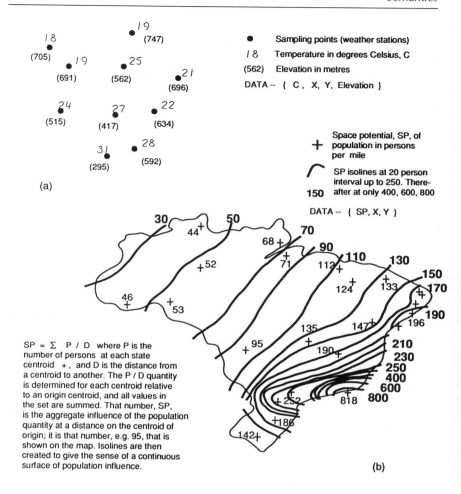

Figure 3.16 Continuous space. (a) Discrete (point) observations of temperature. (b) Aggregation over space using a space potential index.

bounding the point, line, area and volume spatial units.

Natural, physical, or biological phenomena are often clearly continuously variable – temperature, barometric pressure, ocean salinity, soil wetness, slope of terrain and other attributes (Figure 3.16a). At times discontinuities may occur, as in cliffs or weather frontiers, but from a global view they account for a very small portion of all possible locations. Social phenomena may also be thought of as spatially continuous at a particular scale, or can be measured by derived statistics that represent a field view, for example, variations in population density, income levels, aggregate accessibility. Figure 3.16b has examples for a spatial statistical artefact, a continuous surface of population potential, or the integration over area.

As a practical matter, though, the continuously varying attributes are often perceived as discrete elements:

1. As sample points, lines or areas.
2. As isolines and bands between the isolines.
3. As connected pieces.

Sampling, not always scientific, creates a reduced number of locations for carrying the attribute data, so that x, y, and z are a subset of all possible triads of latitude, longitude, and height (Figure 3.17a). A set of geodetic control points, or spot heights is a common technique for carrying data for elevation of terrain above some datum line, that is an accepted horizontal base level related to some sea level definition. The points may be irregularly distributed as in the case of survey triangulation based control points, or regularly distributed as in the case of elevations interpolated from a grid superimposed on existing topographical maps. Lines or areas are at times used as sampling units, as in landscape profiles or plant species surveys using quadrats, respectively.

The contouring of terrain elevation or atmospheric temperature uses the spatial unit of line segments or arcs of equal attribute value (Figure 3.17b). More generally known as **isolines**, this unit partitions space into bands or zones, within which some assumptions are made about the nature of the variation in the phenomena over space – perhaps directly proportional to distance between isolines, or perhaps resulting from a higher order mathematical equation. At times there may be no information provided as to the variability between isolines.

The production of segments, or **piecewise models**, is a common practice for some area phenomena, especially soils, land use, land cover, and other terrain conditions; but also for some social phenomena. In this process, areal units are delimited somehow so that boundaries can be drawn for polygonal entities, producing land use zones, and the like. The attribute information is carried by the area elements, and the boundaries reflect varying assumptions of change between adjacent units. A sharp break or discontinuity is generally assumed, without any possibility of providing a mathematical statement for the differences, and varying assumptions are made about variation within the units. These may be homogeneity, linear, or non-linear variations.

The concepts of fields and **discretization** are also applicable to

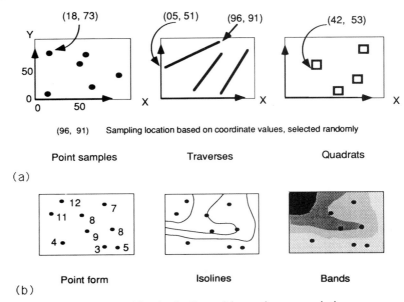

Figure 3.17 Practicalities in dealing with continuous variations.
(a) Some spatial sample types. (b) Continuous field of temperature.

networks. The difference from the case of the complete surface is that x, y, and z have null values in many places: that is, they are not defined off the graph. Otherwise, there can be continuity of values along the links of a graph; sediment load and vehicle densities are two attributes of this kind for hydrology and transportation networks respectively. The discretization can similarly use point or line unit sampling, or segment into particular line elements characterized by homogeneous or other assumed characteristics sufficiently different from neighbours to justify dividing into new units; for example, to recognize variations in pavement conditions.

3.5.2 Isotropicity

A particular aspect of the continuity of phenomena is the differentiation between isotropic and anisotropic (Figure 3.18). An isotropic condition exists in geometry when conditions do not vary with direction. That is, the measure of distance does not vary with direction from a point – a mile is a mile is a mile. In contrast, because some phenomena exist only as channels, for example river or transportation systems or ocean currents,

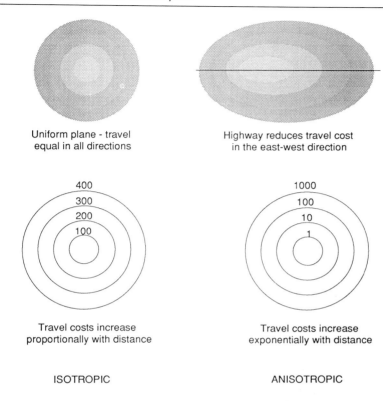

Uniform plane - travel
equal in all directions

Highway reduces travel cost
in the east-west direction

Travel costs increase
proportionally with distance

Travel costs increase
exponentially with distance

ISOTROPIC

ANISOTROPIC

Figure 3.18 Anisotropicity.

the value of space does differ depending on direction, being associated with presence on or off the network.

Most clearly associated with networks, this **anisotropic** condition is seen, also, in the difference between kilometres distance and temporal distance. In geography it is often represented by the measurement of the distances between places by different measures of accessibility, usually expressed and displayed by isoline maps implying continuous variation with directional components, as demonstrated by Figure 3.18.

3.5.3 Discrete and continuous views

We can now categorize phenomena into five fundamental types of unit, following Goodchild (1987). Area objects are differentiated from the connected discrete pieces or units in a plane that make up a tessellation superimposed on continuous spaces; lines are differentiated from line segments. Points are clearly defined in only one sense. The set of non-overlapping units representing discretization of space is often created in a practical sense by producing sets of polygon units from digitized line data; the set of terrain elevation values for all possible locations is often created by spatial interpolation from a lattice of a small number of point locations.

A field view of reality seems to offer advantages in dealing with spatial and attribute data error. That is, it is straightforward to imagine an error band for the values for an attribute measured at a particular point location; but it is difficult to establish error based on discretizations. On the other hand, treating complex and compound objects is not consistent with the continuous surface view of the world; and people tend to think of objects not gradients. At the same time, as discussed more fully later (initially in section 4.5.3), information systems can be more efficient for some purposes if reality is modelled as a continuous map rather than as a set of points, lines and polygons; but there are problems in encoding a very large number of points in space as implied by the field view.

3.6 SPATIAL AND NON-SPATIAL PROPERTIES TOGETHER

Even though at times research and practice may be able to use data only for spatial form or location of objects, much analysis, modelling and querying is based on attributes that are non-spatial. That is, we are interested in the character of space, spatial covariation, location–allocation problems, etc.

Several types of analysis can be recognized on the basis of spatial or non-spatial properties, types of spatial unit, and need to create new spatial units (Figure 3.19):

1. Analysis of the attributes of a single class of entities.
2. Analysis of the attributes of pairs of places.
3. Analysis using both attributes and spatial properties for a single class.
4. Analysis with both for pairs of places.
5. Analysis that involves more than one type of spatial entity, for attributes only.
6. Analysis of attributes and spatial properties for different entity types.
7. Analysis which creates a new class of entity.

The first two types of analysis consist of enquiries for only non-spatial properties. The enquiries may be retrieval from tables of all records for a type of spatial unit that meet a specified condition, for example, land

	ATTRIBUTES ONLY	ATTRIBUTES AND SPATIAL PROPERTIES
SINGLE CLASS OF ENTITIES	Parcels owned by Pierre Leconte	Uses of land owned by Pierre Leconte near to a coastline
PAIRS OF PLACES	Number of air passengers for selected city pairs	Correlation of migration stream volumes with distance between zones
MORE THAN ONE SPATIAL ENTITY CLASS	Correlation of crop yields with data from weather stations	Analysis of health conditions for neighbouring countries
CREATION OF NEW CLASS OF ENTITY	(not applicable)	Noise pollution bands on either side of highways

Figure 3.19 Attribute and spatial data. (After Goodchild, 1987.)

owned by Pierre Leconte, or analysis of all records comparing or correlating two attributes. The difference between the two categories is the need in the second to relate two spatial units in order to create the dyadic unit. (As such, we do not consider creating a dyad to be the same as the seventh category, to be discussed below.) In practice, the query and analysis needs are met by information systems and statistical analysis software packages or routines.

Analysis may involve both spatial and non-spatial attributes. An assessment of political gerrymandering (originally, deliberate and biased boundary drawing to favour one political party over another) might require data for voting splits for different political parties and for the shape of the voting districts. The empirical study of distributions of drug-related violent crimes could benefit from finding the mean geographic centre of low income apartment dwellings. An example for units which are dyads is the correlation of the volume of migrants for the pairs of places with the distance of the migration stream.

At times more than one type of spatial unit may be required, especially if dyads are included. For an example without dyads, the crop yields in sample areas may be correlated with temperature and precipitation conditions at weather stations (point objects). For most spatial interaction studies or modelling the attributes of the pairs of places, for example, the number of migrants between specified origins and destinations of migrants, and the nature of the origins or destinations separately are required. An assessment of spatial autocorrelation could use a statistic comparing the attributes of polygons which are neighbours with those which are not neighbours. This statistic requires access to the area type unit and to the line entity (the boundary for adjacent polygons). Traffic studies may require the combination of point and line entities: for instance, traffic lights are features associated with the lines (streets) and points (intersections).

Some analyses require creation of new entities, either unary or binary. If original data consist only of points in space, then investigations of pairs of places requires creation of dyads. If data exist for points only, it may be necessary to create polygons to which the attributes of the points apply, as in the use of Thiessen polygons to produce postal code areas from the location of post offices. When buffer zones are created for original point or line entities, a new polygon unit is formed.

Also, new instances of one or more spatial unit types may be created in some processes, even if a new class of entities is not created. For example, placing distance based buffer zones around polygons produces more polygons. If the spatial data are contained in a layered database, then overlays produce new areas, lines and points.

Nor are the spatial linkages the only relationships that may be involved. Entities or arbitrarily discretized pieces of continuous space may also be related in functional/behavioural and ownership senses. The first of these includes real world situations like stream flow, the physical connection between entities or the process relationships of changes in the location rather than attributes of entities, such as the encroachment of a barrier island on a lagoon. Ownership relationships refer to grouping of elemental units; for instance, counties, into higher level units, like states or countries.

Studies involving spatial relationships are, therefore, not inherently simple. They may require computations to produce values for spatial metrical properties, or searching to establish topological attributes. While not all operations will be needed in a given case, it can happen that some procedures may not be possible at all with a particular software package because it is not well matched to a particular conceptual model required for analysis. The exceedingly complex spatial reality must be converted to a fixed and somewhat limited form to reside in an information system. As demonstrated here and in previous chapters, there are numerous alternatives, each with its own merits and limitations.

3.7 AN INTRODUCTION TO THE MECHANICS OF SPATIAL DATA ORGANIZATION

Maps have been used for over two millennia by many individuals, cultures, organizations, professions and disciplines to represent knowledge and perceptions of space and its character. The spatial form of buildings is easier to convey in pictures than with words. However, sometimes for convenience, or sometimes because of available instruments, tables of data are often the repository of spatial data. A table is not practically consistent with a field view, but objects or discretizations of continuous space, for some purposes, can be effectively represented in tables of numbers or other symbols. The table by itself has no planned connotation of spatial location, whereas a map by definition contains the position of phenomena. That is, in a table, the ordering of rows and columns is not necessarily important, but it is for a map.

3.7.1 Tables and matrices

Tables, one, two or multiway, linearly organized information, can be used for attributes of different spatial entities, with or without positional information. Entities are, by custom, placed in the rows, and attributes in the columns (Figure 3.20a and b). If the rows contain point, line, area or volume spatial units we have what may be called basic geographic, geological, demographic, or whatever, databoxes. Selections of information reflect different activities or perspectives. In the case of a geographic data matrix, by way of example:

One row represents the character of a place
One column represents the distribution of an attribute
Two or more rows identify the comparative character of places
Two or more columns allow for correlations among attributes
A subset of several rows and columns represents an inventory of
    conditions for regions

In all cases, only attribute data are used, and the order of rows is unimportant. Even if some attributes are inherently spatial, they are included in the table after they are derived or computed from other data for position or topology.

Complex and compound objects may be included in such a matrix organization, for example, the rows could be paths through a network, but for some spatial entities it is more useful to think of a somewhat different type of table. Especially, the **dyadic matrix**, containing one or

SPATIAL UNIT	NUMBER OF PEOPLE	AREA
Argentina	...	...
Brazil		
Chile		
...		
Venezuela	...	...

(a)

SPATIAL UNIT	LENGTH (miles)	TRAFFIC VOLUME	SURFACE TYPE
Road segment 1	4.3	35,000	Concrete
Road segment 2	2.7	18,500	Asphalt
Road segment 3	3.9	17,750	Asphalt
...	...	...	...
Road segment 567	0.6	9.850	Concrete

(b)

DESTINATION ORIGIN	New York	London	Dakar	Rio de Janeiro	Cape Town
New York		3500	3800	X	7900
London	3500		2700	5800	6800
Dakar	3800	2700		3100	4100
Rio de Janeiro	X	5800	3100		3850
Cape Town	7900	6800	4100	3850	

(c)

Figure 3.20 Data matrices. (a) Point or area entities. (b) Line or dyad objects. (c) Dyadic origin-destination matrix.

more attributes for object pairs, arrayed in both rows and columns, should be identified. In this matrix one row represents conditions for places perceived as origins, one column relates to destinations, and one cell has some property relating to the pair of places (Figure 3.20c). Topological properties of contiguity and connectivity are neatly represented in this form, as demonstrated in section 5.5, and mathematical operations appropriate to matrices can be used to derive other pieces of information.

Ordinarily, as demonstrated by Figure 3.21, positional information is organized in simple tabular form. However, line and polygon entities will usually have sets of coordinate pairs of varying number, so that a matching row–column arrangement is not always possible.

3.7.2 Maps

Maps have been used as devices for revealing details of shape of entities as well as position and variations in attributes. A field view can be well accommodated, although some discretization is often practised for portrayal of the continuous surface, albeit not always effectively. Maps can show clearly objects, simple or complex, unitary or compound, but again certain cartographical techniques are employed for effectiveness. Sometimes the imagination of cartographers is challenged where there is a high informa-

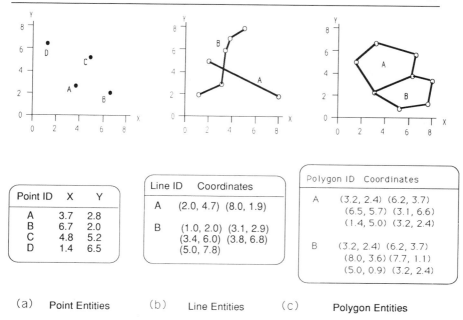

Point ID	X	Y
A	3.7	2.8
B	6.7	2.0
C	4.8	5.2
D	1.4	6.5

Line ID	Coordinates
A	(2.0, 4.7) (8.0, 1.9)
B	(1.0, 2.0) (3.1, 2.9) (3.4, 6.0) (3.8, 6.8) (5.0, 7.8)

Polygon ID	Coordinates
A	(3.2, 2.4) (6.2, 3.7) (6.5, 5.7) (3.1, 6.6) (1.4, 5.0) (3.2, 2.4)
B	(3.2, 2.4) (6.2, 3.7) (8.0, 3.6) (7.7, 1.1) (5.0, 0.9) (3.2, 2.4)

(a)   Point Entities   (b)   Line Entities   (c)   Polygon Entities

Figure 3.21 Tables for coordinates. (a) For point entities. (b) For line objects. (c) For polygons.

tion content. Aeronautical charts (see Figure 3.11) appear to win hands down for the highest density of information presented in map form.

The table and map can, of course, be used for containing information selected or portrayed according to other views. Maps of only topological information can be prepared from data embodied in topographical maps, as discussed in more detail in section 5.1 and illustrated there in Figure 5.1 for a rail transit network. The reverse process may produce several 'maps' from one set of topological information, because the metric information has been thrown away. Maps of topological spatial properties are generally known as graphs because they do not appropriately contain positional information, although this may be present as a spatial reference system for orientation purposes.

In any event, we can regard a **map model** as an alternative form of looking at reality – often a map is placed in the stream of data creation between the real world and data encoding. In this case, the role of the map as an intermediary view should be recognized, and the user should be aware of cartographical conventions of generalization, feature displacement and symbolization.

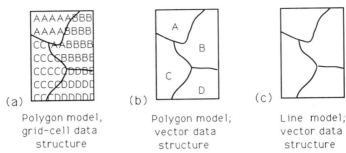

(a) Polygon model, grid-cell data structure

(b) Polygon model; vector data structure

(c) Line model; vector data structure

Figure 3.22 Data model and data structure. (a) An example of a polygon model with a grid-cell data structure. (b) An example of a polygon model with a vector data structure. (c) An example of a vector data structure for a line model.

3.7.3 Data models

The map model has limitations and may indeed constrain the perception of space. The map, predominantly oriented to a static, two-dimensional view of the world, is only one influence on the conception of spatial objects and variations. In a database context of computer based numerical information, the mechanistic organization of data into tables or other forms can be, and generally is, quite different from the map representation. We use the term **data model** to refer to the mechanistic organization reflective of the logical organization of data, an organization which has a semantic basis. The example of tables for holding coordinates represents a Euclidean geometry view of the world.

From a spatial data user point of view, the terms data model, data structure and file structure are often encountered because there are no standard terms across different fields. This is true especially for data model and structure, essentially synonyms for organization, but used differently. The expression **data modelling** is used for a comprehensive set of conceptual tools for organizing data. These are of general utility and can be formally defined, as examined in Part Two of this book for spatial contexts. They include notions of tessellations and polylines. The organization itself may rely on a variety of spatial concepts, like spatial interaction or fractal geometry, that facilitate our understanding of spatial reality.

The **data structure** is a more detailed practical description of spatial phenomena, being concerned not so much with entities like polygons or tessellations, but with tools like run-length encoding for saving storage space for representing phenomena in regular grid cells or chains of coordinates for the boundaries of polygons. While data tables may be used as effective devices for revealing data models, the data structure itself focuses on numerical data organization into lists, matrices or diagrams that are aids in addressing the computing issues of storage requirements and performance. The **file structure** is the other aspect of internal representation within the computer, the particulars of location of data in hardware devices.

Data structure and file structure are not usually of concern to the data user, but if an information system has a data structure of a certain type it may foreclose some kinds of activity for the user. Using Figure 3.22, we first contrast two structures for a polygon data model. Phenomena are seen as polygons, yet on the one hand they are created from squares, and on the other from chains of line elements. Both structures allow operations like finding the area of polygons. Otherwise, comparing Figures 3.22b and c, a vector structure is used for both, but on the one hand data are available only for the boundaries of polygons, and on the other for only the interiors.

3.8 PERSONAL SPATIAL SEMANTICS

Whether the purpose is to find a better way through a city or to understand better the political conflicts in the Middle East, people need to acquire spatial information. The facts may be landmarks or turning points useful for giving directions; they may be images of how places relate to each other. We do not discuss here the process by which people capture information, but we do point to some topics of significance for designing and developing spatial information systems:

1. The personal spatial reference systems that people seem to use.
2. Spatial language.
3. Perception of objects and relationships among them.

Psychologists seem to have established that a child's ability to deal with spatial relationships develops early, perhaps by the age of six or seven years. The ability to work with geometrical concepts seems to come later, although once that facility is developed it appears to be very dominant in some societies. For instance, distances are measured metrically, in kilometres or minutes, and much attention is given to the precise location of boundaries of real property.

Cognitive maps appear to distort shapes but do preserve the relative locations and properties of adjacency and connectivity, although the extent to which any spatial information is adequately handled depends on the amount and type of that information. A knowledge of places is often

Figure 3.23 Cognitive maps – a Northerner's view of how a Londoner views 'The North Country' of the UK. (Based on a map, Figure 1.10, in Gould and White, 1974.)

revealed by drawings from mental maps, although the conclusions about that knowledge are affected by the ability to draw it on paper (Figure 3.23). It appears, in general, that cognitive maps emphasize topology rather than geometry; are azimuthal rather than Cartesian; have non-linear scales for distances; and are object oriented rather than focused on continuous variations.

From this we might conclude that Euclidean geometry frameworks are not necessarily optimal for spatial information systems users, but that people feel more comfortable visualizing relationships among point, line, area or volume objects in different spaces. How much precision is needed in locating objects is more a matter of the purposes in having data in the first place (for example, inventory of real estate), but, for many purposes, topological graphs or highly generalized metric maps may be sufficient for visual operations by users. Also, users may wish to have several maps, at different scales, or one map with a variable scale. Teachers of geography and other spatial fields have long used these techniques in presenting concepts to students.

Certainly, though, efforts should be made not to present spatial facts in misleading fashions, as has happened in the use of Mercator or other

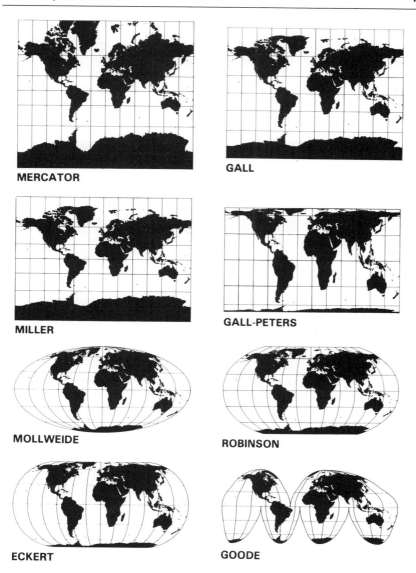

MERCATOR

GALL

MILLER

GALL-PETERS

MOLLWEIDE

ROBINSON

ECKERT

GOODE

Figure 3.24 Alternative views of the world. Rectangular and nonrectangular projections will preserve only some of the basic properties of true shape, area, distance and direction. (Reprinted with the permission of The Cartographic Unit, University of Wisconsin, Madison. See the reference Robinson, 1990, cited in Chapter 4.)

direction-preserving map transformations to show political units of the world (Figure 3.24), leading to misperceptions about the relative sizes of countries or continents, in particular, the large size of Greenland, or the shortest way from Oslo to Tokyo.

Cultural factors may condition users to particular forms or ranges of language for dealing with objects in space. The Hawaiian Islands residents use a radial system for referring to the location of places. The words 'up' and 'down' may mean different things in varying places, bearing no particular reference to north and south as 'top' and 'bottom', based again on a dominant culture of making maps with the North Pole at the top. The concept of direction embodied in northwest is not simply defined, for it may mean objects in a straight line from an existing point, or it may mean everything above and to the left of a single line running northeast to southwest, or the coordinate system quadrant above and to the left of the vertical and horizontal axes.

Some fields that deal with spatial phenomena place much emphasis on visual impressions of landscapes – landscape architecture, civic design and urban planning. Travellers through city streets or river rafters on the Colorado River encounter varying scenes, judged personally in terms of a sequence of images, along paths which themselves have many spatial properties. Building styles may clash; some features may not be seen; others may have unclear boundaries. Space, the world in which we live and experience objects and their form and character, may be seen at different levels: physical, activity based, perceptual and symbolic. In some specific spatial problem solving tasks, the 'real' entities are the symbols of a landscape or the mental painting of a series of views of the walls of the Grand Canyon.

A better design of interfaces between person and machine for spatial information systems requires that we know more about personal reference systems, how spatial data are processed cognitively, and if language has any consistencies across cultures or other subgroups of people.

3.9 SOME OTHER ASPECTS OF SPATIAL INFORMATION

While this chapter has emphasized the logical organization of phenomena in space, let us refer also to some other matters related to the characterization of spatial information. They include:

1. The quality of data, including identification and measurement of error.

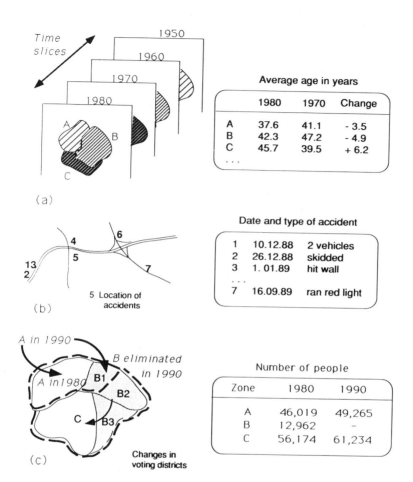

Figure 3.25 Different treatments of time. (a) Phenomena observed at different time periods. (b) Time of occurrence of specific events. (c) Modifiable spatial units.

2. The amount of data.
3. The forms and sources of data.
4. The treatment of the time dimension.
5. The impact of scale.

Here we discuss, briefly, only the quality and time aspects.

3.9.1 Data quality

Error can arise for attributes, topological properties and positioning information. Measuring instruments, physical or conceptual, can be biased, leading to inaccuracy, that is, the property of not producing the real value. **Precision** of locational or attribute measurements refers to the fineness of measurement: distances can be measured to the nearest kilometre, metre, centimetre or millimetre, for instance.

We hope that we do not displace objects from a true location, but this sometimes happens. Recently, for example, the United States Geodetic Office published position offsets for places in the USA that resulted from the use of a new geodetic datum. The recorded location may be not only an inaccurate x or y coordinate, but may cause topological relationships to be erroneous. For instance, a point feature (a school, say) may be placed on the wrong side of a line feature (the street) on which it is located, when using only the coordinate information. Such situations may arise from bad original data for either the point or line feature, but can arise from certain computational procedures in spatial information systems.

The view of spatial phenomena can affect the treatment of error. In particular, an error term can easily be visualized for the continuous surface view by adding, perhaps via a normal statistical distribution concept, a measure for the range of possible values for the positional items, x, y, and z, as appropriate. At least in theory the true value is empirically verifiable. However, for the object-oriented view, error is more complicated, as it has to be defined for complex objects, sets of line segments, contour lines, etc.

3.9.2 The time element

Today commercial spatial information systems do not directly address the matter of time (Figure 3.25). However, time may be included in several ways:

1. As a basis for recording events or attributes.
2. As an attribute of an entity with unchanging spatial properties.
3. As a framework for observing changes in the spatial entities.

In the first case, phenomena or characteristics are observed at varying points of time, perhaps at regular intervals, as for most official statistical censuses (Figure 3.25a). Maps can be produced at different points in time for a single theme, or for several phenomena; different layers in a database can represent observed conditions for different times, for example satellite imagery every twenty-eight days, or population distribution at decennial census intervals, or even surface conditions of different parts of the earth in different geological time periods.

For the second case, data may consist of the date at which events occurred. Thus, a set of data can have a changing number of instances of an entity. For example, as new highway accidents occur, there are additional data records for these events, and the data table contains an item for the date of occurrence (Figure 3.25b). Time can also appear as an attribute as a duration measure. Tabulations of demographic conditions often include data for population growth over a period of time.

For the third item, the focus is on a changing object. Physical features of the earth's surface and subsurface change over time. The construction of new highways changes topological relationships among places. The area of parcels of land changes over time as they are subdivided. Time–space interactions like these may necessitate use of special data organization methods, or representation in different versions in a database. For example, for the subdivision, attributes may be passed from parent plot to new lot; new boundaries are created but some are the same as before; and the date of creation of the new lot is an important attribute for tax purposes. At the least, the question of time discretization has to be addressed for most phenomena; continuous time records are very costly for most purposes.

Definitions may vary at different points in time, if only for attributes of entities which themselves do not change over time. Unemployment or poverty rate definitions are classic examples; the definition of metropolitan areas varies nationally in time, as in the USA, as well as cross-nationally. The recording units themselves may change in time. This occurrence of **modifiable areal units** is especially troublesome for social scientists using social, economic and demographic data from national censuses for spatial units in urban areas. The difficulty may be overcome by the creation of comparable spatial units, but generally involves a process of allocating attribute data for one set of delimited units to another (Figure 3.25c).

A semantic view of entities postulates that spatial phenomena have behavioural attributes, of which change in time is one important condition. While change of attributes through time can be handled by defining new attributes, many phenomena also change locations. Highways are realigned; terrain is modified by major storms, avalanches, and the like; and governments change the boundaries of statistical reporting units. A field view is particularly useful for handling change in time, because new values for an attribute are readily observable at a later point in time for the coordinate points in space. Time changes are more difficult to handle via an object orientation, because new objects have to be created, as in the case of modifiable census units mentioned above, and spatial relationships may change.

3.9.3 Intensional and extensional data

As we examine more fully later in this book, spatial phenomena are sometimes thought of in extensional terms but are represented in databases in **intensional** form, or vice versa. In the second of these, the entity is represented by only a few data compared with what could be used. For example, a river is approximated by a set of numerical parameters instead of by a very large number of vertices. For the intensional state then, we deal with a small amount of data explicitly, and store the rules, procedure or method for obtaining other information. For example, a straight line can be constructed from an ordered series of points or may be created via a mathematical formula (Figure 3.26).

In contrast, in a database in which information is presented by words and numbers, as is often the case for non-spatial phenomena, for example, a bank transaction or a paragraph of text, data are almost always given in an **extensional** way. In this regard the concept of scale, as embodied in ideas of resolution and precision, has different interpretations. For spatial data the precision of recording the positional coordinates may vary depending on purpose, but for data like the amount of an invoice in pounds sterling or the balance of a bank account in Swiss francs every digit is important and cannot be generalized.

3.9.4 General discussion

In a practical sense it may be difficult to ascertain the semantics of situations because of the difference between real and artificial objects, and because different users view a single context differently.

The material on images (photographs, films, digital satellite data) is hard to deal with. Aerial photographs contain shadows that mask detail; maps produced from satellite data may use false colours. Measurements from pictures of landscapes have to recognize and possibly adjust for camera angle, elevation and distance. In addition, different users have different perspectives, so that the data required for a particular purpose, for example transportation planning, varies with whether the user is a traffic engineer, safety analyst or road builder.

Because it is extremely difficult to give full descriptions with words, we

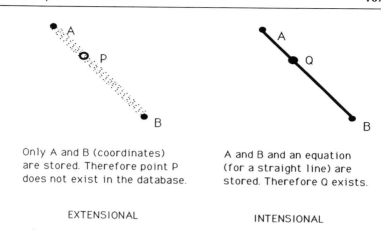

Only A and B (coordinates)
are stored. Therefore point P
does not exist in the database.

A and B and an equation
(for a straight line) are
stored. Therefore Q exists.

EXTENSIONAL

INTENSIONAL

Figure 3.26 Intensional representation for a line.

resort to other means such as numerical measurements, mathematical equations, or annotated pictures, drawings or photographs. Numerical representations of spatial data are characterized by many conditions: entities are multi-faceted; entities may be represented by different spatial unit types at different scales; they may have inexact boundaries; they may differ depending on what data are available; they may be represented by rules rather than by empirical data; they may be modelled as continuous space; they may have attributes correlated in space and changing in time; and they may be accessed via both properties and spatial relationships.

It is unlikely that all ways in which spatial entities may be represented or linked in a study can be stored in a digital or analogue database. But there are good reasons, semantic and procedural (as discussed in Chapters 4–6), for storing some fundamental spatial relationships. Generally there is benefit from storing topology properties, because these may represent user views as matched by data organization. In contrast, distance between point places, even though there may be many, is a derived attribute, not a conceptual approach. But storage and representation are not enough; semantic integration is more the goal than data organization.

In having to deal with spatial relationships explicitly, the entity approach goes beyond the map model, which handles them implicitly. But, as for the map model, so with the semantically oriented non-map model, the recognition of entities depends on scale and purpose of observation. Fundamentally, the question of import seems to be not that of trying to deal with the very large number of ways in which specific instances of spatial data can be related, but, as for map design, to recognize the importance of need or purpose.

It is here that we recall earlier comments about data and information. Not only do our varying views of the real world lead to a need for specific data and information; but information itself is a commodity with varying usefulness to individuals or organizations.

The issues before the designers of spatial information systems are challenging. How should complex reality be effectively and practically dealt with? How should uncertainty and ignorance about some phenomena be handled? How should different basic spatial recording units be defined? We have addressed the conceptual issue of how to organize material at the semantic level. Subsequent chapters develop the theme of data organization and present techniques for conceptualizing the complex real world. Formal descriptions are needed because the meaning of many terms is not clear, and ambiguity exists when some feature is not explicitly recorded. Spatial relationships should be able to be derived in a consistent manner even if not explicitly recorded. Perhaps most importantly, we will be able to reason about and from the formal descriptions of spatial relationships.

3.10 BIBLIOGRAPY

For more details on cartographic concepts and map projections, the reader is referred to both Muehrcke and Snyder. An extensive early treatment of mental maps is provided by Gould and White. A comprehensive coverage of accuracy of spatial data and databases is that of Goodchild and Gopal.

Berry, Brian J. L. 1964. Approaches to regional analysis: a synthesis. *Annals of the Association of American Geographers* 54: 2–9.

Bunge, William. 1962. *Theoretical Geography. Lund Studies in Geography.* Series C, General and Mathematical Geography, no. 1. Lund, Sweden: C. W. K. Gleerup.

Clarke, Keith C. 1990. *Analytical and Computer Cartography.* Englewood Cliffs, New Jersey, USA: Prentice Hall.

Gaile, Gary L. and Cort J. Willmott. 1984. *Spatial Statistics and Models.* Dordrecht, Netherlands: Reidel.

Goodchild, Michael F. 1987. A spatial analytical perspective on geographic information systems. *International Journal of Geographical Information Systems* 1(4): 327–334.

Goodchild, Michael F. and S. Gopal (eds). 1989. *Accuracy of Spatial Databases.* London, UK: Taylor and Francis.

Gould, Peter and Rodney White. 1974. *Mental Maps.* Harmondsworth: Penguin.

Hsu, Po-Siu. 1990. An analysis of spatial structure data in landscapes for geographic information systems. *Proceedings of the Fourth International Symposium on Spatial Data Handling, Zurich*, Switzerland, vol. 2, pp. 888–897.

Langran, Gail. 1989. A review of temporal database research and its use in GIS applications. *International Journal of Geographical Information Systems* 3(3): 215–232.

Laurini, Robert and Françoise Milleret-Raffort. 1989. A primer of multimedia database concepts. In Robert Laurini (ed.), *Multi-media Urban Information Systems*. Urban and Regional Spatial Analysis Network for Education and Training, Computers in Planning Series, vol. 1, pp. 7–75.

Mark, David M. and Andrew U. Frank. 1989. Concepts of space and spatial language. *Proceedings of the Auto Carto 9 Conference, Baltimore*. Falls Church, Virginia, USA: American Society for Photogrammetry and Remote Sensing/American Congress for Surveying and Mapping, pp. 538–556.

Morehouse, Scott. 1987. ARC/INFO: a geo-relational model for spatial information. *Proceedings of the Auto Carto 7 Conference, Washington, DC*, USA, pp. 388–397.

Muehrcke, Phillip C. 1986. *Map Use, Reading, Analysis, and Interpretation*. 2nd end. Madison, Wisconsin, USA: JP Publications.

National Committee for Digital Cartographic Data Standards. 1988. The proposed standard for digital cartographic data. *The American Cartographer* 15(1): 21–28.

Snyder, John P. 1987. *Map Projections – A Working Manual*. US Geological Survey Bulletin P-1395. Washington, DC, USA: US Government Printing Office.

Tomlin, C. Dana. 1990. *Geographic Information Systems and Cartographic Modelling*. Englewood Cliffs, New Jersey, USA: Prentice Hall.

Part Two

Geometries for Spatial Data

4
Geometries
Position, representation, dimensions

Theodor Topol landed in the wrong part of the desert and could not find where he was. Heather Rose tried to explain to her twelve year old pupils that Greenland was not bigger than South America. Ifan Norse measured the distance from his home to the ski slopes by pedometer. Lee Kim prepared to dig a hole in a forest at a location for which his friend provided the latitude and longitude. Bette Mounds tried to recreate the map of land parcels of the seventeenth century from knowing the names of neighbouring landowners. John Hudson used a large computer monitor to visualize a new skyscraper in New York. Art Santander created visual models of landscapes using dragons. Doriana Corona thought she could easily measure the perimeter of Sicily and its islands. Ben Almondbread re-invented dust, snowflakes and trees while Pedro Barrador wondered how to sweep dust.

In Part Two of the book we turn our attention to some basic principles for organization of spatial data and representation of spatial entities. To a large extent drawing on different parts of the field of geometry, the mathematical study of the properties and relations of lines, surfaces and solids, we provide a formal foundation for many practical aspects of spatial information systems design or evaluation. Matters of positioning, representation, and spatial relationships are dealt with in Chapters 4, 5 and 6, respectively. Chapter 7 presents procedures for manipulating spatial data and objects and transforming between different representations. The part concludes with a review of different ways of using spatial data, providing a link back to the first part of the book.

Chapter 4 begins with a discussion of methods for positioning objects in spatial reference systems, local and global. Particular attention is paid to the questions of measuring distance and facilitating access to spatial data. The chapter also has a main theme of different approaches to the representation of artificial and natural objects, especially by contrasting the Euclidean, topological and fractal geometries.

4.1 DIFFERENT GEOMETRIES

In the context of spatial information systems there are needs for dealing with many aspects of space:

1. Two or three dimensions.
2. Planar or non-planar situations.
3. Continuous or discrete referencing.
4. Isotropic or anisotropic conditions.
5. Forms and distances that change over time.
6. Measurements of qualitative or quantitative properties of space.
7. Smooth or rough objects.

Firstly, though, a few words to set the stage. **Spatial** indicates pertaining to space, the void in which material elements exist. Geometry is that branch of mathematics which deals with spatial quantities and the shape of spatial forms. **Shape** (or **configuration**) refers to the structure of those forms. **Form** is the mode of arrangement related to function, or the appropriateness and effectiveness of purpose. For example, the hexagonal pattern of the bee's honeycomb results from the application of both a principle of economy of subdividing space and a principle of stability of structure achieved by the hexagon shape.

Mathematicians have developed many systems and tools for referring to spatial objects, for describing the form of objects, for representing entities and for measuring distances. There are geometries for clearly demarcated entities like buildings, geometries for phenomena that do not have smooth boundaries; geometries for plane or curved surfaces; and geometries for the relationships among entities.

Descriptive geometry is the measurement of properties of objects in space or relationships between objects, that enables the true lengths, angles, lines of intersection and other elements to be determined by graphical means. Topological geometry relates to the relationships of components of form. Fractal geometry deals with the fragmentation and non-smoothness of form and the dimensionality of objects. Computational geometry refers to those concepts, principles, and tools used in the numeric, in contrast to the algebraic practice of geometry.

Geometry provides many tools, and formalisms for helping us work with spatial concepts. Consider first that material objects have length, width, and height, and can be placed in a framework, a **spatial reference system**, that allows measurements of those three fundamental properties (Figure 4.1a), although for many purposes people live in and visualize the world as a 'Flatland' (Abbott, 1952). Generally, the three principal axes are orthogonal (at right angles), but in some applications of statistical

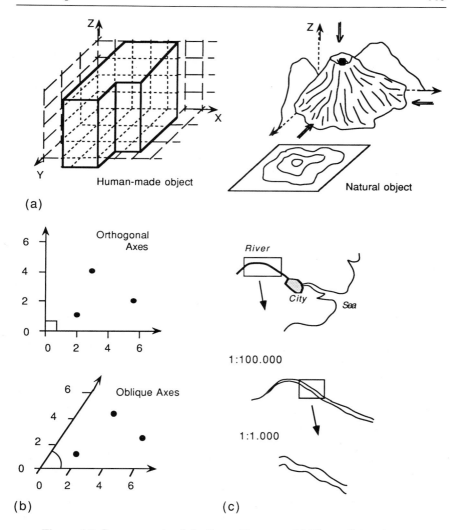

Figure 4.1 Some aspects of dealing with space. (a) Three dimensions. (b) Orthogonality. (c) Scale of observing features.

methods, as might occur in manipulating attribute data, entities may be viewed in oblique coordinate space (Figure 4.1b), reflecting some evidence that abstract measures, the equivalents of the x, y and z spatial coordinate axes, may be correlated.

Secondly, some material objects are not clearly or smoothly demarcated. The water level of a sea varies, giving rise to the use of an average or maximum water level, or an expression of the probability of occurrence of a boundary value, with its spatial manifestation. The line representing the edge of a (wide) highway may be appropriate for a viewing position from a long distance, but a walk along this 'fine line' reveals that it is very irregular at this scale of observation (Figure 4.1c). The field view of spatial objects has this concept of indeterminancy of boundaries, and the **geometry of fractals** provides some ways to represent the unevenness of objects.

Earth scientists may not always be able to use the rigid planar assumptions of solid geometry, and so use coordinate systems with curvature. Of course the earth, as a globe, has sphericity, necessitating measures over a curved surface for true indications, for instance, of distance and area. Map-making, analogue or digital, automated or manual, has long been practised with the aid of conversions from curved to plane surfaces. **Map projections** were devised to allow representation of the spherical world on flat paper.

Material objects in the one, two, or three spatial dimensions have different properties. Shape implies the use of absolute coordinate values as a basis for measuring length, perimeter and subsequent variation from a regular figure like a circle; but many spatial relationships utilize topological properties of connectivity, containment and contiguity. Spatial information systems are often structured on the basis of explicit topology, while some kinds of representations, for example, map models, show those properties implicitly – the human eye is very good at seeing relationships between places and phenomena shown cartographically.

Also, recall from Chapter 3 the discussion of field and continuous perspectives, and the distinction between isotropic and anisotropic. These situations differ in the way they deal with relationships among objects. Taking just the distance from one entity to another, perhaps the simplest concept of spatial relationships (but also quite complex, as discussed in section 4.4), measurements may be constrained to networks, may arise for all possible points in continuous space, and may be different depending on the dimensionality of an object.

4.2 POSITIONING OBJECTS IN SPATIAL REFERENCING SYSTEMS

Positioning objects, fundamental activities of fields as different as geodesy or land surveying and thematic cartography, referring to the real world and map model domains, respectively, involve considerations of:

1. The geometric character of the reference system.
2. Measurement metrics.
3. Types of reference system: Cartesian or polar.
4. Nature of the origin.
5. Discrete or continuous references.

For most purposes, people relate to the length and width **dimensions**, although occasionally, for some people, the height dimension, associated with mountain climbing and skyscraper elevators can be painfully obvious. Today many spatial information systems have data for only the $\{x, y, a\}$, not the z dimension, reflecting origins in natural resources mapping and management, the emulation of map overlay modelling techniques, and, originally, difficulties in programming computers to work with data for three dimensions. Often the z variable is treated as an attribute of objects, not a positioning coordinate.

4.2.1 Continuous space referencing

Within the two- or three-dimensional void, referencing may be Cartesian or polar (Figure 4.2a). In the **Cartesian** case, the position is given by a linear displacement value, as a pair or triplet of values. In the polar case there is again a doublet or triplet, but one measure is distance from an origin, and the others are angular measures, azimuths or bearings. This **polar** type of reference is useful for problems dealing with direction of travel, as a personal reference system, or for mapping phenomena like journey to work or migration that use a spatial dyad as a basic unit. Radar based displays or ground surveys are other examples of the use of distance–direction combinations.

Positioning objects in space generally uses Euclidean (rectangular or plane coordinates) geometry, measuring distances from a specified origin (Figure 4.2b), either global (tied in to the earth) or local, as is generally used for city street atlases (Figure 4.2b). Landscape architects designing the layout of a new ski resort do not need to know (except possibly for predicting weather!) how far they are away from the Greenwich Meridian or the Equator used for global referencing (Figure 4.2c). Information systems for the whole world generally need to tie placements into the geographic reference system of latitude and longitude (Figure 4.2d). Accurate measures for two or three control points can be used to translate from local into geographic coordinates; and spherical coordinate systems may be used for the earth rather than projecting to planar coordinate systems. National map series, such as those in use in the USA

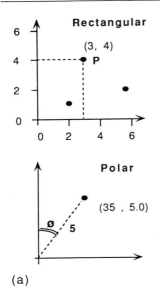

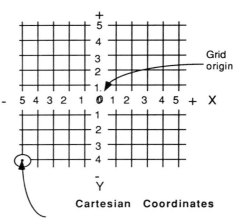

(a)

(b)

Local origin as in street atlas
(Latitude = 35 north, longitude = 75 east)

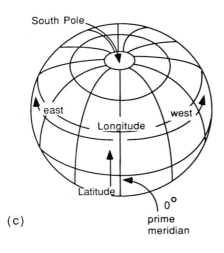

(c)

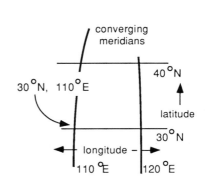

(d)

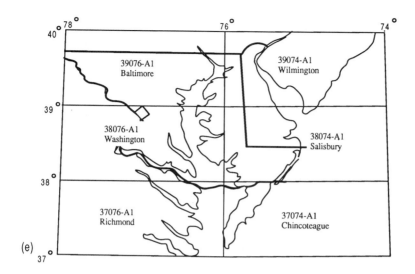

Figure 4.2 Rectangular and spherical coordinate frameworks. (a) Rectangular or polar coordinates. (b) Local origins. (c) Global or geographic referencing. (d) Latitude and longitude. (e) Map sheet divisions based on latitude and longitude.

(Figure 4.2e), often use latitude and longitude to refer to map sheets.

Moreover, **origins** may not be singular, may not be stationary in space, and may be at different elevations (Figure 4.3). Studies of flow phenomena, channelled or not, may require several origin points, for example, the individual towers for air traffic control systems, each usually associated with a polar map (Figure 4.3a). Navigation aids generally work from the current origin of a traveller, meaning that instructions for way-finding, or maps, are based on a succession of reference points (Figure 4.3b). Individual entities may have more than one reference system. For example, a building may be seen from the perspective of its own three dimensional coordinate axes, and by reference to an external frame, perhaps geographic coordinates (Figure 4.3c). Context is not only a personally based view; it is also a real world based frame of reference. In this regard it is interesting to contemplate finding a position in reality, say the (0, 0) latitude–longitude point, compared with its whereabouts on a map.

Features may be located as a relative uni-dimensional distance from some named origin. Civil engineers or transportation planners usually

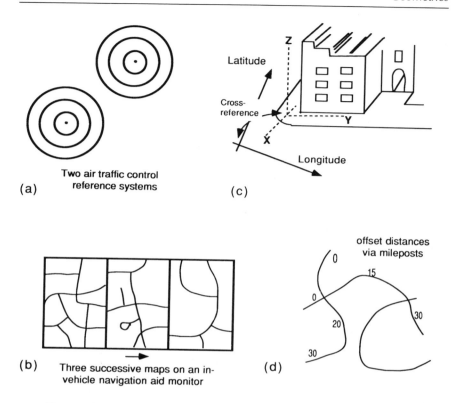

(a) Two air traffic control reference systems

(c)

Latitude

Cross-reference

Longitude

(b) Three successive maps on an in-vehicle navigation aid monitor

(d) offset distances via mileposts

Figure 4.3 Different origins for spatial referencing systems. (a) Azimuthal maps. (b) Moving origin. (c) External and local reference systems. (d) Offset distances.

refer to the location of objects or events on highways as an **offset distance** from a specific origin, perhaps the beginning of a highway, an intersection, or a national boundary (Figure 4.3d). The referencing may not be consistent with identification of objects. For example, a piece of highway with particular properties like the number of lanes or surface condition can be demarcated by how far from an origin the combined attribute of lanes and surface changes. However, a topologically based definition of a one-dimensional object will not recognize that segmentation into pieces based on the attributes.

But, alas, spatial objects move. For one example, we cite the position of land masses of the Earth, perhaps Pangea or Gondwanaland, one billion years ago compared with now. The continental positions are quite different due to plate tectonics. In the modern world the distance between North America and Europe increases every year by a few centimetres. Among other examples, Scandinavia's land mass is rising slightly, and for much of the world there is a steady sea level rise.

Another problem is the existence of earthquakes, implying the movement of pieces of land, and therefore, of landmarks. The question arises for such events, as to whether we consider the landmark coordinates as fixed, or moved. In other words, do we have to change the coordinates in a spatial information system?

Particular entities may have their coordinates obtained from a variety of sources. Ground surveys and mensuration have traditionally been used to establish a control system, or, if it meets certain conditions of accuracy and precision, a **geodetic reference system**. The control system is generally regarded as a set of physical objects, called monuments, put into the ground, and the data describing their positions, which in combination, form a basis for establishing the position of entities on the earth's surface. The geodetic system provides a universal basis for what would otherwise be separate measures. The network of control points is, generally, nationally based for the latitude and longitude dimensions, although national systems may be tied into the whole globe. Sometimes the monuments can be used by travellers to ascertain their positions. For instance, the principal global meridian (zero degrees) is clearly marked at Greenwich Observatory in London, and Sweden maintains a set of monuments marking the location of the polar circle.

Moreover, the officially accepted **horizontal datum**, the standard global framework for establishing latitude and longitude, may change in time. For example, the recent change in the USA from the North American Datum of 1927 to a new one, North American Datum of 1983, caused shifts of up to 100 metres in some coordinates as a result of going to a new spheroid to represent the globe from the ellipsoid used in 1927 which was designed to fit only the shape of the conterminous United States. Elevations or depths are established relative to an accepted sea level. Sometimes different vertical bases have to be used. For example, there is a height difference between sea levels on the Atlantic and Mediterranean shores of France.

Positional data may be obtained in the field by survey methods, by coordinate geometry computations or by satellite surveying. Traditionally, some earth features, especially property boundaries, were located by

metes and **bounds** descriptions. Features were positioned relative to others by distance and bearing, even if at times the base points for measurement were not too fixed, like trees. Today, space-borne transmitters, known as the global positioning system, are increasingly used to pinpoint places on the earth.

Reference to position may not be based solely on the precision implied by a continuously scaled coordinate axis, but can also be discrete. Referencing may also not necessarily involve numerical measures, but can use numerals or words for relative positioning. For example, a house could be positioned as being in one county as opposed to another, or referred to by a (supposedly) unique postal address, as opposed to being positioned as 56,789 kilometres from the intersection of the Equator and the Greenwich Meridian.

4.2.2 Referencing for discrete entities

Discrete referencing, the means of establishing the location of separate entities, other than by coordinates in continuous space, has several possibilities (Muehrcke, 1986):

1. Narrative descriptions.
2. Street addresses.
3. Blocks of earth space.

In many parts of the eastern United States, and in other countries, legal specifications of property have historically been given as a narrative description of boundaries (Figure 4.4a). Descriptions would identify landscape objects, and possibly give some indication of distance. Reference is made to supposedly permanent objects like Farmer Green's barn, the large oak tree or the bridge over the River Exe. Descriptions are as precise as the objects used as landmarks or boundary paths, and as practical only to the extent that these features still exist, and are recognizable, and have not moved over the years. Modern equivalents (Figure 4.4b) use coordinate geometry based direction and distance measures, avoiding statements like 'follow the Crooked River for five miles'.

Street addresses provide an almost universally adopted mechanism of locating discrete properties, such as a dwelling or a plot of land. As with the offset measures for the continuous linear addressing case, location is with reference to some selected, local origin. Particular properties can then be matched to line objects, by indicating address ranges and odd–even numbers associated with particular street segments (Figure

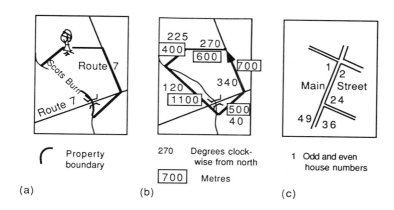

Figure 4.4 Boundary descriptions and street addresses. (a) Description by reference to landmarks. (b) Description by compass direction and distance. (c) Street addresses.

4.4c), or even to zones. The United States Census Bureau DIME system, developed for the 1970 census, was the first large scale practical application of using street address referencing with a topological based identification of objects. Street numbers for left and right sides of streets were incorporated into the basic set of data for the line segments, as well as the indication of the area units (city blocks) on either side of the lines. Street addressing systems are, though, not without problems. Difficulties can arise for numbering streets bordered on one side by a river, and when properties are subdivided.

While data structuring using the topological one-dimensional element is quite common today, street address based referencing has its own practical utility because many data, particularly administrative or taxation type, are recorded by street address. While there may be some shortcomings in precision of location relative to latitude and longitude, and inconsistencies or duplications in naming streets or assigning numbers, the street addressing system is widely accepted throughout the world.

Phenomena may also be referred to by their location in pieces of the earth's surface – perhaps natural entities like an island, or artefacts like postal zones. While there are many possible criteria for creating areal units for referencing, worldwide the most common are: zones of postal services, administrative units, and statistical reporting units for census purposes. As such, these **irregular tiles**, examples of the spatial discretization mentioned in Chapter 3, are useful for accumulating aggregates by particular attributes, and for retrieval purposes using names or other identifiers for those units, although they may lack positional precision.

A **regular tiling** system, that is a subdivision of space into areas equal in shape and size, may also be used for reference to places on the earth. Most street atlases provide tables of street names and a number or code letter for the row and column in which the street can be found. Some countries, for example, Britain and Sweden, have at times compiled statistical information for grid cell units, tied into a geographic reference system by being produced by subdivision of large kilometre square blocks. Referencing makes use of code numbers combining a row and column identifier; for example, the British Ordnance Survey headquarters in Southampton has the address of SU 387148, based on the set of latitude–longitude blocks making up the national grid. Spatial information system databases created from existing map sheets may use the particular sheets for reference purposes or for simplifying management of a large database.

The USA provides two interesting examples of the practical use of discrete reference units. Part of the country has real estate tied into a planar public land survey system, that is based on real geographic coordinates (Figure 4.5a). Lands were subdivided in the eighteenth century by initial reference to a principal meridian, and a selected parallel as a baseline for east to west measurements. Square blocks of six miles per side were identified (Figure 4.5b), labelled as townships, and then hierarchically subdivided into thirty-six sections, four quarter sections, and even to sixteen ten acre units within each 160 acre quarter section. Spatial information systems using this basis of land recording are in existence, and have special features of referencing to the geographic coordinates, and making adjustments for the fact that the geometrically square descriptions are in fact not square because the length of a degree of longitude gets smaller with increasing north latitude. In some parts of the USA a checkerboard pattern of public and private ownership exists within this framework (Figure 4.5c).

The postal code, one of few universal reference systems useful for spatial data, is ostensibly associated with areal units. In reality, the basic unit is the postal delivery route, a concept associated with streets, linear features. Many countries have postal codes reflecting several levels of organization, as we illustrate with an example from The Netherlands

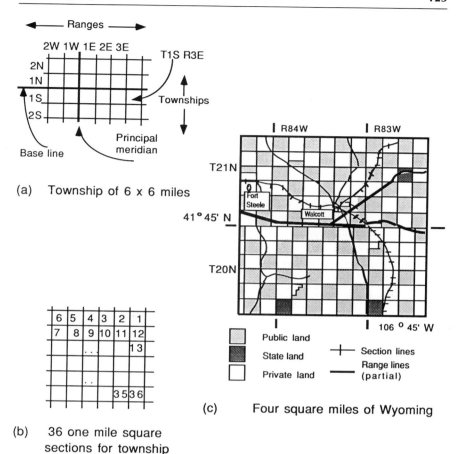

(a) Township of 6 x 6 miles

(b) 36 one mile square sections for township

(c) Four square miles of Wyoming

Figure 4.5 Discrete reference units. (a) Townships of the US Public Land Survey System. (b) Sections. (c) A portion of south-central Wyoming, USA. (Figure 4.5c is redrawn from a portion of Wyoming Public Land Use Maps, Number 8, Fort Steele, published by the Bureau of Land Management, US Department of the Interior, Cheyenne, Wyoming, USA, 1979.)

(Figure 4.6). Similarly to other countries, here the postal code is hierarchical, referring to large territorial divisions, then streets, and routes. The last-named often follow only one side of a street, and often do not break at road junctions as shown.

For several purposes, in later discussions about access to and retrieval of spatial data, line and area entities will be referred to by some limiting

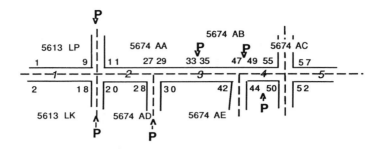

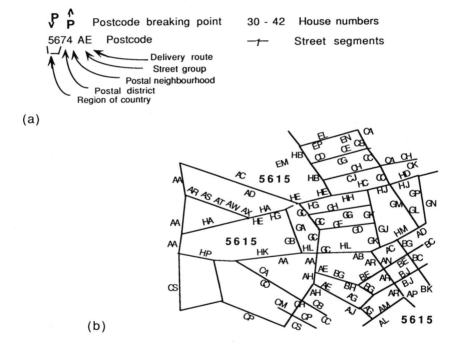

(a)

(b)

Figure 4.6 Postal codes. (a) Postal codes concept. (b) A portion of the postal codes of the city of Eindhoven, The Netherlands. (From Van Est and Scheurwater, 1983 with permission.)

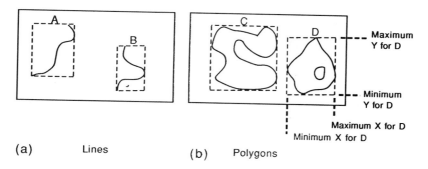

Figure 4.7 Minimum bounding rectangles. (a) For lines. (b) For polygons.

conditions. Most usually these are the coordinates for the ends of lines or the furthest **extents** of lines or polygons in orthogonal x- and y-axes, as shown in Figure 4.7. The extents established by the minimum and maximum coordinates establish the corners of boxes known as **minimum bounding rectangles**. This means that entities will often be represented by two kinds of data – the specifics of their location and/or form, and their positional extent.

By way of summary, we use the term **locator** to refer to any device for indicating relative or absolute position in some spatial reference system. Locators may be Cartesian or polar coordinates, distances from specified but arbitrary origins measured along lines, through space, in one or more dimensions; they may be discrete addresses, or special locational codes known as Peano and Hilbert keys (section 4.7).

4.3 GLOBAL REFERENCE SYSTEMS

The domain of three-dimensional space is very important for the development of global databases such as is presently contemplated and discussed for environmental management and research into global warming and its impact. A discussion of some aspects of global reference systems will allow us not only to mention some interesting practical developments but also to illustrate some of the fundamental matters to do with curved surfaces.

4.3.1 Global referencing

The basic concept of global referencing (Figure 4.2d) consists of using a set of two coordinate values for the position on the real (curved, but flat) surface. Latitude is generally measured north and south from the equator by positive and negative values as a y-axis; longitude is measured east or west by positive and negative numbers of x from an arbitrary prime meridian, usually the Greenwich Meridian but not necessarily. Indeed, in the 1880s there was much political activity before agreement was reached to use the line passing through London, and several countries today have their own prime meridians for mapping purposes.

Currently there exists in space a set of satellite based electronic signal generators that can be used to very precisely and accurately pinpoint a location on the earth's surface. This **global positioning system** (GPS) developed by the United States government, allows accurate spatial referencing in continuous space to a precision of just a few metres, or about the width of an average city street, or even better as electronic equipment improves. It is possible to give every hectare of the earth a unique address, although it still remains quite impractical to collect data for the attributes of such spatial units.

There are several discrete referencing systems for the world. A graticule of grid lines provides a coordinate measurement frame but can also be seen as establishing a tiling system. However, the blocks are not square because of the curvature of the earth. At the poles the width is collapsed to a point – the tiles are then triangles. Reference may be made to blocks of space representing countries, an irregular tiling; map sheets or polygons created by two bounding lines of longitude and latitude, almost perfectly regular tilings; or by the use of special regular geometrical forms suitable for partitioning curved surfaces, as we demonstrate in section 4.3.3.

4.3.2 Map projections

World maps in an atlas provide a spatial referencing system, but the same pieces of the earth can look different according to the map projection used for the map (Robinson et al, 1984). A map projection is a device for producing all or part of a sphere on a flat sheet. Some projections preserve distance or true direction, some maintain correct shapes, and others preserve the property of areal size. Map projections can be studied from the point of view of:

1. The form of the surface used for the projection.

2. The particular viewing origin and the standard points and lines used.
3. Spatial properties, preserved or distorted.
4. The number of points used to transform from a sphere to a flat surface.
5. The formulae used in mapping from a sphere to a noncurved surface.

In manner of construction projections have been classified first of all based on the form of the piece of paper theoretically wrapped around or otherwise touching the globe, principally the three types of cylinder, cone, and disk (Figure 4.8). In the third of these developable surfaces, assuming touching, the chosen graticule and earth features are mapped from the sphere to a disk tangential to the globe at one point, possibly a pole as in the illustration (Figure 4.8a), or any other point suitable as the origin. The cylinder object may be wrapped around the earth in different locations, but touches only along a circle (Figure 4.8b); while the cone is tangential to the surface along the path of one circle (Figure 4.8c). In all cases there is a single line of points (or a single point in the case of the azimuthal surface) for which coordinate values are correct to scale, as suggested by the small circles in the figure.

The projection surfaces may also cut the earth surface; such **secant projections** will have two principal axes, as shown for the conic projection, thereby reducing average distortions compared to the tangential form. The secant type is more common for conic rather than cylindrical or azimuthal projections. The developable surface may also be oriented in different ways, perhaps being aligned with parallels or meridians, or possibly being oblique to them, as shown. Cylindrical projections that are oriented to meridians rather than parallels are known as transverse projections.

Different projection origins may be used. Illustrated only for projecting to a disk (Figure 4.8d), we can imagine a base of infinity, represented by parallel projection lines, or an origin at the pole opposite to the centre of the azimuthal projection, or located at the centre of the globe. The particular point of origin will cause the graticule of equal degree increments for latitude to be represented by different spacing, as shown.

For these azimuthal projections (those with a common centre), parallels may be spaced according to different conditions. (Meridians are always radiating straight lines.) If the linear scale is uniform the projection is equidistant; if equal area is preserved then the spacing between successive parallels of equal degrees increment will be slightly less. The orthographic has a similar pattern, but the distance between successive lines diminishes more rapidly than in the case of the equal area type.

Planar projections may be further classified as to the particular properties preserved. Not all properties can remain undisturbed in a single projection. Some guarantee one factor; others are compromises. The existence of many particular projections is not surprising, given the four different properties, distance, direction, shape, and area, that may be preserved, and their many uses. Engineers, meteorologists and navigators are particularly interested in distances and compass directions; ecologists, geologists and geographers often need area preserving maps.

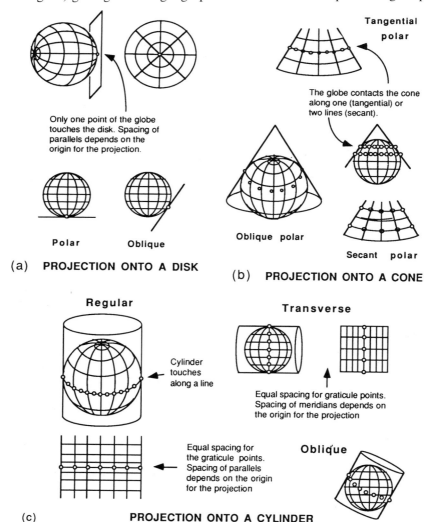

Tangential polar

The globe contacts the cone along one (tangential) or two lines (secant).

Only one point of the globe touches the disk. Spacing of parallels depends on the origin for the projection.

Polar   Oblique

Oblique polar

Secant   polar

(a)   **PROJECTION ONTO A DISK**

(b)   **PROJECTION ONTO A CONE**

Regular

Transverse

Cylinder touches along a line

Equal spacing for graticule points. Spacing of meridians depends on the origin for the projection

Equal spacing for the graticule points. Spacing of parallels depends on the origin for the projection

Oblique

(c)   **PROJECTION ONTO A CYLINDER**

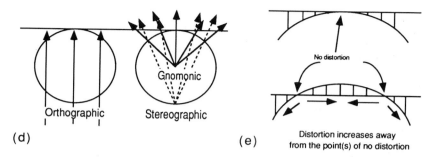

Figure 4.8 Map projections. (a) Cylinder developable surface.
(b) Disk for azimuthal projections. (c) Cone developable surface.
(d) Origins for projection to a plane. (e) Concept of distortion.

Choices are not always easy; compromises may be necessary, and misleading impressions may result. Recently, professional societies of cartographers and geographers in the USA adopted a formal resolution urging an end to the popular use of rectangular maps like Mercator, Gall, Miller and Peters, preferring other projections in order to show the earth as being round, to minimize severe size distortions of large areas of the earth (Figure 3.23), and to avoid representing most distances and direct routes incorrectly.

4.3.3 Some examples of global systems

The more reference points used for the **transformation** (Figure 4.8e) the closer is the representation to the real surface. The representation is more accurate the closer to the reference points, so that the selection of particular (standard) parallels and meridians for certain kinds of projection is important. Some transformations project onto a single sheet of paper with a rectilinear graticule; other projections allow curved lines. In any event, only a few points are true; others are interpolated, being deficient in some regard as to scale. In practice this means that features are represented by some selected known points and rules (the mathematical equations) to establish the coordinates for other points. In a spatial reference system sense, therefore, we have intensional data, a point to which we return later.

In order to provide some standardization throughout the world, particular reference systems have been devised and are in use. Many countries of the world have adopted the Universal Transverse Mercator (UTM) grid system. This is a projection based on wrapping a cylinder around the poles and it divides the earth into sixty longitudinal six-degree zones, covering the earth from north pole to south pole and having about 670 kilometres width at the Equator (Figure 4.9a). A zone six degrees wide is usually divided into 10 kilometre sections for reference purposes. Points are positioned on a northing scale measured continuously from zero at the equator; and with reference to a zone central meridian with an easting of 500,000 metres. Large countries such as the USA have many zones but these provide advantages of the reduction in distortions that the relatively narrow bonds afford longitudinally.

Distortion is absent along two meridians, slightly west and east of the central one, which has a scale set at 0.996 of those lines. However, the developed surface of several transverse zones will have gaps, causing difficulties in working across zones.

Although not global in coverage, the USA uses a system of planar rectangular coordinates, the State Plane Coordinate System, to reduce distortions in mapping an area of continental proportions. Based on either the Lambert Conic or Transverse Mercator projection (Figure 4.9b), it is similar in intent to the Universal Transverse Mercator but has a differently arranged graticule and measurements in feet not metres. Originally devised for permanent recording of the location of original land use survey monuments, it is now used extensively for state mapping purposes. It has 120 zones and a distortion error of no more than 1 foot in 10,000. Both projections used for this system are conformal, that is, they preserve the shape of (small) areas; and provide for an easy interchange of coordinates between grid and geographic reference systems. Other countries have similar national grids but do not usually need as many zones as the USA or other large countries.

Developing a **standard discrete georeferencing system** is a little more challenging because of the properties of curved spaces. While irregular shaped units can be used, and often are (for example, countries), there is interest in a system of regularly-shaped tiles spanning the whole globe. A set of cubes, or octets, may be useful for architects and structural engineers, and, as we see later, for numerous applications in spatial information systems, but this regular geometric figure, an example of a regular tessellation, is not so good for partitioning the surface of a sphere.

Geometric structure is apparent in the three-dimensional world – stacking of oranges in a tetrahedral cluster, patches of black and white on footballs or beachballs, or faceted geodesic domes. The faceted spherical surface of a football (not American!) or basketball consists of mostly hexagonal patches, but also pentagonal patches set at predetermined positions. It can be shown mathematically that only hexagons, or their equivalent, triangles, will not work if curved lines are not allowed in the

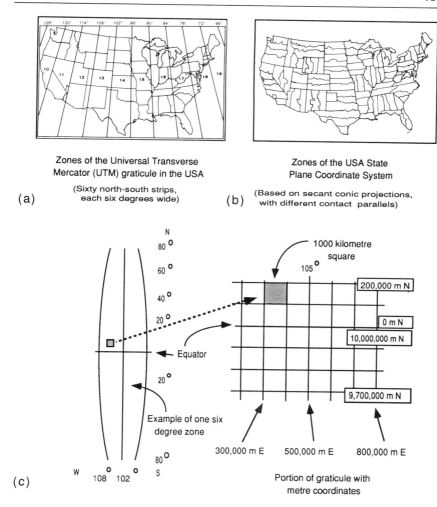

(a) Zones of the Universal Transverse Mercator (UTM) graticule in the USA
(Sixty north-south strips, each six degrees wide)

(b) Zones of the USA State Plane Coordinate System
(Based on secant conic projections, with different contact parallels)

(c)

Figure 4.9 Reference systems for global extents. (a) Universal Transverse Mercator. (b) USA State Plane Coordinate System. (c) Detail for the UTM graticule.

graticule or wireframe model. A set of triangles initiated from an octahedron model for the earth has recently been proposed (Dutton, 1989; Goodchild, 1987) as an effective global reference system. The intent is to have:

1. Basic units almost equal in size.
2. Basic units almost equal in shape.

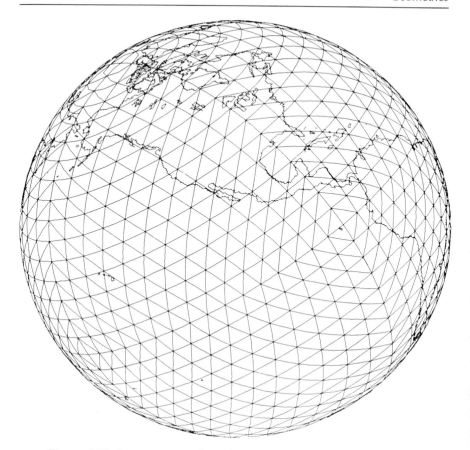

Figure 4.10 A quaternary triangular mesh for a globe. The Level 4 system of triangles viewed orthogonally from over the Pacific Ocean. (Reprinted from Goodchild and Shiren with the permission of the National Center for Geographic Information and Analysis, Santa Barbara, California, USA.)

3. A set of units for which true adjacencies for neighbouring elements are preserved.

However, the shape and size preserving properties cannot be met without compromise at one time in any projection from a curved to plane surface. Unlike a set of squares, the proposed **quaternary triangular mesh** (Figure 4.10) meets these requirements, also allows consistent georeferencing at a variety of scales, and has a structured addressing scheme (section 6.5).

4.4 THE FUNDAMENTAL ELEMENT OF DISTANCE

Distance, or the measurement of length, as encountered in determining relative positions on the ground or attributing size properties to spatial objects, is a building block so important in spatial information systems that it deserves some extra treatment. There are many aspects to distance:

Metric for measurement
Type of geometry
Planar or spherical
Simple or composite
Singular or accumulated
Isotropic or anisotropic
Distance in a graph
Number of dimensions
Precision and accuracy

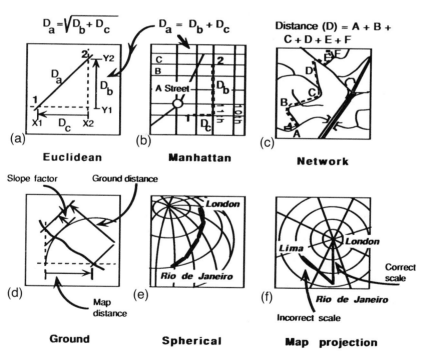

Figure 4.11 Different concepts of distance. (a) Euclidean.
(b) Taxicab. (c) Network. (d) Ground distance. (e) Spherical.
(f) Projective.

Distances in planar Cartesian coordinate systems use the data for coordinate pairs, often computing the straight line distance as the length of a diagonal of a right angled triangle (Figure 4.11a). However, there are other possibilities, of which the most intriguingly named is the Manhattan, taxicab or city block distance (Figure 4.11b). This computes distance by summing the lengths of the sides of the triangle. Algebraically, it is based on absolute distances rather than on the squared distances for the straight line between two points based on the Pythagorean theorem.

The distance measures may be made over a surface or may be computed for the assumed real channels of movement. Real **network distances** or driving routes (Figure 4.11c) are almost always greater than the direct straight line distance. A continuous field model for spatial entities facilitates obtaining direct point to point distances, but will underestimate the reality by ignoring channels for and barriers to movement. Discrepancies between direct distance and 'real' distance will tend to be higher if networks for movement are sparse or irregular, but effective distances should also consider the relative costs of movement on networks (of different kinds) as opposed to simple overland journeys.

Determination of accurate **non-planar distances** requires elevation coordinates so that measures may recognize gradients of the earth's surface. On-ground distances will also be underestimated if the terrain conditions are not considered (Figure 4.11d). And, as noted earlier, false impressions or data will be obtained for global distances if wrong map projections are used (Figure 4.11e and f).

In the human world the effort to move over the earth's surface is often thought of in terms of cost, either in time or money. Transportation planners, urban geographers, and others have developed many measures of **spatial impedance**, and cartographers have devised various techniques to represent the relative positions of places in different metrics. Indeed unless a single composite arithmetic measure is designed for combining different metrics, it is difficult to deal with the inherent conflict of placement of points according to different measures.

Sometimes distance is dealt with as **proximity zones**, generally for simplicity assuming an azimuthal base (Figure 4.12a). For example, distance bands may be created for several market centres to show the relative ease for a farmer to get to each. A comparison of values for each distance zone at a given point may be achieved by overlaying the different sets of circles.

Accumulations of distance from one point to reach all others in a set are useful for producing maps of accessibility (Figure 4.12b). An apparently simple task of determining if manufacturing plants have, on the average, located at greater distances from a port, requires obtaining

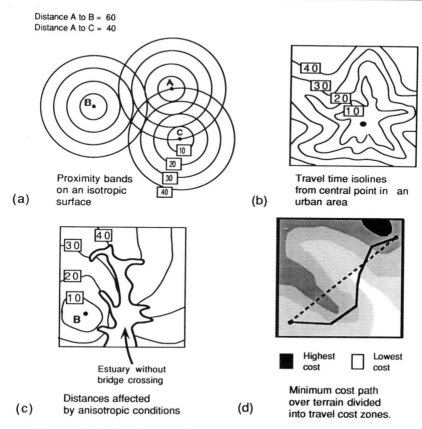

Distance A to B = 60
Distance A to C = 40

(a) Proximity bands on an isotropic surface

(b) Travel time isolines from central point in an urban area

(c) Estuary without bridge crossing
Distances affected by anisotropic conditions

(d) Highest cost Lowest cost
Minimum cost path over terrain divided into travel cost zones.

Figure 4.12 Proximity and accessibility. (a) Proximity zones. (b) Travel time isolines. (c) Distance affected by barriers. (d) Terrain minimum cost path distance.

accumulated mileages through highway networks at different points in time, and then summarizing statistically as a standard distance (Gillies, 1989). Accumulations over distances are also encountered in visibility studies. Imagine standing in a lookout tower in the Alps, or on the viewing platform of the Eiffel Tower. In the respective cases, one could count the number of trees or buildings in different distance bands at, say, one, two or three kilometre increments. Any object in the viewing area may also be seen, or not seen, from many places, depending on obstacles, elevation and surface orientation.

In the anisotropic case, **barriers**, permeable or impermeable, create distortions to the otherwise even patterns (Figure 4.12c). For non-

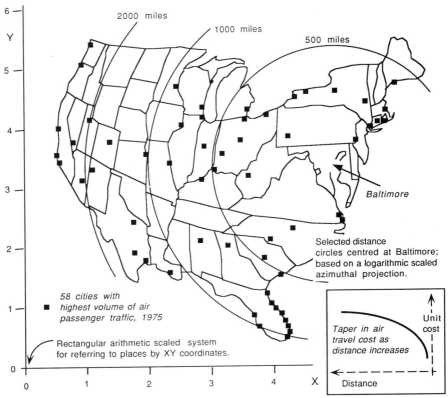

Figure 4.13 Logarithmic azimuthal representation of relative distances from Baltimore, Maryland. The symbols indicate urban areas with substantial air traffic volumes with Baltimore, Maryland. The Cartesian grid is overlaid on a map base of an azimuthal logarithmic scaled graticule centred on Baltimore. The rectangular grid is used to obtain Cartesian coordinates for mapping purposes. (Based on material originally appearing in Thompson and Murphy, 1975, cited in Chapter 16.)

penetratable barriers, movement stops completely, as at a road construction barrier, or at railroad tracks, or, in some cases, at national frontiers. Permeable barriers allow some movement, perhaps representing a real condition of limits to the number of vehicles able to pass a particular point or journey on a selected route, or via an extra cost added to the numbers representing accessibility. They therefore are similar to aggregations of several weighted segments, each associated with a chunk of terrain of different spatial impedance (Figure 4.12d).

Anisotropic conditions require different techniques for expressions of the **relative distances** separating places. For example, the declining cost of per unit travel rates as distance increases, well known in transportation studies as long haul economies, may require a logarithmic or non-linear scaling for distances, as shown for a map of major air travel destinations for originations in Baltimore (Figure 4.13). The map uses a logarithmic azimuthal grid centred on Baltimore–Washington International Airport.

For many purposes, it is desirable to indicate the separation of places as cost, effort, time or personal travel, or other measures. Indeed, several **scales** for relative distance may be needed; mental maps often use something like a logarithmic scaling. An equally difficult problem, though, is when we deal with more than one origin for measurements. The polar map is fine for Baltimore, one place, but, imagine a map showing relative distances among a set of places. Special mapping techniques are necessary, as shown by the sets of isolines in Figure 4.14 for distances from three places at one time. It can be shown mathematically that some distance representations cannot be achieved in map form; we must then resort to dyadic matrices as shown.

The underlying geometry and the unevenness of the world distort measures of distance. **Accessibility** is inherently a multi-faceted concept. Distances measured from maps, or digital data obtained from planar representations without corrections, can be longer than in reality, depending on the map projection underlying the map. Sometimes, as in the example of walking to school, the temporal distance is most appropriate. While a Euclidean base is normally assumed for measuring distances, at times other geometries may be more appropriate. A 'Manhattan' or city block distance is based on absolute distances, not the root of the sum of the squares for a right angled triangle. Sometimes distance may be conveniently measured topologically, not using coordinates at all; and sometimes we have complicated multi-faceted relative distances with which to deal.

4.5 COORDINATES AND SPLINES: THE REPRESENTATION OF LINES

Not only are there interesting questions about distance, but there are practical issues about the representation of one- and two-dimensional spatial elements such as:

1. The amount of generalization in the representation of lines.
2. The mathematical form of the representation for lines.

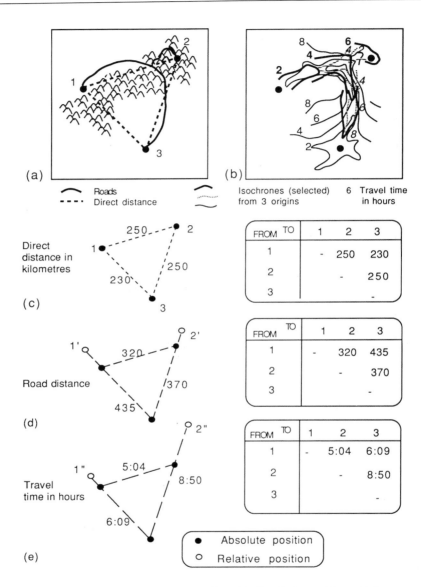

Figure 4.14 Multi-origin accessibility. (a) Geodetic positions of three places. (b) Isochrones for travel time from three places. (c) Direct planar distances. (d) Road travel distances. (e) Travel times.

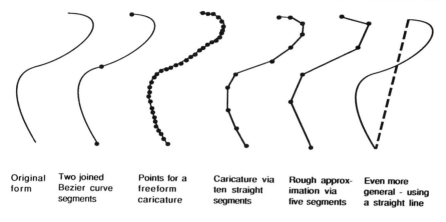

| Original form | Two joined Bezier curve segments | Points for a freeform caricature | Caricature via ten straight segments | Rough approx- imation via five segments | Even more general - using a straight line |

Figure 4.15 Representation of a line.

3. The need for minimizing the amount of data.
4. The amount of regularity in the entity.
5. The accuracy and precision of representation.

Simply put, an irregular line joining fixed and known end points, can be assumed to be straight, as complete as the original data, or represented by some condition between these two extremes (Figure 4.15). The reduction in the number of pieces of information needed to be able to draw the line can be effected in different ways, systematic or otherwise. Of several aspects of generalization, we are here concerned with smoothing and line thinning.

4.5.1 Line simplification

Cartographers often use procedures of line simplification that systematic-ally eliminate certain points from those originally recorded or captured for a linear feature. Ideally, a representation should both preserve the main element of shape, and recognize topological properties. That is to say, for a set of boundary lines, as in the case of the countries of South America shown in Figure 3.13e, it is important to use the topological junctions (line intersections) as inviolate points. Then the undulations and indentations in the shape of the line between the topological end points may be altered depending on how much of the original shape is important for a given purpose. Map users may feel more comfortable if the recognizable features are preserved (in another context, for example,

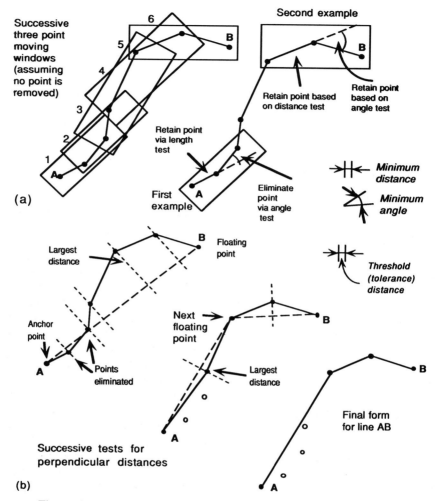

Figure 4.16 Two different procedures for line generation. (a) Three point moving window (after Jenks via McMaster, 1987). (b) The Douglas-Peucker algorithm (Douglas and Peucker).

that the Florida peninsula is not removed), even if they are not necessary for a given specific analytical purpose.

The simplest procedures for line generalization consist of retaining only a fraction of points at systematic intervals: for example, every fifth point is kept and all others discarded. This procedure does not guarantee shape preservation. Out of the very large number of procedures that try

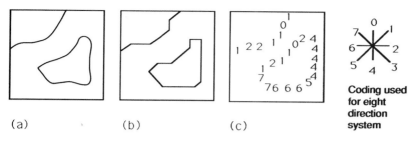

Figure 4.17 Chain encoding for lines. (a) Original lines (for example, portions of contours). (b) Straight segments for eight direction encoding. (c) Numeric codes.

to preserve shape, especially the major turning points, we illustrate just two to demonstrate certain basic principles.

First of all, for an already selected line feature, we can simplify by removing points that are too close to neighbours or have a small angle of separation between vectors (McMaster, 1987). Either of these conditions may be examined separately, or they may be combined in one procedure (Figure 4.16a). In this process a three point moving window can be applied to a line from a topological (or arbitrary) origin. The first three vertices caught in the window are examined with regard to two criteria (Figure 4.16a). One is the distance of a vertex from the previous vertex – if that distance is too short then the second point can be removed. The second criteria relates to the existence of turning points. If the angle between a projected line joining the first two points and the line from second to third is small, then the third point can be discarded. This process continues for every succeeding set of three points in the window.

Perhaps the most well known procedure (Douglas and Peucker, 1973), used in several commercial software systems, uses a slightly different approach to retain the main shape features of a line (Figure 4.16b). The Douglas–Peucker procedure first joins the beginning and ending vertices of a line feature by a straight line, and then examines perpendicular distances to the individual vertices. Those that are closer than a selected threshold distance can be removed. The point furthest away is selected as a new end point for repetition of the process until there are no points closer to a line than the threshold.

In addition to generalizing by point elimination techniques, lines may be represented as a series of compass orientations for successive straight line segments, known as **chain codes**. If lines are constructed on a regular arithmetic grid, the dominant direction may be coded using one of several cardinal directions (Figure 4.17). Assigning integers to four, eight, sixteen or other regular divisions of a circle, as in a compass rose, line features can be indicated by a series of numerical codes showing that general direction of the line. Assuming eight directions, a series of 0 values would represent a line going from south to north, and a mixture of the values two and four would represent a staircase going towards the bottom-right corner of the map.

More precise coding is possible with more directions. Actual coordinates are needed for only one point, either the beginning of a polyline or an arbitrary point for a closed line; and special techniques can be used for junctions. However, it is more difficult to compare or to apply alterations to lines generalized by this incremental type of encoding than it is to lines represented by coordinates.

4.5.2 Smoothed lines

With the same objective of reducing the number of pieces of information, lines may be represented as a single curve, or as a combination of **smooth curves**. Essentially, a smooth curve is fitted, according to certain criteria, to a shape for an irregular polyline or polygon represented by a set of specific points, generally passing through all or possibly just a selection of the specified points (Figure 4.18a). In the case of a line feature, the curve may cross intermediate segments on its route from one end to the other. In the case of a polygon, separate curve pieces generally are fitted to the distinct line segments, such as the five for the polygon in Figure 4.18b. Once this is done, the coordinates for intermediate points can be estimated by interpolation, as shown in Figure 4.19. Of course, if only point data are available, then the same process can be used to fit a curve as opposed to generalizing one.

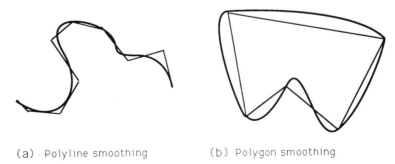

(a) Polyline smoothing (b) Polygon smoothing

Figure 4.18 Line smoothing. (a) For a polyline. (b) For a polygon.

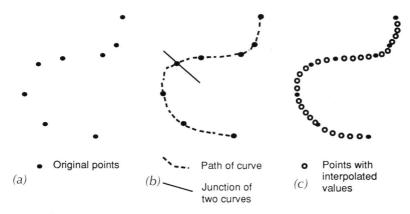

(a) • Original points

(b) `---.` Path of curve

        `___` Junction of
        two curves

(c) o Points with
    interpolated
    values

Figure 4.19 Interpolation of points on a line. (a) Original data. (b) Cubic polynomial fitted to a succession of four points. (c) Drawing a curve using very short straight segments between points interpolated.

Important criteria that may be established for an ideal curve are:

1. It should pass through specific end (or mid) points.
2. It should be tangential to only one point on an individual line segment, or cross at only one point.
3. It should have no discontinuity where separate curves join when making up a series to represent a complex polyline or irregular polygon.
4. It should have compact mathematical representation.

Line smoothing may use statistical averaging techniques or mathematical equivalents to the French curve, the flexible drafting tool used by graphic artists and engineers. This **spline**, made of thin flexible plastic, can (with practice) be made to fit specified key points. If the line is represented mathematically instead of mechanically, it is generally of the form $y = f(x)$ where f stands for function of, or $x = f(y)$, depending on which orientation we prefer or a particular line demands (Figure 4.20a). This function may produce or reproduce a bending line or a straight line as a limiting case.

In practice, lines (or surfaces for the three-dimensional context) are often represented explicitly by curves from the polynomial family, that is, equations having terms of high orders of x, like x^2 (Figure 4.20b), or raised to third and fourth powers (Figure 4.20c). The cubic polynomial, having the form $y = a + bx + cx^2 + dx^3$, is often the preference for fitting curves to successive runs of four points, estimating the numerical

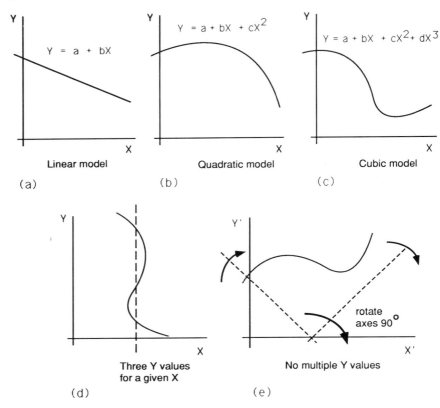

Figure 4.20 Polynomial functions. (a) Linear. (b) Quadratic. (c) Cubic. (d) Problem of several *y* values. (e) Rotation of axes.

coefficients of intercept and slope by least squares statistical procedures. A curve with only four numerical coefficients assures a perfect fit. The polynomial functions are also used for generalizing trend surfaces for attribute data or for modelling surfaces for artificial objects.

However, there is a major difficulty with this functional form approach – for each *x* there should be only one *y* coordinate, meaning that within the range on the *x*-axis there can be only one *y* value. In that many lines do not meet this condition (Figure 4.20d), alternative procedures are necessary. While transposing the *x* and *y*, or rotating the axes can produce satisfactory results in most cases (Figure 4.20e), the preferred alternative is to use a **parametric form** of representation. That is, instead of using a single equation to describe the curve we can use two equations that are explicit in a parameter referred to by *t*. For instance, for a line connecting two points (Figure 4.21a), we can use the expressions:

$$x = tx_2 + (1 - t)x_1$$

$$y = ty_2 + (1 - t)y_1$$

If we set $t = 0$, then we get x_1, y_1; and for $t = 1$ we obtain x_2, y_2. If we increase t from 0 through 1 then we generate all points for the line segment joining the point x_1, y_1 to x_2, y_2. The somewhat more complex case of parametrically creating short arc lengths for creating a circle is shown as Figure 4.21b.

The incremental changes can be determined by equations in this case, but not every curve has a simple generating algorithm. There is, though, a family of curves, including B-spline, Bézier and Hermitian spline that have a basic mathematical form of

$$x(t) = a + bt + ct^2 + dt^3$$

$$y(t) = e + ft + gt^2 + ht^3$$

If elevation, z, is also needed:

$$z(t) = p + qt + rt^2 + st^3$$

where the parametric variable varies between 0 and 1. As with the

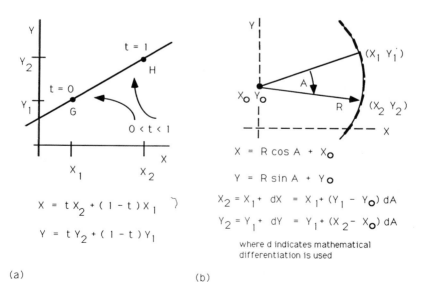

(a)   (b)

Figure 4.21 The parametric form. (a) Line. (b) Circle. (Figure 4.21b is based on material in Harrington, 1987).

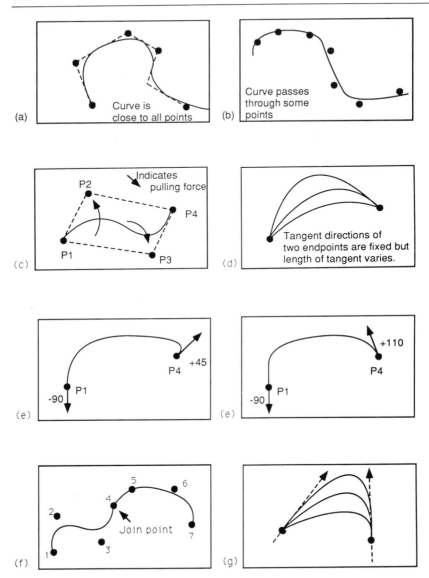

Figure 4.22 Spline and Bézier curves. (a) Spline for six points. (b) Spline for seven points. (c) Concept of control points for spline curves. (d) Fixed directions of tangents. (e) Variable directions of tangents. (f) Two Bézier curves joined. (g) A set of Hermitian curves. (Based on material in Foley et al, 1990.)

functional form, here the cubic order is preferred for the balance it provides between computational complexity and user flexibility.

In these equations a, b, c, d, e, f, g, h, p, q, r and s are special coefficients depending on the coordinates x_1, y_1, z_1, x_2, y_2, z_2, x_3, y_3, z_3, x_4, y_4 and z_4 of the four points: V_1, V_2, V_3, and V_4. Starting from points we can write:

$$x(t) = A(t)x_1 + B(t)x_2 + C(t)x_3 + D(t)x_4$$
$$y(t) = E(t)y_1 + F(t)y_2 + G(t)y_3 + H(t)y_4$$

If elevation is also needed:

$$z(t) = P(t)z_1 + Q(t)z_2 + R(t)z_3 + S(t)z_4$$

in which the $A(t)$ to $S(t)$ are special cubic polynomial functions called **blended functions**, which combine the influences of the several end and control points (Figure 4.22). With different influences from the end and intermediate points, many composite paths may be determined.

Generally, the curve-fitting process is guided by several conditions:

1. The number of points of influence on the path of the curve.
2. The number of points the curve passes through.
3. Tangents at the origin and at the extremity.
4. If continuities are desired at the location where the different pieces of the curve come together.

The control may be exercised by prior specifications, or done iteratively visually. The curvature for a particular line segment is influenced by the **control points** and the conditions to be met where pieces join together. In any event, using cubic polynomials for reasons already stated, as Figure 4.21 illustrated, the smoothness of a specified sequence of curves is established by setting the tangents of two curves at a common point to be equal, in mathematics called first order continuity (Figure 4.22f). The particular shape of the cubic polynomials varies as different points exert more influence.

It is important to be sensitive to the existence of several types of spline curve. We briefly review the salient characteristics, in turn, of the Hermitic curve, natural and B- (for basic) splines, and Bézier curves. Generally speaking, natural splines, models for the physical splines used by draftspersons, have forces pulling them in directions of control points, in a sense like springs, just as weights attached to the flexible metal strips pulled them in various directions (Figure 4.22c).

Hermitian curves use the conditions of tangents at the two end points to direct the path of the curve. The nature of the bend in the polynomial will reflect the direction and length of the tangent vectors, as shown (Figure 4.22d and e). Bézier curves, invented by Pierre Bézier of the

Renault car company for designing the outer panel of cars as a smooth surface, are generally similar in concept, but can also be effective at higher order polynomials, and are easy to split into several other Bézier curve pieces (Figure 4.22f). In the few spatial information systems that provide smoothing tools, apparently for graphical purposes and not for reducing data storage, it is the Bézier one that is the most commonly encountered.

4.5.3 Some realities of line and polygon representation

In practice, several spatial information systems oriented toward mapping features at a high precision offer tools for line simplification by weeding out points, often using the Douglas–Peucker algorithm. The chain codes are rarely used, and smoothing facilities are more common for computer drawing or mapping purposes than for spatial information systems, although as noted above, a few commercial products of the latter type do provide users with capabilities of line smoothing using Bézier curves.

The different representations resulting from different degrees of generalizing are well illustrated by the public domain data sets World Data Bank I and II, containing digital coordinate data for major features of the world, especially coastlines. The two versions, different in the number of points captured by a digitizing process, represent natural features curvatures quite differently. The more detailed of the two, World Data Bank II, also provides a good example of the impact of line generalizing to different degrees. Figure 4.23b shows the loss of detail at steps of different minimum point separation distances.

Line simplification procedures aim to reduce the amount of data while retaining a high level of information content. The data reduction may

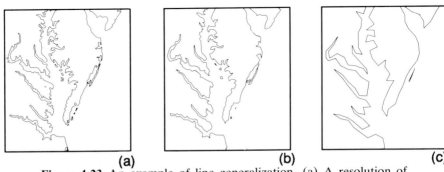

(a) **(b)** **(c)**

Figure 4.23 An example of line generalization. (a) A resolution of 0.01 inch; (b) A resolution of 0.03 inch; (c) A resolution of 0.10 inch. (Extracted from Figure 4 in Anderson et al.)

serve to reduce storage space needs, and reduce access or drawing time, but will be achieved at some cost of loss of precision and occasionally of form of entity.

Overall, different forms of representing irregular linear features or polygon boundaries are better or worse depending on the uses of the data and the computer resources available. Using coordinates is simple but requires data for many short lines if the features are smoothly curving. The success of generalizing by eliminating some details by point elimination or chain coding depends on the kind of curvature in the lines as well as how much reduction might be wanted in data volume. Natural entities like coastlines and rivers may well be represented by successions of spline curves, but they are unlikely to be as useful for boundaries of administrative units or railroads. On the other hand, complicated curves can be represented by only a few pieces of data, the parameters of the equations, rather than by a long string of coordinates.

4.5.4 Intensional and extensional representation of objects

In general, recalling some comments of Chapter 3, the individual entities may be represented in practical terms in two contrasting ways. For example, a straight line can be constructed from an ordered series of points or may be created via a mathematical formula. The arc or spline type of entity implies the latter case. So we recognize these two situations:

1. The extensional character.
2. The intensional character.

For the **intensional** state, we deal with a small amount of data explicitly, and store the rules, procedure, or method for obtaining other information. That is, then, for the river the coefficients of a mathematical equation are stored along with the end points of the line feature. A more simple example is the representation of a triangle by three vertices and rules to compute segment lengths and/or angles (Figure 4.24). As we demonstrate later, we are not restricted to geometric objects, but can also deal with chunks of space in this way.

For the moment, we summarize that intensional data require:

1. The storage of some privileged elements of data for the object.
2. The specification of a rule for generating all possible elements, the generative rule.
3. The definition of a rule for testing if an element, for example, a line segment, is a member of the object, perhaps a polygon.

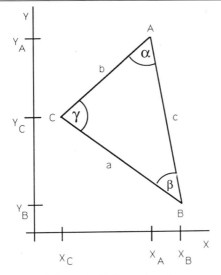

A, B, C coordinates are stored in the database.

Rules are used to create the side lengths and angles for the triangle, for example:

$$a^2 = (X_B - X_C)^2 + (Y_C - Y_B)^2$$

or, by the law of cosines:

$$a^2 = b^2 + c^2 - 2bc \cos \alpha, \text{ and}$$

$$\alpha = \arccos \left((b^2 + c^2 - a^2) / 2bc \right)$$

Figure 4.24 Intensional representation—rules for a triangle.

The third aspect is required for, among other things, ascertaining if a particular coordinate position falls inside an object (say a polygon) or outside, as well as on the boundary. Thus, as we discuss more fully later, there is a way to circumvent the practical difficulty in storing the infinite number of points that lie inside a polygon.

4.6 FRACTALS: A WAY TO REPRESENT NATURAL OBJECTS

So far we have assumed that sharp boundaries or smooth shapes exist for real entities. This assumption reflects a map model or geometric bias rather than a necessarily appropriate model for nature. In spatial information systems we have two kinds of entity to model:

1. Natural earth features such as terrain and coastlines.
2. Person-made objects like buildings and roads.

Fractal geometry has scope enough to represent more adequately than Euclidean geometry real world entities that are not smoothly formed, as is the case with most natural objects. For example, a tree would be modelled in the latter domain by sets of connected cylinders, cones and other regular geometric primitives; but fractal geometry approaches the situation by a different route giving us more acceptable shapes, especially for the leaves and branches.

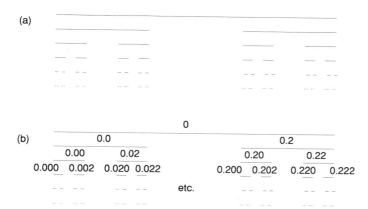

Figure 4.25 Cantor dust. (a) Generation of the dust. (b) Encoding method.

4.6.1 Creation of fractal objects

Fractal geometry (Mandelbrot, 1982) is an attempt to synthesize various mathematical works at the turn of the twentieth century. The word **fractal** implies properties as in fraction or fragmented; in essence fractal geometry has ideas of fragmentation and self-similarity.

Even though objects may be rough or irregular, that is fragmented, they may at the same time have some similar semblance of shape or pattern when viewed from different distances. Self-similarity is symmetry across different scales; there are patterns within patterns. Or, as Mandelbrot says, fractals are geometric shapes that are equally complex in their details as in their overall form. An appreciation for these concepts is best approached by seeing how particular types of object can be created, before demonstrating the use of fractal geometry for natural landscapes.

Consider a line segment split into three equal parts with the middle section discarded (Figure 4.25a). An indefinitely long continued process of splitting the line segments will in the end produce a set of very small, aligned segments called **Cantor dust**. Since a point can be defined as a line whose length is tending to zero, we can say that we end with a set of aligned (Cantor dust) points, the number of which could be very hard to count.

Such a process of **pattern generation** has two components:

1. An initiator (for Cantor dust this is the line segment).
2. A repetitor (two other segments) or pattern generator.

An encoding method will serve to distinguish the different 'points'.

Figure 4.26 The Von Koch curve.

Identifying the initiator as 0, at the first subdivision we have pieces 0.0 and 0.2, but no 0.1 which is the gap in the line (Figure 4.25b). The digit 1 never appears in the encoding, which proceeds to produce numbers as shown. At the second step there are four segments, encoded as shown; at the next step there are eight, and so on, always encoded by only digits 0 and 2.

At step N, the set consists of 2^N segments, and each segment has a numerical code with N ternary digits after the 'decimal point'. The segment length is $l3^{-N}$, where l is the unit length (perhaps in centimetres) of the initiator, here a length of 1. Thus, for the third level of application of the **recursive** process, if the original line was 4 units, then each piece will be $\frac{4}{9}$, that is just under half a unit. For the repetitor, its self-similarity ratio r is $\frac{1}{3}$, identifying the subdivision of the original line into three pieces.

A somewhat more interesting pattern results from the use of the **von Koch curve** which is similar in construction to the Cantor dust except that the middle piece of the split line is replaced by two joined segments with the same length, making an equilateral triangle (Figure 4.26). After applying the repetitor for a while, the pattern takes on a curved appearance.

It is easy to imagine that, at the infinitum, there will be produced a continuous curve formed by the succession of small angles (Figure 4.26). Even if this curve is continuous, it has no tangent or derivatives because a tangent is defined by a line corresponding to the limit of a chord between two points taken in the curve. If we are at a certain level of resolution we can define a chord and take its limit; but if we change the resolution the chord angle alters as we make new fractal segments along the von Koch curve.

If the line segment initiator of the von Koch curve is replaced by an equilateral triangle, a snowflake is produced by the recursive process (Figure 4.27). At step one, the initiator is replaced by a six-pointed star corresponding to a polygon with 12 sides. At step two there are 48 sides;

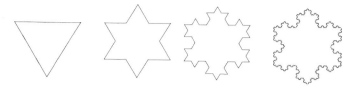

Figure 4.27 The Von Koch snowflake.

at each step the number of sides is multiplied by four. So, for an initiator perimeter of length 1, the perimeter becomes $1(\frac{4}{3})^N$ which number evidently tends to infinity, although the area tends to a finite limit. The self-similarity ratio is again $\frac{1}{3}$.

An initiator and repetitor as shown in Figure 4.28a will, by self-similarity, produce new smaller islands (Figure 4.28d). A tree form, perhaps also a river system, is produced as shown in Figure 4.28e. Many other patterns can be produced via fractal geometry, but this group of dust, snowflakes, islands and trees apply to most spatial information systems. Polygons created via fractal geometry, called **teragons** (a term coming from the Greek word *teras* meaning a dragon), may be produced in a deterministic way, as with a von Koch curve, or stochastically, using a procedure outlined below.

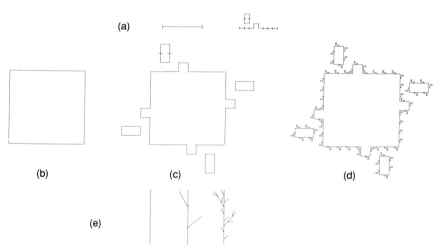

Figure 4.28 A few other fractal objects. (a) Initiator and repetitor for the island generator. (b) Starting the island generator. (c) Second step in generating islands. (d) Third step. (e) Example of a fractal tree.

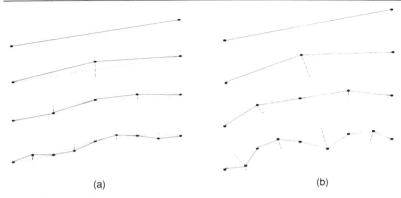

Figure 4.29 A stochastic interpolation for a line. (a) Displacement of the middle point along the *y* axis. (b) Displacement of the middle point along the segment perpendicular bisector.

4.6.2 Stochastic fractals

So far we have produced only regular objects. To create more 'natural looking' shapes, that is, involving variety in the sense that the leaves of a tree species may generally have the same form, but are individually different, we use randomization. The **fractional Brownian model**, a family of one-dimensional Gaussian stochastic processes of value in analyzing time series for natural phenomena, incorporates a curve which is self-similar.

This kind of fractional movement is difficult to simulate. An approximate solution is given by the formula:

$$y_{\text{new}} = \tfrac{1}{2}(y_1 + y_2) + u\sigma_0 2^{-lh}$$

where *u* stands for a random number, σ_0 is the parameter for a Gaussian (normal curve) distribution, *l* is the level of recursivity, and *h* is a fractal parameter specifying the roughness of an object. For example, for coastlines or terrain generation, this is often near 0.8. This formula means that instead of having the pure middle of a line segment one-half of the addition of the two *y* values, we add a small error term which has a Gaussian distribution, and fading with the level of recursivity. In other words, the stochastic quantity is less important when we increase the level of recursive splitting.

In practice, an acceptable approximation of the Brownian motion (curve) can be achieved by using a recursive subdivision process and this formula. Considering a line segment as an initiator, the repetitor is created (shown in Figure 4.29a) as follows. At the first step we have a

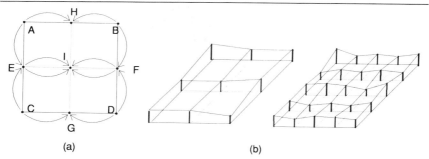

(a)   (b)

Figure 4.30 Stochastic interpolation for terrain. (a) Procedure. (b) Repeated application of the procedure (after Fournier, et al., 1982).

linear interpolation; afterwards we take the middle and we move it a little bit, so giving two segments. In sequence, we take the middles of these two segments and move them a little according to the distribution noted above. A pseudo-random number based on the Gaussian distribution is generated and used as a distance in the y-axis to displace the middle point of the line segment. In this way, the equation provides a stochastic interpolation, as discussed in section 7.1.1.

A slight variety of this process displaces the point along the segment mediator (Figure 4.29b), the line perpendicular to the line joining the two points at the ends of the starting line segment. Some authors prefer to move the middle along the perpendicular bisector, and some prefer movement along a line parallel to an axis. For the mediator the formula is more complex but truer.

The process can also be used for two-dimensional contexts, to produce simulations of terrain (Figure 4.30a). Starting from points A, B, C and D we generate midpoints on each side of the original rectangle (or square), and then get I from the midpoints E, F, G and H. Gaussian displacement values are again generated, producing an effect as shown (Figure 4.30b). Continuing this process to small line segment lengths will produce a terrain simulation from the non-systematic irregularities in the height values (Figure 4.31). The surface modelling, varying according to the values of the h parameters, can be made to produce rough or smooth terrains as illustrated. As h increases towards 0.9, the surface roughness lessens and the surface approaches a plane surface.

4.7 SPACE-FILLING CURVES AND DIMENSIONALITY

Data processing and storage may be more economical if less information can be used to meet the same requirements. Thus area units may be

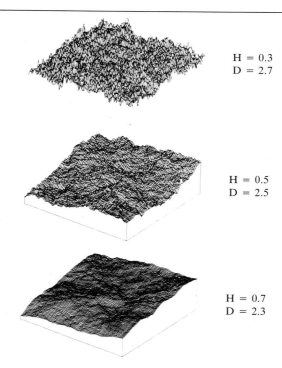

H = 0.3
D = 2.7

H = 0.5
D = 2.5

H = 0.7
D = 2.3

Figure 4.31 Terrain representation using the fractional Brownian process. This example shows the three surfaces that are outcomes of different H parameters. As H increases so the surface roughness lessens, and the fractal dimensionality tends towards that for a planar surface. (Extracted from a figure in Goodchild and Mark (1987), and reprinted with the permission of The Association of American Geographers, Washington, DC, USA.)

represented by a centroid, a zero-dimensional object or by parametric curves. A data reduction can also occur if objects could be positioned in only one dimension rather than two or three. Indeed, the question also arises as to how many dimensions a material object has.

4.7.1 Paths through space

The matter of dimensionality is encountered in spatial information systems in different ways. In an earlier chapter we discussed the identification of material entities as having zero, one, two or three dimensions. It also arises in terms of addressing systems. Indeed there are procedures to represent relative locations of the two- or three-dimensional

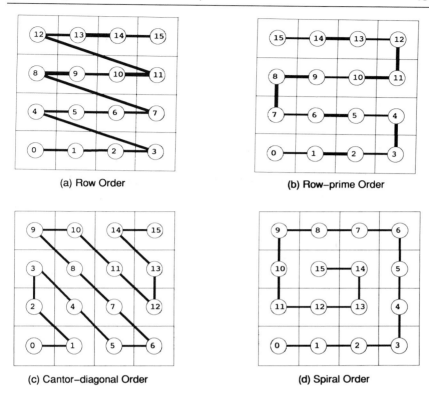

Figure 4.32 The concept of paths through space. (a) Row order. (b) Row-prime order. (c) Cantor-diagonal order. (d) Spiral order.

kind by a one-dimensional system. For example (Figure 4.32a), we can imagine we can refer to the sixteen adjacent map sheets in a single order by traversing from one to others along rows from bottom to top.

There are different orderings, that is **one-dimensional paths**, through the two-dimensional tiled space. If we think of a regular distribution of nodes for the tiles, then we have to solve a location problem of the routing kind, except we are not constrained by a network. Paths could zigzag, could go along a row in one direction and in a reverse direction in the next row, like a bidirectional computer printer (Figure 4.32b), or could follow a path that reduces the total distance of travel through going to as many immediate neighbours as possible, and having a small number of longer connections, or could have diagonal or spiral forms (Figures 4.32c and d).

A good sequential ordering should have certain properties that provide

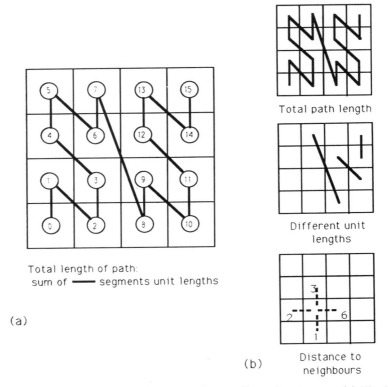

Total path length

Different unit
lengths

Distance to
neighbours

Total length of path:
sum of ——— segments unit lengths

(a)

(b)

Figure 4.33 Measures for curves in two-dimensional space. (a) The N
curve. (b) Measures.

some conveniences in single dimension addressing for two- or three-
dimensional sets of regularly shaped tiles. The path should pass only once
to each tile in the two- or N-dimensional space, and neighbours in space
should be adjacent on the path. The path should be useable even if there
is a mixture of different sized spatial units, and should work equally well
in two or three dimensions, and for connecting to adjacent blocks of
space. In reality there is no ideal path; there are just orderings with some
of these properties.

The row order, the kind used for building up a television picture, has a
longer total path than the row prime sequence in which every other line is
traversed in a reverse direction, and has several places in which
neighbours on the path, for example 3 and 4, are not adjacent in space.
The diagonal and spiral orders are like the row-prime sequence in having
the immediate adjacency property. However, the diagonal sequence

mixes up corner and side joins, while the spiral order terminates in the middle, making it impossible to connect to other blocks of space.

Another path (see Figure 4.33a), represents a Z or N, depending on the direction of viewing; this path mixes a few longer jumps with mostly steps from one tile to an immediate neighbour. Because it joins, sometimes at edges, and sometimes at corners, it has three different lengths.

A comparison of different paths (for a given resolution) can use several measures (Figure 4.33b):

1. Total length of the path.
2. Variability in unit lengths, where unit length is the distance from one point on the path to the next in sequence.
3. The average distance on the path from tiles to their four neighbours in space.

For an example of a given block of space divided into sixteen tiles, the row order has a total length of approximately 22 (four rows of horizontal lines of length 3, plus three diagonal lengths each of $\sqrt{10}$. The row prime and spiral have shorter lengths, each of 15. The Cantor order, made up of unit lengths and the 1.414 for the hypotenuse of triangles with two equal unit length sides, is approximately 18, and the N path has a length of about 20.

The third measure represents how easy it is to reach neighbours in space via the different paths. For example, for the tile identified as point 9 in the row order map, it has neighbours with path order values of 5, 8, 13, and 10, giving four distances of 4, 1, 1, 4, a total of 10 (average = 2.5). Other distances sums for this one tile are 10 for row-prime, 16 for spirals, 14 for diagonal, and 12 for N order. Comparative averages for the central block of four squares are shown in the table below along with other properties of a sixteen-tile mosaic.

Path type	Length (approximate)	Variability	Average distance
Row	22	2	10
Row-prime	15	1	10
Diagonal	18	2	18
Spiral	15	1	13
N	20	3	12

A more complete appraisal could include the distances for corner and edge cells. Other considerations might be important in addition to path lengths, such as the point or entry and exit from the mosaic.

4.7.2 Space-filling curves

More especially we talk about space-filling curves rather than paths through space. These curves are special fractal curves which have characteristics of completely covering an area or volume. While they have a topological dimension of two, their fractal dimension is two when filling an area, or three when completely occupying a volume space.

If we define a point as a zero-dimensional object, theoretically speaking it is not possible in the spirit of Euclidean geometry to find a one-dimensional curve passing through the infinity of points. However, if we think of a point as a (small) two-dimensional square the side of which tends towards zero, then, in the spirit of fractal geometry, it is possible to find a curve filling a two-dimensional space. So now the curve is defined as a sort of ribbon, the width of which tends towards zero. Similarly, at three dimensions, a point is defined as a small cube for which the side length tends towards zero, and the curve as a three-dimensional curve, a sort of spaghetti noodle for which the width also tends to zero.

Consequently, thinking of paths in space now as space-filling curves, lines that pass to all possible points in space, we see they should have the following properties:

1. The curve must pass only once to every point in the multi-dimensional space.
2. Two points that are neighbours in space must be neighbours on the curve.
3. Two points that are neighbours on the curve must be neighbours in space.
4. It should be easy to retrieve the neighbours of any point.
5. The curve corresponds to a bijective mapping from a multi- to a one-dimensional space.
6. The curve should be able to be used for variable spatial resolution, that is, a mixture of different-sized 'points'.
7. The curve should be stable, even when the space becomes very large or infinite.

In reality, we do not possess such ideal curves, but there are some with valuable properties for our purposes.

The original space-filling curve was exhibited in 1890 by the Italian mathematician Giuseppe Peano (Peano, 1890). A later variety, now known as the **Peano** or **N ordering** (Figure 4.34a), facilitates retrieving neighbours, and although neighbouring points in space are not always neighbours on the curve, they generally are. It is also possible to deal with different resolutions as shown, and the curve is stable. The **Hilbert**

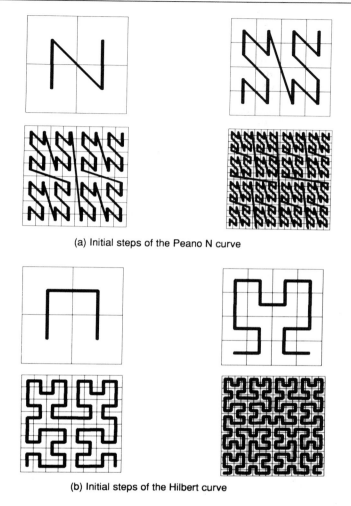

(a) Initial steps of the Peano N curve

(b) Initial steps of the Hilbert curve

Figure 4.34 Initial steps of the Peano and Hilbert curves. (a) Peano N. (b) Hilbert.

curve, sometimes known as the Peano–Hilbert curve, passes through all points in a set by means of single length steps only, having the form of a Greek letter π (Figure 4.34b). This curve meets most of the conditions noted above, but does not provide an easy way to retrieve neighbours and is not stable, as described below.

In practice, the encoding of space-filling curves uses one coordinate, called a **key** by most practitioners, to stand for two or more coordinates.

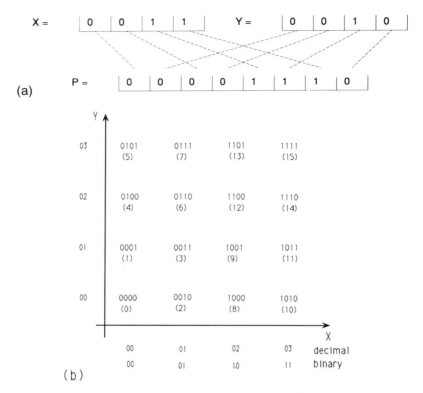

(a)

(b)

Figure 4.35 Obtaining the Peano key by bit interleaving. (a) The bit interleaving for *x* and *y*. (b) Keys for a block of sixteen spaces.

For the Peano N curve the keys are easily obtained; it is quite complex to get the keys for the Hilbert curve, as noted below. For the former, the binary digits for the *x* and *y* values are interleaved, as shown in Figure 4.35a. For example, *x* = 3, and *y* = 2 in binary are respectively 0011 and 0010. The key, *p*, thus is 00001110, corresponding to the decimal number 14. A set of keys for a block of sixteen squares is shown in Figure 4.35b. (There is a special arithmetic called tesseral arithmetic that serves to look after the peculiarities of mechanisms such as bit interleaving.)

It is a little more awkward to deal with real and negative coordinates compared with the case of integer coordinates just described, but the idea is the same. With four digits after the decimal point for *x* and *y*, we get eight digits after the decimal point in the Peano key, for example, for *x* = 11.0101 and *y* = 10.1011, *p* = 1110.01100111. If there are ten digits after the decimal point, the Peano key codes will have twenty digits. Conversely, whatever is the even number of binary decimal digits of a number, p, it is always possible to change it to two other real numbers, one by taking digits with even rank to get x, and odd rank digits to get y. So, this mechanism allows us to cover space with any resolution. For negative coordinates the preferred solution is to transform the values to positive coordinates by origin shifting, that is, geometric translation.

To demonstrate the awkwardness in using **Hilbert keys**, we present one algorithm (Mark and Goodchild, 1986). We will use H(X, Y) to denote the Hilbert key of x and y.

PROCEDURE H to X, Y (H, X, Y)

Unsigned H;
Integer X, Y;

```
Begin
      base − 4 P[8], Q[8] /* handles grids up to 256 × 256 */;
      binary X[8], Y[8];
      integer N;
      integer i;
      N = 8;

/* first construct vector P from H and make another copy Q */
for i := 1 to N do
begin
      P[i] := (H mod 2 ** (2 * (8 − i + 1)))/2 ** (2 * (8 − i));
      Q[i] := P[i];
end

/* next rearrange the bit pairs in Q */
for i := to N − 1 do
begin
      if P[i] = 0 then
      for j := i + 1 to N do
      begin
          if (Q[j] mod 2) = 1 then Q[j] := 4 − Q[j];
      end
      else if P[i] = 3 then
          for j := i + 1 to N do
          begin
          if (Q[j] mod 2) = 0 then Q[j] := 2 − Q[j];
          end
end

for i := 1 to N do
```

```
begin
    switch (Q[i])
    begin
        case 0 : X[i] := 0; Y[i] := 0; break;
        case 1 : X[i] := 1; Y[i] := 0; break;
        case 2 : X[i] := 1; Y[i] := 1; break;
        case 3 : X[i] := 0; Y[i] := 1; break;
    end
    /* finally assemble X and Y from their binary vectors X[ ] and Y [ ] */
    X := 0;
    Y := 0;

    for i := 1 to N do
    begin
        X := X + X[i] * 2 ** (2 * (8 − i));
        Y := Y + Y[i] * 2 ** (2 * (8 − i));
    end
end
end
```

Generally, the ordered paths have similar shapes at different scale levels (Figure 4.34). They are said to be **self-similar**, that is the entity is called self-similar if any part of it, after enlargement, cannot be distinguished from the whole object. However, the particular place of a point or tile in the sequence for a particular curve type may not be consistent across scales. While the N curve does have such **stability,** as revealed by the coded numbers (Figure 4.36b), this is not so for the Hilbert curve (Figure 4.36d). That is, for the Peano case, if the space is extended by doubling each side for the block of four quadrants, we see that the order of squares is not perturbed. The square numbered 1 keeps its same number; it is stable.

Space-filling curves have two principal, practical uses in the domain of spatial information systems. Firstly, they provide some efficiencies in scanning operations, either hardware devices or searches through datafiles. Secondly, they are used as spatial indexes, simplifying two dimensional addressing as single dimension addressing. The United States Bureau of the Census uses Peano keys for this purpose, and the Canadian Land Information System, often regarded as the first working geographic information system, used the concept of orderings under the label of Morton sequences (Morton, 1966). Orenstein (1986) calls the two-dimensional reverse Peano order the Z-order. (We avoid reference to the Z-order, recognizing that *z* is usually used to indicate the third space dimension. Different forms of curve may be more or less appropriate for particular circumstances; but, in general, it appears that the Peano and Hilbert orderings are more robust, that is better performers over a range of requirements.

4.7.3 Dimensionality

Recent advances in this area of fractal geometry have allowed us to model natural objects, as shown earlier, but also give us different ways to think about dimensionality. For example, the length of a coastline can

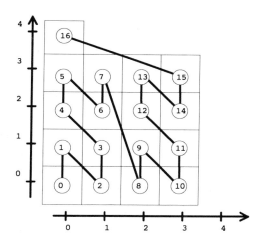

(a) Peano N space–filling ordering

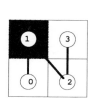
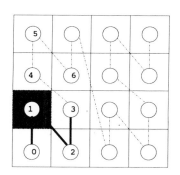

(b) Ordering stability with the N order

Figure 4.36 (Continued overleaf)

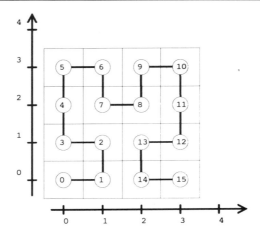

(c) Hilbert space–filling ordering

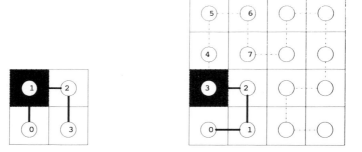

(d) Ordering instability when space extension with Hilbert ordering

Figure 4.36 Ordering stability for the N and Hibert orderings. (a) Peano N space-filling ordering. (b) Ordering stability with the N order. (c) Hilbert space-filling ordering. (d) Ordering instability when space extension with Hilbert ordering.

vary depending on scale, ranging from an apparently infinitely high length (assuming no limit to the precision of measurement) to a very short distance if we highly generalize the shape. It is interesting that fractal geometry can give us measures of the dimensionality of objects that are different from those we expect from Euclidean geometry. The fractal dimension tells us how densely a phenomenon occupies the space in which it is located. It is not dependent on the measurement units used or alteration of the space by stretching or condensing.

Different objects may be identified by a particular fractal dimension. The **fractal dimension** of many, but not all, entities can be obtained by the formula (from Mandelbrot, 1982):

$$d = \log n/(\log (1/r)) \qquad [\text{or } \log n/\log s]$$

where n is the number of pieces in the repetitor, r is the self-similarity ratio, and d is the fractal dimension. Alternatively, s, the scaling factor, the inverse of the self-similarity ratio, can be thought of as the number of pieces that an entity is split into. In the case of Cantor dust, we have $n = 2$ and $r = \frac{1}{3}$, that is log 2/log 3, giving $d = 0.6309$. For the snowflake, $n = 4$, $r = \frac{1}{3}$, giving $d = 1.2619$. For the island generator $n = 16$, and $r = \frac{1}{8}$, producing $d = \frac{4}{3} = 1.333$. For the space filling curve for the plane $d = 2$.

We can imagine a continuum where a value of close to 0 would mean an entity is close to a point, a value of 1 means like a line, and if it is near 2, it is like an area. The island generator falls somewhere between a line and an area. In this way, we reach another definition of fractal – an entity is called fractal when its fractional dimension exceeds its topological dimension. The Cantor dust 'dots' topologically have a zero dimensionality, whereas they have a fractional dimension of 0.6309, closer to a 'line'. A smooth line has a dimensionality of 1.0 (that is, log 3/log 3), but a line formed from the von Koch triadic snowflake has a fractional dimension of greater than 1.

Imagine a string of virtually no width (finer than a hair) being fed into a box. As more and more string is pushed into the box it fills up the entire (three-dimensional) space. Or better still, look at a ball of string from a hundred metres away, and then from a distance of one foot. If we define a point as a three-dimensional box whose side length tends to zero, then we can have a curve (piece of string) filling a three-dimensional space. That is, a one-dimensional object at a limiting state can take on a **higher order dimensionality**, like a solid three-dimensional object, depending on perception. Or, an item of infinite length can fit in a finite space.

Similarly, a smooth line will have a dimensionality of 1, but an irregular line has a higher value, certainly greater than 1.0. The idea of a continuous magnitude of dimensionality provides a basis for comparing different representations of a given phenomenon, as in rivers shown on maps of different scales, or for assisting cartographers in developing principles for map generalizing. For coastlines the mean fractal dimension is $d = 1.2$ and the stochastic parameter $h = 0.8$, whereas for terrain, d is about 2.3 and $h = 0.8$ (Figure 4.31). Smoother objects have smaller d

numbers, and dimensionality values of 1.5 or 2.5 are more appropriate for highly indented lines and rugged terrain, respectively. Such natural objects are not pure fractals, in the sense of having a constant dimensionality, but are approximately so, in the sense of being represented by a non-integer dimension.

Overall, fractal geometry has promise for some of the requirements of spatial information systems. Two-dimensional stochastic interpolations (discussed more fully in Chapter 7) are useful for terrain modelling; one-dimensional applications are for coastlines or boundaries of entire continents. As noted earlier in this chapter, fractal geometry gives us a way to indicate the dimensionality of specific objects and structures, and underlies the concept of space-filling curves. Fractals may also be used for cosmetic embellishments for cartographic and other graphic displays, for example, image generation.

4.8 SUMMARY

Spatial data are quite special compared with data usually stored in databases. We have to deal with an infinite number of points in space, not a fixed number of entities like bank transactions or people in a survey. We have both intensional and extensional forms of representation; and there are varying degrees of precision of representation. We also have, and will continue, to use geometry deterministically. That is, we are interested in the representation of particular rivers, terrain or buildings, not in discovering laws about all rivers, all buildings, or all terrains, the realm of statistical geometry.

Ignoring the trivial case of point entities, our discussion of geometries can be summarized primarily as a need to be sensitive to various ways of representing line entities. These one-dimensional objects in Euclidean geometry may be represented by:

1. 'Real' position, using Cartesian or polar coordinates.
2. Some generalization of the vertices for the line feature using procedures that depend on coordinates.
3. Topology, ignoring geometry.
4. Parametric representation.
5. Fractal geometry.

Some data are discarded if only generalized versions are maintained in a database. But the data volume can be lowered by a reduced number of coordinates, or, better still under some conditions, by the parameters of functions or equations, or the initiators and generators for fractal representations. Coordinate-based representations are associated with digitizing or line-following forms of data capture. Procedures of simplification and data reduction usually are applied to those data, unless regular geometric figures are the entities.

Otherwise, detailed geometric representations are dealt with by stored data for positional coordinates. Absent good reasons for representing phenomena intensionally, that is by rules, as for regular geometric figures, then devices are needed for adequately and effectively recording position extensionally. The many spatial databases existing today are characterized by Euclidean measures, and projective geometry, as well as by extensional data.

Continuing a theme first developed in Chapter 3, spatial entities can be viewed from several perspectives regarding dimensionality. The spatial units can be dealt with in terms of Euclidean dimensionality, or thought of via fractional geometry. For some needs in spatial information systems applications the reduction in dimensionality associated with space-filling curves is important.

Working at a global scale necessitates certain sensitivities and practices in the use of map projections to avoid excessive and misleading distortions, but also makes it challenging to create spatial referencing schemes suitable for a sphere, and ways to store very large volumes of data.

The use of Cartesian coordinates in a digital environment necessitates, too, an appreciation for errors that may arise for data for position, and distance, especially when approximations are used, no matter how appealing. Whether a reflection of computer machine precision or a user's choice of measurement needs, the limitations in recording are best thought of via interval geometry rather than as a continuously refined measurement concept. Scale of observation, embodied in the concepts of resolution and precision, means that not all the digits for a positional coordinate are important.

The broad extension of this concept of precision of representation takes us to a position of recognition that for many phenomena boundary lines are artefacts, that a field oriented view can be more appropriate than an entity orientation, and that fractal geometry has value for dealing with natural objects. In the end, for these and other situations discussed in the next chapters, we must not think only through the minds of Euclid or Descartes, but also via Peano and Mandelbrot.

4.9 BIBLIOGRAPHY

For more extensive treatments of Euclidean geometry, the reader is referred to Gasson (1983) and Gatrell (1983). A comprehensive survey of line generalization is that by McMaster (1987). Elaborations on spline curves are provided by Foley *et al.*, and by Bartels *et al.* In addition to the treatises by Mandelbrot, fractal geometry concepts are covered extensively by Barnsley. The book by Clarke provides the cartographic perspective on spatial objects, oriented especially to vector geometry; while either Robinson et al. or Muehrcke provides an extensive review of spatial referencing systems and map projections.

Abbott, Edwin A. 1952. *Flatland: A Romance of Many Dimensions*. 6th edn. New York, USA: Dover Publications.

Anderson, Delmar, James L. Angel and Alexander J. Gorny. 1977. WDBII: *Content, Structure and Application*. Paper presented at the First International Advanced Study Symposium on Topological Data Structures for Geographic Information Systems. Harvard University, Cambridge, Massachusetts, USA.

Angel, Edward. 1990. *Computer Graphics*. Reading, Massachussetts, USA: Addison-Wesley.

Armstrong, Marc P. and Brian T. Dalziel. 1989. Digitizing and distance estimation errors in spatial analysis: sinuosity and network configuration effects. *Proceedings of the International Geographic Information Systems Symposium*, Baltimore, Maryland, USA, pp. 75–84.

Barnsley, Michael. 1988. *Fractals Everywhere*. San Diego, California, USA: Academic Press.

Bartels, Richard H. *et al.* 1987. *An Introduction to Splines for Use in Computer Graphics and Geometric Modeling*. Los Altos, California, USA: Morgan Kaufmann.

Bell, Sarah B. M. 1986. Tesseral addressing and arithmetic – practical concerns. *Proceedings of the Workshops on Spatial Data Processing using Tesseral Methods*. Swindon, UK: Natural Environment Research Council, pp. 11–16.

Butz, A. R. 1971. Alternative algorithm for Hilbert's space-filling curve. *IEEE Transactions on Computers* (April): 424–426.

Clarke, Keith C. 1990. *Analytical and Computer Cartography*. Englewood Cliffs, New Jersey: Prentice Hall.

Committee on Map Projections. 1989. Geographers and cartographers urge end to popular use of rectangular maps. *The American Cartographer* 16(3): 222–223.

Davis, John C. 1973. *Statistics and Data Analysis in Geology*. New York, USA: Wiley.

Diaz, Bernard M. 1986. Tesseral arithmetic – overview. *Proceedings of the Workshops on Spatial Data Processing using Tesseral Methods*. Swindon, UK: Natural Environment Research Council, 1–10.

Douglas, David H. and Thomas K. Peucker. 1973. Algorithms for the reduction of the number of points required to represent a digitized line or its caricature. *Canadian Cartographer* 10(4): 110–122.

Dutton, Geoffrey. 1989. Planetary modelling via hierarchical tessellation. *Proceedings of the Auto Carto 9 Conference, Baltimore*. Falls Church, Virginia,

USA: American Society for Photogrammetry and Remote Sensing/American Congress for Surveying and Mapping, pp. 462–471.

Dutton, Geoffrey. 1990. Locational properties of quaternary triangular meshes. *Proceedings of the 4th International Symposium on Spatial Data Handling, Zurich*, Switzerland, pp. 901–910.

Epstein, Earl F. and Thomas D. Duchesneau. 1990. Use and value of a geodetic reference system. *Urban Regional Information Systems Association Journal* 2(1): 11–25.

Foley, James D. *et al.* 1990. *Computer Graphics, Principles and Practice*, 2nd edn. Reading, Massachussetts, USA: Addison-Wesley.

Fournier, Alain, *et al.* 1982. Computer rendering of stochastic models. *Communications of the Association for Computing Machinery* (6): 371–384.

Gasson, Peter. 1983. *Geometry of Spatial Forms*. Chichester, UK: Ellis Horwood.

Gatrell, Anthony. 1983. *Distance and Space*. Oxford, UK: Clarendon Press.

Gillies, Charles F. 1989. The motor freight system and port of Baltimore, Maryland 1950–1980: thirty years of spatial change. Thesis, University of Maryland, College Park, Maryland, USA.

Gleick, James. 1987. *Chaos, Making a New Science*. New York: Penguin.

Goldschlager, L. M. 1981. Short algorithms for space-filling curves. *Software, Practice and Experience* 11: 99–100.

Goodchild, Michael F. and David M. Mark. 1987. The fractal nature of geographic phenomena. *Annals of the Association of American Geographers* 77(2): 265–278.

Goodchild, Michael F. and Yang Shiren. 1989. A hierarchical spatial data structure for global geographic information systems. NCGIA Technical Paper 89-5, National Center for Geographic Information and Analysis, University of California, Santa Barbara, California, USA.

Harrington, Steven. 1987. *Computer Graphics: A Programming Approach*. 2nd edn. New York: McGraw-Hill.

Jenks, George F. 1981. Lines, computers, and human frailities. *Annals of the Association of American Geographer* 71(1): 1–10.

Kennedy, Hubert C. 1980. *Peano: Life and Works of Giuseppe Peano*. Studies in the History of Modern Science, No. 4. Dordrecht and Boston, USA: Kluwer Academic.

McMaster, Robert B. 1987. Automated line generalization. *Cartographica* 24(2): 74–11.

Mandelbrot, Benoit B. 1982. *The Fractal Geometry of Nature*. San Francisco, California, USA: W. H. Freeman.

Mandelbrot, Benoit B. and J. W. Van Ness. 1968. Fractional brownian motions, fractal noises and applications. *Society for Industrial and Applied Mathematics Review* 10(4): 422–437.

Mark, David M. and Michael F. Goodchild. 1986. On the ordering of two-dimensional space: introduction and relation to tesseral principles. *Proceedings of the Workshop on Spatial Data Processing Using Tesseral Methods*. Swindon, UK: Natural Environment Research Council.

Meixler, David. 1983. Peano keys. Paper presented at the Spatially Oriented

Referencing Systems Association Forum. 10–14 October 1983, College Park, Maryland, USA.

Morton, G. M. 1966. A Computer-oriented Geodetec Database and a New Technique in File Sequencing. IBM Corporation, Canada – Ontario Report. Ontario, Canada: IBM.

Muehrcke, Phillip C. 1986. *Map Use, Reading, Analysis, and Interpretation.* 2nd edn. Madison, Wisconsin, USA: JP Publications.

Orenstein, Jack A. 1986. Spatial query processing in an object-oriented database system. *Proceedings of the Association for Computing Machinery, Special Interest Group, Management of Data, Washington, DC, USA*, pp. 326–336.

Peano, Giuseppe. 1890. Sur une courbe qui remplit toute une aire plane. *Mathematische Annalen* 36(A): 157–160.

Peano, Giuseppe. 1957. edn. La curva di Peano nel 'formulario mathematico'. In Giuseppe Peano, *Opere Scelte di Giuseppe Peano.* Rome, Italy: Edizioni Cremonesi, vol. 1, pp. 115–116.

Poelstra, T. J. 1990. Frankie goes to town, or how to capture reality in urban areas. In Nicos D. Polydorides (ed.), *Data Acquisition for Spatial Information Systems.* Athens: Urban and Regional Spatial Analysis Network for Education and Training, Computers in Planning Series, vol. 7, pp. 56–66.

Quinqueton, Jean and M. Berthod. 1981. A locally adaptive Peano scanning algorithm. *IEEE Transactions on Pattern Analysis and Machine Intelligence* (4): 403–412.

Robinson, Arthur, *et al.* 1984. *Elements of Cartography.* New York, USA: John Wiley.

Robinson, Arthur H. 1990. Rectangular world maps – No! *Professional Geographer* 42(1): 101–104.

Sierpinski W. 1912. Sur une nouvelle courbe continue qui remplit toute une aire plane. *Bulletin de l'Academie des Sciences de Cracovie*, Serie A: 462–478.

Snyder, John P. 1983. Map projections – used by the US Geological Survey. *US Geological Survey Bulletin*, P-1532. Washington, DC, USA: US Government Printing Office.

Van Est, Jan and Jan Scheurwater. 1983. The geographic base register and a geographic base file: the use of post codes as segment identification. Paper presented at the SORSA Forum, 10–14 October; College Park, Maryland, USA.

Van Gelder, B. H. W. 1990. Global Positioning System: state of the state. In Nicos D. Polydorides (ed.), *Data Acquisition for Spatial Information Systems.* Athens: Urban and Regional Spatial Analysis Network for Education and Training, Computers in Planning Series, vol. 7, pp. 36–42.

Veregin, Howard. 1989. A taxonomy of error in spatial databases. Technical Paper 89-12, National Center for Geographic Information and Analysis, University of California, Santa Barbara, California, USA.

Warntz, William. 1964. A new map of the surface of population potentials for the United States, 1960. *The Geographical Review* 54(2): 170·184.

Wilson, Alan G. and Michael J. Kirkby. 1975. *Mathematics for Geographers and Planners.* Oxford, UK: Clarendon Press.

5
Topology
Graphs, areas, ordering

Freddie Zip delivered the mail to the boxes on his highway route. Freda Johnson missed her ride home to school because she did not realize there were two crossings of Ridge Road and Crescent Avenue, and waited at the wrong place. Inge Svenssen needed to map the vacant spaces in her forest zone. Gerry Mander tried to tame a tiger to help him persuade his political clients that they could do their own legislative district map drawing on their own personal computer. The French police could not apprehend Carlos Coca because he disappeared into a house in Spain even though they completely surrounded the house. François Reclus was not quite sure to which authority to pay taxes because his kitchen was in one city and his bedroom in another.

Relationships between places and spaces are examined in this chapter from the point of relative positions. Concepts from the mathematical areas of topology and graph theory are valuable for revealing the spatial structure of entities seen as points, lines, areas and solids, after the details of geometry are stripped away. After reviewing some properties of graphs for point and line elements, this chapter then introduces concepts for dealing with area objects by means of topology. Spatial properties of adjacency, connectivity and containment are examined conceptually, and then an introduction is given to data models for spatial relationships.

5.1 NETWORKS AND GRAPHS

Space can be viewed as a theoretically infinite set of points with observations possible for one or more attributes or spatial properties anywhere. But some phenomena are not defined for all points in this continuous space, or cannot correctly be regarded as samples from an infinite set. For instance, for studying rivers and transportation, we deal only with a subset of positions, constrained by the predetermined location of the associated zero-dimensional objects. In other words, the position of the nodes is the dominant element.

Formally, we define an isotropic condition as the property of

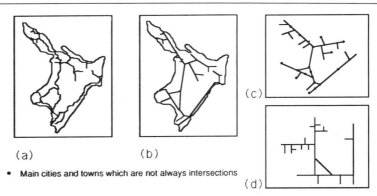

(a)   **(b)**

• Main cities and towns which are not always intersections

Figure 5.1 Graph for a rail transportation network. (a) Map based on geographic coordinates. (b) Staightening of links. (c) Removal from context to emphasize spatial structure. (d) Vertices and edges.

identicalness in all directions. Anisotropic refers to the existence of properties that differ with direction in space. Physical networks, man-made or natural (Figure 5.1 and Figure 3.11), in two or three dimensions represent an anisotropic state. Virtual networks or channels exist for air masses, ocean currents and air routes, among other examples. Many scholars and practitioners work with phenomena of this kind, for example, transportation planners, air traffic controllers, water and sewer service maintenance crews, marine biologists, hydrologists and geo-morphologists.

5.1.1 Graphs

The combination of line segments, the network, may be viewed as a graph (Figure 5.1). Assuming we begin with a map, showing the exact location of the transportation features, as for the railways of New Zealand, we will generally throw away the line shape information, the compass orientation, and the line lengths, concentrating on the **structural components**, the junctions and connections. The principal element is connectivity, but orientation, the recognition of a begin and end direction to a line, is sometimes important. Several steps of generalization and extraction may be involved, as shown.

The resulting graph has only a few basic elements (Figure 5.2):

1. The intersections or end points of lines, usually referred to as vertices or nodes, depending on context.

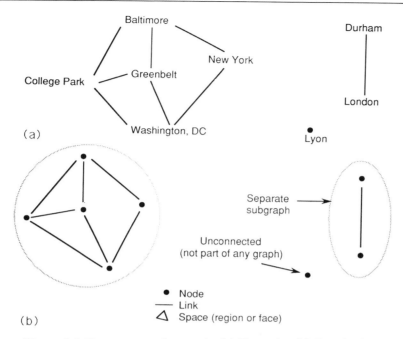

Figure 5.2 Components of a graph. (a) Example. (b) Terminology.

2. The lines, generally called edges, or links, but sometimes arcs or chains.
3. Separate single links or disconnected sets of lines, called subgraphs.
4. Vacant spaces (also referred to as faces or regions) between or outside edges.

Geometrically speaking, a vertex is a unique point position which has neither size nor dimension, and an edge, a non-intersecting curved or straight line, with ends that terminate in two, not necessarily different, vertices. If there are no unconnected edges, that is, no subgraphs, then every edge must be associated with two vertices, and every edge divides two vacant spaces. Of course, each vertex can have a varying number of edges, from at least one, representing a terminal vertex. Vertices can be connected in many different ways, subject to constraints imposed by the existence of subgraphs, and whether a vertical dimension is allowed. Connections may be **planar**, that is all crossings occur in the plane, or edges may cross without producing an intersection, the non-planar case.

For a three-dimensional context, there are virtual or concrete lines that do not meet, and the component of the space or void between edges is

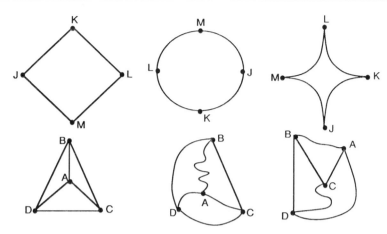

Figure 5.3 Two sets of isomorphic graphs.

harder to visualize. Ground transportation systems are examples of **non-planar graphs** if they have over- or underpasses, bridges or tunnels. In nature, subterranean river systems can be viewed as three-dimensional graphs, while aircraft flight paths are virtual three-dimensional graphs in the sense that there should not be any physical intersections (plane collisions). For a solid complete three-dimensional object, sometimes termed a cell, the graph consists of a series of edges bounding faces, plane, concave or convex, and converging, at least three in number, at the vertices. A solid object also has an inside, and an outside, but no connections to other objects. In a case of joined solid objects any edge may have several facets converging at it.

In that graphs are devices for dealing with certain kinds of spatial relationships, it is possible to use them to reveal structural similarities, such as comparing linkages by different modes of transportation, or changing patterns of connections over time. Two graphs are said to be **isomorphic** if there is a one-to-one correspondence between their vertices and their edges. That is, in other words, the conditions of connectedness and adjacency correspond, even though shapes may be quite different (Figure 5.3).

Graphs may have loops or circuits or cycles (Figure 5.4a). If they have none, they are called **tree graphs**, or just trees, for example, a star-shaped pattern of connections in a local area electronic communications network. Structures that have no circuits but which have directed edges, or **directed acyclic graphs**, are very common for sewerage and some water supply systems. Graphs with **circuits**, that is, cyclic graphs, have at least one vertex connected to itself without the need for traversing one edge in both directions. Highway and air transportation systems have many circuits; river systems generally are tree graphs, except in the case of braided streams. Both acyclic and cyclic graphs may be directed or oriented, as shown by arrows for directions of movement.

The network configurations represent different conditions; these range from simple patterns such as all vertices singly connected to a central node, or singly connected to a peripheral node, to complete linkages, that is, having all possible connections. Some special patterns of connection are the spanning tree, in which each vertex must be connected to at least one other vertex, the travelling salesman configuration, for which there must be a way to return to a starting point without retracing an edge, or a radial pattern (Figure 5.4b). Conditions of accessibility for the set of edges, and for particular nodes may vary substantially for different configurations. Some represent economies of building; others, like the minimal spanning tree represent economies of effort in use with costs of use associated with the edges. We can analyse the graph to reveal the set of links that has a minimum cost of connecting all places.

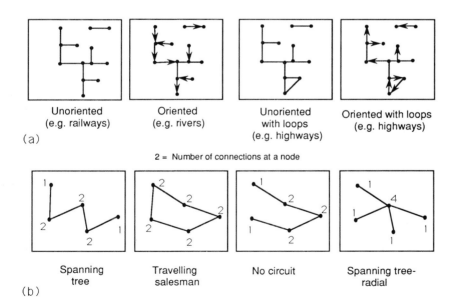

(a)

| Unoriented (e.g. railways) | Oriented (e.g. rivers) | Unoriented with loops (e.g. highways) | Oriented with loops (e.g. highways) |

2 = Number of connections at a node

(b)

| Spanning tree | Travelling salesman | No circuit | Spanning tree-radial |

Figure 5.4 Some types of graph patterns. (a) Oriented and un-oriented graphs. (b) Patterns of connections for the same set of nodes.

5.1.2 Properties of graphs

Various measures have been devised for revealing properties of particular graphs. In addition to counts of the number of vertices, edges and subgraphs, there are measures for properties of connectivity that help in answering questions like:

1. What is the relative level of accessibility of the different cities served by air routes?
2. How will travel be affected if a new bridge is built over a hitherto uncrossed estuary?
3. How many alternative ways might I travel through a city street system from my home to workplace to avoid traffic jams or road works existing on my usual route?

The measures relate to particular elements, the vertices or edges, or the entire graph.

Besides comparing graphs by their counts of edges, vertices and circuits; it is possible to devise derived and normed measures as useful descriptive tools. The particular vertices may be compared by counts of the number of edges (total, or in or out if these are different), often being referred to as the **degree** of a vertex. In the example, Figure 5.4, for a skeletal highway system, it is clear there are major differences in accessibility to nodes. The conditions of connectivity for the whole network may be summarized by a count or average nodal degree, or may use measures based on edges, like the minimum or maximum number of edges for a given number of vertices; the number of edges relative to the maximum possible number; the number of circuits relative to the maximum possible; the ratio of edges to vertices.

Not only are there such descriptive measures for networks, but, as made apparent by the Swiss mathematician Euler a few centuries ago, graphs have some special properties, representable by a single equation. For the case of two-dimensional graphs, following the style of Wilson (1985), we have:

$$V + F = E + S$$

where V and E indicate vertices and edges, and F denotes the spaces between a cycle of edges and S is the number, called the Euler number or genus (often denoted G), that provides numerical balance to the equation. This number also varies depending on whether or not the outside region around a graph or set of polygons is counted as a face. As demonstrated in Figure 5.5a, if it is, then S is 2; otherwise it is 1. These **Euler** (sometimes called Euler–Poincaré) **equalities** specify a constant

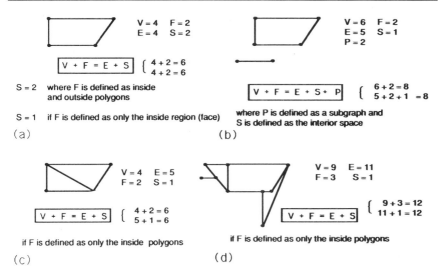

Figure 5.5 Demonstration of the Euler equality for the planar two-dimensional case. (a) Simple case of four connected vertices. (b) Example of subgraphs. (c) The addition of one edge for existing vertices. (d) An example of a more complex graph.

relationship for a set of links on a plane surface. That is to say, as demonstrated in Figure 5.5, the number of vertices and faces is dependent on the number of edges.

Adding edges for non-existing vertices add equal numbers of edges and vertices, thus balancing the equations (Figure 5.5d). An additional nonconnected edge (Figure 5.5b) will add two vertices and one edge and one subgraph, again balancing the two sides of the equality. Additional edges for existing vertices do not change the count of vertices, but do add more loops, as shown (Figure 5.5c); that is, one extra circuit (face) for each extra edge, again providing balance to the equation.

In dealing with graph representations of natural or person-made phenomena, we have both elemental level entities, the vertex, and edge, but also several compound entities. The combination of edges and vertices gives us the graph, or network; the selection of certain edges produces a **path**, which may then be selected as a route. A change of scale may necessitate use of a **hypergraph**, as when both intercity and intracity transportation features need to be portrayed.

Edges may be put together in sequences, or chains, as in finding the shortest route from one place to another, or as naturally represented by a river drainage system. An **oriented path** is an ordered set of edges from a specified origin to a specified destination vertex. A path may be non-oriented, reflecting an ordered set of connections, without specifying origin or destination ends. If the graph is more than a tree, then different paths may exist for any pair of vertices depending on the number of circuits (loops), which reflects how generally connected the graph is. An oriented cycle, a path from a vertex to itself, is made up by a set of edges arranged in sequence. Consequently, in working with paths, the topological property of sequence (of edges) is necessary, along with the property of connection.

Moreover, there are many different contexts for laying out paths between nodes. As demonstrated in Chapter 17, we may place paths over terrain to avoid steep gradients, to circumvent hostile land types like swamps or lakes, or to take advantage of relative costs of different forms of transportation. In a graph sense, the nodes representing end points may be fixed, but a practical solution may require that a particular edge be subdivided to recognize points standing for the facets of land of particular terrain slope, land cover or freight rate, respectively.

Note, too, that the space within the set of edges making a circuit is not regarded as an object with particular attributes of interest. Identified for purposes of mathematical completeness, as in the case of the Euler equation, the area element can serve a useful role; but for the application of graphs to real world phenomena, the vertices and edges are the entities of concern. For example, with reference to Figure 5.6, we attach no significance to what lies within the triangles between the air routes, in contrast to the case of discretized space for land use. For graphs,

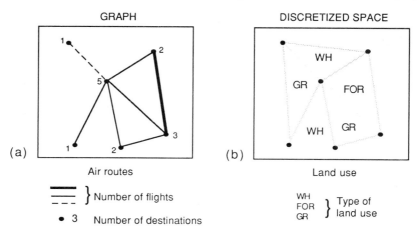

Figure 5.6 Graphs, areas, and attributes. (a) Attributes for nodes and links of a graph. (b) Attributes for polygon spaces.

attributes can be associated with edges, to produce weighted graphs, for example one showing the number of planes travelling on particular routes, and the time taken to travel each route. Vertices may have measures like the total number of air passengers, and both edges and vertices may have spatial properties derived from the graph components as illustrated earlier, and again in section 5.3.2.

In the history of spatial information systems the two-dimensional units were given more prominence at first in comparison to one-dimensional entities, reflecting applications oriented to natural resources and thematic mapping for population data. More recently, spatial analysis involving networks and the interest by public utility companies in computer based inventories has led to a greater attention to line phenomena. Graphs not only are valuable tools for representing structure in physical entities, but are also used for dealing with connections among other phenomena or abstract entities. Patterns of interpersonal communication among people can be modelled as networks. The intellectual building blocks for conceptual models are often shown by graphs (see Chapter 9); connections among apparently disparate pieces of information are implemented in software like *HYPERCARD* (Chapter 16) through the concepts of buttons (nodes) and links (edges).

5.2 GRAPHS AND AREAS

Recall that contiguity of polygons is another topological property, and recall that one type of discretization of space produces a complete set of polygonal entities without holes or overlaps. Here we turn our attention to concepts and tools that treat the spaces within networks as well as the vertices and edges. A set of polygons representing, say, the three countries of Great Britain, comprising vertices, edges and spaces, can be treated as a planar **topological map**, provided that there are no overlapping regions, and that there are no line crossings that do not produce vertices.

5.2.1 Digital line graphs

In its preparation of digital encodings of topographic maps the United States Geological Survey devised in the late 1970s the concept of the **digital line graph** (DLG). The hydrological, transportation, land use and other phenomena contained in those maps were viewed as point, line or

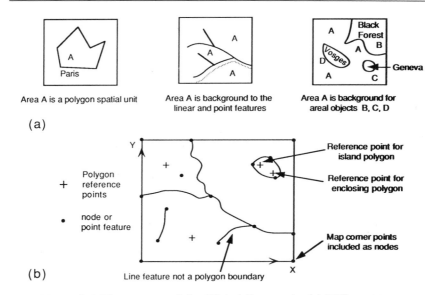

Figure 5.7 The concept of the Digital line graph. (a) Different types of areal spatial unit. (b) One instance of a line graph. (Based on material in US Geological Survey, 1986.)

area objects, as Figure 5.7a shows. Several kinds of area phenomenon were recognized:

1. Polygonal units of data, such as administrative areas.
2. The area forming a background to point or linear objects.
3. The background area for polygon spatial units.

The polygons were seen as a set of boundary edges; and lines were viewed as having beginning and ending vertices, called **nodes** in the context of topology (Figure 5.7b). Polygons have single special reference points associated with them, not necessarily contained within the boundary of the polygon, acting as a surrogate, or polygon seed, to which attribute data for the areal units may be associated (Figure 5.7b). Individual point features were seen as degenerate nodes, that is, lines of zero length.

Somewhat earlier, the Census Bureau of the USA had used concepts from the field of topology to organize the data for the street segments of urban areas, via the **line segment** or *DIME* structure as used for the creation of digital files, officially called the GBF/DIME (Geographic Base File/ Dual Independent Map Encoding) System (Figure 5.8). Designed to help produce maps for field enumerators and relate addresses to streets, and later to support cross-references to demographic statistics for city blocks and other units, this topological encoding, using an underlying graph theory concept, focused on the city street segment as the principal graph edge. Streets, transportation features and other map elements were broken into straight segments for which address ranges, and identifiers for areas on left and right were encoded, as well as the nodes at beginning and end of the line segments (Figure 5.8b).

Both of these practical applications used concepts from the mathematical field of topology. Geometrically speaking, this perspective treats the polygons as **connected cells**, like in a bee's honeycomb. Most primitively, the set of cells is a **tessellation** of (irregular) triangles, where each cell has only three edges. In the natural world, where we represent phenomena as discrete units such as soil types, we usually have area entities with more than three edges, and generally we do not encounter vertices with more than three edges converging. For example, for the conterminous states of the USA there is only one vertex where there are more than three edges – the Four Corners point where Arizona, New Mexico, Utah and Colorado touch (Figure 5.9), and where many people

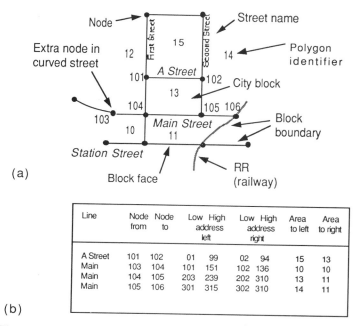

(a)

Line	Node from	Node to	Low High address left		Low High address right		Area to left	Area to right
A Street	101	102	01	99	02	94	15	13
Main	103	104	101	151	102	136	10	10
Main	104	105	203	239	202	310	13	11
Main	105	106	301	315	302	310	14	11

(b)

Figure 5.8 The DIME geocoding concept. (a) Portion of an encoded street map. (b) Sample tabulation showing basic data items for the DIME concept. (Based on material in US Bureau of the Census, 1970.)

Figure 5.9 The example of the states of the USA.

seem to wish to have photographs taken of themselves in an unflattering spread-eagled pose.

Incidentally, the map of the fifty states of the USA can be viewed as a graph of connections of boundaries, but has the special cases of several pieces separated by water (Alaska, Hawaii) and a hole for the District of Columbia (Figure 5.9). Alaska and Hawaii may be conceived of as being connected via telecommunications, but are otherwise not physically part of the graph; whereas the federal government territory, the District of Columbia, is a polygon which does not have three corners – only two states, Virginia and Maryland, touch the District of Columbia. While we can observe adjacent polygons in this case, the situation does not completely meet the formal requirements for a tessellation or topological map, and is more correctly a connected graph.

5.2.2 Topological consistency

For planar graphs, recalling that each edge has two vertices and separates two voids, we now present the concepts used by the Census Bureau of the

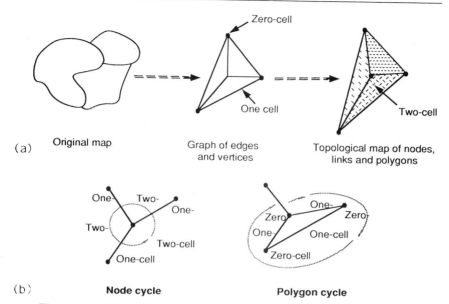

Figure 5.10 Topological cells. (a) Maps and graphs. (b) Cycles for nodes and polygons.

United States. We also cite the different but often used terminology adopted by that organization for a topological representation of a planar set of polygons. That is, for a complete tessellation, we have a group of five conditions for the vertices, edges, and areal **cells** (Figure 5.10a):

1. Every ONE-CELL has two ZERO-CELLS.
2. Every ONE-CELL has two TWO-CELLS.
3. Every TWO-CELL is surrounded by a cycle of ONE-CELLS and ZERO-CELLS.
4. Every ZERO-CELL is surrounded by a cycle of ONE-CELLS and TWO-CELLS.
5. There are no intersections that are not ZERO-CELLS.

The zero-, one or two-cells correspond to the geometric entities of points, lines and areas, but have definitions particular to the context of existence in a tessellation. Thus, isolated points will still be points, whereas points that are topological junctions are zero-cells (nodes). Disconnected or unrelated linear features are not one-cells; this term refers to the bounding edges of polygons.

The first two conditions, which state that every line that is a one-cell has two ends and two associated areal spatial units, also represent a special condition of the duality of zero- and two-cells for the edges. The third and fourth conditions refer to the completeness of the mixture of entities: a cycle around every vertex should encounter a succession of edges and spaces; a cycle around a polygon should consist of a series of alternating edges and nodes (Figure 5.10b). The fifth condition means that the geometry of the set of polygons does not contradict the topology: if two lines cross, then there should be a one-cell defined. Otherwise, a non-planar condition exists, there is an error in position of one or more lines, or there is incorrect topological data encoding. This subject of error detection is developed fully shortly.

Not all features on cartographic maps are topological. The curvature of lines, the shape of polygons, and the labels for places are not. Moreover, several different maps may have the same topological structure. As Figure 5.3 demonstrates, shape can vary, and the absolute, but not relative, placements of points can differ. The properties of relative position are said to be **invariant** under change (deformation) in the base. For example, the writing on an inflated balloon alters shape as the balloon is stretched and twisted, but still retains its pattern. Lengths and angles can change, but four elements must not be altered:

1. Incidences (two nodes per line, or lines at nodes).
2. Intersections.
3. Adjacencies (neighbours for areal units).
4. Inclusions (points in polygons).

These properties are referred to as the map properties that are invariant under continuous deformation of a surface, the map base. Notice that tearing and puncturing can alter some of these in particular cases, but stretching will not do so. This state of affairs proves quite valuable for map transformations of certain kinds and for substantiating checks on the integrity of data.

The topology may not necessarily be logically internally consistent. For instance, an edge may be missing for two vertices, or there may be no occupied area within three edges, as in, by analogy, a badly treated umbrella that has a panel of cloth missing. We define a **topologically consistent map** (or database) as having all spatial entities projected upon a plane surface, with no freestanding features, and having complete topology. That is, the five rules identified above are obeyed. Completeness of inclusion means that there are no isolated non-connected points and that all lines are parts of boundaries of polygons. The completeness of incidence means that lines intersect only at points associated with the ends of lines. These consistency conditions may be seen as constraints for the integrity of a database of cartographic information.

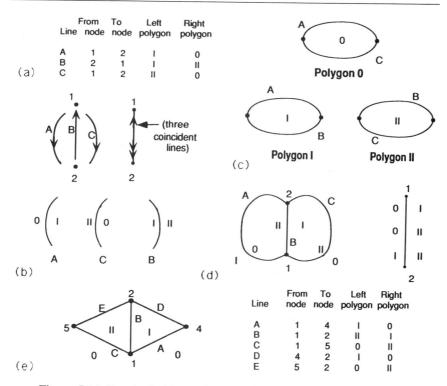

Figure 5.11 Topological inconsistency. (a) Data for three one-cells. (b) Graphs for three one-cells. (c) Only two-cell data are used. (d) Ambiguous situation. (e) Unambiguous data.

A few examples demonstrate the occurrence of inconsistency or ambiguity in a topological structure (Figure 5.11). Consider the case of two polygons, I and II, apparently touching at one common edge, and having two other topological one-cells, as recorded by the data table (Figure 5.11a). Each of the lines has two nodes and two polygons, including the exterior region, 0. But if we had only zero-cell information the graph would consist of three arcs as shown, or all coincident (Figure 5.11b); if we had only two-cell information, then the next diagram results. Working with only polygon boundary lines again produces ambiguity, as occurs even if we use both node and polygon information (Figure 5.11d). We have forgotten that a two-cell has a minimum of three sides and three vertices. For the example, the complete set of data must have records for five edges and four nodes, as shown (Figure 5.11e).

So we can raise questions about a map of phenomena, addressing certain kinds of spatial relationship implied by the map. Of what zero-, one- or two-dimensional elements does the map consist? Which areas do particular lines separate? Which points terminate which lines? For particular vertices, which lines are incident, and for particular areas, which lines bound? Is an area completely bounded by a set of lines, and is the set of lines and areas for a particular point complete? What is the set of areas adjoining a given area?

Creating a topological structure from geometrical information is not that simple in practice. Imagine tracing map features as part of a digitizing process in which the way individual elements are connected is not recognized. Today most spatial entity oriented software takes this 'spaghetti' of lines and builds a topological structure from it. Consequently, the software must know where lines cross in order to establish vertices, must create all the separate edges, must then assemble ordered sets of edges to make regions, and must associate edges with vertices. The topological information is deduced from the metric data in this process, or it can otherwise be provided by the creator. In either case, errors can arise.

5.3 ERROR IDENTIFICATION

Various tools have been devised for checking digital encodings representing planar enforced surfaces. The US Census Bureau has a series of **edits** to assist it in identifying errors in its 1990 *TIGER* files (and for the 1980 *DIME* files). Some map conditions are best seen and corrected manually, but some can be more easily dealt with by computer routines.

5.3.1 Possible conditions in digital maps

A summary of possible conditions that may result from preparing digitized versions of maps is provided by Figure 5.12. Typically, hand digitizing produces some over- and undershoots of lines beyond other lines or a map edge; some edges or nodes may be completely omitted or duplicated, and line shapes may be bad. Whether or not some conditions are errors depends on external requirements or validity. For example, an overshoot could represent a cul-de-sac in a street system, or an unconnected edge might be a railway line that does not join any other line.

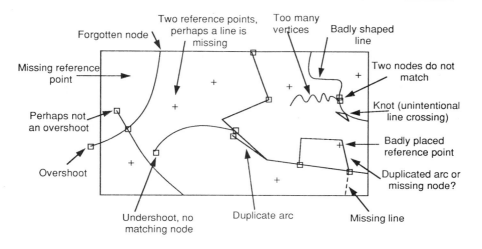

Figure 5.12 Some conditions for digitized maps.

Some geometric conditions are:

1. A node is missing or misplaced.
2. An edge is missing or misplaced.
3. An edge has a bad shape or too many (or too few) points on it; or coordinates are missing or incorrect.
4. A node has more than one position.

Some topological conditions are:

1. Unconnected edges exist.
2. A polygon has a gap between two edges, that is it is not closed.
3. Duplicate arcs are present.
4. A polygon has more than one or no reference point associated with it.
5. A node has only one or two edges, rather than at least three.
6. A polygon may be missing.

Such topological conditions may arise from imperfect geometry, as when there is a line undershoot, with or without a node identified on a nearby line; or when a node is misplaced. At times, unless external data are available for checking, it may not be easy to validate whether a condition is inherently geometrical or topological, or indeed, whether or not it represents an error (Figure 5.12). Topological errors will have to be uncovered by automated spatial data editing procedures or visual map inspections; or they may be dealt with via other techniques associated with database management systems, a topic for Chapter 15.

For a planar, complete surface, a topological map will have at least three edges intersecting at topological junctions. Nodes with only one or two edges (sometimes referred to as dangling and pseudo-nodes respectively) represent singly connected edges or artificial points necessary for allowing duplicate arcs between the same pair of points (as needed, for example, for the boundaries of the District of Columbia), or for dividing one line based on different attributes. But they may represent a failure to close a polygon by reason of an undershoot in digitizing. Generally speaking, it is relatively easy to detect errors for areal spatial units; but sometimes it is not straightforward. For instance, a missing polygon in a cadastre may not be clearly visible from a plotted map that shows boundaries for all neighbouring parcels of land.

Line or point entity phenomena are not so easily dealt with when detecting errors other than by graphical means. For example, power transmission lines may cross each other without any connection, or they may meet at substations. A planar graph will need to recognize intersections in both cases, and differentiate between the two – but this is not possible directly using only spatial data. An indirect method is to provide additional information that says even though a topological junction exists, it is not possible to travel from all edges to all others at the intersection.

Individual point entities, not connected to other phenomena as part of graphs or topological maps, may be positioned on the wrong side of line entities or polygons, possibly because data come from different sources. For example, an obstacle to air navigation, like a radio tower, on a map may be placed on the wrong side of a major visible ground feature like a highway, causing problems for flight via visual rules. Topology and geometry can be in conflict not only at line intersections, but for other spatial relationships and at the boundaries of adjacent tiles or map sheets.

5.3.2 Some procedures for checking for errors

The Euler equation provides an initial simple tool for checking, relying as it does on only counts of the basic elements in a graph or map (Figure 5.14). If the Euler number test fails, then a graph is not topologically internally consistent. On the other hand, there is no guarantee that the reverse is not the case because there can be compensating miscounts. For example, one fewer node and one extra edge will balance in the equation (Figure 5.13f). In the case of Figure 5.13b compared with 5.13a, a vertex is not recorded in a data table; or, as for 5.13d and 5.13c, there may be an undershoot without dangling nodes, or an undershoot with one

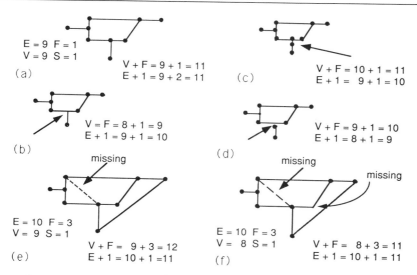

$$E = 9 \quad F = 1$$
$$V = 9 \quad S = 1$$
(a)

$$V + F = 9 + 1 = 11$$
$$E + 1 = 9 + 2 = 11$$

(c)

$$V + F = 10 + 1 = 11$$
$$E + 1 = 9 + 1 = 10$$

(b)

$$V = F = 8 + 1 = 9$$
$$E + 1 = 9 + 1 = 10$$

(d)

$$V + F = 9 + 1 = 10$$
$$E + 1 = 8 + 1 = 9$$

missing

missing

missing

$$E = 10 \quad F = 3$$
$$V = 9 \quad S = 1$$
(e)

$$V + F = 9 + 3 = 12$$
$$E + 1 = 10 + 1 = 11$$

$$E = 10 \quad F = 3$$
$$V = 8 \quad S = 1$$
(f)

$$V + F = 8 + 3 = 11$$
$$E + 1 = 10 + 1 = 11$$

Figure 5.13 Use of the Euler equality for detecting topological inconsistencies. (a) Original graph and data. (b) One vertex not encoded. (c) Double encoding for one vertex. (d) One edge is not connected. (e) One edge is missing. (f) One edge and one vertex are missing.

dangling node. In Figure 5.13e the missing line reduces the edge count by one, although the number of vertices and regions is correctly recorded.

Various possible situations are shown systematically as Figure 5.14 for both polygons and graphs. Notice the variation in the S number for the different cases; of course, if S is incorrectly specified, then the equality is inherently incorrect, even before other quantities are examined.

Topological editing is based on the five rules presented on page 187. A surface can be represented as smooth and continuous if there are positive answers to a series of questions. Is every one-cell incident with two zero-cells? Is every one cell incident with two two-cells? Is every two-cell bounded, and is the neighbourhood of a zero-cell equivalent to a disc? Particular routines for topological editing make use of some of these properties identified earlier. For example, there should be a sequence of edges and nodes around a polygon, and any labelled node should occur only twice. The assessment, excluding the line intersecting which is done computationally, requires a mixture of encoded data and checking routines. For example, using line segment encoding (as with the *DIME* structure) the first two questions are answered by records which contain data for the from- and to-nodes and left- and right-hand side polygons,

(a) Examples for polygons

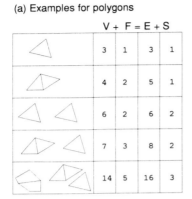

	V	+ F	= E	+ S
	3	1	3	1
	4	2	5	1
	6	2	6	2
	7	3	8	2
	14	5	16	3

(b) Examples for graphs

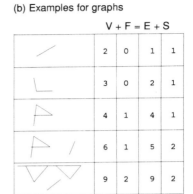

	V	+ F	= E	+ S
	2	0	1	1
	3	0	2	1
	4	1	4	1
	6	1	5	2
	9	2	9	2

Figure 5.14 Examples of Euler equalities for graphs and polygons. (a) Polgyons. (b) Graphs.

allowing the dual incidence to be determined, while the third and fourth questions are addressed by algorithms, as now described.

A polygon chain routine can detect certain kinds of errors, and a node chain cycle routine can detect others (Thompson *et al.*, 1982). Thus an error in orientation of lines can be detected by a **node chaining edit** which recognizes the to- and from-designations (Figure 5.15a) – the cycle of one- and two-cells does not produce a chain of repeated polygon labels. The third record should be CB, not BC, reflecting the inadvertent switching of the direction of the line. A node edit using both orientation and adjacent polygon data operates as shown in Figure 5.15b. The edges at a specific node are collected, and then gathered so that orientation is correct; this is done by writing out duplicates for the original three lines, with the from- and to-node designations reversed. If there is a chain through the polygon labels, then no error is detected, but if one polygon is badly labelled, or if another, non-adjacent polygon has the same label, then the chain will not work correctly. Thus, additional checks are needed beyond both the counting to see if each edge has two nodes and the adjacent polygons because it is possible to make labelling errors.

A **polygon chaining** procedure works in a similar way to the node chaining edit, undertaking a cycle around particular polygons looking at the succession of one- and two-cells. This procedure looks for missing or redundant nodes; the node chaining procedure looks for missing or redundant lines. The United States Census Bureau uses a series of such automated checks on its digital boundary data files (Marx and Saalfeld, 1988). An umbrella edit (node chaining) looks for the correct cycle of ribs and panels. An elementary cycle edit (polygon chaining) looks to see if an

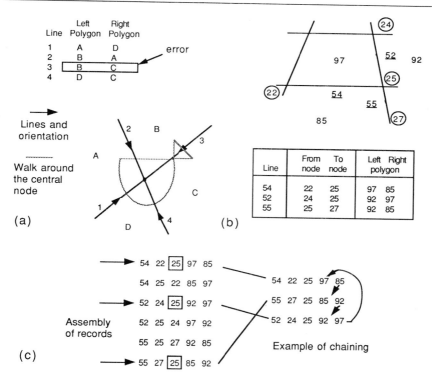

Figure 5.15 Node cycle edit. (a) Node cycle edit concept. (b) Example. (c) Chaining procedure.

area unit has a complete boundary. An intersection detection routine is used to look for intersections that are correctly coded as topological junctions. Using an example from the US Census Bureau (Figure 5.16), we emphasize that there can be different conditons and, therefore, different solutions, represented by one instance. The geometric and topological resolutions produce different maps.

Of course, the structural checks apply to only internal consistency. While everything appears to be fine with the logical conditions of the graph or topological map, there may still be errors in correctness. The shape of some lines may be badly drawn. The **external validity**, the correlation of data and reality, must also be checked. Some polygons may have been overlooked – suppose that the counties of Yorkshire and Lancashire were joined as one unit? However, very little of this type of problem can be automated easily, unlike the checks for internal consistency.

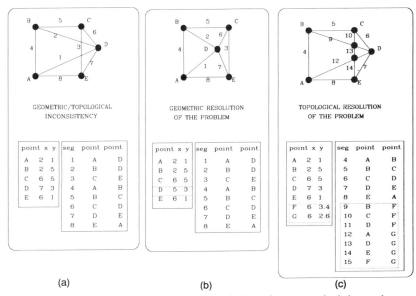

Figure 5.16 Resolution of a topological and geometrical inconsistency. (a) Original conditions. (b) Geometric resolution of the problem. (c) Topological resolution of the problem. Nodes are here called points; seg refers to lines. (From Marx and Saalfeld, reprinted with the permission of the US Bureau of the Census, Washington, DC, USA.)

The design and range of functions available in software for capturing spatial data for entities seen as points, lines and areas are important for aiding the efficient production of error-free databases, a problem which will be re-examined through the concepts of integrity constraints in Chapter 15. Systems with automated procedures based on topological concepts are likely to be more effective than those that rely solely on comparisons of original and digitized maps. Furthermore, because of human imperfections in the hand-digitizing process, it is helpful to have tools that are adaptable to the different levels of error that can be tolerated. For example, undershoots measured in hundredths of millimetres, but nonetheless causing a common node to be represented by two different coordinate values, may be dealt with in the digitizing process by automatic snapping of the nodes together.

The varying **tolerances** that may be invoked as editing controls in building spatial databases address conditions for the different spatial units (Figure 5.17). Line generalizing may accept differences between vertices greater than a certain distance, or angle, depending on the weeding procedure used. Nodes can be snapped together, according to another tolerance item for the acceptable distance apart of the different representations of the same node. Line overshoots can be adjusted or not, according to a dangle tolerance. And a test applied to all features, a fuzzy tolerance, works globally, sometimes with undesirable or unforeseen results, on the distance between points, whether they are nodes or line vertices. As such, this tolerance, by the value assigned to the smallest distance acceptable, establishes the resolution for entity oriented spatial geometric data. Tolerances can be established for other spatial contexts; we have presented them here in some detail only for irregular tessellations.

5.4 POLYGONS AND AREAS

Recalling our earlier discussion (Chapter 3), we know that polygon entities are encountered in dealing with many natural and man-made phenomena. Polygon data may represent naturally occurring entities like lakes or forest stands or person-made entities like administrative units, or they may be discretizations of continuously varying phenomena. Area units are most often assembled from line data, and, less commonly, made from point data. In these cases they generally take the form of irregular tessellations with no hierarchical ordering of polygons at different levels. As such, they do not deal explicitly with the property of containment. For

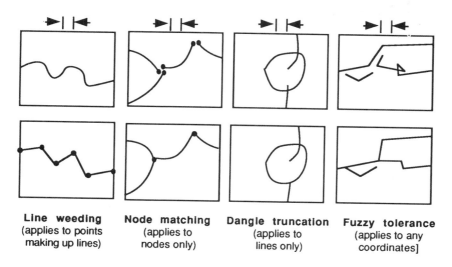

Line weeding
(applies to points
making up lines)

Node matching
(applies to
nodes only)

Dangle truncation
(applies to
lines only)

Fuzzy tolerance
(applies to any
coordinates]

Figure 5.17 Some spatial tolerances.

pedagogical purposes, we treat two-dimensional units in two sections, based on whether or not containment is a factor.

5.4.1 Types of areal spatial unit

We separate area phenomena that have no overlap, no holes and no isolated pieces from the additional features of enclaves and exclaves. Referring to Figure 5.18, we define a **simple polygon** as a single unit of space bounded by three or more lines, generally having an irregular shape, and containing no holes. More generally, we can use the concept of a **cell** to refer to any unit of space, regular or irregular, bounded by three of more lines and not containing holes. A polygon is **complex**, sometimes referred to as an area, if it contains one or more holes. A **region** is made up of two or more separate polygons (or cells), not necessarily connected although, in the field of geography, the term region has other definitions. Any point position is in only one space (polygon), unless it lies on a boundary or in an overlap of two or more polygons or areas.

It is not just a matter of recognizing that enclaves and exclaves can occur, it is also a matter of completeness geographically. Many countries or parts of countries have land separated partially or completely by water. How does one establish contiguity across intervening water? Is Alaska adjacent to the state of Washington? As Figure 5.19 suggests, conceptually there are different boundaries for administrative and physical representations. In the case of political redistricting in Maryland in 1981, for instance (Thompson and Slocum, 1982), it was necessary to assign a special code to contiguity relationships across the Chesapeake Bay. Topological and geographic views can be in conflict at times.

Much of the difficulty associated with polygons is practical, not conceptual. That is to say, it may be necessary to intersect many polygons

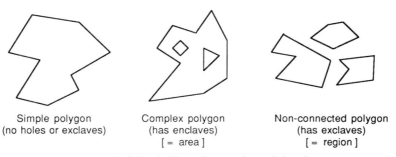

Simple polygon	Complex polygon	Non-connected polygon
(no holes or exclaves)	(has enclaves)	(has exclaves)
	[= area]	[= region]

Figure 5.18 Definitions for areal spatial units.

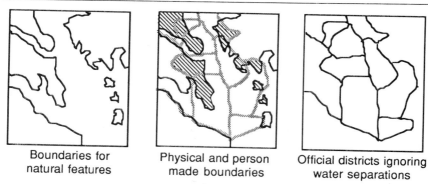

| Boundaries for natural features | Physical and person made boundaries | Official districts ignoring water separations |

Figure 5.19 Different versions for encoding areal phenomena.

to create homogeneous units; it may be desirable to combine several similar units to make a more-or-less homogeneous larger unit, or they may have to be created from point or line data. For example, land parcels are both a basic geometric two-dimensional unit and a useful statistical unit; yet homogeneous terrain units are made up from separate polygonal units for a mixture of soil conditions, slope, land use and other variables. While the basic areal spatial units for which attribute data exist may satisfy many user requirements, they may not be the most **basic spatial recording unit**. So we define additional objects: a basic statistical unit is the smallest unit for which attribute data of a given kind are collected. This may be the housing unit, as in the case of demographic data, or field point samples aggregated to represent areas of space. As is the case with census data, reporting units may not be the same as data collection units in order to protect individual privacy.

Another fundamental matter is the extent to which there is a one-to-one relationship between the spatial units and the primitives (Figure 5.20). For the simplest case, a one-to-one matching (such as watersheds which are separate water management districts), the respective topology properties are compatible. Because each natural unit is matched with only one administrative entity, there is also a matching between the boundaries and the adjacency and connectivity topology data. Even if the spatial units at a given level are assembled from separate pieces, for example the European Community and European Free Trade Association blocs, with twelve and four countries respectively, the larger unit boundaries match up. The larger unit of the particular trading block is a combination of the separate pieces, none of which overlap into the other. On the other hand, there may be disjointed regions, in which the semantic unit, the region, for example A in Figure 5.20b, is made up of two separate pieces; or there may be overlapping areas (Figure 5.20c).

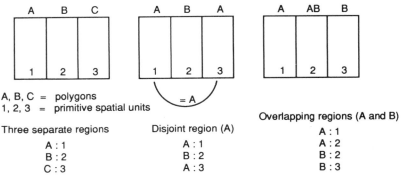

(a) **One-to-one relationship** (b) **One-to-many relationship** (c) **Many-to-one relationship**

Figure 5.20 Primitive spatial units and polygons. (a) Three areal units. (b) Disconnected region. (c) Two overlapping regions. (Based on material in Kirby, and used with the permission of the Cambridge University Press.)

5.4.2 Containment and coincidence

We have just described a scenario about the property of **containment**, that is the inclusion of one spatial entity within another. A point may be in different areas; lines may be in different areas or parts of boundaries of different polygons; and areas produced by overlap lie within different polygons. Such situations arise not only for administrative or other forms of territoriality, but also for natural phenomena and for linear spatial units (Figure 5.21). A boundary of a county may also be a river or a boundary for electoral constituencies. An edge comprising part of a land parcel boundary may be part of several real features, possibly a river or road, or the parcel itself.

Different spatial situations that arise for sets of spatial units that are completely or partially **ordered** by inclusion are:

1. There is a perfect hierarchy of small units combining at several levels to make larger units, with no overlaps and no holes, as in the case of election precincts combining to create legislative districts.
2. There is incomplete nesting in which, for at least one level of spatial subdivisions, some units are split.
3. Areal or linear subdivisions may be put into different larger units for different purposes.

Man-made areal spatial units frequently exist in a vertically structured organization. Political units, for example, on a worldwide scale, begin with nations, which are subdivided into a few large territories like the

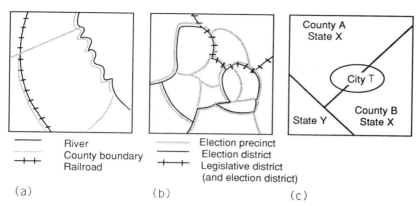

Figure 5.21 Examples of inclusion relations. (a) Line coincidences for river/boundary and railroad/boundary. (b) Common boundaries for election precinct and district, and for election precinct, election district, and legislative district. (c) City split by a county boundary.

republics of the USSR, and then into progressively smaller administrative territories. Generally, but not always, there is a one-to-many relationship of the larger unit to the smaller constituent pieces. Sometimes exceptions occur, as in the case of the city of Takoma Park near Washington DC – this city lies in two counties of the State of Maryland (Figure 5.21c). On the other hand, the Vatican City in Rome is not in any hierarchy at all.

While administrative hierarchies with only a few exceptions have a property of **nesting**, that is boundaries are coincident (Figure 5.21b), there are person-made spatial units that are dominated by overlapping, that is have only partial ordering. The market areas of consumer products, for example, regional newspapers, can have major areas of **overlap**, meaning that the one small piece of territory falls into more than one larger unit.

These examples demonstrate that some types of spatial organization are not easily handled in a straightforward way by only the concept of a plane surface. A single map showing non-nested many-to-one relationships looks rather messy (newspaper circulation); several maps are needed (one for each newspaper), or non-map representations can be used. Hierarchical tree graphs are frequently used for such situations (Figure 5.22). They can show the overlap or nesting quite well, although, in practice, cases with many units might require very large pieces of paper to show all relationships. Here, for simplicity, the polygons are built up from regular blocks, but irregular geometric figures are more natural.

As a practical matter, the processes of spatial subdivision or its converse, spatial unit aggregation, work with basic spatial units at several

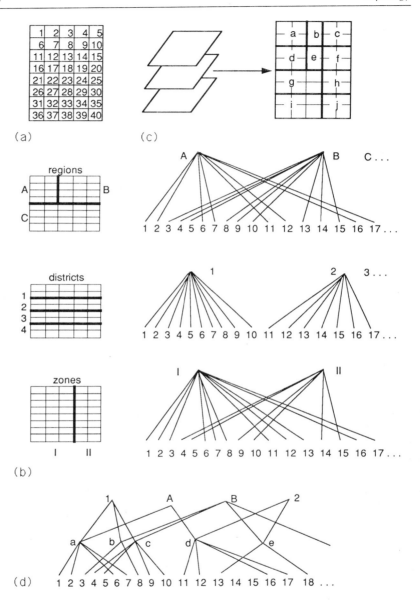

Figure 5.22 Ownership ordering for areal spatial units. (a) Areal cells. (b) Ownership trees for three sets of territories. (c) Common geographic units. (d) Ownership ordering. (Based on Meixler and Saalfeld, 1985.)

levels. It is possible to model the containment relationships, otherwise referred to as ownership ordering, so that the lowest level has all individual pieces, the smallest spatial unit, or two-cell (in the terminology of the US Census Bureau). For example, as Figure 5.22 shows, two-cells are combined to create zones, regions and districts at a higher level (Figure 5.22b). There is no ordering for districts, regions and zones, so the sequence at the higher level does not mattter. In this case, the basic units are established prior to combinations being made. If a subdivision process is taken, then the units at the lowest level will be produced by identifying the units produced by overlaying the district, region and zone maps, as shown in Figure 5.22c. These, the minimum number of units necessary and sufficient for producing the several combinations for districts, regions, and zones, are entered in the tree diagram as a new layer as shown in Figure 5.22d.

So (after Visvalingham *et al.*, 1986), we distinguish among (Figure 5.23):

1. Largest common spatial units, resulting from overlaps.
2. The smallest units (cells) which are combined to produce the common spatial units.
3. The spatial primitives (points and lines) which make up the smallest areal units.
4. The spatial units for which data are recorded (the basic spatial recording unit).

In theory, there can be as many cells as there are largest common spatial units. In layered databases these will be the same if the lowest level common spatial units are produced by polygon overlay intersections. In theory, too, the cells establish the resolution, the minimum scale, at which the phenomena are to be observed. In practice there may be economies if the data analysis works at the level of common spatial units rather than cells. For example, for its 1990 census, the US Census Bureau has about ten million two cells; it is hoped that there will be many fewer largest common spatial units.

At the same time, a building block structure may facilitate data aggregation. Population recorded for individual households, as illustrated for houseboats in Figure 5.23, is ordinarily grouped to data reporting units to preserve privacy of information. Other thematic attributes, or events, such as traffic accidents, thunderstorms, or homicides associated with basic spatial recording units, can be aggregated to special purpose zones like streets or telephone service areas.

The treatment of linear features parallels that of areal units. The largest common spatial units are those where there is coincidence, as for

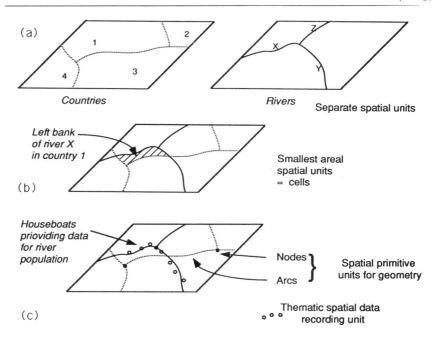

Figure 5.23 Types of spatial unit according to ownership. (a) Two examples of separate thematic units. (b) Combination of polygons for two themes. (c) Geometric and thematic primitives.

the part of the northern boundary of Durham County, England, which is also a river, and the limit of an electoral district. There may not be any one-dimensional cells, but the lines will certainly be made up of other entities, either primitive segments, or represented by mathematical equations. **Coincidence encoding** for line objects can be a valuable check on data integrity, avoiding inconsistencies that tend to arise with separate digitizings, although there are limitations. For example, a boundary line between nations may remain stable even as the river on which it is based changes over time.

Coincidences of two-dimensional spatial units may be approached from the perspective of **boundaries** as well as the interiors of the areal units, especially to be able to more effectively deal with holes inside spatial units (Kirby *et al.*, 1989). That is, the geometric and phenomenal parts are separated, and only the boundary line information is used. With reference to Figure 5.24, the enclaves and islands require boundaries for both insides and outsides, as shown by the double lines around the entities. The containment hierarchy of boundaries can be seen from the

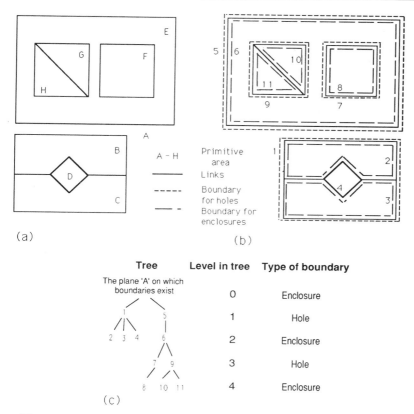

Figure 5.24 Containment encoding using boundaries. (a) A set of primitive areas. (b) The boundaries of the primitive areas. (c) The tree representing the containment hierarchy of the boundaries. (Reprinted from Kirby, 1989 with the permission of the Cambridge University Press.)

tree diagram. This approach requires a data table for holding the boundary data as well as tables, discussed earlier, for the lines, nodes, and intermediate points. This organization contrasts with others (see also section 5.5) in which there is no explicit coding of the boundaries, but in which other devices, such as special tabulated numerical codes for lines, or artificial connectors called stalk arcs joining separated polygons are used.

Moreover, there may be artificial precision associated with the boundaries. Thus, for climatic regions, based on data for continuous surface concepts of gradients of temperature and precipitation conditions, boundaries are arbitrary. In other cases boundaries may represent the equi-probability of membership of more than one area unit, so that the concept of a membership function is more appropriate than discretization.

If data organization is conceived as separate maps or data layers of polygons for different themes, then it is desirable to be aware of conditions for both boundaries and interiors. Many combinations exist for these elements, regarding the topological properties of touching or overlap. For boundaries two area units may be completely separate, touch at one point, partially overlap, or be identical (perfect overlap). For interiors, one unit may be inside the other, or the reverse, or there may be overlap regarding the one or more holes. It is important to note that exclusion is only a semantic property: unlike inclusion it cannot be computed.

Different questions as to containment may need to be answered. We may wish to know the polygons included in a given piece of arbitrarily delimited territory; or which polygons overlap with any piece of that territory. For example, respectively, which regions of Europe fall entirely in the circulation area of the *Manchester Guardian* newspaper, or which regions include a portion of that circulation area? Other queries may involve specification of size of areas, or coincidence of boundaries. Different data arrangements and content may be necessary to adequately deal with all possibilities.

5.5 DATA FOR SPATIAL RELATIONSHIPS

In working with digital databases it is important to recognize that some geometric or topological properties may not be directly encoded, having to be derived from other data that are. It is thus important to identify what topology is **explicit**, that is to say, in a practical sense of looking at tables of numbers, what is directly recorded.

Some forms of spatial data organization do not encode any topology properties. The **polygon model** (appropriate for lines and point features too) has tabulations for the coordinates of each area unit taken separately (Figure 5.25a). More generally, this approach has a one-to-one correspondence between a map feature, point, line or area, and a record in a table. Entities are geometrically defined, but no spatial relationships are recorded. So touching polygons or connected lines are revealed only in visual form when the entities are drawn. Common boundary line coordinates are listed twice. A related structure, the **points list** or dictionary, also does not separate individual polygons; it still does not

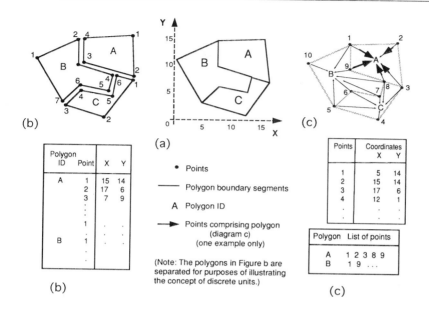

Figure 5.25 Spatial data models without explicit coding of topology properties. (a) Data for three polygons. (b) Polygon model concept. (c) Points list model.

record adjacency or connectivity directly, but it avoids duplication of feature coordinates (Figure 5.25b). The polygon or line units are made up by indicating which points are necessary; the table for points has only coordinate data.

Such **non-topological data models** are those in which only positional information is recorded. The recognition of individual spatial units as separate unconnected elements, characterized the early data organization for digital cartographic data. This concept was implemented in the polygon model of the *SYMAP* software, developed primarily at Harvard University in the late 1960s, the first computer thematic mapping package to gain international fame.

Historically, the first topologically explicit data model was the **line segment** structure (Figure 5.26a), devised and used by the US Census Bureau for the 1970 census. This *DIME* format had a basic record for line segments of from-node, to-node, left polygon, and right polygon, and associated information like the coordinates of the nodes, and names of the polygons. There were no intermediate ('shape') points; consequently, a single record could conveniently handle both geometry and topology, although this meant that major turning points along lines had to be

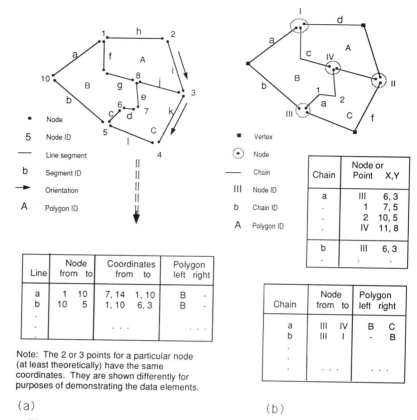

Figure 5.26 Explicit coding of topology. (a) Line segment encoding concept. (b) Chain format concept.

recognized as nodes as well as the street intersections. Moreover, only line segment features were directly addressed. Other entities, particularly polygons, had to be assembled (often quite laboriously) from the data encoded in the line segment records.

Encompassing additional properties or spatial units directly generally involves a structure of several tables of data. A **chain model** (Figure 5.26b), sometimes known as the *POLYVRT* structure, used by the *ODYSSEY* system of Harvard University from about ten years ago, first explicitly recognizes intermediate points along lines, as well as the line end points. Data are stored in different tables for the chains, as for the line segments in the *DIME* model, with from-node, to-node, left polygon, right polygon; but unlike the *DIME* arrangement, also for nodes, for

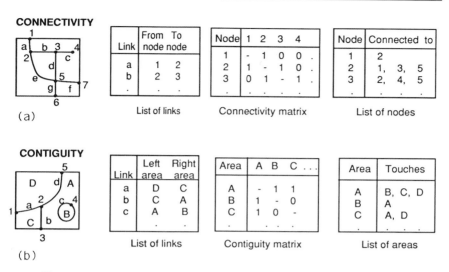

Figure 5.27 Recording data for connectivity and contiguity.
(a) Connectivity. (b) Contiguity.

coordinates; for points, for coordinates; for polygons, and for the chains making up the boundaries. The US Geological Survey's original digital line graph concept was encoded in the form of a set of tables, for nodes, with (x, y) coordinates; for areas, with coordinates for centroids; and lines, with starting and ending nodes, and coordinates for them, and left- and right-hand side areas.

Connectivity (Figure 5.27) is readily handled by a link-node list, showing from- and to-node labels, and a separate table for the coordinates of the lines. But it can also be dealt with by a connectivity matrix and separate node coordinate list. **Contiguity** is directly addressed by a table of lines with data items for left and right polygons, and a separate table for the coordinates for the lines (Figure 5.27); or by a contiguity matrix or list, with separate data for coordinates for the polygons, either as polygons, or sets of lines. Lists may be better than matrices for some purposes, such as for connections in skeletal graphs and contiguity relations for contour lines. In both cases, the matrices would be very sparse, containing many null entries. However, some arithmetic operations are performed more easily on matrices, better than on lists.

Area features can be represented by polygon–line topology, using tables for the lines that make up the polygons, and the coordinates for the lines. Special codes may be used to identify the orientation of the

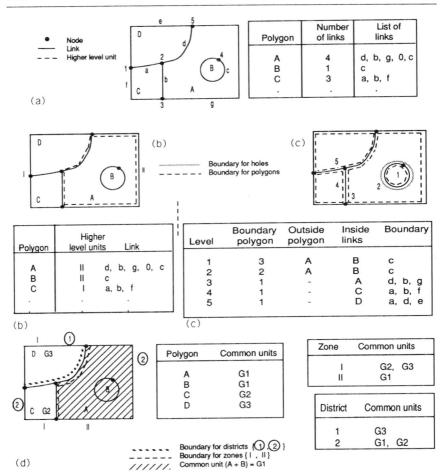

Figure 5.28 Representing polygons. (a) Islands encoded via a list of links. (b) Hierarchy shown by a label. (c) Polygons and islands represented by boundaries. (d) Encoding for common spatial units.

topology edges, and the existence of inside and outside boundaries if enclaves exist (Figure 5.28a). Node features can be tabulated as separate records with data items for lines that converge, as well as coordinates. As discussed earlier, containment has to be dealt with somewhat differently if it is to be dealt with directly, rather than indirectly, via special codes for boundaries. A boundary oriented structure, shown as Figure 5.28c, adds records for boundaries to those for edges, nodes and coordinates. A hierarchical organization for handling the several layers of polygon units

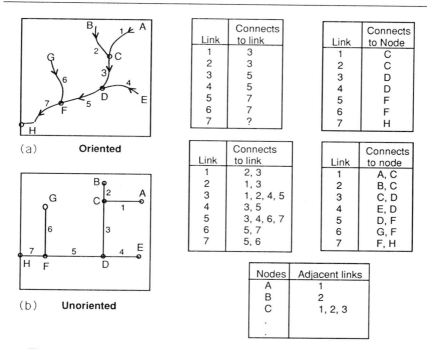

Figure 5.29 Topological encoding for linear entities. (a) Link lists for oriented graphs. (b) Link and node lists for unoriented graphs.

that can be built up in a nested fashion from common units is undertaken by labelling of different level units, portrayed in simplified form as Figure 5.28b. And the hierarchical approach used by the Census Bureau, as noted earlier via Figure 5.22, by contrast is handled as in Figure 5.28d.

Several possibilities exist, too, for graph entities. Each link in a river or sewer or highway network could have a data item of the link before it and the link after it (Figure 5.29); or the system could be modelled as links which have nodes which have links. The *DIME* structure's duality for nodes provided a base for moving through a network by searching for links with matching node numbers. If the link end points do not have unique labels, then matching can be based on the coordinates of the end points of edges in the graph, as, indeed, in the *TIGER* system, a new structure devised by the US Census Bureau for its 1990 census, as described later in section 8.6.3. Alternatively, and better for most purposes, the organization could have tables for the edges adjacent to the nodes, the nodes at the ends of the links, and the coordinates for the edges.

5.6 SOME OTHER CONSIDERATIONS AND SUMMARY

The structural properties of graphs and planar surfaces exist, minus any information for the attributes of the point, line, area or volume objects, or for their positions. Depending on purpose, the geometric data may be necessary, the topological properties may be required, or, for many reasons, both are essential. The metric information is used for:

1. Drawing maps of edges, nodes, and surfaces.
2. Differentiating between duplicate edges.
3. Locating intermediate points on edges.
4. Measuring distances, perimeters, shapes, or areas and volumes.
5. Determining positions of entities.

The requirements for certain categories of retrieval of data for spatial entities are met only by positional information, although this does not have to be based on Euclidean geometry. There are several ways of spatial addressing as covered in both this and subsequent chapters.

Topological information is valuable for accomplishing tasks involving relative positions of entities or for dealing with spatial relationships, although it is not necessary because visual representations, maps, contain data to establish those properties. Topology must be encoded explicitly, however, to permit numerical spatial data problem solving and reasoning, or, at times, to facilitate some data editing operations. The topological data are used for:

1. Undertaking analyses that need information for connectivity of line elements.
2. Undertaking procedures that utilize data for sequences of line elements.
3. Determining the character of adjacent area units.
4. Automating some error detection procedures.
5. Facilitating searches in neighbourhoods.
6. Facilitating data retrieval for associated elements.
7. Making data updates more feasible by separating the metric information from the structural.
8. Reducing limitations inherent in the use of only metrical information.
9. Facilitating aggregation of primitive spatial units into larger units.
10. Providing a basis for automation in map matching and transformations.
11. Facilitating spatial reasoning.

The two kinds of spatial data may not be completely recorded in spatial information system databases. Not all topological properties may be present at all; some topologically structured databases may deal with only one, or not all, properties explicitly; and not all situations can be easily handled topologically. For example, discontinuous surfaces or non-planar graphs require special treatments. It may arise that geometry is required at one level, and topology in another, for a layered concept for a database. For example, when several polygon type layers are overlaid, only the combined topology is retained, not the new geometry. And, from a database point of view, topological data are redundant because they can be derived from the geometric data.

The topological data model may have disadvantages for cartographic production. Having to deal with nodes and links can disrupt smoothness in digitizing activities, and the short lines associated with a graph structure (compared to unbroken features in a map model) can increase map drawing time because of many starts and stops. Unfortunately, a misleading distinction is sometimes made between cartographic databases and topological spatial databases implying that topology properties are not necessary for the former. On the contrary, thematic mapping, especially polygon shading, can be more effective if the boundary lines between adjacent polygons with similar attribute values are known. And some topological properties are valuable for editing cartographic data used only for base mapping, even for line features.

In any event, it is better to distinguish between geometric and topological spatial data rather than cartographic and topological databases. In a **geometric** (pseudo-cartographic) **model**, entities can be manipulated completely singly, not being assembled from separate disconnected pieces. For example, a long highway, such as Route 1, can be identified and stored as a single linear feature rather than requiring retrieval of a long set of separate line pieces created by the topological junctions at intersections with other highways. The topological model does not allow direct reference to logical complex objects – these must be identified at the stage of data input.

Topologically structured data can be manipulated algebraically and arithmetically to produce valuable derived statistics for revealing graph or surface properties, or facilitating error detection. Many people favour the **topology model** because it directly addresses entities in a phenomeno-logical sense, that is, as to what exists. At the same time, it does this imperfectly because the visible objects may be polygons and networks, rather than the encoded primitives of points, lines and areas. However, many spatial relationships cannot be (easily) computed and must be explicitly recorded. Some relations can be computed from just metric information, like distance and intersection, but some may more easily be dealt with non-metrically. Topological properties are expensive to produce from geometry, and are not needed for certain tasks.

In summary, returning to the general subject of spatial relationships, geographers have known and worked with many concepts like containment, neighbours, and distances for a very long time. Today, various scholars in several fields are developing a more formal basis for understanding spatial relationships using the three categories of:

1. Metric relations (distance, direction).
2. Topological relations (connection, orientation, contiguity).
3. Order relations (inclusion).

Metric relationships are based on (usually Euclidean) geometric measurements for entities located in a two- or three-dimensional void. Topological spatial relationships are based on the notion of a continuous surface of space and concepts of boundaries, and interiors of cells. While inclusion properties can be dealt with topologically, they are directly addressed by the order relations via concepts like partially ordered sets.

Moreover, a strong foundation can be provided by use of the principles developed in various branches of mathematics, geometry, graph theory and topology. While we have taken an intuitive approach to these matters, nonetheless the practical implementations for topologically structured spatial data draw upon the formalisms of those fields to great advantage. Formal descriptions of phenomena based on mathematics are very valuable for clearly, concisely and consistently defining spatial relationships.

5.7 BIBLIOGRAPHY

For more extensive treatments of graph theory, the reader is referred to Harary, among others; while the book by Haggett and Chorley (1972) is a comprehensive review, albeit dated, of network analyses for both social and physical phenomena.

Broome, Frederick R. 1986. Mapping from a topologically encoded data base: the US Bureau of the Census example. *Proceedings of the Auto Carto Conference, London, UK*, vol. 1, pp. 402–411.

Corbett, James P. 1979. *Topological Principles in Cartography*. Technical Paper No. 48, US Bureau of the Census. Washington, DC, USA: US Government Printing Office.

Environmental Systems Research Institute. 1987. *ARC/INFO Users Guide*. Redlands, California, USA: Environmental Systems Research Institute.

Frank, Andrew U. and Werner Kuhn. 1986. Cell graphs: a probable correct method for the storage of geometry. *Proceedings of the Second International Symposium on Spatial Data Handling, Seattle,* Williamsville, New York, USA:

International Geographical Union Commission on Geographical Data Sensing and Processing, pp. 411–436.

Frank, Andrew U. and Bruce Palmer. 1986. Formal methods for the accurate definition of some fundamental terms in physical geography. *Proceedings of the Second International Symposium on Spatial Data Handling, Seattle.* Williamsville, New York, USA: International Geographical Union Commission on Geographical Data Sensing and Processing, pp. 583–599.

Haggett, Peter and Richard J. Chorley. 1972. *Network Analysis in Geography.* London, UK: Edward Arnold.

Harary, Frank. 1969. *Graph Theory.* Reading, Massachussetts, USA: Addison-Wesley.

Kainz, Wolfgang. 1989. Order, topology and metric in GIS. *Technical papers of the Annual Convention of the American Society for Photogrammetry and Remote Sensing/American Congress on Surveying and Mapping Annual Convention, Baltimore, Maryland, USA,* vol. 4, 154–160.

Kainz, Wolfgang. 1990. Spatial relationships – topology versus order. *Proceedings of the Fourth International Symposium on Spatial Data Handling, Zurich, Switzerland,* pp. 814–819.

Kirby, G. H., M. Visvalingham and P. Wade. 1989. Recognition and representation of a hierarchy of polygons with holes. *The Computer Journal* 32(6): 554–562.

Marx, Robert W. and Alan J. Saalfeld. 1988. Programs for assuring map quality at the Bureau of the Census. *Fourth Annual Research Conference, US Bureau of the Census, Arlington, Virginia.* Washington DC, USA: US Government Printing Office.

Meixler, David and Alan Saalfeld. 1985. Storing, retrieving, and maintaining information in geographic structures. A geographic tabulation unit base (GTUB) approach. *Proceedings of the Auto Carlo 7 Conference,* March 1985. *Washington DC,* USA, pp. 369–376.

Meixler, David and Alan Saalfeld. 1987. Polygonization and topological editing at the Bureau of the Census. *Proceedings of the Auto Carto 8 Conference, Baltimore, Maryland, USA,* pp. 731–738.

Nyerges, Timothy L. 1989. Design considerations for transportation GIS. Paper presented at the *Annual Conference of the Association of American Geographers, Baltimore, Maryland, USA.*

Peucker, Thomas K. and Nicholas R. Chrisman. 1985. Cartographic data structures. *The American Cartographer* 2(1): 55–89.

Saalfeld, Alan. 1987. Stability of map topology and robustness of map geometry. *Proceedings of the Auto Carto 8 Conference, Baltimore, Maryland, USA,* pp. 78–86.

Sauvy, Jean and Simonne Sauvy. 1974. *The Child's Discovery of Space.* Harmondsworth, UK: Penguin.

Slocum, Terry A., Robert J. Hanisch and Derek Thompson. 1984. The design and applications of a cartographic data base for redistricting. *Geo-Processing* 2: 151–176.

Thompson, Derek and Terry A. Slocum. 1982. A geographic information system for political redistricting in Maryland. *Proceedings, Applied Geography Conferences* (College Park, Maryland, USA) 5: 73–87.

Thompson, Derek *et al.* 1983. The Maryland Reapportionment Information System. Department of Geography, University of Maryland, College Park, Maryland, USA.

US Bureau of the Census. 1970. *The DIME Geocoding System.* Census Use Study Report Number 4. Washington, DC, USA: US Government Printing Office.

US Geological Survey. 1986. *Digital Line Graphs from 1:24,000-Scale Maps.* National Mapping Program Technical Instructions Data Users Guide 1. Reston, Virginia, USA: US Geological Survey.

Visvalingham, Mahes, P. Wade and G. H. Kirby. 1986. Extraction of area topology from line geometry. *Proceedings of the Auto Carto London Conference*, pp. 156–165.

Werner, Christian. 1985. *Spatial Transportation Modeling.* Scientific Geography Series, vol. 5. Beverly Hills, California, USA: Sage Publications.

White, Marvin S. 1984. Technical requirements and standards for a multi-purpose geographic data system. *The American Cartographer* 2(1): 15–26.

Wilson, Peter R. 1985. Euler formulas and geometric modelling. *IEEE Transactions on Computer Graphics and Applications* 5(8): 24–36.

6
Tessellations

Regular and irregular cells, hierarchies

Wong Lee prepared some map overlay models to predict the effect of acid rain on natural vegetation using 100 hectare square parcels of land. Carmen Adobe proposed to replace the grid pattern of streets by a new form with less dangerous intersections. Kiki Vroomanoff interpreted some LANDSAT images as part of her study of vegetation change in tropical regions. Wendel Christoff saw settlements in human space as sets of points within hexagons of different sizes and shapes. Eva Sanchez could not understand why or how the marketing department of her company created postal zone boundaries artificially. Geo(rge) Control used a network of geodetic control points to establish surface elevations. And Olivoushko and Karinoushka, who platonically love a dodecahedron (or maybe an octtree), go in search of the pyramid at address 003201320110230.

The theme of polygons as spatial units is extended in this chapter to the general topic of tessellations, sets of regular or irregular pieces of space accounting for all of that space. After presenting properties of geometrical figures, the discussion turns to the role of regular four-sided figures, in planar and hierarchical varieties. Irregular tessellations are then presented, especially the triangulated irregular networks which have several interesting properties.

6.1 MOSAICS, TESSELLATIONS AND LATTICES

The discretization of space, a topic introduced in Chapter 2, takes many forms. Discretization is the process of partitioning continuous space, either lines or areas, into pieces. This process is referred to as tiling in the case of two-dimensional spatial units, or segmenting in the case of lines. **Mosaics** or tiles of zones of irregular and varied shape and size may be created via various processes. Originally conceived as polygons with four equal corners, as implied by the Greek word *tetara* or Latin word *tessella*, mosaics also can be made up of other regular geometric figures or a mixture of different types. Mosaics of quadrilaterals can be seen in combining photographs or paper map sheets to make larger, seamless, documents, or arithmetic graph paper with a square grid can be used to record the character of the earth represented on maps in large or small, consistently shaped and sized pieces.

6.1.1 Tessellations

Tessellations (sometimes in the computer graphics field called meshes), are sets of connected discrete two-dimensional units (Figure 6.1). Tessellations may be regular or irregular in geometry. A regular tessellation (for example, grid squares) is an (infinitely) repeatable pattern of a regular polygon (two-dimensional figure) or polyhedron (three-dimensional figure). This cellular decomposition implies every point in space is assigned to only one cell. An irregular tessellation is an (infinitely) extending configuration of polygons or polyhedra of varied shape and size. Irregular planar tessellations may also be regarded and represented as topological two-cells. A tessellation provides a way to deal with the occupance of space, in contrast to dealing with identifiable entities like lakes or towns. However, some person-made phenomena occur as tessellations, for example land ownership parcels.

In practice, **irregular** tessellations are used in areas such as:

Zones for social, economic, and demographic data
Administrative or political units
Surface modelling using triangles
Partitions of a large geographic database
Man-made phenomena like land parcels
Irregular sampling of a continuous spatial distribution
Wireframe modelling of buildings

Regular tessellations are encountered in:

Image data from remote sensing
Data compilations from maps by grid squares
Organization of map libraries
Data generated by photogrammetric systems as lattices of points
Uniform sampling of a continuous spatial distribution

The former have the advantage (discussed earlier) of being able to be processed generally as topologically structured vector data. However, their creation may be based on doubtful assumptions for the discretization process, or involve lengthy and cumbersome computations if different tessellations are overlaid, as discussed in Chapter 8.

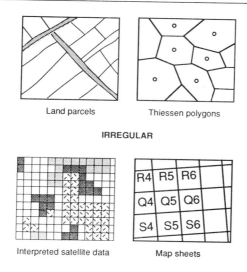

Land parcels   Thiessen polygons

IRREGULAR

Interpreted satellite data   Map sheets

REGULAR

Figure 6.1 Regular and irregular tessellations.

For the context for this chapter's discussion, we must remember that the underlying semantic questions regarding what is in (this) space or at (this) point and what is the location of (this) object are not in a one-to-one relationship with geometric form. An object can be positioned in Cartesian coordinate space, or it could be referenced by location within a hexagon. Grid squares are often used to record the nature of space, but the square is just the extreme, perfectly symmetrical case for all tessellations. From the larger set of all possible tiling figures, we can also employ irregular polygons which may represent real objects or may be used for descriptions of the contents of space.

6.1.2 Lattices

If we use point entities arranged over space in a regular, perhaps square or hexagonal, pattern, we talk of lattices rather than tiles (Figures 6.2a and b). The lattices may be seen as the intersections of the grid lines or as the centres of a set of squares (Figure 6.2c). Or, they may consist of a quinconcial pattern made up of the points and middle for a set of squares. Originally the word lattice referred to thin strips of wood or metal, crossing diagonally (Figure 6.2d), and capable of forming different patterns or networks.

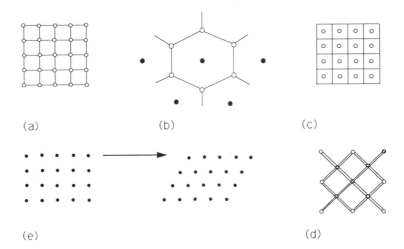

Figure 6.2 Lattices and tessellations. (a) Intersection of grid lines. (b) Hexagonal arrangement of points. (c) Centres of square cells. (d) Lattice frame. (e) Transformation from square to hexagonal pattern.

If the question, What is here?, is answered from the framework of a tessellation (say a set of squares) or a lattice of points, then different conditions arise. For the regular units, the attributes for a grid cell are an average (or other summary statistic) of conditions over that space. If a point framework is used, then the concept of average may be inappropriate. The zero-dimensional entity is best seen as a unique position at which some entity can be directly observed. Yet, the instances in a set of points may at times carry data for a summation over a continuous surface, as in the concept of aggregate accessibility space potentials. In this case, though, there is no implication of a discrete areal unit being represented by a point; the data represent influences of an aggregation of other points in the entire set.

Moreover, the regular tessellations may be seen as different patterns of points. The patterns of the sets of points may change via application of a linear transformation, for example, tilting the y-axis, just as a lattice of wood pieces anchored by pins at the crossings can be changed into different forms by pushing or pulling in one or the other coordinate axis. Thus, sets of rectangles or parallelograms can be made from squares (Figure 6.2e) by remapping the centroids of the squares. And a hexagonal arrangement of points can become a square grid of points or vice versa. There is not a one-to-one relationship between grid squares and a square lattice, for a square mosaic can be changed into a set of equidistant points

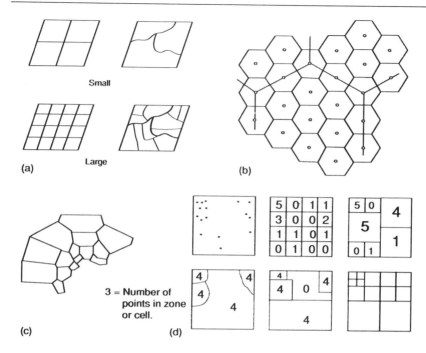

Figure 6.3 Variable resolution and scale. (a) Different scales. (b) Different scales for hexagons. (c) Variable resolution hexagons. (d) Varying resolutions at one scale for point data.

arranged in a hexagonal pattern (or other forms), although there is a duality of zero (lattice points) and two cell (tessellation polygons) units.

6.1.3 Scale and resolution

The basic spatial units of a tessellation can vary in size, shape, orientation and spacing. Also, for a given theme, several tessellations can exist, representing different scales of observation, or aggregations or subdivisions from one initial scale of observation (Figure 6.3). For regularly-shaped two-dimensional units the structuring is akin to a pyramid; for irregular shapes, the new larger units produced by combining smaller units, as in creating school districts from census tracts, will not have a shape consistent across the different scales. Varying scale is seen for lattices in different point densities per unit area.

The basic unit may vary in size for a given theme for a given scale. In this way the recording unit may better match the size properties of entities or the diversity of the phenomena in space. These concepts are met in theoretical formulations like the different sizes of hexagons for central place theory (Figures 6.3b and c), a set of propositions which, among other things, addresses the question of the size and shape of market areas for services provided from towns. It also arises for practical applications like surface modelling triangulations, as discussed in section 6.6. Irregular tessellations generally have implicitly **variable spatial resolution** (Figure 6.3c), but regular tessellations with variable size units for one theme have to be produced from the smallest scale by subdivision, as is seen in the process of creating quadtrees by hierarchical decomposition (section 6.3).

Tessellations, mosaics and lattices are not only to be seen as spatial units for recording data, but also as devices for facilitating access to databases for continuous space. We have already seen this use as a basis for space-filling curves. While any piece of the earth can conceivably be used for addressing and retrieval, regular tilings have some advantages for this purpose as is demonstrated later in this book.

6.2 THE GEOMETRY OF REGULAR TESSELLATIONS

Of the many regular geometric figures, from 3 to N sides, the square, triangle and hexagon are those currently most likely to be generally encountered in spatial data contexts. Some, like the pentagon, appear in special circumstances. The square has some practical advantages historically, being the unit used in graph paper and mechanical printing devices. The hexagon has utility in theory and practice where adjacency is an important property or where packing of geometric figures into space is important. The triangle has a property of being the most primitive polygon, that is, it cannot be further subdivided into a different figure, and it can readily have varying shape based on angles or length of sides conditions. The regular polygons all are characterized by identical lengths of sides, and equal angles subtended by lines from vertices to the centre of the figure; and only the triangle, square and hexagon regular figures, unlike circles, can fill a plane completely.

Comparisons of the value of different tiling units are best made on the basis of certain standard properties. In examining the geometry of regular tilings we will consider:

shape
adjacency

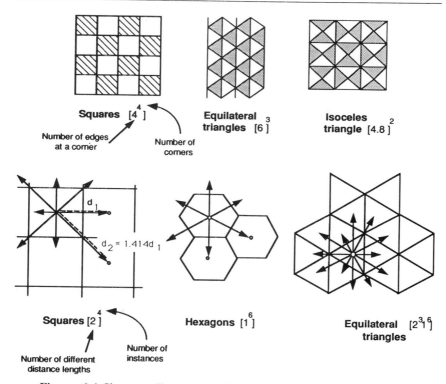

Figure 6.4 Shape, adjacency, and connectivity for some regular tessellations.

connectivity
orientation
self-similarity
decomposability
packing properties

Shape is the geometry of the figure's different number of sides, albeit that particular instances of a particular type of figure may have different shapes, where there is no tight restriction on the length of sides or the angles, as for triangles. Shape for regular figures is identified by the number of edges at a vertex and the number of vertices (Figure 6.4). For example, the hexagon has six instances of three edges meeting at corners, and the isoceles triangle has one vertex with four edges, and two vertices with eight edges converging.

Adjacency refers to the conditions of touching for repeated instances of the basic figure. The contiguity may be at the sides (edges), or corners (vertices), or both (Figure 6.4). This situation varies from the sides only for hexagons, to both vertices and edges for triangles and squares or rectangles. Metric measurements can be used to distinguish different figures – the adjacency distance is the distance between the centroid of one tile and a neighbour, and the adjacency number is the number of different adjacency distances. The hexagon has one such distance, six in number (one for each equidistant neighbour). The square has two, and the equilateral triangle has three different distances.

Related to adjacency, but considering only the conditions at the intersections, for some applications, like transportation, the hexagon has the **connectivity** advantages of having junctions at only three edges, unlike the four for a regular grid, as in an American-style city block street system (Figure 6.4). For most purposes, though, the polygon space adjacency is of more interest that the intersection connectivity patterns.

The tile **orientation**, defined as the base relative to coordinate axes, is uniform for the hexagon and square, but not so for the triangle as not all instances of triangles in a tessellation can have the base parallel to either the x or y coordinate axes (Figure 6.5). Tiles with the same orientation can be transformed into each other by translations, that is, by shifting along the x- and y-axes, without necessitating rotation or reflection.

If the basic figure shape is similar to that produced by aggregations to a higher level, then the property of **self-similarity** exists (Figure 6.5). This is

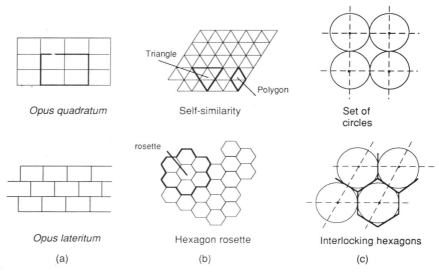

Opus quadratum	Self-similarity	Set of circles
Opus lateritum	Hexagon rosette	Interlocking hexagons
(a)	(b)	(c)

Figure 6.5 Self-similarity and packing. (a) Rectangles. (b) Triangles and hexagons. (c) Packing of hexagons.

unlimited for regular figures which have each side on an infinitely straight line composed entirely of edges. This condition is seen clearly in the case of the tiles touching at corners (*opus quadratum*) as opposed to the bricklayer's scheme (*opus lateritum*) with only horizontal unbroken lines (Figure 6.5a). However, only two tessellation types, the square and equilatral triangle, with identical side lengths, and two tessellation figures with unequal side lengths, the isosceles and the 30–60 degrees right angle triangles, meet this condition.

It is also of interest to examine changing conditions if the figures are **decomposed** into smaller units or combined to larger units. The square and equilateral triangle can be decomposed into units of the same shape, but the hexagon cannot; it can be split into six equilateral triangles. Indeed, as Figure 6.5b demonstrates, the aggregation of seven hexagons produces a rosette, while the aggregation of equilateral triangles produces polygons if pairs are taken, and higher-level triangles if four basic triangles are combined. While the hexagon is not decomposable with self-similarity, it can be used at different scales, rather as a lattice pattern, without nesting (Figure 6.3b). Only the square can be decomposed into smaller or assembled into larger tiles of the same shape and orientation, and be compatible with the orthogonal axes of Cartesian coordinate systems.

The overall pattern of how the repeated regular figures **pack space** is an important condition in engineering for properties like stress, and is also important for living organisms in their effectiveness in the use of space. A structurally stable situation (Figure 6.5c) is demonstrated by the relationships of a set of circles at right angles, at oblique angles representing rotation of one of the two coordinate axes, and then displacement to oblique axes where the vertical spacing is half the horizontal.

Essentially, the figures are locked together at this position. If squares or hexagons are substituted for circles, the same principle is at work, but it can be shown that the hexagon is more economical in the occupance of space. For example, there are gaps left if circles are used, but not if hexagons are used. This packing condition appears in the fields of geography and regional economics in central place theory, a set of propositions governing patterns of settlements and trade areas; in the insect world it is represented by the honeycombs of bees.

It is not surprising, though, on the basis of this review of properties, that the square figure dominates the world of regular tessellations for spatial information systems for two- or three-dimensional entities (except if a single sphere is to be partitioned). The square has simple and valuable conditions of equality of sides, decomposability, and stability for orientation and aggregation. So the square appears in the form of arrays of picture elements, grid cell data recording, lattices for elevation data, and electronic data capture and displays. The hexagon is more complex although it has nice properties of edge touching and economy of packing space. In theory it has had many uses, but not in practice. However, recently it has made an appearance as the data unit for one particular spatial database. The triangle has advantages of being the most primitive polygon, and does appear in several practical contexts, although in the form of irregular rather than regular triangles, as discussed in section 6.6.

6.3 FIXED SPATIAL RESOLUTIONS: REGULAR CELL GRIDS

The commonly encountered organization of data into an array of regular cells, almost always, but not necessarily, square, is regarded as practically straightforward, but conceptually does not explicity deal with questions of the type, Where do I find this particular entity? The basic unit of the square handles both location, geographic or arbitrary, and the attributes of the location. Often known as raster encoding of earth features, this cell array approach contrasts with the vector object-oriented encodings, having some practical advantages, as well as some limitations.

6.3.1 Data encoding

Some fundamental considerations to be examined regarding grid-cell data organization are:

1. The existence of an *a priori* fixed resolution.
2. The method of determining what attribute is in a cell.
3. The level of measurement used for the attributes.
4. Limitations in recording precisely point or line features.
5. The absence of recording coordinates for features.
6. The lack of explicit topology.
7. The flexibility in undertaking many operations easily.
8. A match with a field view of the world.

Clearly, the focus on space rather than on features can produce problems in some cases (Figure 6.6). If a high degree of precision is required for capturing data for point or line features, a very fine **resolution** may be required, at the possible expense of having a very large amount of data. If it is not required, then the entire grid cell is used to

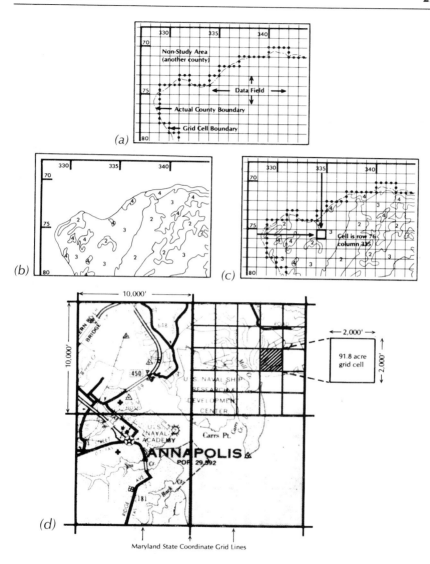

Figure 6.6 Grid cell data encoding. (a) The MAGI system grid for encoding. (b) Example of polygonal data. (c) Grid overlay. (d) Resolution. The encoding unit is the 91.8 acres block made from the 2,000 feet grid lines on state base maps. (Taken from Maryland Department of State Planning, 1981, Figures 1 and 3. Reprinted with the permission of the Maryland Office of Planning, Baltimore, Maryland, USA.)

record that a piece of highway or boundary exists in part of the cell, possibly causing problems upon reconstruction of an entire line feature from several cells. It is possible that some features are not recorded at all at a given scale of cell size.

Figure 6.6, taken from the Maryland Automated Geographic Information (MAGI) System created in 1974, shows the details of phenomena relative to the chosen grid side length of 2,000 feet, a resolution of 91.8 acres (Maryland Department of State Planning, 1981).

Some difficulties arise because of the nature of the mappable entities; others occur as a result of the **rules used for encoding** data into the grid cells (Figure 6.7). Three choices exist: a rule of dominance, a rule of importance, and encoding of only what is at the centre of a cell. In the first case, a single feature like a well or river segment, is identified as occurring in a cell, no matter how much of the space it occupies. But for polygons, a decision is made, based on the proportion of the cell area, as to the dominating category. For example, if three types of land use are seen to occur in a particular cell, then the highest one is recorded, and the others are not represented (Figure 6.7b). To retain more information in a case like this, then secondary, tertiary, quaternary (or even more) levels of incidence are recorded (Figure 6.7c). Similarly, for point and linear entities, while a rule of dominance would record, say, only primary highways, a rule of order of importance would also record highways of lesser rating.

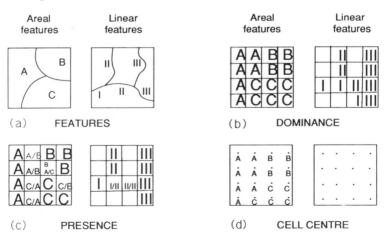

Figure 6.7 Rules for encoding. (a) Data for areal and linear features. (b) Dominance—encoding of only the principal attribute value. (c) Presence or importance—often recorded as multivalued attributes. (d) Based on middle of cell (shown only for areal features).

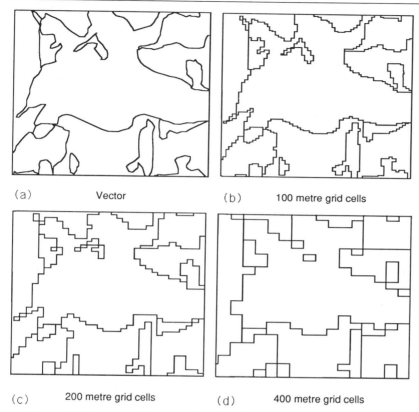

(a)   Vector   (b)   100 metre grid cells

(c)   200 metre grid cells   (d)   400 metre grid cells

Figure 6.8 Some deficiences of grid cell data representation. Contiguity and connectivity relationships may be altered, sometimes substantially. (a) Vector representation. (b) Encoding as one hundred metre square cells. (c) Two hundred metre square cells. (d) Four hundred metre square cells.

Avoiding the issue of dominance or importance is done by recording data values for only grid cell centres, but much information may be omitted in this way (Figure 6.7d). If this procedure is used, then the data are in effect a lattice, and no judgements can be correctly made about the occupance of space. Lattice representations are better for continuously variable quantities like temperature or precipitation measurements, or elevations above or below sea level.

Generally, grid-cell representation is used for categorical data, either binary or multinomial, although it is possible to record **scalar quantities** (interval and ratio scales) in cells. Indeed, there have been examples, as in the UK and Sweden, of recording demographic data by grid cells. The practicality of doing this rests upon the availability of disaggregated data, that is for individual houses or addresses, which can be accumulated for the regular tessellation units. At the same time, it is interesting to note that lattice representations are valuable for data from remote sensing for some purposes, rather than the rasterized version based on pixel classification techniques. Each coordinate position on the earth can have scalar values for the multi-dimensional spectral signature, using several spectral bands.

The encoding of point and line feature data has limitations, subject to the grid-cell size used, in the ability to show the locations and shapes precisely. Boundaries may appear rather blocky, a point may occur near the edge of a cell, some data may disappear, and line features that are close together may be shown as touching (Figure 6.8). Thus, some aspect of spatial relationships may be violated. Of course, the precision of representation is a function of the size of the unit in the regular tessellation. A fine **spatial resolution**, defined as the minimum distance that can be recorded, and measured from cell centre to centre, produces many more grid cells with which to work. Whether the trade-off between precision and amount of data is worth it depends on various factors, not the least important of which is purpose.

6.3.2 Spatial properties

Spatial relationships are implicit in the data, but with only a few exceptions do the software systems for grid cell data allow direct handling of relationships between entities. Metrical distance relations along orthogonal axes are readily derived, although not without limitations, from the differences between x and y coordinates (row and column values), and other distances are computed as the hypotenuse of right angled triangles. While axial and diagonal lines can be measured by cell increments without loss of precision, this is not so for other hypotenuses. Thus, perimeter measures for areal objects or lengths for linear features may be severely erroneous. Shapes can be approximated by observing orientations of lines of cells relative to those of neighbours, assuming that the resolution is good enough to produce sufficient unambiguous data.

Metrical spatial properties can be readily dealt with by square cell tessellations, bearing in mind the resolution implied by the cell size. Distances are harder to obtain for hexagons, because the centres of the cells are arranged symmetrically along three coordinate axes, but properties based on neighbours are much easier because there are no corner adjacencies. Distances and adjacencies are more awkward for sets of equilateral triangles, because they have six triangles touching at corners and three different distance values for each triangle centre.

Graph properties of connectivity have to be explicitly recorded for cells containing nodes, and details of direction of edges leaving nodes (necessary in order to create a precise graph) must also be encoded. Consequently, the desirable situation is a combination of cells and vector representation of point and line elements within cells. Details for points and lines within a grid cell could be retained by particular coding techniques, or by linking the grid cell to other data by reference to the grid cell identifier. Some techniques have been devised to record information explicitly in a grid framework for line and polygon entities. Thus, the directional coding presented in section 4.5 is sometimes used for boundary line or linear feature representation.

Topological properties are only indirectly dealt with by the regular tessellation forms, because they do not treat phenomena as entities. Of course, there are properties of adjacency and connectivity among cells, but qualitative spatial relationships for entities must be derived from the data that are recorded. The topological properties of non-connectivity may be violated, such as the apparent joining of highways or touching of inner and outer boundaries of polygons. The graph property of existence of holes can be determined by calculating the distance to the nearest edge of a polygon (a set of cells).

Cell encoding forces pattern to be inferred from the data values for adjacent cells, but it does facilitate the process as it directly establishes neighbours and neighbourhoods. If cells are square, then eight touching pairs exist in the immediate ring around a cell; other situations may take in other cells, such as the neighbours of a block of four cells. Yet, unlike for topological encoding, the identification of neighbouring regular cells provides little information unless the attributes of the cells are considered, such as in creating zones by finding cells of a particular attribute value.

However, many operations, logical or arithmetic, are quite straight-forward when working with grid cell data. Area measures simply involve counting squares, not computational geometry; distances involve sub-tracting, and can be readily ascertained even on diagonals. Aggregations can be made within proximity zones, and map overlay modelling using attribute information essentially does nothing more than compare attribute values for identical grid cells, as demonstrated in more detail in section 8.3. At the same time, vector geometry may have an advantage in that much of the space is void: for regular tessellations, the space is 'full'.

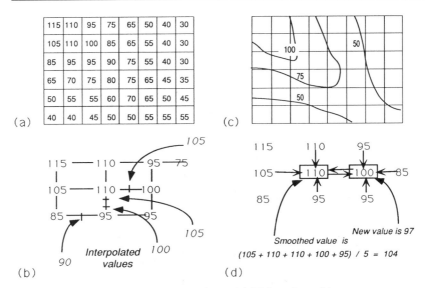

Figure 6.9 Surface representations. (a) Values for grid square centres or lattice points. (b) Mesh-based interpolation. (c) Contour lines. (d) Smoothing lattice or cell data.

6.3.3 Surface modelling from lattices

The **elevation dimension** is readily manipulated and displayed by means of data for cells, although it is usual to record such data as a lattice (Figure 6.9a). Such data, often referred to as digital elevation models because they are used for interpolating or extrapolating values for places in addition to the original data points, generally are themselves produced from other data by a process of gridding. Each cell or point carries data for an elevation or depth relative to a horizontal datum.

Much use has been made of grid cells or lattice-point data for representing fields of varying quantities across space, whether continuous or broken. The mesh form of the data, whether the values are for grid line intersections or cell centroids, lends itself to simple linear interpolations along the *x*- and *y*-axes (Figure 6.9b). The outcome is often simply an isoline map, as shown in Figure 6.9c. Proximity surfaces are easily created from the mesh data by accumulating distances in (circular) zones from specified points, or in bands out from line or area objects. Smoothing of attribute values for blocks of touching cells, five or nine, as demonstrated in Figure 6.9d, provides a means of generalizing the spatial attributes which are in scalar forms of measurement.

Row ID	Column ID	Attributes A B C ...
1	1	16 5 ...
.	2	19 8 ...
.		
.		

(a)

Layer Row ID ID	Column ID	Attribute value
A 1	1	16
. 1	2	19
B 1	1	5
. 1	2	8
.		

(b)

Layer ID	Zone	Attribute value	Cells row col.
A	1	19	1, 2
	.		.
	2	16	1, 1
	.		.
B	1	5	1, 1
.			

(c)

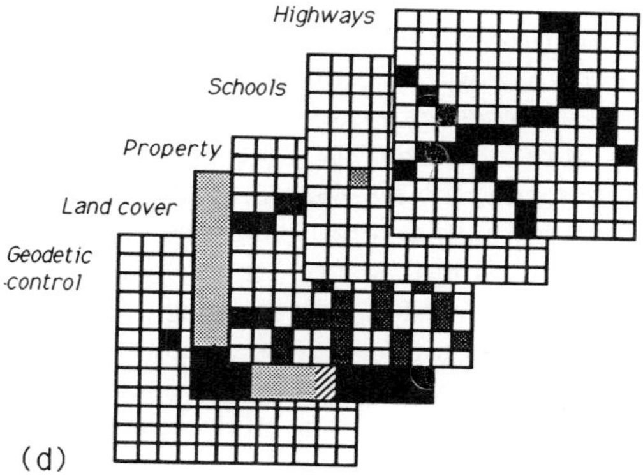

(d)

Figure 6.10 Structuring grid-cell data. (a) Grid-cell encoding. (b) Layer or matrix organization. (c) Clustering encoding. (d) A layer-encoded version of Figure 1.3.

6.3.4 Structures for grid cell data

Just as there are numerous varieties of organizing geometric and topological properties in data tables, so there are different ways of structuring regular tessellation data (Figure 6.10). The common form for a basic tabular record of data for grid cells consists of a row coordinate value, a column coordinate value, possibly a cell identifier, in addition to the row–column combination, a series of values for attributes found in a cell, possibly a code value pointing to another table of data, a reference to the identifier for the nearest cell on the sides or corners. If only one attribute is of interest, then the basic table consists of a row identifier, column identifier (jointly making a space locator) and one data item: an organization reflecting each grid cell as an individual entity.

The records, that is one per grid cell, can be listed sequentially in row order or column order, or, taking advantage of computer storage formats known as arrays, the entire **matrix** of cells can be stored (Figure 6.10b). The matrix format may also be used if there are several data items for each grid cell. The term **raster** (from the German *raster*, meaning screen), while often used interchangeably with grid cell data, implies, though, an ordering through the whole matrix in line scan form, that is, one row followed by the others, as in the set of parallel lines making up a television picture. Another variety **clusters** observations in groups of like character, and then provides coordinates for each pixel (Figure 6.10c). Conceptually, a grid encoded database looks like Figure 6.10d.

Although often associated with a map layer concept, the overlay arrangement is not a requirement for regular tessellations as the grid-cell or pixel form suggests. However, the matrix and clustered forms are designed to work with thematic layers, meaning that the attribute data are not recorded in sequence for each and every cell. While the layering architecture is utilized for both entity-oriented and tessellation encoding of phenomena, the two types certainly do treat attribute data differently. For entity-oriented representations, attributes are separated from the spatial information in most cases, whereas for regular tessellations the positional and attribute data are associated.

Spatial aggregation can be achieved quite well for tessellations by combinations of adjacent cells; spatial disaggregation can be dealt with by splitting cells. Aggregations might be necessary to facilitate generalization from one resolution (geographic scale) to another. For example, the number of points in four cells can be added to get the incidence for a set of four cells, or, if the attributes are scalar, an average could be obtained. Combinations might be necessary to produce the grid cell equivalent of homogeneous regions or administrative districts. However, awkwardness arises for shapes other than the square figure. Hexagons can be subdivided into only the triangle regular figure, although this is a straightforward process; equilateral triangles may be subdivided into other triangles, but they will not always be equilateral. Triangles can be combined into parallelograms, or polygons, or hexagons depending on how many are joined; and hexagons produce rosette shapes when seven are joined together.

It is not surprising, therefore, that the square cell is used for aggregating or disaggregating across scales. The resulting form, known as the **pyramid model** (Figure 6.11a), provides a **multiple scale representation**, with spatial units constant for a given scale. Used often in image processing, it provides a means to remove or hide detail in order to focus on structural components like general shape, or to reveal details of form that might be hidden by too much generalization. It is designed for rapid detection of global (overall, not earthly) features in a complex image. An

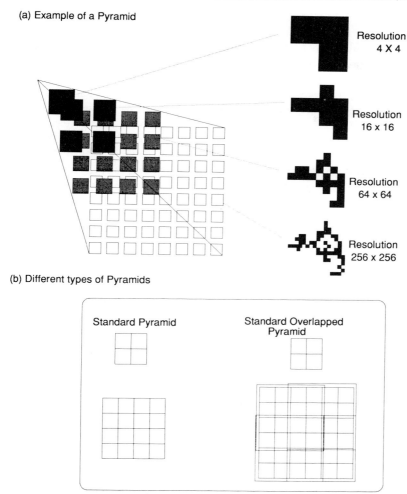

Figure 6.11 Pyramid form of organization. (a) Example of a pyramid. (b) Different types of pyramids.

ordinary pyramid has blocks of four cells combining to larger cells at higher levels, without overlap. A standard overlapped pyramid has 50 per cent overlap for adjacent blocks (Figure 6.11b).

A hierarchical representation has uses in image processing: for browsing at different scales, for simplifying mapping, for matching data collected at different resolutions, and for access at different levels. In this way, the grid cell becomes a handy spatial indexing tool, rather than a storage unit, for space filling curves.

6.4 VARIABLE SPATIAL RESOLUTION: QUADTREES

Geographers and cartographers have for a long time used various analogue devices to represent the unevenness of spatial distributions or the need to show more detail for congested areas. Thus, atlases have maps at different scales, and sheet maps have insets. Theoretically, the geometry of central place theory has equal-sized hexagons only if there is an underlying isotropic surface, that is, a uniform distribution of people on a plane. Otherwise, the hexagons are distorted in shape as their size is altered to represent a fixed number of people, giving rise to a tessellation of irregular hexagons (Figure 6.3c). Cartograms are often used to represent variable densities of phenomena in space.

In spatial information systems, this concept of variable spatial resolution is encountered in data models using regular spatial units. Phenomena conceived as occupying space may be aggregated into square blocks of the same or different sizes. Consider a distribution of houses (Figure 6.3d) represented as dots, and possibly represented by irregular polygons or fixed size grid cells. Aggregating the houses to rectangles containing an equal number of houses produces units of different sizes. If all space is to be covered by squares, then several possibilities exist, although it is very difficult in this case to directly draw the map of four squares, each containing four people.

The concept of variable spatial resolution implies varying sized units at a given resolution level. The choice of the shape of the units is a different matter; as discussed earlier, there are practical advantages in using the square figure. The square is particularly handy if the process of creating the blocks of varying size is one of decomposing space from a general level to more detail. It is relatively easy to do this with squares. For example, as Figure 6.12 illustrates, a polygon can be successively approximated by sets of blocks at different levels. If the process involves systematic splitting of space in two-dimensional space by a rule of four, then the structure is known as a **quadtree**, one type of hierarchical data

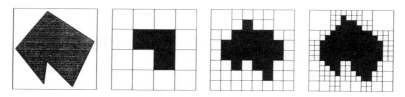

Figure 6.12 Successive approximations of a polygon by a quadtree. This method assumes approximation by default, that is, each square must be fully occupied by the black object.

model. A three-dimensional equivalent is known as an octtree because it involves an eightfold splitting.

Recalling an earlier discussion of rules for encoding grid cell data, in regard to Figure 6.12, we must mention that the approximation of the polygon has employed a rule of complete coverage by the presence condition (black area) for a given quadrant at a particular level. A different rule, sometimes known as approximation by excess, would encode blocks only partially covered by the black object. In this case, only the top-left, top-right and bottom-left quadrants would be left white.

The hierarchical arrangement (Figure 6.13) is often charted as a tree diagram, showing the four blocks produced at each level of subdivision of the squares. A polygon recognized as a collection of cells with like attribute, for example urban land use (Figure 6.13a), is in this way represented by cells which have an identical numerical or colour code (Figure 6.13d). The detailed geometric form is picked out by the combination of units of different sizes. The empty space, provided it coincides with a square at a higher level, does not need to be represented by multiple small squares. The attribute data condition is shown at the different tree levels by the (conventional) black and white coding.

More extensive attribute coding can be accomplished in like manner. The different types of urban land use (Figure 6.13c) are encoded by the set of quadrants at levels 2 and 3, as tabulated in Figure 6.13e. A complete listing, including the space blocks outside the urban polygon, will appear in the form of Figure 6.13f. This time there is an extra table column showing the area represented by each quadrant. Reference to the individual blocks may use one of several possible locational reference schemes. For the moment we illustrate numerical locational codes, referring to the sequence of NW, NE, SW and SE blocks by the numbers 0, 1, 2 and 3. Thus, 321 stands for the SE corner at level 1, the SW at level 2, and the NE at level 3.

Subdivision continues until either a predetermined limit of resolution is reached, or all detail in the phenomenon is accounted for (as in Figure 6.12). The minimum-sized spatial unit is sometimes designated by ε (epsilon). If the digital data are made from source documents containing much curvature detail for linear features or boundaries of polygons then, as a practical matter, the decomposition process will be terminated at a given resolution, perhaps because of computer resource restrictions. The hierarchical organization allows spatial resolution to vary with the complexity of the phenomena being recorded.

Different kinds of data can be treated in this hierarchical subdivision fashion, to a greater or lesser degree of success. Ideally, we would like to be able to:

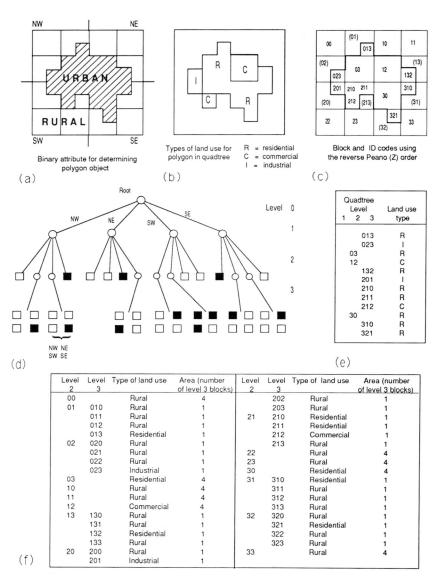

Figure 6.13 The quadtree data organization. (a) Binary attribute for determining polygon object. (b) Types of land use for polygon in quadtree. (c) Quadtree blocks and identifiers based on location codes. (d) Hierarchical organization. (e) Table of attributes for the urban polygon. (f) Complete table of data for entire area.

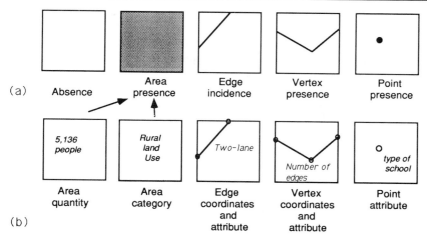

Figure 6.14 Different types of quadrants. (a) Types of quadrants. (b) Some possible attribute data.

1. Treat point, line, and area data in the same way.
2. Capture metrical details for entities.
3. Facilitate various kinds of operations.
4. Deal with different ways of measuring attributes.
5. Have consistent locational referencing.

The simplest form of quadtree recognizes the presence or absence of an attribute in space, whether point, line, or area. Computer scientists usually refer to the **binary incidence** representation as colour coding, using black and white to indicate presence and absence. A cell could contain a scalar value, or a pointer to sets of attributes under the condition that the cell is a lowest geographic unit. Thus cells may be used to represent point data, such as cities, where each cell contains one city; or linear features, say water pipes, where each cell contains a segment of a pipe or a junction of several water lines. So we may define (Figure 6.14a) an attribute presence quadrant, an absence quadrant, an edge quadrant, a vertex quadrant and a point quadrant.

In the standard form, the geometry of edges and points is not retained, only incidence. However, as for fixed resolution regular tessellations, additional information can be encoded for cells (Figure 6.14). In the case of edges representing polygon boundaries or graphs, this could consist of the x and y Cartesian or the polar coordinates to establish where the edges cross the boundary of a cell, or for vertices or points, the exact coordinates for a point within a cell.

If, for linear features, the incidence is only a vertex with one or more

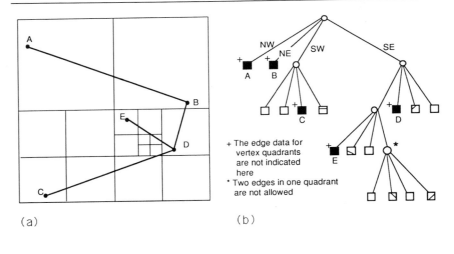

Quadrant ID	Vertex label	Vertex coordinates	Edge coordinates	Edge attributes

(c)

Figure 6.15 Quadtrees for line data. (a) Map and quadrants for line objects. (b) Tree and encoding rules. (c) Data items record.

graph edges or a piece of an edge, then a quadtree representation will look like that in Figure 6.15. Various possibilities for encoding exist; rules must be established before the quadtree database is created from the original data.

Unconnected points may be handled in different ways. A regular figure decomposition process (Figure 6.16) could produce squares from the orthogonal coordinate space by subdividing using both x and y, with varieties depending on whether or not all four squares at a given level of decomposition were recognized (the MX quadtree) or not (the PR quadtree). The second type requires coordinate information to establish position within the block; the former does not, representing the point at a corner of the cell.

Various locational reference schemes are possible, and are indeed used for meeting different requirements. One simple scheme is to consistently order the four blocks at each level in a NW, NE, SW, SE sequence, using data in the record for a tree node to point to the four nodes at a lower level if such exist (Figure 6.17a). Numerical coding representing the NW, NE, SW and SE by integers as noted earlier could be used (Figure 6.17b); some coordinate values could be used, or a space path could be employed to simplify movement through the entire set of cells, without using actual

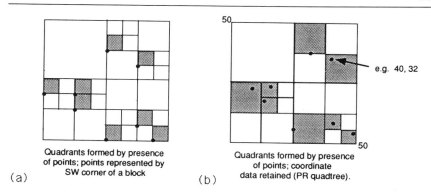

Quadrants formed by presence
of points; points represented by
SW corner of a block

(a)

Quadrants formed by presence
of points; coordinate
data retained (PR quadtree).

(b)

Figure 6.16 Quadtrees for point spatial units. (a) Point presence, no precise location. (b) Points and positional data.

coordinate values. Referring to Figure 6.17, coding using row and column identifiers would require more data to be stored than for a locational coding scheme, using the NW, NE, SW, SE orientations, while a Peano N path has single dimension addressing and has stable numbering across different levels of resolution. Thus the larger blocks in the quadtree would be represented by fewer positional pieces of data than the number of blocks (Figure 6.17c), and the final table would contain items for the Peano key and quadrant size, often the number of smallest size pixels on the side of the square block.

Node level	Nodes at next level NW NE SW SE	Node type	Node attributes

(a)

Locational code	Size of quadrant	Node type	Node attributes

(b)

Peano key	Size of quadrant - side length	Land use
0	2	Rural
4	2	Urban
8	2	Rural
.	.	.
.	.	.
.	.	.
28	1	Rural
29	1	Rural
30	1	Urban
31	1	Rural
32	4	Rural
48	4	Urban

///. Land use (urban)

(c)   4   Peano key

(d)

Figure 6.17 Locational coding for quadtrees. (a) Compass orientation locational coding. (b) Space path locational coding. (c) Peano ordering of quadrants. (d) Quadtree table for Peano ordering.

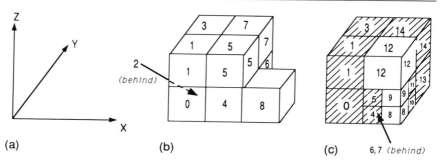

Figure 6.18 Octtree representation. (a) Axial labels for Peano-N ordering. (b) One level numbering sequence for the Peano order. (c) Two levels numbering and an example of colour coding.

The general properties and principles for quadtrees are applicable to the three-dimensional variant, the octtree, used to some extent for geologic modelling and representing three-dimensional solids. Figure 6.18 illustrates the arrangement and numbering of the octant blocks.

Hierarchical decompositions may be undertaken on the basis of the **empirical information** to be encoded and stored in contrast to the regular subdivisions so far discussed. The latter are data independent; the former are data dependent. For example, a distribution of point features, such as cities, may be subdivided into rectangular, rather than square blocks, on

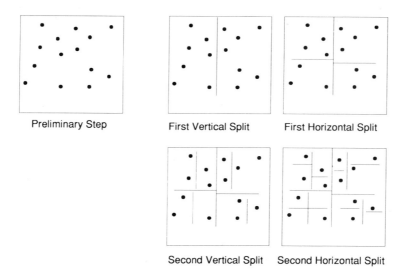

Figure 6.19 Splitting scattered points into a 4-tree.

the basis of alternating x and y axes (Figure 6.19). A similar process can produce two or four branches at each step. Thus, the empirical information, the exact position of the points, governs the data structuring, not a fixed-grid scheme. The binary subdivision, which is one of a group of **K-dimensional (KD) trees**, is generally regarded as superior to the point data quadtree for operations done in sequence.

The quadtree and related structures, clearly based on a tessellated discretization of space, provide semantic value by their recognition of varying density of incidence of phenomena in space, and can deal with both vector and raster data. The hierarchical structuring cleverly addresses spatial variations at different scales, it offers the valuable adaptability property to empirical conditions and with good locational referencing provides a basis for efficient spatial access and indexing. As discussed more fully in section 8.4, Boolean operations such as union, intersection and difference are easy to perform, whereas translation, rotation and scaling are not.

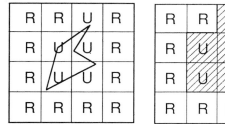

Sixteen cell resolution

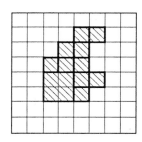 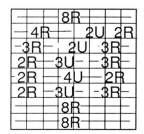

Sixtyfour cell resolution

(a) (b) (c)

Figure 6.20 Impact of different cell-encoding methods on data quantity. (a) Array of grid squares. (b) Quadtree blocks. (c) Run-length encoding.

In general, the hierarchical tessellations are regarded as offering benefits in the reduction in the amount of space needed to store data for phenomena. We contrast the more extensive grid cell encoding with the quadtree, and another device, the run-length encoding (Figure 6.20). The first of these records data for each cell, demonstrated here for two different resolutions. The quadtree will use a smaller number of spatial units as produced by the hierarchical subdivision; the run-length encoding reduces data storage by recording runs of like conditions for rows (or columns) as shown. The degree to which the space-saving methods reduce storage depends primarily on the amount of homogeneity in the mapped data. The extremes are a perfectly uniform landscape, for which the quadtree block is best, or a checkered pattern in which each cell is different from all its neighbours. In this case, there is no particular advantage in using the two space-saving techniques. Alternative data storage schemes like linked lists are preferable for sparse matrices.

The hierarchical structures may be differentiated on the basis of types of data, the principles guiding or governing the decomposition process, and the type of spatial resolution. However, because they are based on regular spatial units, they also have advantages and limitations associated with the use of grid squares. Particularly, there are limitations in dealing precisely with point and linear features, and in not explicitly addressing topological spatial properties.

6.5 HIERARCHICAL TESSELLATIONS FOR A SPHERE

A particularly interesting and challenging need is a hierarchical structure of cells for covering the entire world. As discussed in section 4.3, a global database can be structured as a set of triangles. While there are many ways to subdivide the surface of a ball, at one level, some of the geometrical properties of particular figures may have practical short-comings for a single or multiple scale representation scheme for the earth.

A globe can be divided into a single tessellation of only triangles that at one level have five corner neighbours, that is the icosahedron three-dimensional model. However, it is not possible to anchor the graticule to all important global features, the pole, equator and meridians. The dodecahedron, a set of pentagons, while having a nice property of only touching at edges of cells, is even worse for not having hemispherical symmetry, being flat at the top (or bottom) and pointed at the diametrically opposite pole. Out of the five convex regular polyhedra shown in Figure 6.21, and known as the **Platonic solids** (Gasson, 1983),

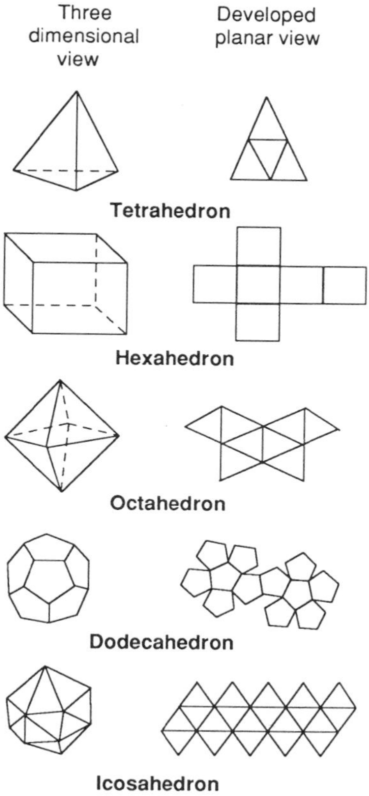

Three dimensional view	Developed planar view	Edges per face	Faces per vertex	Vertices	Edges	Faces
Tetrahedron		3	3	4	6	4
Hexahedron		4	3	8	12	6
Octahedron		3	4	6	8	12
Dodecahedron		5	3	20	30	12
Icosahedron		3	5	12	30	20

Figure 6.21 The Platonic solids.

the icosahedron is the most nearly spherical because it has the largest number of vertices, edges and facets.

The tetrahedron, or pair of these that make the octahedron, is better for fitting to the requirements of polar symmetry and for mapping vertices along the equatorial plane. Six anchor points, each with four triangles meeting, correspond to the north and south poles, and the 0°, 90°, 180° and 270° subdivisions around the globe (Figure 4.11). The eight initial triangular facets are then subdivided into a set of regular triangles (Figure 6.22a), providing global referencing for areas closely similar in shape and size, and facilitating hierarchical referencing to detailed positions on the earth. Within each principal triangle, four equal subdivisions are referred to by 0, 1, 2 and 3; at greater resolution levels there will be extra digits, using this same scheme of four numbers (Figure 6.22b). The numbering

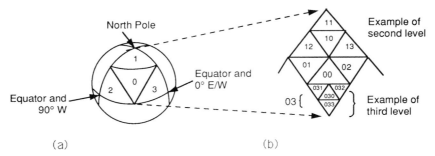

Figure 6.22 Triangular mesh coding. (a) One octahedron face with four level one triangles. (b) Triangular subdivision tiling and locators. (Based on Dutton.)

scheme used for the tetrahedron can also be regarded as providing indexing keys for access to each cell.

6.6 IRREGULAR TESSELLATIONS BASED ON TRIANGLES

Two types of irregular tessellation have valuable properties for spatial information systems: triangles and proximal polygons. They both represent variable spatial resolution at a given scale and can be dealt with hierarchically although, at the moment, there are few practical applications of this variety. A real need that demonstrates the value of a set of triangles is the representation of earth surface terrain conditions. It is generally thought that, at least visually, it is preferable to break up a surface into triangular facets rather than squares or other polygons. In order to create areal units from only point data, a technique of creating proximal polygons is often used.

6.6.1 Proximal regions

For the second of these needs, consider a distribution of administrative offices in space. We can argue for locating them so that their territories are demarcated such that the people in every household living within them travel to their nearest centre (Figure 6.23). As such, the polygons are sometimes called proximal regions, or are often called **Thiessen** or **Voroni polygons**. If any people in a household within the designated proximal zone then choose to travel to a different centre, they bear an additional cost by increasing their travel distance above an optimum minimum based on the nearest centre.

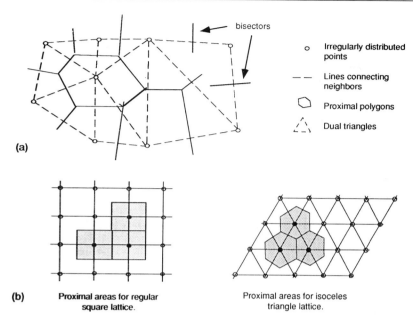

Figure 6.23 The concept and construction of proximal polygons. (a) Proximal areas and duals. (b) Regular proximal polygons.

The polygons are created by subdividing lines joining nearest neighbour points, drawing perpendicular bisectors (sometimes called mediators) through those points, and then assembling the several polygon edge pieces out of those lines, as shown (Figure 6.23a). The concept of the proximal area, sometimes used as a standard for evaluating equity issues for travel to administrative centres or public service facilities, is known in mathematics as the **Dirichlet domain**. This space encompasses a set of points closer to a given point than to any other points in the set. With reference to Figure 6.23b, note that the domain takes on different forms for varying point patterns, regular or otherwise.

6.6.2 Triangulation

The irregular triangulation for surface modelling is a somewhat more involved concept than creating proximal areas because it is oriented to line features as well as points. As Figure 6.24 suggests, using triangles to represent terrain, a more realistic representation will be achieved if the spatial data units recognize natural surface changes in slope, at peaks,

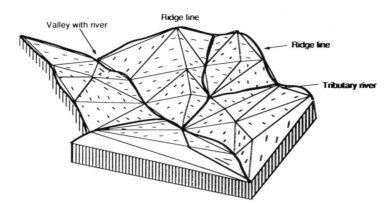

Figure 6.24 Modelling terrain: concept of a triangulation closely following the major terrain features.

pits, passes, ridge lines, saddle points and course lines or discontinuities, rather than just be fitted arbitrarily. A set of triangular facets can be created to meet these conditions by having triangle edges fall along approximations of ridges and river channels, and having their corners located at control points with exact known coordinates from earth surveys, or at river confluences, or at peaks or depressions of terrain.

Ideally we would like to have:

1. The triangle corners match important turning points in the terrain surface.
2. The important linear features be represented by triangle edges.

The process of triangulation has three stages:

1. Choosing the data points.
2. Connecting points to create triangles.
3. Storage of necessary and additional desirable information.

These are usually followed by the interpolation of elevation values for other points in space, although the triangles may be used for other purposes.

Assuming for the moment that the z variable is terrain elevation, but noting that, in principle, other phenomena can be treated in the same manner, then original data for height may come from several sources of different spatial structure. Data may be lattices of heights in the form of digital elevation models (themselves often created from digitized contours), irregularly distributed spot heights, contours or a mixture. It is important to know the spatial distribution and whether data are point or

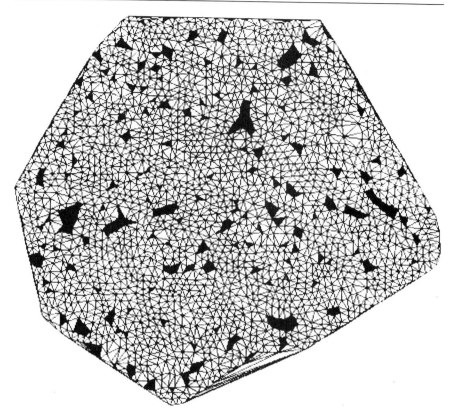

Figure 6.25 An example of a triangulated irregular network (TIN) for a high density of data points. The black areas indicate triangles with virtually no slope. The hull of outer points is not convex in this example.

linear. It is also most important, especially for terrains, to recognize natural breaks of slope, and key landscape features like coastlines, course lines, ridges and peaks. Estimated heights or surface representations will be more reliable if there is an even spatial distribution of information, except that in areas of no relief, then fewer sample points are needed.

Whatever the original form of data, the triangulation method uses x, y, z coordinate triads, fitting a set of irregular triangles to all data points, and then interpolating intermediate values of z from the known values at the corners of the triangles. This **triangulated irregular network** (TIN) therefore is a tessellation model (Figure 6.25) applied to known positions, or, at least a subset of them. Because estimates of height will be more reliable the closer they are to the original data points, it is better to use

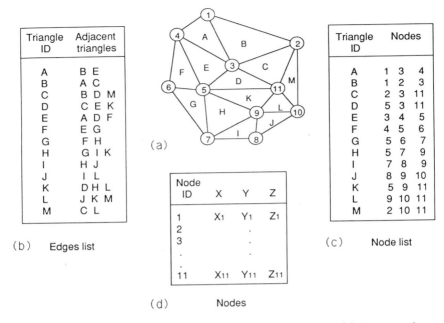

Triangle ID	Adjacent triangles
A	B E
B	A C
C	B D M
D	C E K
E	A D F
F	E G
G	F H
H	G I K
I	H J
J	I L
K	D H L
L	J K M
M	C L

(b)   Edges list

Triangle ID	Nodes		
A	1	3	4
B	1	2	3
C	2	3	11
D	5	3	11
E	3	4	5
F	4	5	6
G	5	6	7
H	5	7	9
I	7	8	9
J	8	9	10
K	5	9	11
L	9	10	11
M	2	10	11

(c)   Node list

Node ID	X	Y	Z
1	X_1	Y_1	Z_1
2		.	
3		.	
.			
.		.	
.			
11	X_{11}	Y_{11}	Z_{11}

(d)   Nodes

Figure 6.26 Data for a triangulated irregular network. (a) Annotated triangulation map. (b) Edges list. (c) Nodes list. (d) Nodes.

triangles as close to equilateral as possible, although this condition will have to be compromised if triangle edges have to follow course lines, ridges or shores. It is especially important to avoid long narrow triangles, such as might occur when using data from widely spaced contours with many points on each contour line.

Thus the triangular tessellation consists of:

1. A set of points carrying elevation data.
2. A set of lines consisting of pairs of points, joined by straight lines.
3. A set of triangles, having triplets of x, y, z coordinates.
4. Adjacency relations for the edges of the triangles.
5. A list of triangles in which particular edges are included.
6. The triangles in which particular nodes are contained.

As such the triangular tessellation combines topological and geometric information, some items of which are illustrated in Figure 6.26. Various derived data are computable as needed or computed and stored for the following topological elements:

1. The gradient and aspect of edges of triangles.

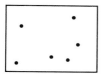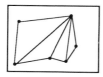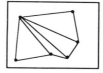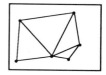

(a) A set of points and three instances of a triangulation of these points. The third is closest to the ideal of equilateral triangles.

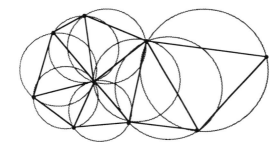

(b)

Figure 6.27 The Delaunay Triangulation. (a) Alternative triangulations. (b) An instance of a Delaunay triangulation. (Based on Saalfeld.)

2. The planar and surface area of the triangular facets.
3. The slope and aspect across the facet.

Naturally many sets of triangles could be fitted to a set of points (Figure 6.27a). A triangulation that produces **Delaunay triangles** is generally the preferred relatively straightforward method, producing triangles with a low variance in edge length. This type of triangle, based on a proximal distance criterion, is defined by the condition that the circumscribing circle of any triangle does not contain any point of the data set inside it, as in Figure 6.27b (Saalfeld, 1987). The Delaunay triangles are duals to the Thiessen polygons discussed above (Figure 6.23); and the polygon centre is the centre of the circumscribing circle.

Triangulated tessellations have a number of useful features for spatial information systems. The triangles can be treated as irregular polygons; the tessellation exhausts all space; there is planar enforcement; they are appealing spatial units that appear to provide acceptable models of certain kinds of terrain surface. At the same time, their creation is computationally demanding; there are many possible triangulations for any set of points; and they can miss important aspects of surface morphology unless the edges are constrained to fit major breaks of slope.

They are not the only data models conceptually possible or actually used in estimating surface height values from a set of spatially distributed points. If contour data exist, elevations can be interpolated for points lying within the isolines. A uniform distribution of points, a lattice form or intersections of grids, facilitates estimation of heights of points along the grid square sides, but generally the regular distribution of point values is itself estimated from an irregularly spatially distributed set of data by a process known usually as **gridding**. In simple form, this consists of using one or more original data points found in a neighbourhood of the grid intersection. More discussion is provided in section 7.1.

Absent much empirical evaluation of different surface representation techniques, it appears from logical grounds and some experimental studies that irregular tessellations can recognize important surface conditions, provide data for topological properties, produce reliable interpolated values by passing the surface through known data points, and allow for different scale representations. Gridding procedures producing regular tessellations do not usually recognize data points, do not provide explicit topological information, and are not adjusted to known conditions like breaklines.

In the end, though, choices must be made on the basis of **purpose and type of terrain** being modelled. Some hydrological simulation models work reasonably well with gridded data; subsurface depth estimations from limited information can be done better by grid data for there is usually no information as to natural break conditions. Simple, regular, nearly plane surfaces are better handled by grids, but dissected fluvial landscapes are apparently successfully modelled by triangulated irregular networks. Glaciated landscapes may indeed be best represented by neither technique, but instead by fractal geometry.

6.7 IMPLICATIONS FOR SPATIAL INFORMATION SYSTEMS

Evaluations of the merits and limitations of different ways of organizing spatial data or representing phenomena are not straightforward because the success of one approach depends on the requirements to be met. Indeed, at this stage in our presentation of concepts, it is inappropriate to do more than raise the issue because we have not yet dealt systematically with processing operations applied to the data. Instead, let us summarize some of the properties of different classes of representation, and identify some criteria and issues that should be considered in general.

Regular tessellations structures clearly are oriented to the characteristics of space, not the character or location of spatial objects. They, therefore, do not deal directly with topological spatial relations, although geometric character and relations can be incorporated into cellular units. Depending on resolution, cell representations will not record curvilinear features or irregular polygon entities as well as explicit coordinate encoding or parametric expression. Irregular tessellations do use metrical information for partitioning space, but for some phenomena are (sometimes gross) approximations. Discretization, which can be applied to linear or two- and three-dimensional phenomena (perhaps on the basis of attribute data, not just spatial as in the case of topological encoding), produces many connected elements in conceptually whole entities.

Generally, there is no conceptual reason for not combining geometrical and cellular representations of line, area or volume objects. In practice, there are several possibilities. Point spatial units can be stored in grid cell or quadtree blocks or rectangular blocks produced via binary or quaternary subdivision. Polygons can be represented as bounding rectangles which in the case of an irregular tessellation, will produce many overlapping rectangles, for which R-tree structures are possible, although not too suitable. Other uses of rectangles for representing and retrieving spatial entities is discussed in Chapter 7.

Moreover, in any representation directly treating entities rather than space, assumptions are made and approximations occur. The latter may include smooth curves fitted to irregular boundaries, non-smooth features represented by fractional geometry, and changes in spatial dimensionality. The entities may not fit together well in planar tessellations or graphs. A hierarchical organization may be necessary, for both entity and cellular approaches, to deal with varying spatial scales, identifying the smallest common spatial units, and for dealing with containment.

Structural properties may be more important than metric conditions, so that graph and topological map organization become more important than capturing precise coordinates. Only some entity-oriented data models deal explicitly with topological properties of adjacency and connectivity. On the other hand, regular cellular structures deal explicitly with access to entities in space, although different varieties of the same approach have different properties.

Tessellations and entity-oriented representations are alternatives to direct treatment of continuous variabilty in space. Not restricted to natural phenomena, surfaces avoid the arbitrariness of many discretizations, although the surfaces can have discontinuities. Concepts such as gradient, dominant orientations of flows, accumulated accessibility, spatial differentiation and geodesic paths can be worked with empirically via lattices or irregular distributions of points; discretizations may be quite inappropriate. However, in many problem domains several spatial properties may be necessary, perhaps requiring a mixture of network, tessellation and field space concepts.

Changes in hardware technology for spatial data capture, storage, processing and output have led to dominance of one form or another of data structuring at different times, although today the great range of devices available provides much flexibility in encoding and storing data. Much of the early period of computer mapping, in the late 1960s, and the time of the birth of working computer based spatial information systems, was dominated by grid-cell data concepts and practice, reflecting the limited power of computer processors, and the relative ease of use of printer-oriented output devices, but also a realization of the expense of storing large quantities of data.

The 1970s saw a swing to vector representations through the appeal of line drawing pen-plotters and vector graphics display monitors. Concepts of topological structuring were refined in the 1970s by geographers and cartographers, even while computer scientists were developing the ideas of hierarchical data structures for regular cell spatial data. Triangulated irregular networks and planar topological encoding got much attention in the early 1980s at a time when automated scanning equipment was much more expensive than digitizing devices. Now much more use is made of scanners for data capture, and commercial software for spatial information systems includes quadtree implementations.

Even as the technology dimension becomes background, and the mathematical foundations for topology structuring, hierarchical encoding, and so on become more well known, the cognitive dimension is coming to the fore. It appears that an entity orientation is more natural than thinking of regular partitions of space, but there are gaps in our knowledge about how people process spatial data, the languages used, and the spatial reference systems with which they are comfortable. Mathematical formalisms and high-speed computers may not help at all if people do not know what spatial relationships are.

6.8 BIBLIOGRAPHY

A principal set of references for further reading on hierarchical data structures are the books by Samet. Geometrical principles for tessellations are covered by Gasson (1983).

Bell, Sarah B. M. *et al.* 1983. Spatially referenced methods of processing raster and vector data. *Image and Vision Computing* 1(4): 211–220.

Berry, Brian J. L. 1967. *Geography of Market Centers and Retail Distribution*. Englewood Cliffs, New Jersey, USA: Prentice Hall.

Brassel, Kurt E. and D. Reif. 1979. A procedure to generate Thiessen polygons. *Geographic Analysis* 11(3).

Diaz, Bernard and Sarah B. M. Bell (eds). 1986. *Spatial Data Processing Using Tesseral Methods*. Swindon, UK: Natural Environment Reseach Council.

Dutton, Geoffrey. 1990. Locational properties of quaternary triangular meshes. *Proceedings of the Fourth International Symposium on Spatial Data Handling, Zurich, Switzerland*, pp. 901–910.

Gasson, Peter. 1983. *Geometry of Spatial Forms*. Chichester, UK: Ellis Horwood.

Gold, Christopher M. 1989. Spatial adjacency – a general approach. *Proceedings of the Auto Carto 9 Conference, Baltimore*. Falls Church, Virginia, USA: American Society for Photogrammetry and Remote Sensing/American Congress for Surveying and Mapping, pp. 298–312.

Holroyd, Fred. 1990. Raster GIS: Models of raster encoding. Paper presented at the GIS Design Models and Functionality Conference, Leicester University.

Ibbs, Tony J. and A. Stevens. 1988. Quadtree storage of vector data. *International Journal of Geographical Information Systems* 2(1): 43–56.

Lam, Nina S-n. 1983. Spatial interpolation methods: a review. *The American Cartographer* 10(2): 129–149.

Maryland Department of State Planning. 1981. *The Maryland Automated Geographic Information System*. Baltimore.

Morehouse, Scott. 1987. ARC/INFO: a geo-relational model for spatial information. *Proceedings of the Auto Carto 7 Conference, Washington, DC*, pp. 388–397.

Morehouse, Scott. 1989. The architecture of ARC/INFO. *Proceedings of the Auto Carto 9 Conference, Baltimore*. Falls Church, VA: The American Society for Photogrammetry and Remote Sensing/American Congress for Surveying and Mapping, pp. 266–277.

Peucker, Thomas K. *et al.* 1978. The triangulated irregular network. *Proceedings of the Digital Terrain Models Symposium of the American Society for Photogrammetry/American Congress on Surveying and Mapping, St Louis, Missouri, USA*, pp. 516–540.

Peuquet, Donna J. 1979. Raster processing: an alternative approach to automated cartographic data handling. *The American Cartographer* 6: 129–139.

Peuquet, Donna J. 1988. Representations of geographic space: toward a conceptual syntheses. *Annals of the Association of American Geographers* 18(3): 375–394.

Piwowar, Joseph M., Ellsworth F. LeDrew and Douglas J. Dudycha. 1990. Integration of spatial data in vector and raster formats in a geographic information system environment. *International Journal of Geographical Information Systems* 4(4): 429–444.

Saalfeld, Alan. 1987. Triangulated data structures for map merging and other applications in geographic information systems. *Proceedings of the International Geographic Information Systems Symposium, Arlington, Virginia*.

Samet, Hanan. 1984. The quadtree and related hierarchical data structures. *ACM Computing Surveys* 16: 187–260.

Samet, Hanan. 1990. *The Design and Analysis of Spatial Data Structures.* Reading, Massachussetts, USA: Addison-Wesley.

Samet, Hanan. 1990. *Applications of Spatial Data Structures: Computer Graphics, Image Processing and GIS.* Reading, Massachussetts, USA: Addison-Wesley.

7
Manipulations

Interpolations, geometric operations, transformations

Billy Jackson wants a method to know how to interpolate geological layers between two borehole measures. Christegonde de l'Aigliere wishes to know whether she is inside or outside Amazonia. David Marcowitz wants to go to the very centre of Russia. Maria Schwarzenberg looks to find where she crosses all country boundaries when flying from Berlin to Tokyo. Min Wong wishes to convert her digital remote sensing data to polygons and lines. José María Gallegos plans to build a quadtree out of a digitized map. Jan van Delft wants to update a map of highways in the Netherlands from new air photographs. And Krading Gingging is fixated by a pixilated key-ring.

The purpose of this chapter is to alert the reader to some issues about the manipulation of spatial data and objects, especially the use of computational geometry. When dealing with geometric information, it is necessary to have several procedures to handle spatial entities. This chapter first provides a rapid introduction to the many ways in which to manipulate geometric objects, particularly for undertaking interpolation, finding line intersections, ascertaining if a point is inside a polygon, or finding a centroid for an areal unit. We concentrate on actions which use positional information, but illustrate some which also require topological data.

The chapter also presents a framework for spatial data transformations, including geometrical, topological and between different representations. Most operations which require attribute data for spatial objects are dealt with in Chapter 8; these include algorithms for spatial analysis, using graph data, such as routing and path finding, or polygon attributes, like districting. Spatial map overlay modelling, whether using cell data or polygons, is also dealt with in the next chapter.

We do not aim to demonstrate all categories of spatial data manipulation, nor provide comparative algorithms for specific requirements. Readings in the bibliography, especially Preparata and Shamos (1986), provide extensive treatments of these topics. We believe it is more important to demonstrate the varied ways in which spatial data may have to be dealt with, and the flavour of the approaches that may be used.

7.1 INTERPOLATION AND EXTRAPOLATION

Concentrating on positional information, we present in this and the next section some aspects of geometric inference and computational geometry useful in the domain of spatial information systems:

1. Interpolation and extrapolation.
2. Line and segment intersections.
3. Point-in-polygon operations.
4. Determination of centroids.
5. Operations on polygons.

In these discussions some practical considerations also will be mentioned.

7.1.1 The interpolation and extrapolation concept

For situations in which we have limited information, possibly a planned scientific sample for selected points, lines or areas, or simply just what spatial units are available in a practical sense, it is often necessary to interpolate or to extrapolate in order to get more information, that is, to 'guesstimate' a new value or set of values. We have already alluded to this need in discussing triangulated area networks, and it also arises in allocating values or attributes from one or a set of polygons to others created by overlap; or from one or a set of linear features to others created by coincidence.

For now, we consider only **point-oriented interpolation**. That is, data for one or more points in space are used to produce estimated values for other positions at which it is not possible to record values directly. Several interpolation approaches are demonstrated in Figure 7.1. The horizontal axis is assumed to be a distance scale, although it could also be time or values of an attribute. In sequence the methods are:

1. Nearest value interpolation by which we confer the value of the nearest data point.
2. Linear interpolation, in which a simple straight line based on two points is used.
3. Spline interpolation based on three or more points.
4. Stochastic interpolation based on a pseudorandom number generator and a fractal parameter and several points.
5. Model-based interpolation.

Estimation could also use a set or an average of nearby data points rather than just the nearest. Proximal interpolation in the form of

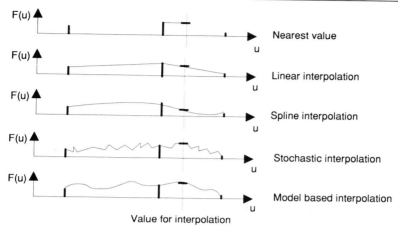

Figure 7.1 Several possibilities for interpolation.

Thiessen polygons has already been encountered in section 6.6.1. Spline interpolation was presented in section 4.5.2, and fractal interpolation in section 4.6.2.

When a more specific simulation of the phenomenon of interest is known, perhaps as a result of a mathematical model or scientific theory, it must be used in order to confer more confidence in the estimation of new values. Thus, a trigonometric function could be used to interpolate from a cyclic progression if a theory suggested a phenomenon occurs with repeated patterns. For every method, a certain level of precision must be established, and each procedure has some inherent limitations of data accuracy.

When we are outside the range of data points, we speak about **extrapolation**. As for interpolation, several kinds of estimation can be utilized (Figure 7.2):

1. Nearest value extrapolation based on the last data point or an average of neighbours.
2. Linear extrapolation based on the two last data points.
3. Spline extrapolation based on several data points.
4. Stochastic extrapolation based on fractal parameters.
5. Model-based extrapolation.

Again, the last is perhaps the more prudent approach and provides more confidence in the results. When extrapolating, as for interpolating, a certain level of precision must be given, although in particular contexts it may be difficult to establish how much error can be tolerated.

In the foregoing context, represented by the diagrams, only one spatial

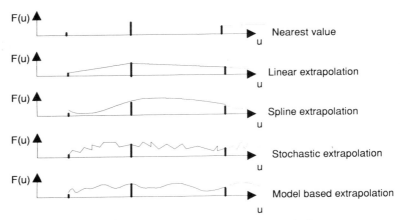

Figure 7.2 Several possibilities for extrapolation.

parameter (coordinate axis direction) was used in the interpolation or the extrapolation. However, there are more usual situations; for example, in terrain modelling, when we estimate new values for a two-dimensional surface based on neighbouring elevation values. Sometimes, we have to interpolate in one coordinate dimension, and extrapolate in the second coordinate dimension. For this type of surface modelling, the term **geometric inference** is sometimes used, meaning that we are in a complex situation characterized by:

1. Two or three space coordinate dimensions.
2. Various methods for selecting data points.
3. Either interpolation or extrapolation or both.
4. Selection of the type of procedure (linear, spline, model based, etc.).

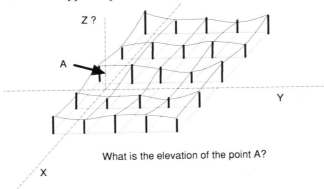

Figure 7.3 Inferring a new elevation point in a gridded terrain.

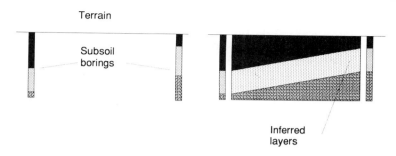

Figure 7.4 Inference of geological layers from borings.

This field-oriented approach is illustrated assuming the estimated variable z is elevation above sea level, or depth below a vertical datum. However, while other attributes, like temperature or average income can be imagined, the predicted attribute must be in the form of scalar values, not as categorical or ordered data.

Digital terrain or digital elevation models which give the observed or recorded elevation for some **privileged points**, for example along a grid or along contour lines, are often used in order to infer values for some new points (Figure 7.3). In this case, the observed values are at a regular grid spacing, and interpolation occurs along both coordinate axes, perhaps using a mathematical model like a high order polynomial. Another example, from geology (Figure 7.4), is the inference of the subterranean layer structure from borings. This time there is two-dimensional estimation, but also, in practice, the inferences are made from very few points. In general, geological interpolation or extrapolation is very difficult because there is usually very little recorded data, and often considerable real world variability even if mathematical models might be considered as appropriate for a given context.

7.1.2 Some practicalities

Sometimes the privileged points are irregularly distributed in space, and the elevations are estimated for the intersections of grid lines or a lattice of points. In this context it is most important to make a prudent selection of data points. Conceivably, all point samples could be used to obtain a value for z for positions at which no observed values exist. At the other extreme, the estimate can be based on only one known value, the nearest in space. The data points used can be chosen on the basis of several criteria:

1. The number of data points to be used for each lattice position.
2. The distance away from the lattice position.
3. The location within a compass direction away from the lattice position.
4. The interpolation or extrapolation procedure used.

The nearest value, straight line and spline interpolations use a restricted number of data points, but even so, each one of these could itself be an average of a set of points. In any event, if a procedure can use a large number of points of observed values, then the decision as to how many to use is based partially on the extra accuracy benefits that may arise relative to the effort needed to process more data. For example, for a set of 3,100 points representing counties of the USA, an estimate for a location in California is not likely to be much influenced by conditions in Florida, represented by about 100 points at a considerable distance from California.

Generally speaking, nearby data points are given more weight than those further away, and there may be a limit to the spatial range used. Sometimes only immediate neighbours are used unless some compass directions are not represented in the set of data. In this case, say, for four or eight compass sectors, each must contain at least one point of data even if it is positioned a long way from the lattice position, in order to mitigate possible effects of spatial bias.

Consequently, in geometric inference it is important to be sensitive to the facts that:

1. The number of sample points may be an unrepresentative spatial distribution.
2. There is a condition of spatial autocorrelation; that is, places closer together in space tend to be more similar than different.

The first statement means that attempts must be made to cover all space in a study area with a high enough sample of data points to reflect its nature. Otherwise some natural variability may not be detected, or the estimated values will be systematically biased or have an unacceptably high range of error. The second, the condition of **spatial autocorrelation**, implies that nearby data points receive the highest weight in averaging or modelling type interpolations, or that points beyond a certain distance can be eliminated. One modelling technique developed by mining geologists, **kriging**, specifically uses information for the spatial auto-correlation, the numerical measurement of the degree to which points different distances apart in space are alike in regard to an attribute.

Interpolation of an attribute over space will naturally be more reliable if more good data are used, but the outcome is also affected, often substantially, by the procedure used. Dealing only with point interpolations still, important considerations are:

1. If the procedures are local or more extended spatially.
2. If they produce an outcome that honours the input data.
3. If the reality being modelled is broken or has barriers.

Some techniques are local in orientation, that is, they are applied systematically to different parts of a study area, using data limited to one point or gathered from a neighbourhood, as for proximal mapping and piecewise spline patches. Other techniques fit a continuous surface to a set of points scattered over space, as with trend surface analysis using functions like polynomials or sines, or consist of integrative measures of accessibility or space potentials. Such functions use all data points, that is they are **global interpolators**. Otherwise, they may use data from a large area, referred to as regional interpolators. Local processing has advantages of being faster because fewer data are needed, but global operations may provide better results under some conditions.

Some procedures will produce estimated values for the positions of observed data that are not the same as those observed values. Exact interpolation produces a surface that honours all known values, but some procedures fit a global or regional trend that is not constrained to the local circumstances. Kriging and B-splines ordinarily are exact interpolators; polynomial trend surfaces and moving averages are not.

As noted in earlier chapters, the interpolation may be made to recognize natural conditions like breaks in slope or barriers. The latter, usually either linear or areal in geometry, include impenetrable features like geological faults, some political boundaries, or inhospitable terrain; or permeable barriers like landscapes of different ease of transportation. Geometrically speaking, trend projections would have to be truncated or weighted, respectively, to recognize these two contexts, in order to produce abrupt or gradual changes. The latter may be easier to deal with in practice, but many situations are contrary to unbroken continuous space, if only because of the existence of anisotropicity.

Spatial interpolation may also be desired or necessary in cases of linear or areal phenomena. Examples encountered earlier include comparisons of spatial units with different boundaries or extents at different points in time, or an allocation of attributes such as the number of voters per constituency to census tracts. In addition to proportional area techniques, areal interpolation can use the intermediate steps (as discussed in section 7.3.4) of representing polygon data by sets of grid cells and then producing a continuous surface.

7.2 BASIC OPERATIONS ON LINES AND POINTS

Several of the most common needs in spatial information systems, access to phenomena encoded as entities, or determination of overlaps that result from map overlays or geometric window searches, utilize just a few basic, but very important, operations on the spatial data primitives of points and lines.

7.2.1 Line intersections

There are several contexts in which it is necessary to find the intersection of line entities. Not only is it necessary for determining topological junctions from separately digitized line features, or for finding where a line meets a map edge, but also for solving the important polygon overlay problem. We can differentiate between two contexts: one in which lines are of infinite length, and may or may not cross; the other in which there are straight line segments terminating in nodes or vertices and may or may not intersect. For representation, there are many mathematical expressions to obtain the locus of a line.

To represent a straight line in the two-dimensional Cartesian space, a **linear equation** is generally used:

$$ux + vy + w = 0$$

Some people prefer $y = a + bx$ where a is the intercept and b the slope. Unfortunately, this technique produces some difficulties in representing vertical or horizontal lines parallel to the coordinate axes, in particular because the slope in this case is zero.

For representing segments, we need to know the end-points x_A, y_A and x_B, y_B (Figure 7.5a) and the line equation. Using the linear equation cited above, where

$$u = y_B - y_A$$

$$v = x_A - x_B$$

$$w = y_A x_B - x_A y_B$$

and, considering another line, passing through points C and D, so the coordinates of the intersection are:

$$x_i = \frac{(y_C x_D - x_C y_D)(x_A - x_B) - (y_A x_B - x_A y_B)(x_C - x_D)}{(y_B - y_A)(x_C - x_D) - (y_D - y_C)(x_A - x_B)}$$

$$y_i = \frac{(y_A x_B - x_A y_B)(y_D - y_C) - (y_C x_D - x_C y_D)(y_B - y_A)}{(y_B - y_A)(x_C - x_D) - (y_D - y_C)(x_A - x_B)}$$

These equations can constitute the **point-in-line rule** which is used in Chapter 10.

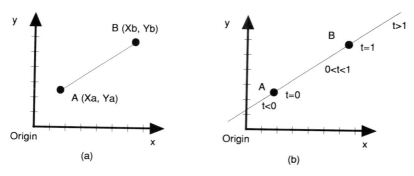

Figure 7.5 Segment represented by end point coordinates and parameter. (a) Segment represented by end-point coordinates. (b) Segment represented by end-point coordinates and parameter.

7.2.2 Segment intersections

It is rare in geomatics to have to deal with an infinite line. Generally speaking, we have line segments and for representing them we must have:

1. The line equation.
2. The end point coordinates.
3. A way to refer to only between those two end points.

The linear equation solution given above is not very convenient for these conditions. Another possibility is to use the so-called **parameter expression**, that is, have a parameter t varying from 0 to 1 for the relevant portion of the line (for example, Figure 7.5b), and take a value of 0 at one end and increase to 1 at the other. So we have:

$$x = x_A + t(x_B - x_A)$$

$$y = y_A + t(y_B - y_A) \qquad \text{where } 0 \leqslant t \leqslant 1$$

Concerning the intersections of two segments (Figure 7.6), suppose that we have two segments AB and CD. Using t and s to denote parameters, the corresponding equations are as follows. For the first segment:

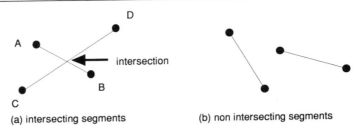

(a) intersecting segments   (b) non intersecting segments

Figure 7.6 Intersection of segments. (a) Intersecting segments. (b) Nonintersecting segments.

$$x = x_A + t(x_B - x_A)$$

$$y = y_A + t(y_B - y_A)$$

For the second segment:

$$x = x_C + s(x_D - x_C)$$

$$y = y_C + s(y_D - y_C)$$

At the intersection the parameters t and s are given by:

$$t = \frac{(x_C - x_A)(y_C - y_D) - (x_C - x_D)(y_C - y_A)}{(x_B - x_A)(y_C - y_D) - (x_C - x_D)(y_B - y_A)}$$

$$s = \frac{(x_B - x_A)(y_C - y_A) - (x_C - x_A)(y_B - y_A)}{(x_B - x_A)(y_C - y_D) - (x_C - x_D)(y_B - y_A)}$$

provided that $0 \leqslant t \leqslant 1$ and $0 \leqslant s \leqslant 1$.

In order to compute the intersection for two line segments, we have two possibilities:

1. To compute the intersection point of both supporting lines and to check whether the point lies in the segments.
2. To use the parametric equations, checking whether both parameters, t and s, range from 0 to 1.

This second method constitutes the **point-in-segment rule** which will be encountered again in Chapter 10. Sometimes known as the point–vector form, this representation for a straight line has valuable properties of being able to be used for lines and line segments, and does not require special treatments for vertical or horizontal lines parallel to coordinate axes.

For either infinitely long lines or abbreviated line segments, identification of lines that may cross can conveniently take advantage of a simple initial procedure that compares the coordinates for the corners of enclosing rectangles. (This concept is discussed in section 4.2.2.) If the bounding rectangles do not overlap at all, then the lines contained within them cannot possibly intersect. If there are many lines or segments, then preliminary sorting in ranges of x and y coordinates can help the process of finding possible intersections, even if enclosing rectangles are not used.

7.2.3 Point-in-polygon procedure

An important geometric operation, known as the point-in-polygon procedure, is to ascertain if a discrete point, or the end or vertex of a line, lies within a particular polygon. Of the several algorithms that exist, the most well-known is based on the half-line (Jordan) theorem. Let there be an (x, y) point and a polygon (for example, Figure 7.7). In order to know whether the point is inside or outside the polygon, we generate a half-line starting from this point and extending to beyond the range of the known extreme coordinates for the polygon. We then count the number of intersections of this half-line, a line projected in only one direction from the point, and the polygon boundaries. When this number is odd, the point is inside the polygon, and when even, the point is outside. The procedure incorporates a stage of counting the line intersections. In order to facilitate this procedure we ordinarily use half-lines parallel to the x-axis.

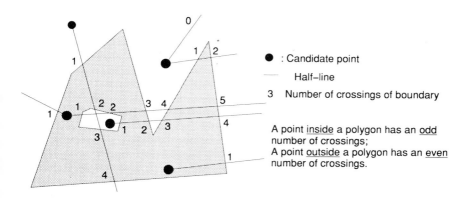

Figure 7.7 Illustration of the half-line theorem (point-in-polygon rule).

The procedure works well for so-called special cases of island polygons, polygons with holes or concave polygons. Some decision rule is needed, though, if a point falls exactly on a boundary line, or when it falls exactly on a vertex, or when a segment is collinear to the half-line. For all these situations, when the point originates from a mouse device, we usually move it slightly. When the half-line is parallel to one Cartesian coordinate system axis the algorithm for counting intersections is simpler, as explained below in the discussion of the segment-in-polygon algorithm.

In order to speed up this point-in-polygon algorithm, especially when the polygon has many edges, the minimum bounding rectangle, also called an extent, is used. Indeed, it is easier to check whether or not a candidate point is inside or outside an extent rather than for a complex polygon. When the point is outside the enclosing rectangle, it will be outside the polygon. When it is inside, though, it is necessary to apply the Jordan theorem. This speed-up procedure is usually used in database retrieval since the probability of a point being inside is very low. This algorithm will be used for the **point-in-polygon rule** (discussed again in Chapter 10).

Generally speaking, the problem is not to find whether a point lies inside or outside a single polygon, but to determine as to which spatial unit of a tessellation the point belongs, that is a **point-in-tessellation test**. A possibility is to use repeatedly the point-in-polygon algorithm for each polygon; but this solution is very time consuming. Suppose one has 1,000,000 polygons, then there is an average of 500,000 tests to perform. In other words, the search time, the time taken to find the polygon in which the single point lies, is linear; that is, it is proportional to the number of units. However, in order to undertake this task more rapidly, two acceleration techniques exist:

1. To evaluate the extent before the entire polygon is tested.
2. To organize polygons hierarchically.

The **hierarchical organization** functions as follows. Suppose we wish to know in which town in the USA an x-y point, perhaps representing a natural hazard, belongs. A naive algorithm is to apply the point-in-polygon test to all polygons representing towns. Another possibility is first to test the states, and when a state is found, to test a county, and so on. If there is even just one level of hierarchy, the number of separate tests is substantially reduced. The N polygons can be regrouped into $\sqrt{N}$ group of $\sqrt{N}$ towns, so the average number of tests is $\frac{1}{2}\sqrt{N} + \frac{1}{2}\sqrt{N} = \sqrt{N}$. Compared to the complete testing situation for our example of 1,000,000 polygons, now we have but 1,000 tests to perform instead of the expectation of 500,000. We can thereby expand the number of hierarchy levels. For example, at three levels the average number of tests is 150, at four levels there are only 20 tests, and so on.

Point-in-polygon testing occurs for several types of need in spatial information systems work. It arises most commonly in comparing the position of a point entity relative to one or more polygons, for example, a school within a county, or a television tower in an airport zone. In a database query context it arises when the cursor is used to pinpoint a geographic location for obtaining information about a polygon drawn on the monitor screen.

The general concept of point-in-polygon can be extended to also cover point-in-volume, point-on-line or line-in-polygon. The last of these requires more tests, for there are two end points of a line segment to be evaluated, and a line segment could lie outside, completely inside or partially inside a polygon (section 7.3.1).

7.2.4 Centroid definition

At various times we need to represent a polygon (or a line) by a single point, perhaps for choosing a position for map labelling purposes, or perhaps for simplifying distance measurements to other objects. Let us imagine we wish to determine a position, called a **centroid**, which is approximately in the middle of a polygon. There are several definitions, but no one is substantially better than others:

1. Defined from vertices.
2. Obtained as a statistical bivariate median or the centre of gravity.
3. Computed as the centre of an enclosing or enclosed rectangle or of an enclosing or inscribing circle.
4. Obtained as the peak value of a surface fitted within the polygon.
5. Chosen intuitively.

The first solution is to define centroid coordinates (x_c, y_c) as the **average of all vertex coordinates** (Figure 7.8a):

$$x_c = (1/N) \Sigma x_i$$

$$y_c = (1/N) \Sigma y_i$$

As seen in Figure 7.8, when a polygon is very rugged on one side, in other words, the number of vertices is very high in some parts, the vertex centroid shifts towards those vertices. In order to avoid this condition, instead of having a vertex centroid, the centre of gravity can be employed (Figure 7.8b), or a centre can be computed arithmetically from the corners of an enclosing rectangle or circumscribing circle, or from an inscribed circle or other regular figure. Of course, additional computational steps are necessary if centres are determined for fitted figures compared to just using coordinates data as is the case of the average of vertex coordinates.

The previous definitions hold when the polygon has no big **concavities**. In fact, in this case, a vertex or a gravity centroid can be outside the polygon and we have to move it into the interior (Figure 7.8c). Similarly, when the centroid falls inside a hole within the polygon, then it again should be moved. It is in cases such as these that a visual interactive approach to positioning the central point can be useful, or a computational procedure can be used. A very crude algorithm to calculate a centroid in this context is first to compute the centre of gravity and then, using the half-line method, test whether this point is inside or outside the polygon. If so, this point becomes the centroid; otherwise the point is moved along the line which has the longer segment forming the intersection with the polygon.

7.2.5 Some spatial statistics based on point data

Distances between two points may be readily obtained from the Pythagoras' theorem:

$$\text{Distance} = [(x_B - x_A)^2 + (y_B - y_A)^2]^{\frac{1}{2}}$$

where A and B designate the two points separate in space, or at the ends of a straight line. The formula is readily extended to other metrics, especially the Manhattan distance in which the two axial distances are summed as in:

$$\text{Manhattan distance} = [|x_B - x_A| + |y_B - y_A|]$$

That is, using absolute differences, the length between points in the two axial directions.

If distances are measured along lines or links made up of more than one straight segment, then we have:

$$\text{Length} = \sum_{i=1}^{N-1} [(x_i - x_{i+1})^2 + (y_i - y_{i+1})^2]^{\frac{1}{2}}$$

For distances from points to lines or polygon boundaries it is necessary to find first the closest line segment and then measure distance. This process may require testing against each vertex on a line, especially if it is

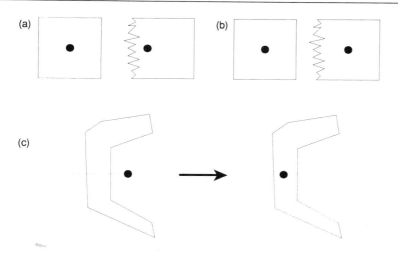

Figure 7.8 Definition of centroids. (a) Examples of centroids. (b) Examples of centroids defined as centres of gravity. (c) Moving an outside centroid.

complex, in addition to using an intersection algorithm to yield a point which is not a vertex.

The nearest neighbour statistic, a common measure for point entity distributions, requires that distances from a point to its nearest neighbour, and possibly second and higher order neighbours, be known before computing the summary statistic based on all points in the distribution. Used at times for evaluating the uniformity in a set of points prior to using geometric inference, the statistic compares the observed distribution in space with a theoretical distribution based on a process producing a random spacing pattern.

Distance measures are also involved in the computation of various spatial statistics related to the centroid or median position in two space dimensions. Ring and sector counts are based on distances from a bivariate centre, and a bivariate standard distance deviation, akin to the standard deviation, is a summary statistic of some utility.

7.3 SOME OPERATIONS FOR POLYGONS

Comparisons of different discrete polygons may require overlap of their boundaries, overlapping a rectangle or other geometric figure, or comparing a line with a polygon boundary or interior. The first situation constitutes operations which are very common in polygon overlay, particularly unions and intersections. The second represents the box clipping problem. Other operations on single polygons viewed as separate entities include area computation, creating boundaries of buffer zones inside or outside a polygon and parallel to its boundary, and measuring shapes.

7.3.1 Intersection of lines with polygons

A common situation comprises the intersection of a segment with a polygon, or in the opposite way, the computation of what part of the segment lies inside a polygon (Figure 7.9a). In order to determine the segment intersection, we can compute the intersection of the straight line for the segment and the line segments for the polygon boundaries. If there is no intersection with the boundaries, there is no intersection of the segment and the polygon. If there are intersections of the straight line and the polygon, we can determine what parts are inside by a repetitive use of the Jordan theorem.

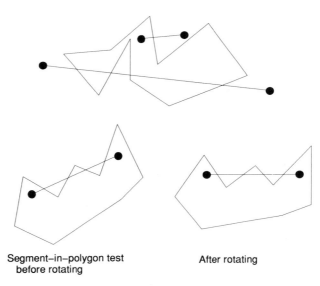

Segment–in–polygon test   After rotating
before rotating

Figure 7.9 Intersection of segments with a polygon. (a) Intersection of segments with a polygon. (b) Solving the segment-in-polygon problem using the Jordan theorem.

For checking whether a segment is totally included in a polygon, it is not sufficient to test end-points because the polygons can have strange concavities intersecting this segment. Obviously we cannot examine the infinity of segment points to see whether they are all inside or outside the polygon. In order to facilitate the testing and to re-use the result of the Jordan theorem, a nice step is to rotate the polygon so that the segment under study is parallel to an axis (Figure 7.9b). After this, it is easy to count the number of intersections only by comparing coordinates. If the number is different from zero, we know that a part of the polygon is inside and a part is outside. By re-using the half-line procedure on the end-points when the number of intersections is zero, we then know whether the end-points are inside or outside, and, consequently, whether the totality of the line segment is inside or outside.

7.3.2 Union and intersection of polygons

One of the major needs and challenging problems in spatial information systems is to compute the difference, union and the intersection of polygons. For example, considering two polygons A and B (Figure 7.10a), we need to find their union; that is, their joint extent (Figure 7.10b), and their intersection, the common area (Figure 7.10c). Drawing on Preparata and Shamos (1986), who present several union and intersection algorithms, we present only possibly the simplest way, based on the **slab technique**. Each polygon is divided into parallel slabs, usually horizontal for the convenience of parallelism with the coordinate axis, as shown in Figure 7.11, created by drawing lines through the polygon vertices. This procedure creates trapezoids which are easy to compare. Figure 7.12 gives some varied examples to illustrate this method. When the edges are not intersecting inside the slabs, the comparison is

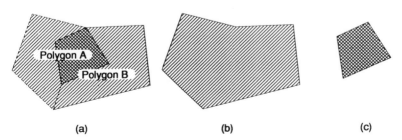

    (a)   (b)   (c)

Figure 7.10 Union and intersection of polygons. (a) Two intersecting polygons. (b) Union of A and B. (c) Intersection of A and B.

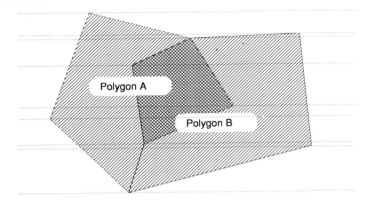

Figure 7.11 Splitting two polygons into slabs.

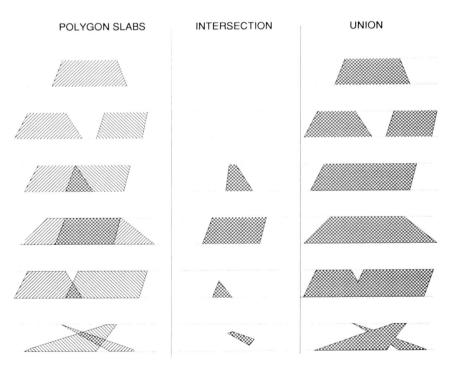

Figure 7.12 The method of slabs for finding the union and intersection of two polygons.

straightforward; otherwise, as in the last row in the diagram, some other slabs can be created passing through these intersection points.

7.3.3 Area computation

Area computation sometimes uses the trapezoid idea although a procedure using geometric cross-products is usually preferred. Lines perpendicular to the x-axis (alternatively, the y-axis could be used) are dropped to it from the vertices of polygons. This process creates a set of overlapping figures, with four line segments: the base on the x-axis, two vertical lines and an edge of a polygon. After the area of each trapezoid is computed, they are summed, including subtractions for the lower pieces. Enclaves and exclaves can be handled by this procedure by using encoded data showing the gaps or separate pieces of the polygons.

However, the classical way in computational geometry is to use the **geometric cross-product** to compute areas, and mixed products for volumes. Here we provide the procedure for only areas. For computing the area of any polygon with or without holes, if we have in total N vertices ordered in a counterclockwise sequence from 1 to N, the area is given by:

$$\text{Area} = \tfrac{1}{2}\left(\sum_{i=1}^{N-2} (x_i y_{i+1} - x_{i+1} y_i) + (x_N y_1 - x_1 y_N)\right)$$

This process, established via vector algebra using the cross-products, is more rapid than the trapezoid decomposition. The dot product device is also useful for obtaining angles between vectors, and for concatenating boundary lines for polygons using a centroid as origin for the vectors.

7.3.4 Areal interpolation

An additional interesting situation that arises for polygons is the allocation of attributes from one polygonal representation to another. This need often arises in comparing census demographic and other data for different years for spatial enumeration units that have changed between censuses. Assignment of values is easy if there is a nested hierarchy of polygon units because systematic aggregation and decomposition can be used. But, if there is overlap, then allocations have to be made for the pieces created by overlapping the polygons. The areal interpolation, a variant of point or linear based interpolation, or the transfer of attributes from one area spatial unit basis to another, may be based on any of the following:

1. Proportionate areas, assuming uniform densities of phenomena.
2. Other, correlated data.
3. A continuous surface representation of an attribute.

Not only is it necessary to create new polygons and undertake area measures by procedures described above, but it is also necessary to allocate the attribute values associated with each original polygon to the new piece. The simplest procedure, based on proportional areas, finds the area of a new polygon resulting from the intersection operation, and then allocates the attributes of the two original polygons proportional to the percentage area occupied by the new polygon. If the fortunate occurrence of another attribute distribution matching the new polygons arises, then this other variable can be used to estimate, assuming a statistical correlation exists for the two attributes.

Alternatively a field orientation, rather than a discretization implied by the use of polygons, may be a more appropriate method. In this case a surface estimation procedure will have to be used first, followed by aggregation of point estimates taken from that continuous surface. The different possibilities take advantage of the concept of a regular grid or lattice as an intermediate step.

Using an example of human population distribution, one method allocates the housing density (computed from number of houses divided by polygon area) for each areal spatial unit to its centroid, interpolates from these irregularly distributed points to a set of grid cell centroids, reconverts to the number of housing units from cell density, and aggregates to the zones for the second set of irregular areal units. An alternative procedure assumes that there will be no stepped function for the global surface and that the sum of attribute values in a zone must equal the given total. So, a cell's value is replaced by an average of its neighbours, and the sum of all cell values is adjusted to the total of the original zone. An iterative process will lead to convergence on the control values for all zones.

7.3.5 Shape measures for polygons

The shape of landscape features is an important part of the pattern recognition techniques used in image processing. Shape properties are sometimes recorded for natural phenomena for scientific purposes, especially through an approach that relates form to function, as for different types of glacial deposit or classification of coastal barrier islands. Shapes of land cover of different kinds can be quite important for wildlife, for agricultural practices or for the layout of recreational and parkland areas. For example, long thin strips of open space can assist the movement of animal spaces through otherwise inhospitable terrain.

If only because of the concept of 'gerrymandering' associated with political redistricting, it is worthwhile to examine the measurement of shape properties of polygons. It is for supporting allegations of racial or partisan bias in drawing boundaries of new electoral districts that statistics for shape regularity or indentedness are commonly encountered. (The term gerrymandering is based on the name of Governor Gerry of Massachusetts and the form of the salamander, a shape perceived by some challengers to resemble a rather strangely formed district.)

Shape and compactness measures may utilize sets of parameters defining particular properties of areal objects, or may use data for sets of line segments fitted to the boundaries. For the former, area, perimeter, longest axis, shortest axis, centroids, radius of circumscribing or inscribing figures are encountered – shape indices are computed from two or more of these basic measures. The difficulty in producing a single measure reflects the multi-faceted nature of shape, involving conceptually distinctive elements of elongation, compactness, puncturedness and fragmentedness.

We present just two methods to illustrate the underlying computation (Figure 7.13). Firstly, the perimeter of a polygon, obtained by summing the lengths of line segments, is compared with the perimeter of a circle of the same area as the polygon. Although different classes of shape can produce the same statistical value, the measure does pick out the irregularity in figures. Secondly, again comparing with a regular figure of a circle, the departure from the perfect form can be detected by the length of radials drawn from the centre of the areal unit, as illustrated. This procedure is computationally more involved than the other, because line intersections must be determined, and requires finding a good centroid.

7.3.6 Polygon clipping

A clipping situation occurs commonly when displaying or retrieving spatial entities within a defined area like a rectangle. When one has a rectangular window on a screen or on another graphic device, it is necessary to know what part of the object (say, a house) has to be displayed on the screen. Similarly, when using enclosing rectangles for overlay purposes, particular polygon or line objects may be cut. Three situations arise, illustrated for the display context (Figure 7.14):

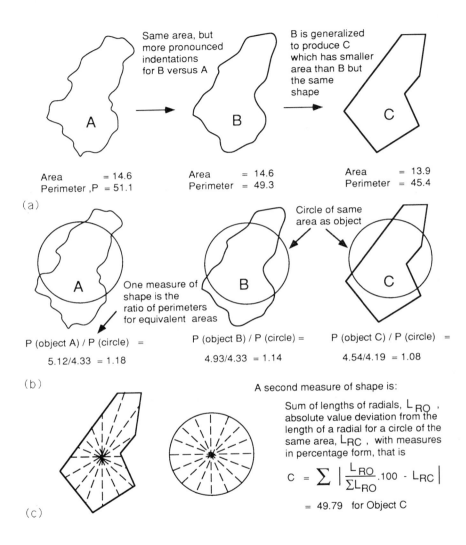

Figure 7.13 Measures of compactness. (a) Influence of generalization. (b) Compactness I Index: Simpler index based on comparison with an equivalent area circle. (c) Compactness II Index: More complex index requiring line inter-section computations. (From Thompson et al, reference cited in Chapter 8.)

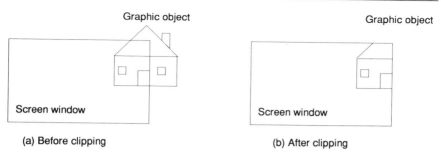

(a) Before clipping   (b) After clipping

Figure 7.14 Clipping of an object by a rectangular window.

1. If the object is completely inside the window, it must be totally displayed.
2. If the object is totally outside the window, there is nothing to be done.
3. If the object is overlapped by the window, the part to be drawn must be computed and displayed.

In order to give just a flavour of a clipping algorithm, only the problem of segment clipping will be presented. In this algorithm, the two-dimensional space is divided into nine subspaces by projection of the rectangle bounding lines, and Boolean codes (binary digits) are conferred to the segment end points based on location relative to the two horizontal and vertical axes.

The first bit must be set (to 1) when the extremity is above the top line, the second when below the bottom line, the third when more at the right than the right line, and the last more at the left than the left line (Figure 7.15). Once this encoding is done, we test all extremities and the clipping must happen; in other words, we can compute the segment part to be displayed. As an output of this algorithm, the segment can be totally displayed, removed or partially displayed.

In the situation of clipping a polygonal area, it is necessary to follow both the edges of the area and those of the window, requiring an algorithm a little more complex than that just illustrated.

7.3.7 Buffer zones

One other class of spatial operations includes the creation of boundaries, inside or outside an existing polygon, offset by a certain distance, and parallel to the boundary. Referred to, respectively, by the terms **skeleton** and **buffer zones**, these new units require distance measures from selected points on the boundary of the polygon (Figure 7.16a). The skeleton is

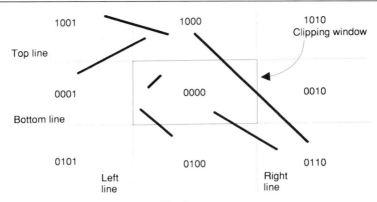

Figure 7.15 Clipping a segment.

akin to contracting the polygon by moving straight line segments in, parallel to their original position, and the buffer has lines moving outwards. Skeletons (Figure 7.16b) have some value in assisting labelling operations for polygons or edges; buffers are much used in spatial analysis and modelling.

For example, as demonstrated in Figure 2.18, buffers are used to establish critical areas for analysis or to indicate proximity or accessibility conditions. In contrast to the simpler world of proximity values for grid cell data, for vector data, computational geometry operations are required to establish buffer zones. If the concept of gradation in accessibility is required, then a sequence of buffers at selected increasing distances must be created. Similarly, for a zone within two specified distances, two new buffer boundaries have to be determined (Figure 7.16c). If a vector data representation is required, then this buffering operation is best done as a batch or background process because of the time it can take. In contrast, the equivalent operation for regular tessellations can be done much faster because it does not need to employ time consuming computational geometry operations.

7.3.8 Polygon overlay process

The matter of polygon overlays as a whole is perhaps the most challenging computational requirement for vector type spatial databases. We have already shown the need for some line segment intersection procedure; areal measures are often required too, and new spatial units have to be created. The general process of overlay to create new polygons consists of:

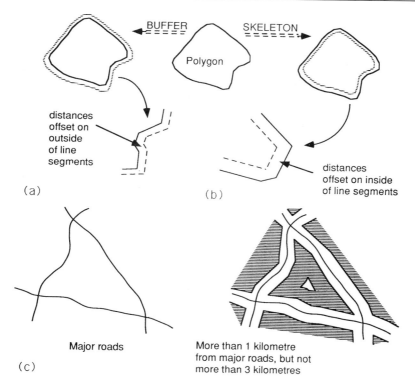

Figure 7.16 The concepts of buffers. (a) Buffer zones. (b) Skeleton zones. (c) Example of a buffer zone for major roads.

1. Identifying line segments, preferably having topology.
2. Establishing minimum enclosing rectangles for the polygons.
3. Ascertaining if line segment(s) of one polygon are inside a polygon of the overlay map by a point-in-polygon process.
4. Finding intersections of segments representing boundaries.
5. Creating records for new line segments and their associated topology.
6. Assembling the new polygons from the appropriate line segments.
7. Relabelling polygons and reallocating the attribute data if appropriate.

Taking a somewhat simple case as an example (Figure 7.17), after rectangle overlap is found, then a point-in-polygon test procedure can be used to determine if line segments fall in particular polygons. After the vertex of intersections is computed for the selected lines, the left- and right-side identifiers are applied from the original segments to the new segments; and lines making up the new polygon can be assembled. For a

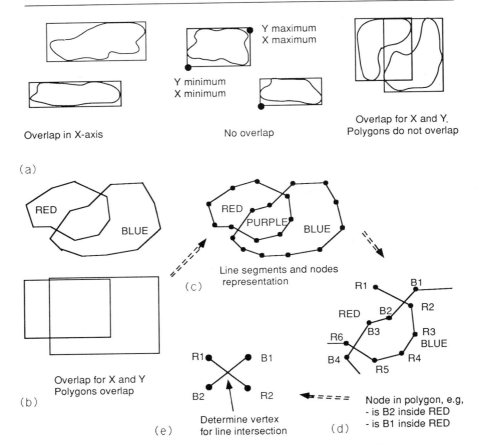

Figure 7.17 The polygon overlay process. (a) Overlap conditions for enclosing rectangles. (b) Example of overlapping polygons. (c) Node and line segment representation. (d) Point-in-polygon testing. (e) Finding vertex for intersecting line segments.

completely enclosed unit, a point-in-polygon test can be used to see if the single node lies within a larger polygon or not, unless boundaries data for ascertaining containment are explicitly encoded.

While there are ways to speed up the entire process, it is still computationally demanding of resources when there are many polygons. The time taken is influenced by the number of polygons, the number of line segments, and the complexity of the lines. Additional complications may be the need to resolve if lines touch or cross, to assess the reliability and validity of small polygons or sliver polygons created via the overlay process, and the uncertainties of use of imprecise boundaries or tolerance values.

Sliver polygons (Figure 7.18a) must be examined and eliminated, perhaps by removal of the longest shared boundary, perhaps by a line midway between the others. Slivers may be detected, other than visually, by being small areas; having elongated shapes, as indicated by a large value for the perimeter-to-area ratio; or having only two boundary lines. Unreliability in location of new boundary lines may cause a conflict with other spatial entities, especially point features, which might now be positioned on the wrong side of a line. Such a change could have disastrous results, if, for example, a major obstacle to aircraft is shown on the wrong side of a highway. Or, one boundary, for example a river or political unit, may be unalterable in the interest of preserving 'truth'. For line entities, coincident features may have representations which have some discrepancies (Figure 7.18c).

It is also difficult to deal with the identification of containment for new polygons created by overlay processes. This is illustrated in Figure 2.17 as part of a process to create critical zones at one thousand feet from water. The buffer polygon holes were not detected directly as holes. Without special coding techniques, it is possible to detect these only visually and then take corrective action by modifying the tabulated data designating position within a buffer zone.

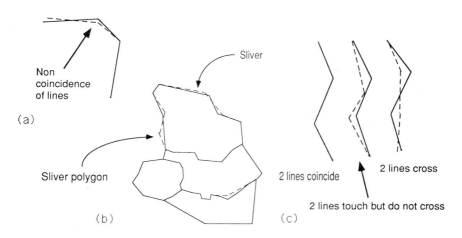

Figure 7.18 Polygon or line discrepancies. (a) Noncoincidence of lines. (b) Sliver polygons. (c) Some possibilities of line coincidence.

From the point of view of building a digital database, the question arises as to whether line intersections, especially for polygon overlaps, should be precomputed and stored for future use, or computed only when needed. The former is better for improving the integrity of the data, but may not be possible, especially for buffer zone polygons, if the new polygon phenomena are not known in advance. A query mode of use of a spatial information system can produce many unanticipated needs for buffer zones.

7.4 SPATIAL DATA TRANSFORMATIONS

There are many changes in the state of information in the process of creating and using a spatial information system database. There are changes in media as going from paper map to digital encoding to display on a monitor. There are changes in form, as when going from tables of numbers to graphical representation. There are changes in representation of entities, as when going from vector geometry to grid cells. Some alterations occur in conceptualization; others in digital encoding; others in product creation.

We may consider at least the following kinds of **data transformation**:

1. In changing the dimensionality of an object.
2. In recording the position of entities.
3. In generalizing spatial entities.
4. In the map space and reference system used.
5. In symbolization.
6. In data structure.
7. In type of representation.
8. In measurement scale for the attributes.
9. In allocating attributes to divided or aggregated spatial units.
10. In deriving topology from geometry.

Not all of these are discussed explicitly here; some items have already been encountered in earlier chapters, and symbolization is not discussed at all.

The procedures to make changes are not always based on an algebra – some transformations cannot be expressed mathematically. It is also most important to know if a procedure can be reversed in order to recreate an original. However, some procedures discard data that would be necessary for restoration, as in the case of line generalizing, or going from geometry to topology representation without keeping the coordinates.

Conceptual model	Representation				
	FROM:	TO: CELL	POINT	LINE	POLYGON	VOLUME
	POINT	D	Ideal	S	S	S
	LINE	D	R	Ideal	S	S
	POLYGON	D	R	R	Ideal	S
	VOLUME	D	R	R	R	Ideal

Ideal = optimum match
Cell = regular figure, e.g. square or triangle or cube
D = possible distortion and imprecision
S = possible substitution and symbolisation
R = information reduction

(a)

From Encoding	To Encoding			
	CELL	POINT	LINE	POLYGON
CELL	Scale change	Make lattice	Boundary estimation	Areal interpolation
POINT	Encoding	Interpolation	Triangulation	Thiessen polygon
LINE	Encoding	Collapse (e.g. contours)	Generalization	Link chain
POLYGON	Encoding	Centroid	Profile	Distort shape

(b)

Figure 7.19 Transformations among spatial unit types. (a) From concept to data representation. (b) Examples for changes between representations. Volume objects are not shown.

7.4.1 Changes in dimensionality

There are many instances of dimensional changes for spatial entities, as Figure 7.19a demonstrates. First of all, real world features conceived as points, lines, areas or volumes may be represented by matching dimensions or by a regular tessellation. Notice that we do not allow a cell conceptualization for the real world, only as a representation and that we show only simple entities, not complex units like networks. For most purposes the direct topological match is used, although this depends on the scale at which data are likely to be needed. For instance, if there are no requirements for studying conditions within an urban area, then it can be encoded as a point object. Representation at a lower dimension implies information reduction, which can conceivably be regarded as information loss under some circumstances. One approach is to retain data as originally encoded and generalize as needed.

For the same categories of spatial unit plus regular cells, there are twenty-five possible conversions among entity representations, as illustrated for most cases in Figure 7.19b, where the rows represent the from-state, and the columns the to-state. Some conversions are not of serious interest, such as polygon-to-line, but most are likely to be encountered in spatial information systems processing. The diagonal represents no change in topological definition; but it does include some transformations, for there may be scale, geometric form or data structure changes for a particular category. For example, the diagonal includes line generalizing operations, polygon boundary simplification, and reduction or enlargement in cell size.

Transformations are not necessarily single; there may be a sequence of changes – perhaps point data are extracted from contour lines, and then used as a basis for triangulation. Not only is there information discarding, but it may become difficult to appreciate the changes in implicit error along the chain of processing.

In addition, alternative versions of one or more objects may have different dimensions, or a change in state may occur over time in the representation form. For instance, a street network can be regarded as a set of polylines or a set of areas, one and two dimensions respectively. In this case, a change in dimensionality implies the coexistence of two different representations.

7.4.2 Changes in position

Conversions not involving dimensions of entities or their representations include many types of changes to positional information: map projections, geometric rotation and scaling, rubber sheeting, tin sheeting and conflation. Including topology for comparative purposes, Figure 7.20 demonstrates particular properties that are preserved or not for transformations likely to be encountered in spatial information systems. In addition to topological transformations, the non-Euclidean category includes affine and projective. The latter involve length and angular distortion, the former only angular change, in that parallel lines are kept parallel. This condition is, of course, also true for rotation, translation and scaling, but these changes also have other properties, as shown.

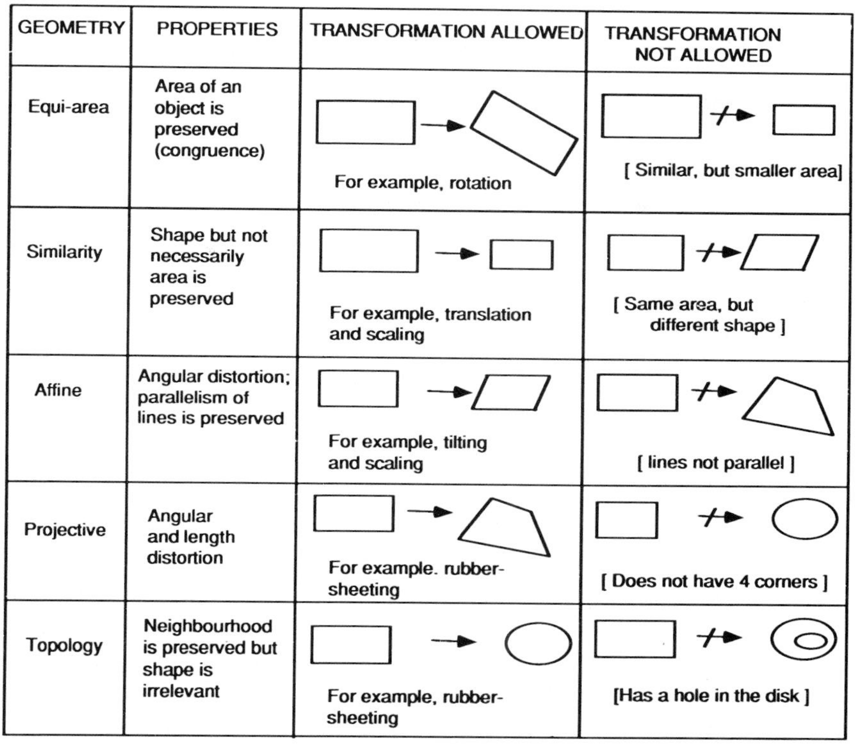

GEOMETRY	PROPERTIES	TRANSFORMATION ALLOWED	TRANSFORMATION NOT ALLOWED
Equi-area	Area of an object is preserved (congruence)	For example, rotation	[Similar, but smaller area]
Similarity	Shape but not necessarily area is preserved	For example, translation and scaling	[Same area, but different shape]
Affine	Angular distortion; parallelism of lines is preserved	For example, tilting and scaling	[lines not parallel]
Projective	Angular and length distortion	For example. rubber-sheeting	[Does not have 4 corners]
Topology	Neighbourhood is preserved but shape is irrelevant	For example, rubber-sheeting	[Has a hole in the disk]

Figure 7.20 Types of geometric transformations. (From Taylor, 1977, with permission.)

All three types – topological, projective and affine – are often labelled as **rubber sheeting** (Figure 7.21). While this term has many popular connotations, it technically implies the possibility of continuous distortion without tearing or sharp breaks. **Tin sheeting** implies a more rigid structure, as when triangles in a triangulation meet along edges without a smooth transition (that is, there is no derivative continuity). **Conflation** (Saalfeld, 1988) implies neither *per se*, but, in its focus on matching objects from two representations via graph concepts and measures, may use both forms of changes in relative position to achieve its goals. The stretching implied by rubber sheeting often is activated on a map subdivided into triangles as manageable pieces.

Area preserving conversion is rotation or reflection, maintaining angles as congruent after either process. As shown in the diagram, **rotation** is undertaken by a trigonometric operation, where the angle specified is the angular change in orientation of the figure or set of entities. Object shape

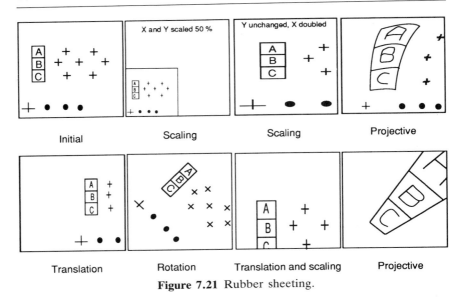

Figure 7.21 Rubber sheeting.

similarity is preserved too in both **scaling** and **translation** operations; but in the former case the object size is enlarged or diminished, and in the second case its position in the reference system is changed by a shift along one or more of the x-, or y- or z-axes. These affine transformations from one Euclidean coordinate base to another, shown in more detail in Figure 7.21, are effected by a series of arithmetic or trigonometric computations. As a group, the affine transformations stretch or shrink the axes, thereby changing distances and angles, but they preserve properties of collinearity and parallelism in lines.

In the **projective** change, the entity maintains its figure definition, that is, number of points and sides, but one or more of them can change in length and/or orientation as discussed earlier, for instance as the topic of global map projections. The geometry, that is, the coordinate distances, is changed, without altering the topological spatial relations. In other words, there is a one-to-one relation between vertices in the prior and resultant states. For example, a square can become a rectangle or an irregular quadrilateral, but not a triangle (Figure 7.20). The process may be automated if it is possible to use equations for the transformations; otherwise a computer assisted intuitive procedure is used. The triangle is often used as the base spatial unit for some kinds of automated rubber sheeting. The spatial adjustments may be global, in the sense of applying to an entire source document, or just partial, in the sense of correcting in only selected locations. If modifications are local and few in number, then graphics editing of one document (in the foreground) against another (background) can be done interactively.

In addition to the map projections category, rubber sheeting is encountered in matching one map representation to another, in graphic designing, and in registration of air photographs or satellite images to global geographic coordinates. For example, the geometry is not easily known for air photographs, or for ground photographs taken from moving vehicles, especially if using a fish-eye lens camera (Poelstra, 1990). In such cases referencing is achieved by matching a few fixed points (like road intersections) which are very clear on the images to the same control points on a surveyed map, and then adjusting other features on the images to fit those on the 'true' document.

7.4.3 Conflation

The matching of features on two documents may use a procedure of conflation that involves more spatial properties than mathematically based rubber sheeting. The appropriate matching of phenomena, perhaps photograph to map, or one map source to another map source for the same area, or documents at different points in time, may require change in relative positions as well as geometry. Visual, intuitively-based corrections can, for example, insert a new highway that produces new topology by comparing the two documents; but affine or projective geometry changes would be insufficient to undertake the task correctly.

Some automation can be achieved by comparing the graph representations of the two documents. Consider, for example, the highways shown on a topographical map of a certain date (Figure 7.22a), with those shown on an air photograph of a more recent date (Figure 7.22b). The latter reveals additions such as bypasses, straightening of some segments, or even road closures creating dead-ends. Counts of the number of edges and nodes are a starting point to reveal differences between the two maps, or other measures like change in the number of circuits, or lengths of chains. Absolute large positional displacements of nodes may also be detected structurally.

Patterns of connections may also be quite valuable in tracking down discrepancies. Nodes with missing edges will have different degrees of connectivity, unless there are compensating discrepancies (Figure 7.22d). In the example (Figure 7.22c) nodes 8 and 15 have degree of three based on the prior state, map B, and four based on map A. Examination of the general orientation of links for the nodes, possibly portrayed via compass

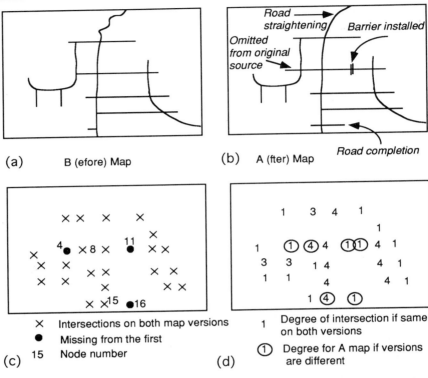

(a)   B (efore) Map   (b)   A (fter) Map   *Road completion*

×	Intersections on both map versions
●	Missing from the first
15	Node number

(c)

1	Degree of intersection if same on both versions
①	Degree for A map if versions are different

(d)

Figure 7.22 Conflation. (a) Map B. (b) Map A. (c) Intersections for both maps. (d) Degree of intersections for the two maps. (Based on material in Saalfeld, 1988.)

rose-type diagrams, as demonstrated by Saalfeld (1988), might provide clues as to missing edges or new links. Road straightening by itself will not be detected by devices using topological data, as indicated.

Metrical displacements, basically shape distortions or errors in area measures, may be detected by various techniques using positional data. Point position discrepancies will be shown by comparisons of coordinate pairs or triads if the height dimension is also used. Line differences will be revealed by the amount of area enclosed by the two versions of lines or by statistical analysis of patterns of point coordinate differences for the ordered array of points. That is, the pattern of discrepancy between two lines might show a bias towards one side or the other, or simply a large or small overall deviation. The directional bias can be picked out by examining the runs of deviation above or below a line element.

7.4.4 Changes in topology

Conversions for primarily topological properties for spatial entities includes not only the building of topology out of individual feature encoding, for example, from the polygon data structure, or from spaghetti-digitized data; and the reverse, of going from topologically structured form to graphic entity structure, but also changes for the topology itself. Creating topology properties for line and polygon entities from geometry, that is, lines not known as related to others, involves numerous steps, presented here in only summary form since many operations have already been discussed:

1. Finding line intersections.
2. Splitting lines into segments and labelling.
3. Identifying junctions, and labelling all nodes.
4. Applying a node-matching procedure where necessary.
5. Snapping nodes to line segments.
6. Assembly of line segments for nodes, and undertaking a node-cycle edit.
7. Assembly of line segments to create an envelope polygon.
8. Assembly of boundaries for individual polygons, and undertaking a polygon-chaining edit.
9. Assigning identifiers to polygons, and determining centroids.
10. Creating minimum-bounding rectangles for polygons.

Alterations of the dimensionality of objects include some cases already discussed: creating Thiessen polygons from point data, forming buffer zones about lines, representing polygons by centroids, and creating gridded data from points.

7.5 TRANSFORMATIONS BETWEEN REGULAR CELLS AND ENTITIES

Perhaps because of the influence of technology more than anything else, much captured spatial data are in the form of arrays of square or rectangular data units and may need to be changed into entity representations. For some analytical purposes it is better to structure data as regular tessellations, rather than sets of polygons, lines or points. In any event, there is much interest in having the capability in spatial information systems to convert between raster and vector representations.

7.5.1 Change to regular cells

Transformation to regular cells from points, lines or polygons, already introduced in section 6.3, is perhaps easier than the reverse process. Logically straightforward, the **rasterization process** detects the occurrence of entities in the cells of the chosen size and shape, recording presence/absence, or other attributes. Basically, features are put in 'pigeon holes' by a scanning procedure, perhaps operating on rows defined by particular coordinate values, or on columns; or the combination, the two-dimensional array; or by reading lists of coordinates for points or lines. Visual encoding, the predominant practice originally for direct cell encoding from maps, is a much easier process, albeit having some practical limitations as discussed in Chapter 6.

Automated procedures may involve different types of operation:

1. Transferring Cartesian coordinates for point-and-line-entity vertices into the matrix of predetermined resolution and known positional values.
2. Use of a single scan line (row or column) or strip of several adjacent scan lines that detects intersections of cell boundaries with linear features, and records how many grid units have been passed up to the intersection.
3. For polygons, after detection of vertices, and, therefore, line segments, using a bidirectional scan traverse that knows when it reaches the edge of a polygon.

In the first situation, the coordinate values for the points or line ends are compared with the range of coordinates for each cell, or the original coordinates can be converted by a numerical factor to indicate into which cell they fall. In the extreme state, the number of cells needed can be determined from the number of coordinate positions in the original data set. Cells intermediate between line end point cells will have to be inferred in this process. Otherwise, to more directly get at the whole length of lines, the crossing of the line by the horizontal and vertical grid lines can be computed.

Except for the corner-touching case, the edge intersection will cause the two adjacent cells to be encoded with a line object. In the case of exact match at the corner intersection, all four touching cells can be encoded with presence of a line feature, or some other decision rule could be used, perhaps by looking at the other cells already encoded. In the case of lines, every cell containing even just a piece of line should be noted, otherwise lines will be incomplete in the raster representation. However, to avoid then creating fat lines in the raster, a rule of minimum line length could be applied to the amount of edge in each cell.

For polygons, the boundaries may be encoded to cells in the same manner as for linear features, except that the attribute data for the polygons must be associated with the cells. Unless multiple attribute coding for individual cells is allowed or not wanted, a decision has to be made as to which attribute will be encoded, perhaps using a dominance rule, although this needs an extra step of computing the proportional area of a cell occupied by the different polygons. The difficulty in working with the attribute data is enlarged if the variable is continuous rather than categorical. Spatially continuous data allocated to a polygon centroid, on rasterization will be applied only to cells with points, unless the data are allocated to polygons or interpolated to a lattice before conversion to grid cells. A spatial autocorrelation function can be applied to the point data, rather than assume spatial constants in local neighbourhoods.

7.5.2 Change from regular cells to vectors

The reverse process, going from grid cells to geometric figures, usually known as **vectorization**, is more involved, partly because it requires several stages and partly because there is no topological structure already encoded in the tessellation. Ideally, we would like to:

1. Transfer topology, that is the connections and adjacencies visually apparent in the grid representation.
2. Transfer the correct shape of objects.

So there is need to vectorize and then recognize entities. For example, for sets of polygons it will be important to follow cells that make up the boundaries, creating vectors, and then assemble the vectors, including detecting junctions, to obtain polygons (Figure 7.23). Along the way, it will generally be necessary to smooth the rectangular edges of cells to eliminate the staircase effect. However, boundary smoothing is sometimes considered an optional cosmetic step.

In general, the procedure consists of the following stages. First of all, cells are coded, perhaps by binary states, so that all vertically and horizontally contiguous cells with the same attribute category can be assembled. Each such group of touching cells can then be assigned a unique label. Next it is necessary to establish the path of boundary lines by identifying the cells with different values on at least one side and then traversing the boundary from a starting point which is the corner junction of three or four cells with different values. The boundary line may be recorded as directional codes using four compass points as shown, or

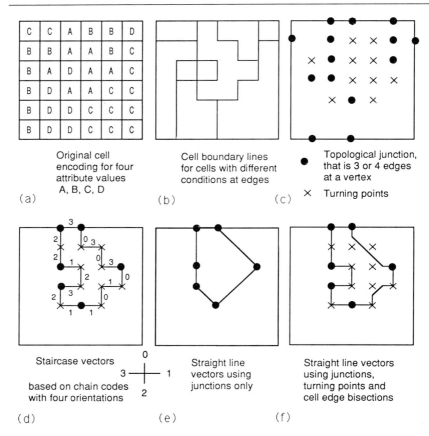

Figure 7.23 Raster to vector conversion. (a) Original cell encoding for 4 attributes (b) Cell boundary lines for cells. (c) Topological junction. (d) Staircase vectors. (e) Straight line vectors. (f) Straight line vectors using turning points.

somewhat smoothed by using eight points. Diagonals may be established, rather than simply joining corner points by straight lines, by bisecting the cell edges or a more complex procedure subdividing the cell edges based on neighbourhood conditions. Part of this process clearly involves topology construction, that is the nodes at the corners of touching polygons; and part of it attempts to create acceptable approximations for the geometry.

The construction of the vector equivalent of raster representations of lines involves line skeletonization and recognition of connectivity. In raster mode, lines have width; geometrically, they are not supposed to. So, we need to obtain a width of only one cell, generally achievable by recognizing and eliminating cells with short lengths and allowing corner neighbours. Different algorithms exist for implementation of this general procedure (Piwowar *et al.*, 1990). Some procedures have an intermediate stage of creating graphs of vertices and edges, and simplifying by eliminating vertices through dilution when the edges are not needed for topology. Subsequently, lines can be smoothed as illustrated above for polygon boundaries.

Vectorization may be well assisted by interactive procedures. Often the smoothing can be effectively judged from the visual display of cells, clearly revealing neighbourhood conditions. Otherwise, fully automated procedures with varying success in building topology and producing acceptable geometric approximations are available. For both rasterizing and vectorizing, judgements of effectiveness can be based on several criteria; we consider accuracy of area measure, perimeter measure and minimization of feature displacement to be as important as considerations of memory space and speed of processing.

7.6 ACCESS TO SPATIAL DATA

Recalling the discussion of space partitioning by tessellations (Chapter 5) and the spatial referencing by continuous and discrete methods (Chapter 4), it is appropriate at this point to consider the important matter of access to spatial data in a database. Generally speaking, in alphanumerical databases, access to information is based on the attributes. In spatial databases we have, in addition, location based access to data.

7.6.1 Access by identifiers and by locators

Access to spatial data will vary according to the data types and structuring. For the moment, however, we consider this matter conceptually. So we may enquire of data according to:

1. The name of an object.
2. A particular single position on the earth's surface or relative to an arbitrary reference system.
3. A specified block of space in the spatial reference system.

If it is possible to associate unique names with spatial entities, then this form of access may be preferred by many users. But not all 'places' have names, and for some features which do have unique names, there is no clear spatial limit for the name. All things stored in a database will have a unique identifier of some kind, so that unambiguous access is always possible; but possibly not in a form comfortable to the user.

Also, different forms of representation may not lend themselves so well to straightforward use of names. For example, regular tessellations will need to have a name associated with a set of cells, either by direct storing of the feature name of each cell in the set, or by a numerical code which is translated to a name by a lookup table. If a cell forms part of several objects, then it will need to have several associated numerical codes or names. For a layered vector data model, names are conceptually attached to features like polygons; for example, lakes, not the vertices and line segments needed to demarcate them. So an additional data type of annotation may be necessary if the software does not allow combination of the primitive spatial units to a level to which names are appropriate.

Earlier discussion (section 4.2) of discrete and continuous referencing pointed to different forms of coordinates and tiling systems. Queries as to the nature of space can be executed by identifying positions or tiles and then providing the descriptive information about the identified position. However, some software systems may not be able to handle all points in space too well, recognizing only entities as encoded; that is, they lack the ability to perform nearness, or point-in-polygon or point-on-line tests quickly or at all.

It is important to be sensitive to a double requirement for accessing spatial objects:

1. Identifiers for entities or their encoded forms.
2. Locators for spatial objects or chunks of space.

Recall that locators may be geometric, either Cartesian or azimuthal, or topological (such as route mile point or postal address), or space-filling (for example, a Peano key). And the locators may be of mixed form, for example a postal code area delimited by coordinates for vertices.

7.6.2 Rectangles and strip trees

At times lines, either one-dimensional objects, or the boundaries for polygons, are accessed and retrieved by reference to the boxes that contain them. Enclosing figures, generally rectangles, but possibly circles or spheres as discussed later, serve to identify the range of values in the x and y dimensions, whatever the particular form of representation for the details of the shape of the line. As discussed in section 4.2.2, lines or

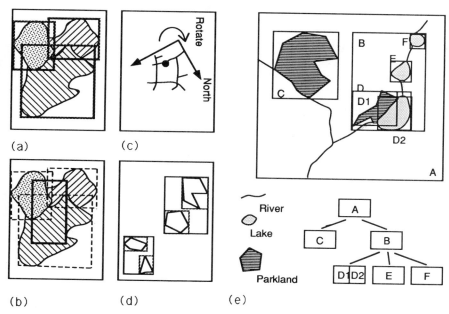

Figure 7.24 Rectangles for polygons. (a) Minimum enclosing rectangles for a tessellation of polygons. (b) Truncation of polygons for search window. (c) Rotation requirement. (d) Rectangle hierarchy for nonconnected polygons. (e) Example of an R-tree.

irregular polygons encoded geometrically can be represented by rectangles drawn in orthogonal dimensions to touch the extreme points of the object in both x and y dimensions. These minimum-bounding rectangles, shown in Figure 7.24a, are useful devices for extracting particular lines or polygons from a set, requiring specification of only the range of x and y, rather than undertaking a spatial search on coordinates for the polygon boundary vertices.

The rectangles (or cuboids), with sides parallel to the two (three) coordinate axes, may also serve to clip entities stored as polygons or lines in the interest of finding what exists within a block demarcated by a particular range of x and y (and z if required). Or the rectangle may represent a larger spatial unit, a map sheet or tile subdivision of the entire database. The fragmentation of entities has limitations, though, such as failure to retrieve an entire object (Figure 7.24b), and the fragmentation can grow to undesirable levels as boxes are made smaller in cases where the amount of data increases. Otherwise a rectangle can be used to access all objects within it, but, again, the chance of retrieving a few objects is correlated with the size of the rectangles relative to the density of empirical phenomena at a given scale.

If the rectangles are bounding rectangles, then the number of rectangles will be equal to the number of entities, and they no longer serve a purpose of simply partitioning space. In the case of a database of different thematic layers, polygons in each layer may be enclosed by sets of minimum rectangles, leading to overlapping of bounding rectangles when searching for all thematic objects in a specified coordinate range (Figure 7.24b). It is, though, easier to compute intersections of overlapping rectangles than to find where irregular figures might cross.

The enclosing rectangle is just one of several **spatial access devices** using regular figures. Some needs may be better met if the enclosing boxes are not dependent on being parallel to coordinate axes, as, for example, when searching for objects in a rectangle rotated to a certain orientation, or a varying angle in the case of vehicle navigation displays in which the compass direction at the top of the map display varies in order to keep the vehicle icon-oriented in the direction of the vehicle (Figure 7.24c). Circles and spheres are insensitive to the rectangular axes of Cartesian coordinate systems, and are also appropriate to searching in azimuthal coordinate space. On the other hand, it is more involved computationally to fit a circle around an irregular polygon.

This type of spatial unit organization can also have a **tree structure** in order to get different spatial resolutions, but does not involve a regular partitioning of space as with the quadtree (Figure 7.24d). Indeed areas void of polygons or other objects can be ignored. Known as **R–trees** (rectangle trees) and illustrated in Figure 7.24e, this organization is preferable for unconnected polygons rather than tessellations (see also section 15.2.3).

Here, as we illustrate, we can reference the two large rectangles B and C, at one level, but to more clearly access the matching parkland within B, we go through two more levels to get to D. Access may be by identifier, or by a locator for the bottom left corner (x, y based). The intent in building the rectangles is to keep overlap to a minimum while still keeping as many features as possible completely within a single box.

Linear features, as well as polygons, can be represented by rectangular cells. One technique, using the **strip tree**, is beneficial for operations requiring search of linear features. It is appropriate for single curves, rather than sets, being produced by successive approximations enclosed in bounding rectangles (Figure 7.25a). The process of decomposition can be stopped when strips are of a predetermined width or height. If long linear features are not split into pieces, their extents will be large, possibly being inconsistent with zooming operations for visual display changes of scale.

In cases like this the low-level clipping routines will have to be invoked.

A strip tree is very nice for representation of polylines at different resolution, as might be done in the case of approximations of a coastline at different scales. Several special conditions may have to be taken into consideration, such as closed curves, by breaking an initial rectangle into two or more separate but connected pieces. A disadvantage is that here the rectangles, which can have varying orientations, are not tied into a given coordinate system (Figure 7.25b). The data tabulated will consist of the strip width and level, a code pointing to the next level block in the strip hierarchy, and geometric and other information, as shown (Figure 7.25c).

7.6.3. Sheets and tiles

The topic of division of space also arises in the creation of databases. Because much spatial data now exists in the form of separate pieces of paper, the digital equivalents are often created as discrete units even if they comprise part of a series, as for most national topographical maps, for example, Figure 4.2. While the concept of a seamless coverage for the joined base maps in digital form is laudable, there are practical matters to be dealt with in order to produce effective and accurate spatial databases

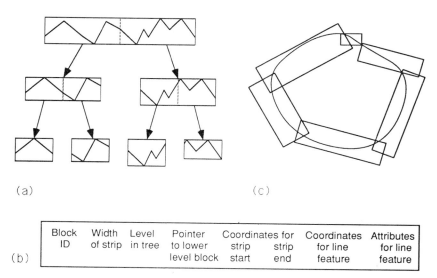

(a)

(c)

Block ID	Width of strip	Level in tree	Pointer to lower level block	Coordinates for strip start	strip end	Coordinates for line feature	Attributes for line feature

(b)

Figure 7.25 Trees for linear features. (a) Strip tree. (b) Strips for an enclosed polyline. (c) Representative data for a strip tree.

(Chrisman, 1990). There are two main types of space partition in this context:

1. The physical subdivisions, generally sheets representing original paper or photograph documents that are to be combined to produce a single cover.
2. Tiles produced as logical units for reasons of user querying or possible management purposes.

A sheetless database should be created so as not to have map edge matching problems resulting from mismatched positions of map features, and will avoid the problems of queries returning bad data because a sheet boundary truncated an object. However, notwithstanding the physical space savings in using logical tiles rather than physical irregular chunks as a major unit to partition a coverage, by devices like a reduction in the number of digits needed for storing coordinates by reducing the range covered by the map area, queries cutting across tiles will still occur. However, while logical partitioning is a good idea, given that user needs may not be well anticipated, there is still a need to provide for operations that cross tile boundaries, for example, to assemble the pieces of a road. A possible device is a double encoding for the spatial units (Chrisman, 1990); and another option is to have overlapping tiles.

7.6.4 Different forms of spatial address

The specification of location for spatial databases is itself not necessarily a simple matter. The conventional indication of position by coordinates does not cover all aspects associated with location. For a good understanding of the quality of locator data in spatial data processing, four elements need to be addressed:

Scale
Resolution
Precision
Accuracy

Scale (denoting the order of magnitude or level of generalization at which phenomena exist or are perceived or observed) and resolution (the size of the smallest recording unit, akin to precision, the fineness of measurement) are traditionally specified separately from the traditional Cartesian coordinate form of indicating position. Thus the scale of observation might be specified by a cartographic ratio like $1:1,000,000$ or by reference to a unit of observation, for example nation. The resolution may be indicated by the size of shortest length to be measured on the ground, or by a fuzzy tolerance value for a digital map.

While Cartesian coordinates may have a length (number of digits) that varies with machine precision (whether integer or real numbers in a digital computer environment), dictated usually by the hardware's word length, this length does not inherently indicate which digits are significant. The numbers themselves say nothing about scale, and carry no guidance as to the accuracy of the measurements. Something akin to the statistician's sampling error measures must be provided separately, as part of the ancillary metadata. Accuracy, the correctness of measurements, in the sense of validation against reality, is not the same as precision, which reveals the detail in recording the observed properties.

Other types of locator can be constructed to convey something of scale, resolution and accuracy as well as position (Dutton, 1991). We have already presented a locational coding scheme for quadtrees. A quadtree address, such as 321 for the block shown in Figure 7.26a, conveys position for a quadrant in the NW, NE, SW, SE sequence, and indicates resolution by the number of digits. Precision in location can be achieved by making the quadrant sufficiently small to encompass the entire object. Accuracy can be conferred by establishing a buffer strip around the object, that is, making the square larger by some quantity. Similarly, a triangular tessellation (Figure 7.26b) coding scheme denotes position, scale and resolution. In the example, the level is shown by the number of digits, and 0, 1, 2, 3 refer in a consistent way to the four

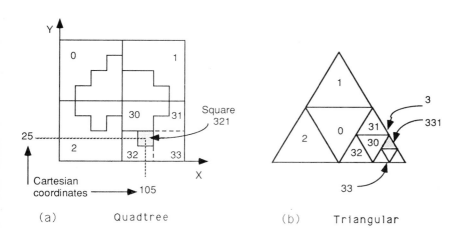

Figure 7.26 Coordinate and tessellation addressing. (a) Quadtree Z-order locational coding. (b) Polyhedral tessellation addresses.

triangles. Using an example provided by Dutton (1991) the Giza Pyramid has an address of 003201320110230. The first zero in this address at the fifteenth level indicates the global octant in the octahedron system discussed in section 6.5.

We shall return to this topic of methods of spatial addressing again later in Chapter 15. For the moment it is important to be sensitive to the multifaceted nature of establishing position of objects in space.

7.7 SUMMARY

The scope of this chapter was not to present all situations, problems, general procedures and algorithms for manipulating spatial objects. We have merely tried to give a flavour, in order to transfer something of an appreciation for data manipulations in spatial information processing. We have accordingly:

1. Emphasized the variety of transformations that may be needed.
2. Presented some methods for access to spatial data elements.
3. Identified the fundamental computational geometry operations.
4. Described basic computations needed for some spatial statistics.
5. Shown the need to both differentiate, and yet deal jointly with, the spatial characteristics of geometry and topology.
6. Established a basis for intensional rules that can be incorporated into the design and implementation of spatial information systems.

The treatment of transformations provides a framework for discussion of alterations of states for conceptual modelling. Later we will treat the topic of multiple representations, but it should be clear already that many different states can exist. The great variety of operations on the spatial data may have challenges for their implementation in software systems because there is not a simple one-to-one match between operation concept and method. Returning to the idea in Chapters 1 and 2 of a spatial information system being a toolbox of operators applied to data, we have in Chapter 7 presented some of the low level operations that may need to be performed in order to undertake spatial analysis or mapping tasks, or reorganize a spatial database. Chapter 8 covers some of the higher semantic order operations utilized in several domains.

We have presented operations involving positional data: scaling, rotation and translation; and distance, length and area measurement. We have discussed more complex needs requiring several basic operations, as in conflation, shape computation, buffering, areal interpolation and polygon overlay. All of these have been dealt with in the domain of encoding of geometry and topology. This arises, in part, because many fundamental operations for regular tessellation representations are much simpler, but we shall come back to this point in Chapter 8. In addition, we have so far paid only scant attention to the attribute data, although, of course, interpolations and extrapolations apply to one or more spatially variable attributes.

It is important to emphasize that the algorithms and methods discussed will be used in data modelling as rules, to be covered in Chapters 10 and 11. Indeed, spatial data representation must embody not only coordinates and topology, but also some geometric semantics as rules. Whatever the form, process or need, the impact of interpolation and extrapolation on spatial information systems is the encoding of a rule or model for deriving new data from encoded initial data. The determination of centroids, line intersections, shapes and area or volume measures, buffer zones and clipping regions, requires geometric operations in the form of rules. The intensional representations may be unavoidable or they may be desired for reasons of data storage economy; but they must be recognized and provided for in addition to complete, explicit, deterministic, extensional encoding.

7.8 BIBLIOGRAPHY

Either Foley *et al.* or Preparata and Shamos will provide the reader with extensive coverage of computational geometry algorithms. The work of Davis provides extensive coverage of spatial interpolation, while Clarke contains several chapters on cartographic data transformation.

Aronson, Peter. 1987. Attribute handling for geographic information systems. *Proceedings Auto Carto 8 Conference, Baltimore*, Maryland, USA, pp. 346–355.
Ballard, Dana. 1981. Strip trees: a hierarchical representation for curves. *Communications of the Association for Computing Machinery* 24: 310–321.
Boyce, Ronald R. and W. A. V. Clark. 1964. The concept of shape in geography. *Geographical Review* 54: 561–572.
Chrisman, Nicholas R. 1990. Deficiencies of sheets and tiles: Building sheetless databases. *International Journal of Geographical Information Systems* 4(2): 157–168.
Clarke, Keith C. 1990. *Analytical and Computer Cartography*. Englewood Cliffs, New Jersey, USA: Prentice Hall.
Davis, John C. 1973. *Statistics and Data Analysis in Geology*. New York: Wiley.
Deichmann, Vive, Michael F. Goodchild and Luc Anselin. 1989. *A General Framework for the Spatial Interpolation of Socioeconomic Data*. Technical

Report, National Center for Geographic Information and Analysis, University of California, Santa Barbara, California, USA.

Dutton, Geoffrey. 1990. Locational properties of quaternary triangular meshes. *Proceedings of the 4th International Symposium on Spatial Data Handling*, July 23–27. Zurich: Switzerland, pp. 901–910.

Dutton, Geoffrey. 1991. Polyhedral hierarchical tessellations. The shape of GIS to come. *Geo Info Systems* 1(2): 49–55.

Foley, James D. *et al*. 1990. *Computer Graphics, Principles and Practice*, 2nd edn. Reading, Massachussetts, USA: Addison-Wesley.

Frank, Andrew U. 1987. Overlay processing in spatial information system. *Proceedings of the Auto Carto 8 Conference, Baltimore*, Maryland, USA, pp. 16–31.

Goodchild, Michael F. and Nina S-n Lam. 1980. Areal interpolation: a variant of the traditional spatial problem. *Geo-processing* 1: 297–312.

Guevara, J. Armando. 1985. Intersection problems in polygon overlay. *Proceedings of the Auto Carto 7 Conference, Washington, DC, USA*.

Mark, David M. 1987. Recursive algorithm for determination of proximal (Thiessen) polygons in any metric space. *Geographical Analysis* 19(3): 264–272.

Olea, Ricardo A. 1974. Optimal contour mapping using universal Kriging. *Journal of Geophysical Research* 79(5): 695–702.

Oliver, Margaret, Richard Webster and John Gerrard. 1989. Geostatics in physical geography. Part 1: theory. *Transactions of the Institute of British Geographers* 14(1): 259–269.

Openshaw, Stan. 1984. *The Modifiable Areal Unit Problem: Concepts and Techniques in Modern Geography*. Norwich, UK: Geo Books.

Peuquet, Donna J. 1981. An examination of techniques for reformatting digital cartographic data. Part I: the raster-to-vector process. *Cartographica* 18(1): 34–48.

Peuquet, Donna J. 1981. An examination of techniques for reformatting digital cartographic data. Part II: the vector-to-raster process. *Cartographica* 18(3): 375–394.

Piwowar, Joseph M., Ellsworth F. LeDrew and Douglas J. Dudycha. 1990. Integration of spatial data in vector and raster formats in a geographic information system environment. *International Journal of Geographical Information Systems* 4(4): 429–444.

Poelstra, T. J. 1990. Frankie goes to town, or how to capture reality in urban areas. In: Polydorides, Nicos D. (ed.) *Data Aquisition for Spatial Information Systems*; Athens, Greece: Urban and Regional Spatial Analysis Network for Education and Training, Computers in Planning Series 7: 56–66.

Preparata, Franco P. and Michael I. Shamos. 1986. *Computational Geometry: An Introduction*. New York: Springer-Verlag.

Saalfeld, Alan. 1988. Conflation: Automated map compilation. *International Journal of Geographical Information Systems* 2(3): 217–228.

Taylor, Peter J. 1977. *Quantitative Methods in Geography*. London, UK: Houghton Mifflin.

Tobler, Waldo R. 1979. Smooth pychophylactic interpolation for geographical regions. *Journal of the American Statistical Association* 74: 519–530.

Unwin, David. 1981. *Introductory Spatial Analysis*. London, UK: Methuen.

8
Spatial Analysis
Attribute data, modelling, integration

The timber company assigned specific tasks to different employees – Freda enumerated the amount of forest according to tree type; Fred obtained statistics for the slope of the land in the forest; Lesley marked forest stands and burn areas on maps, while Leslie computed the volume of timber that could be produced from half-mile wide strips along major trails cut through the forest; Christiane examined the correlation between tree density and slope of terrain, while Christophe worked on the siting of lookout towers in order to encompass all the forest land. Jerry Mander hopes to be elected to parliament; while Bon Chen finally extricates himself from the maze. Marie Pierre found out that her software system could quickly tell her the acreage of land owned by different proprietors, but could not give her figures for the amount of land needed for a new highway and which landowners would be affected.

After a preliminary review of the ways in which attribute data are associated or combined with the spatial data in spatial information systems of different types, this final chapter of Part Two discusses data processing needs of several classes of spatial analysis. Not all categories of analysis are covered, only enough to give a flavour of the many possibilities. Path- and route-finding in planar networks are put in the context of both geometric space and topologically structured spatial data. Map overlay approaches to spatial predictive modelling are considered in the framework of both regular tessellation and irregular polygon data. Then, including some examples of public and commercial models for spatial data, the chapter ends with brief discussions of integrated and multiple representations.

8.1 INTEGRATING THE ATTRIBUTE DATA

Attribute data have been discussed only indirectly during the course of the previous four chapters because of their concentration on spatial data. However, it is important to pay attention also to the descriptive information for places or facilities or landscapes, as probably most uses of spatial information systems pertain to the thematic information. There are three situations:

1. Actions requiring no spatial properties.
2. Actions using only spatial properties.
3. Actions that combine spatial and attribute data.

The first category includes tasks such as naming entities, comparing the income level for households in different countries, finding the composite character of a particular place, or modifying the numerical values for the different categories of an attribute for a set of grid cells. Processing non-spatial data may involve changes among the attributes or particular values, as in converting from one measurement scale to another, or going from continuous to discrete enumeration. It may consist of retrieval of instances satisfying certain conditions, often accomplished by set operations using Boolean logic, such as union and intersection. In any event, there are no changes to the spatial units, there are no new data for spatial properties, and no new spatial units are created.

The second category, operations using only spatial properties, include tasks like finding the names of countries that border Iraq, measuring the area of the burned areas in Yellowstone National Park, USA or determining the total length of sewer pipes underneath the streets of Paris, France. Note that we do not consider the identification of spatial entities by name or character to be contrary to the designation of only spatial properties; the units were selected on the basis of name or entity type, not on the basis of descriptive characteristics.

Tasks that involve combinations of spatial and attribute data (as discussed briefly in section 3.6) include tasks like finding the number of students who live within areas demarcated by no more than thirty minutes driving time on a school bus by the fastest routes to particular schools, or measuring the area of damaged property on coastal barrier islands that have recently suffered storms.

For actions that combine attribute and spatial data, we are concerned for the moment with the ways in which the data may be organized, rather than in producing and discussing a long list of operations. The data for spatial properties can, of course, be **precomputed** and stored as attributes in tables. For example, rather than examine the tables of spatial data for line segments in a graph representation in order to create chains in a network by matching the line end point node labels or coordinates, the links can be stored as attributes of nodes (Figure 5.29). The data for which countries border Iraq can be tabulated as such, rather than requiring computation when needed. The lengths of the perimeters of islands are recorded directly, rather than as sets of coordinates and a rule for obtaining the length.

Many analytical or mapping tasks can be facilitated if spatial

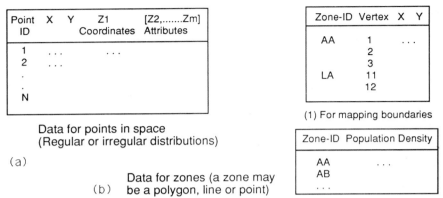

Point ID	X	Y	Z1 Coordinates	[Z2,.......Zm] Attributes
1	. . .		. . .	
2	. . .			
.				
.				
N				

Data for points in space
(Regular or irregular distributions)

(a)

Zone-ID	Vertex	X	Y
AA	1	. . .	
	2		
	3		
LA	11		
	12		

(1) For mapping boundaries

Zone-ID	Population Density
AA	. . .
AB	
. . .	

Data for zones (a zone may
(b)   be a polygon, line or point)

(2) For thematic mapping of attributes

(c)   Georelational Concept

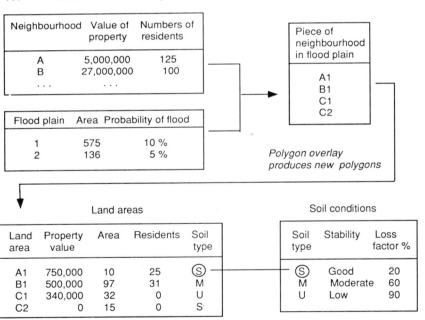

Neighbourhood	Value of property	Numbers of residents
A	5,000,000	125
B	27,000,000	100
. . .	. . .	

Flood plain	Area	Probability of flood
1	575	10 %
2	136	5 %

Piece of
neighbourhood
in flood plain

A1
B1
C1
C2

*Polygon overlay
produces new polygons*

Land areas

Land area	Property value	Area	Residents	Soil type
A1	750,000	10	25	Ⓢ
B1	500,000	97	31	M
C1	340,000	32	0	U
C2	0	15	0	S

Soil conditions

Soil type	Stability	Loss factor %
Ⓢ	Good	20
M	Moderate	60
U	Low	90

Figure 8.1 Contrasted data organization for attributes. (a) Data for points in space (regular or irregular distributions). (b) Data for zones. (c) Attributes data linking for the georelational model (coordinates and topology are not shown).

properties, geometric or topological, are already tabulated. However, query operations, characterized by selection of only a few spatial units out of a total, may require spatial data processing at the time of the query. For example, finding which farms might lose land through construction of a new highway and measuring the land area, for different possible routings of the highway, can be undertaken by computing the intersections between lines on a farm boundary data layer and lines drawn on the screen for different road alignments. In such cases it is most important to have a good organization of the spatial and attribute data so that their connections can be made quickly.

We can see four possible categories of processing in this context (Aronson, 1987):

1. One attribute at a time is processed for spatial units.
2. Several attributes are used in conjunction for a set of spatial units.
3. Attribute and spatial properties are both required for a task but are processed separately.
4. Attribute and spatial data are processed together.

Historically, computer software for working on attributes for spatial units were oriented to just one variable at a time. The thematic mapping programs and packages characteristic of the 1960s and early 1970s were of this type, perhaps making contour maps for one set of elevation data, or producing choropleth maps for one variable at a time. They used data tabulations as shown in Figures 8.1a and b. Composite mapping, developed in the 1970s, allowed operations like adding different themes together before mapping. A map overlay concept and grid cell data encoding (section 8.3) facilitated this type of processing.

As computer software design and programming skills, techniques and tools improved, and as better hardware came along, more challenging analytical and query tasks were performed. Such tasks went beyond the simpler requirements for thematic mapping not entailing use, or computation of spatial properties. Most modern spatial information systems software, including some oriented to mapping rather than analysis, can do both spatial and attribute data processing, although they may have architectures that, on the one hand, separate the two kinds of data, or otherwise do not. One common type is the **georelational**, consisting of different tables of attributes for different kinds of spatial unit and separate tables for coordinates (Figure 8.1c). For the georelational model to perform an operation like overlay of two polygon maps, the spatial intersection is done for units not containing any thematic data. For another type, the spatial and attribute data are handled jointly, as can be readily accomplished with grid cell forms of data, or via other means, as is discussed later in this book.

The ability to perform certain tasks, then, and the efficiency of the performance, depends not only on what data are required but how they are organized. We illustrate this by a series of examples, representing several different categories of spatial problem solving and types of spatial data – for regular and irregular tessellations, and for graphs. This approach puts the processing side of spatial information systems alongside the data organization side. A data processing model is the formalization of the operations to be performed on the spatial data units; a data organization model is the systematic arrangement of the data for the spatial units.

8.2 SOME OPERATIONS FOR PLANAR NETWORK ENTITIES

Recall that transportation and hydrology networks, or the cables and pipes used for water, electricity, gas, sewerage or telephone services, are structurally identical to graphs and networks. For such entities, tasks like the following arise:

1. Find all paths for a huge lorry in a road network among all origins and destinations.
2. Find the place at which service is disrupted because of a breakage or malfunction.
3. Find a best route to pick up more passengers.
4. Find the nearest electricity or telephone pole for a new customer for those services.
5. Locate a service facility on a highway network.
6. Allocate children to their nearest schools defined on the basis of minutes of travel along the city streets.

A primary processing requirement is to create a chain from the line segments of the graph. Navigation through the set of tabular records for line segments could be done by matching end points or their coordinates, or by using a table in which links associated with nodes are stored explicitly. In earlier days of solving routing problems, data were stored as coordinates, not for topological relationships, so that the connectivity conditions had to be derived from the geometry in the original data tabulations. Otherwise, a graph structure better matches the need for quickly finding chains (Figure 5.29), although this requires the tabulation of the spatial impedances – the attributes measuring friction

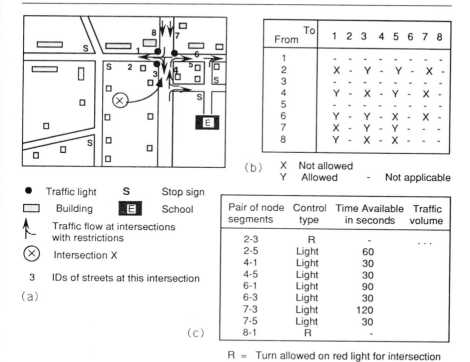

From \ To	1	2	3	4	5	6	7	8
1	-	-	-	-	-	-	-	-
2	X	-	Y	-	Y	-	X	-
3	-	-	-	-	-	-	-	-
4	Y	-	X	-	Y	-	X	-
5	-	-	-	-	-	-	-	-
6	Y	-	Y	-	X	-	X	-
7	X	-	Y	-	Y	-	-	-
8	Y	-	X	-	X	-	-	-

(b)   X   Not allowed
        Y   Allowed   -   Not applicable

●   Traffic light   **S**   Stop sign

▭   Building   **E**   School

� Traffic flow at intersections
with restrictions

⊗   Intersection X

3   IDs of streets at this intersection

(a)

Pair of node segments	Control type	Time Available in seconds	Traffic volume
2-3	R	-	. . .
2-5	Light	60	
4-1	Light	30	
4-5	Light	30	
6-1	Light	90	
6-3	Light	30	
7-3	Light	120	
7-5	Light	30	
8-1	R	-	

(c)

R = Turn allowed on red light for intersection

Figure 8.2 Network turn restrictions. (a) Map of some city streets.
(b) Turn matrix for intersection X. (c) Turn table for street pairs.

along the line segments – needed for some problems incorporating minimizing travel. If these impedances include geometric distance then this can be precomputed.

If measures are needed for the connectivity in the whole network, a matrix form of data organization is better (Figure 3.30). Nodes are recorded as origins and destinations in a two-dimensional table. Matrix algebra can then be used to ascertain conditions of connectivity beyond the direct non-stop connection, by looking at links involving two, three, or more legs in the network.

For graphs with many circuits and multiple link junctions, particularly city street systems, it is insufficient to know that certain links connect at certain nodes. It is at least as important to know if one can pass from one link origin to another without restriction, or if there is a signalling device at an intersection (Figure 8.2a). Consequently, we have to deal with attributes of link pairs or link–node combinations. One device to do this is the turn matrix (Figure 8.2b) which tabulates conditions for every

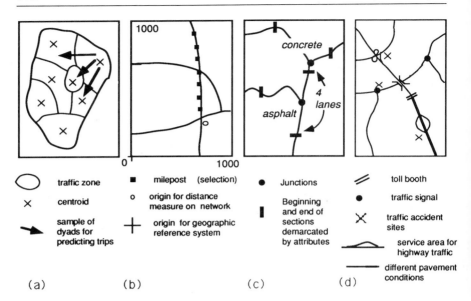

Figure 8.3 Attributes and topological models for transportation networks. (a) Zones, centroids, and dyads. (b) Offset distances. (c) Segmentation of a network based on attributes. (d) Some kinds of observed entities.

possible combination of link pairs at a junction. This concept is especially valuable for dealing with overpasses and underpasses, which are semantically incorrectly recorded as intersections in a planar-enforced representation. The turn table (Figure 8.2c) is convenient for associating attributes with the link combinations. Alternatively, such attributes can be recorded for particular node-link combinations.

There may also be conflicts between geometric and topological contexts for transportation or utilities infrastructure network processing studies or planning (Figure 8.3). Different sources of data for instance, global positioning systems and map sources, may lead a hazard or other object to be placed on the wrong side of a highway. Or some operations may use point data, for example zone centroid-to-centroid distances (Figure 8.3a) rather than graph data. Thus, we have to recognize the existence of a Euclidean coordinate world – there is an infinite number of points to be dealt with. The spatial referencing may be achieved by tabulating entities in continuous or discrete referencing systems, as polar or Cartesian coordinate, or street address, or an offset distance (Figure 8.3b). A point-to-point association is most important for situations like finding the nearest supply point for a new customer for electricity or telephone – the connection may be to the nearest node or to the nearest point on the graph edge representing the electricity or telephone cables.

There may be differences also between a topological view and an attribute view (Nyerges, 1989). The former stresses junctions and connections; the latter is oriented to road pavement, number of lanes and conditions of roadside barriers. **Segmentation** of the links may be necessary to indicate where attributes change (Figure 8.3c). This process can produce a large number of pseudo-nodes, giving rise to additional complexity in processing, that is, much more work to chain all the links together. The alternative of identifying change in attributes, or recording the location of an event like a traffic accident, by a distance from a specified origin or a topological node has the advantage of dealing directly with attributes but at the expense of requiring more tables of data. Yet it also simplifies the process of encoding the basic graph structure; this can be done once ignoring the attributes of the line segments. An associated data type of a route, a collection of arcs with a common characteristic, allows for recording of linear entities without requiring original arcs to be split by pseudonodes where attributes change.

From these and other tasks, it is clear that the transportation domain utilizes a great variety of spatial units, a few of which are shown jointly in Figure 8.3d:

1. Points in space, the positions of traffic accidents or zone centroids.
2. Links, the one- or two-way traffic arteries.
3. Line features, like the edge of embankments or roadside barriers.
4. Points, centroids, that represent areal units, the traffic zones.
5. Points, like traffic barriers or dead-end streets, associated with lines.
6. Node–link combinations, for example stop signs.
7. Link–link combinations, as for city street turn restrictions.
8. A path, a particular combination of links from an origin to destination.
9. A route, the ordered set of links used to go from an origin to destination.
10. Distances of point features like signals or accidents from some origin.
11. Volumes like cut and fill for engineering works.
12. Volumes such as revetments holding up an embankment.

Some of these units are not simple, or at least are not addressed in many general purpose spatial information systems today. These include the path and route units, and turn restrictions. The path and recommended route are difficult because they result from solving a shortest path problem. As such, they produce additional information to be stored, possibly as a binary attribute of inclusion or not in the route, as an attribute of a link indicating the next link in the sequence, or as an attribute of a dyad representing the origin and destination combination.

Most of the challenges are not computational but focus attention on the need for topological structuring of the data. This is essential for some analytical operations, and may require access to several categories of spatial unit. The shortest path problem needs to use both line and node objects, and the link–link entity if traffic turn prohibitions are included. In addition, there are requirements to integrate positional information correctly, so as to preserve correct topology. The exact path of highways may be based on accurate field measurements or on data from global positioning systems which can place a point entity on the wrong (topological) side of a linear feature. Moreover, link–node structured systems are unable to deal with infinite space as represented by a continuous surface; it is not possible directly to manipulate data for positions off the graph, although points can be located on a network by distance from the beginning of a link.

8.3 SOME OPERATIONS FOR GRID-CELL BASED MAP OVERLAY MODELLING

Much spatial analysis has traditionally been done by analysis of spatial distributions of phenomena via map overlays. We illustrate the character of map modelling firstly using regular tessellations, and then, in section 8.5, using topologically structured vector data. Operations on data in grid cell or lattice form are different from those for spatial entities recorded as geometric objects. Computational geometry generally gives way to simpler procedures of counting and attribute value searching and changing. The operations can be thought of essentially as examining one or a combination of attributes, which may be spatial properties, for a given single cell unit or point in a lattice, or for a set of cells in tessellation layers.

8.3.1 Basic operations for grid-cell data

In thinking about manipulation of data in grid cell form it is usual to imagine a series of layers of different map themes (Figure 6.10), for example forest type, distance from a city, and gradient of slopes which are used in some modelling procedures. It is not necessary for the data to be stored in this way, though. Operations may be arithmetic or logical,

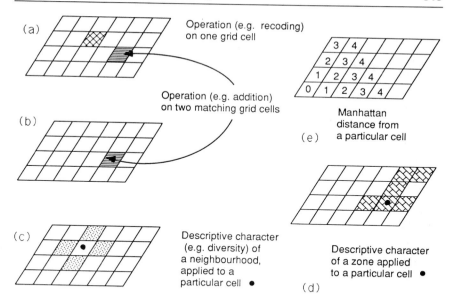

Figure 8.4 Types of spatial contexts for grid cell data operations. (a) Operation on one grid cell. (b) Operation on two matching grid cells. (c) Descriptive character of a neighbourhood. (d) Descriptive character of a zone. (e) Distance from a particular cell.

provided that the attribute data are measured in ways to which those classes of operation are appropriate. So, two layers may be compared for similarity or difference of values or categories for each and, potentially every, cell. Or, arithmetic operations, like adding or multiplying, can be undertaken for individual cells if the numerical data are measured on interval and ratio scales. For example, one layer has elevations at the top of a stratum and another layer has elevations at the bottom: the height of the stratum can be determined from the difference.

Some operations for a single theme may use values for a set of neighbours for a given grid cell, instead of just the individual cells. For example, in smoothing local variabilities of elevation, to get an overall trend, the values of every set of nine cells can be employed rather than just the cell value in the middle of that set.

Operations may be actions for a single unit, one grid cell; or a repetition of the same action for all individual units, the set of grid cells, or a subset of cells making up a neighbourhood or region. Following Tomlin (1990) these different contexts, illustrated via Figure 8.4, may be categorized as:

1. Characterizing individual locations (that is grid cells).
2. Characterizing locations within neighbourhoods.
3. Characterizing locations within zones.
4. Descriptions of spatial properties.

The operations in the first category (Figure 8.4a), dealing with cells divorced from any spatial context, include: reclassification of the stored value, if an alphanumerical code; or rescaling arithmetically or functionally, for example square root or other mathematical operations, if numerical; or sorting on the basis of all values for the one layer, and assigning rank order values. Other actions result from comparing values for the same grid cells for different attributes (layers), perhaps by addition, average, difference, product, maximum, minimum, ratio, variability, cross-classification and so on.

17	12	35	33
11	15	11	34
26	32	34	34
99	26	41	47

RECLASSIFY --
original land use
code values are
changed by grouping
in major land use
categories.

10	10	30	30
10	10	10	30
20	30	30	30
00	20	40	40

SCAN FOR DIVERSITY
-- each 9 cell block is
scanned; number
of different land use
categories determined,
and value applied to
middle of block
of 9 cells.

10	10	30	30
10	10	10	30
20	30	30	30
00	20	40	40

PREDOMINANT
value for land use
category in each of
four cells touching given
cell at edges.
( ) indicates no type
predominant, therefore
the cell value is used.

10	10	30	30
10	10	10	30
20	30	30	30
00	20	40	40

1	2	2	2
3	3	2	2
4	5	4	3
3	4	3	2

10	10	10	30
10	10	30	30
(20)	20	30	30
20	(20)	(60)	(60)

(a)   (b)   (c)

The grid cells may be assigned values to characterize some conditions in an immediate neighbourhood, a block of territory defined by the adjacent cells, either four (Figure 8.4c), if only edge touching is allowed, or all eight neighbours.

Thirdly, new values may be created for each cell on the bases of observed conditions in larger areas, extended neighbourhoods, or for zones identified by the homogeneous region in which the cell falls, on some arbitrarily defined block of cells (Figure 8.4d). All the locations within a zone will receive values for the computed or otherwise derived measure for an attribute in another layer for that zone, for example, average, highest value, diversity, and so on.

For any of these three situations, new data are produced by various operations that generalize or otherwise modify observed values. Figure 8.5 illustrates just a few. Thus, each cell takes on the value for the

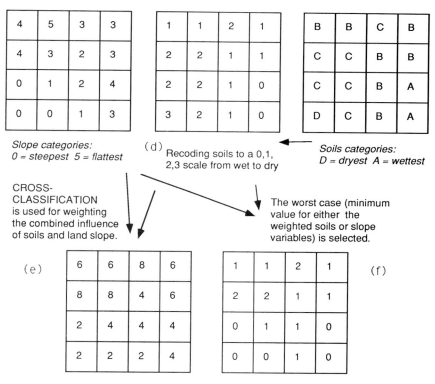

Figure 8.5 Some examples of operations for cell data. (a) Reclassification. (b) Scanning for diversity. (c) Indicating predominant value. (d) Recoding. (e) Cross-classification. (f) Selection of the worst case.

predominant land use in the nine cell area (Figure 8.5c), or the diversity of land use (Figure 8.5b) as indicated by the number of different categories found. After slope and soil conditions are simplified by recoding into just a handful of categories, each cell is given a weight based on the joint conditions of slope and soils, and then the worst case, the minimum value of the two numbers for a cell, is written out as the result.

The fourth category includes the creation of measures for spatial properties, like distance or narrowness of regions. It also includes determination of slope and aspect from elevation data by looking at the difference between a cell's value and that of immediate neighbours. Gradients may also be computed for other variables, like income level or percentage population with college degrees, that are scalar. Immediate and extended neighbourhoods or zones can be examined for spatial properties like length or area of objects, or gradients. Accumulations of properties with increasing distance from a focal point, line or area, can be determined by spreading outwards in distance increments and counting the numerical values for an attribute for the cells falling in different distance zones (Figure 8.4e).

Area and perimeter measures for homogeneous blocks of cells or other sets of contiguous units grouped into zones, perhaps via a special thematic overlay, are obtained, respectively, by cell counts and summing the exterior edges of cells in the zones. Distances, obtained by row and column coordinate differences and application of the Pythagoras' rule for right triangles, are quite easily obtained although, as pointed out in Chapter 6, subject to error. Distances for individual cells to a boundary of a zone or from a linear or point feature can be readily accomplished principally by cell counting operations. Shapes of blocks of cells can be measured crudely by comparing the perimeter of a zone to that length likely for a regular figure, usually a circle, of the same area, and basic spatial statistics like centroids are easily determined from row and column integer values for vertical and horizontal coordinate axes.

Thus, while the concept of the entity and spatial relationships are not natural for grid-cell systems, nonetheless, most spatial properties can be fairly effectively obtained via the attribute data. Polygons are identified by a common code, and a conceptual layer that consists of sets of cells coded for the different units, for example, counties. Or the counties may be shown by pixels coded for the boundaries of spatial units. Point features can have numerical codes referring to data in another table, for example, one containing city names. An overlay of county and city codes can produce the equivalent of a contained-within concept.

8.3.2 Spatial modelling with grid-cell data

The grid-cell form of data encoding facilitates map analysis involving many data items or processing steps. Figure 8.6 serves to illustrate some of the arithmetic procedures commonly encountered in working with cell data for a **map overlay modelling task**. Preliminary planning has produced a flowchart of operations required to produce a single scale of numbers representing the potential for residential development in areas not yet built-up. The data processing consists of a mixture of operations drawn from the four categories noted above.

Proximity to services and highways is represented by incremental distance values, inverted to indicate declining desirability with distance, or based on a minimum distance as in the case of the highways. Original soil categories have been assigned code numbers suggesting their susceptibility to easy construction, especially the soil wetness. Slope gradients have been computed from elevation data. The variety of existing land cover in cells, computed by scanning the neighbourhood, is a measure of the likelihood of finding homogeneous conditions for building. The relative influence of slope and soil types is combined by assigning potential values after a cross-classification operation. New cell values are combined in various ways to obtain the composite final scale for the potential for residential development, as shown.

This procedure has used elementary logical operations in order to apply the renumbering. For example, the logical test logical OR has been applied to two attributes, looking for the existence of either D for the first variable, and 23 for the second, and then assigning a weight of 10 to the result of the selection. Cells with the joint condition of D and 23 are identified by the logical AND operation. A third situation of selecting the complement of the overlap can be undertaken using the logical XOR operation. Complex combinations can be created by suitable logical statements combining more than two attributes.

The ease of undertaking such logical processing (based on **set operations**) with cell data, arises from the use of identical spatial data units (the grid cells), and of simple binary or decimal arithmetic. In a comparison of two attributes, each of which can take on two states (presence or absence) denoted by 1 and 0 respectively, the intersection set operation (logical AND) of the two is the product of the two values 1. The other three possibilities each produce a zero product, when multiplying by zero. The union operation produces values of 0, 1 or 2 for the logical OR operation via addition, also identifying the complement of the union by the zero value.

Much use of grid-cell tessellations for spatial analysis arises from the relative simplicity in comparing two or more spatial distributions, in the manner of traditional map overlays. The comparisons do not necessitate time consuming computational geometry operations to find where lines intersect as is necessary in overlaying geometrically encoded coverages. However, the simplicity of the process rests upon a requirement that cell tessellation layers match spatially. Mapping and modelling are more cumbersome otherwise, perhaps involving the need for scaling, origin

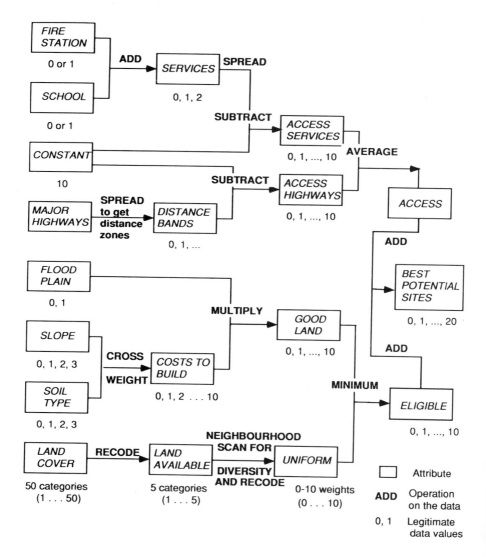

shifting, and rotation, which operations can involve further assumptions or rebuilding databases completely.

If the layers do match, even though there is no explicit encoding of spatial relationships, neighbourhood effects can be ascertained, various spatial statistics can be computed, and surfaces are well represented. Thus, the field oriented approach inherent in raster tessellations has several advantages to put alongside its limitations of not being able to position precisely point and line features nor to accomplish route problem solving easily via a graph representaton.

8.4 OPERATIONS FOR QUADTREE TESSELLATIONS

The quadtree and pyramid varieties of regular tessellations can facilitate operations associated with grid-cell data sets, and may accomplish some actions faster. To illustrate the basic process, consider the combination of the two maps of Figure 8.7. Separate quadtrees are made from the cell encoded data, and then a traversal of each tree, from the top level down, takes values for comparable spatial units and creates a third map. If there is no branch at a given level, then the value found at that level for the first attribute is assigned to each of the four squares at the same level for the other attribute.

In the example the different combinations of soil and slope conditions are obtained by assigning the binary states for each attribute to the sixteen cells. The bottom right quadrant, the fourth branch of the tree in the NW, NE, SW, SE ordering for the diagram, has a homogeneous block of side-length 2 for each attribute, thereby accounting for four small cells with the combined character of low slope and wetland. The

Figure 8.6 An example of map overlay modelling. The example demonstrates a simple model with eight data items, ten different operations, and a few spatial statistics, for producing a single scale of integers representing the suitability of land (as encoded as grid squares) for residential development. The final scale ranges from zero for worst to twenty indicating the best. The modelling process moves from left to right. Original data are used for seven items; the eighth, CONSTANT, a map of uniform conditions, is specially created. Modelling occurs sequentially along several tracks. Several data manipulations occur in parallel; there is one example (cross weighting of soils and slope conditions) of the simultaneous use of data items. Particular spatial operations include CONSTANT, SPREAD for producing distances from specific locations, and SCAN to measure the diversity in a set of eight cells touching a particular cell.

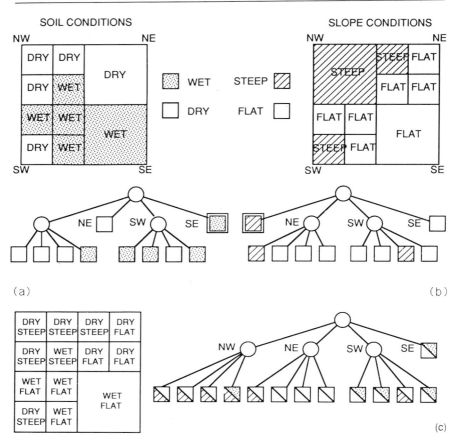

Figure 8.7 Combination of two maps via quadtrees. (a) Soil conditions map and tree. (b) Slope conditions map and tree. (c) Combined map and tree.

one piece of wetland with steeper slope is the fourth cell in the top-leftmost block of four.

The quadtree is especially useful for performing set operations like union and intersection, again doing this by traversing simultaneously two thematic trees, making tests for the attribute coding of the nodes of the hierarchical structure. Figure 8.8 demonstrates the basic process, revealing by intersection the single cell that should be avoided at all costs because of the dangerous combination of the presence of both wetness and steep slope. A similar double tree traversal for union produces the cells meeting the condition of either steeper slope or wet conditions.

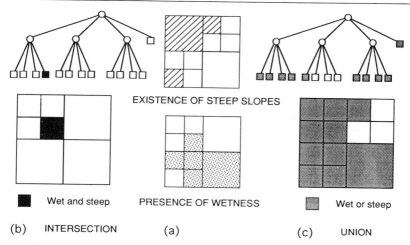

Figure 8.8 Union and intersection for quadtrees for the presence of both phenomena. (a) Original maps. (b) Tree and resultant map for intersection. (c) Tree and resultant map for union. For (b) and (c) tree diagrams the groups of four identically coloured nodes are shown only to facilitate comparisons. The tree branches do terminate at the higher level.

Other traversals could reveal other combinations like the joint occurrence of dryness and flatness.

Area computations are made by counting quadrants of particular conditions and multiplying by the size of the quadrant. Real area unit measures may then be obtained by multiplying by a map scale factor. So, in Figure 8.7, the second level has a block of wetland, and at the third level, there are four smaller cells also with wetland, giving a total area of half of the study area.

Algorithms exist for determining areas of regions made up of neighbouring pieces (Samet, 1990). Because cell adjacency is usually not coded directly in the data structure as in topological models, neighbours are not as easily ascertained. However, concepts of graph theory can be used to ascertain edge and corner neighbours, by a process called **connected component labelling**. This process also underlies the cell expansion device, to produce the equivalent of geometric buffering. However, distance computations are not as straightforward as with individual cell encoding, and the rotate, translate and scaling operations are more difficult than for vector geometry, particulary if the rotation is by any angle other than a multiple of ninety degrees. Of course, the quadtree inherits this condition of non-variance from grid cells; it is a feature of data sets based on arbitrary subdivisions of space.

Changes to the spatial base are more difficult than with vector representations because attribute data are an integral part of the spatial encoding. If translation and rotation are required, possibly to bring tessellations into spatial alignment, then the database has to be recreated (Figure 8.9). The true conditions of the distribution of phenomena may be violated in this process if the translation shift is not a distance equal to the cell size or a multiple thereof, and if the rotation is other than ninety degrees. For irregular cell representations rubber sheeting changes to the geometry can readily be accomplished without affecting the attributes.

Overall, it appears that hierarchical quadtree tessellations are good for searches of single or combined characteristics that might also involve some spatial properties like area and neighbours. Both browsing and specific query tasks can be accommodated. Spatial overlay operations and some kinds of geometric transformations can be efficiently handled by the quadtree structure. Manipulations of the values of attributes, of the kind represented by map overlay modelling, are handled outside the quadtree structure. Also, ascertaining the spatial relationships of containment and nearest neighbours is easily handled for quadtrees, while graph oriented problem solving is not.

8.5 OPERATIONS FOR IRREGULAR POLYGONS AND FOR GRAPHS

Many spatial data mapping, query, analysis or other problem solving tasks pertain to polygon entities. Much of the early development of spatial information systems worked with continuous distributions of phenomena as grid cell tessellations, rather than as discretized polygons, for example for composite mapping and modelling resources or for phenomena like soils and land cover. However, it became apparent that some queries or analyses needed explicit topological encoding, to undertake more effectively tasks like:

1. Ascertaining the type of land use on either side of hedgerows, or the sequence of land-cover along rivers.
2. Amalgamating small areal spatial units to make larger regions.
3. The allocation of attributes for polygons at one point in time to a different set at a later time.

The early and later concepts for organizing attribute data differentiate between simple, single tables and sets of tables, based partly on the way in which the spatial units data were structured. The first keeps the

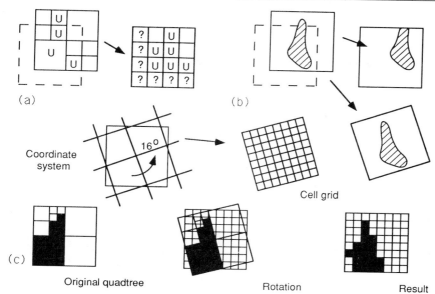

Figure 8.9 Translation and rotation spatial operations. (a) Translation for quadtree. (b) Translation and rotation for object. (c) Rotation for quadtree. The details of the result are influenced by the encoding rule, for example if squares are coloured black if only a portion of the square is occupied. (Figure 8.9c is based on Samet.)

polygon attributes in a different table so as to avoid a lot of repetition because the spatial data are tabulated with lists of records for the vertices within a polygon. The second relates attribute data for polygons by use of a common link (the polygon identifier) associated with the centroid (sometimes called the data seed), while also linking to the line segments that make up the polygon. Neither of these works well for containment, so that a hierarchical organization of attributes is necessitated in this case.

8.5.1 Creating regions

Combinations of polygons based on a single attribute to achieve larger units meeting certain criteria are encountered in many **districting** or regionalizing situations. Regions may be created to meet general criteria of homogeneity, for example sugar producing areas; functionality, such as newspaper circulation, or administration, for instance school districts. The resulting regions may have different spatial properties: nodal, based on a focus, as in newspaper circulation areas; non-overlapping or overlapping, or with or without enclaves and exclaves.

If we take the case of producing territories for political representation, we have a relatively simple case, geometrically speaking, almost always requiring no enclaves or exclaves, no overlap and no territory omitted. In the example for this task, the higher level regions, the legislative districts (the units used for representation), are created from smaller units, the election constituencies (the units used for voting), by adding the latter in order to get to a certain number of people contained within the larger spatial unit.

If the procedure is an intuitive rather than mathematical one, then the requirements for handling spatial data address the properties of contiguity, containment and sometimes shape of the districts created (in case of claims of gerrymandering). A topologically structured database with adjacency encoded is highly desirable for undertaking additions of neighbouring polygons; otherwise the process would have to be based on visual indications of adjacency carried in maps of precincts.

A system developed for political redistricting in Maryland in 1981–1982 (Thompson *et al.*, 1983) demonstrates some of the requirements for a special purpose spatial information system, and further examples of the great variety possible in data organization and encoding. Working from the following set of criteria applicable to the creation of districts for the state legislature, as stipulated in the constitution of the state, or as otherwise deemed appropriate, conditions could be that:

1. Districts should be substantially equal in population.
2. Districts should consist of one piece without holes.
3. Districts should not have unusual shapes suggestive of gerrymandering.
4. Districts should recognize major natural features of the state.
5. Due regard should be paid to boundaries of political subdivisions.
6. Districts should not dilute minority people's voting strength.
7. Boundaries should be drawn to avoid battles between incumbents.
8. Boundaries should recognize communities of interest.
9. Districts should foster voter recognition of representatives.

From this set of legal, political and geographic criteria, without implying which were actually used in the specific case, let us concentrate on just a few with regard to data requirements. Reference to gerrymandering implies the availability of shape statistics, necessitating geometric data processing. The combination of election precincts to larger units is substantially facilitated by knowing neighbours of precincts. Use of the racial balance item would require allocation of data for census tracts to election precincts in cases where boundaries do not match. Dealing with the physical features can be nothing more than making maps of topography. Otherwise, to facilitate a process of interactive district

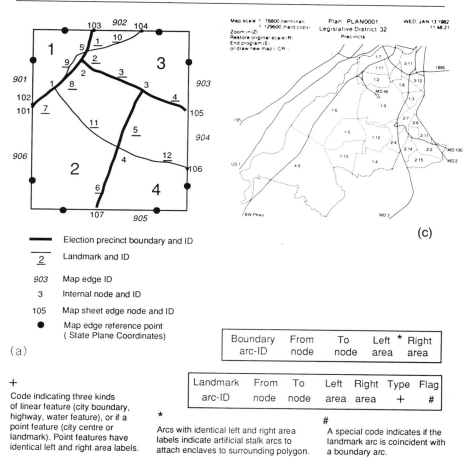

(c)

Election precinct boundary and ID

2   Landmark and ID

903   Map edge ID

3   Internal node and ID

105   Map sheet edge node and ID

●   Map edge reference point
(State Plane Coordinates)

(a)

Boundary arc-ID	From node	To node	Left * area	Right area

+

Code indicating three kinds
of linear feature (city boundary,
highway, water feature), or if a
point feature (city centre or
landmark). Point features have
identical left and right area labels.

Landmark arc-ID	From node	To node	Left area	Right area	Type +	Flag #

*

Arcs with identical left and right area
labels indicate artificial stalk arcs to
attach enclaves to surrounding polygon.

#

A special code indicates if the
landmark arc is coincident with
a boundary arc.

(b)

Figure 8.10 Spatial data model for redistricting. (a) Map encoding.
(b) Data format for districts and orientation data. (c) Example of a
map of a districting plan. The map shows the overlay of highways to
provide orientation. Election precincts are shown within one
legislative district that was created as part of a districting plan.

creation by looking at maps on a monitor, it is helpful to have orientation
information such as major highways, rivers, shorelines or other
landmarks.

   The design concept recognized the primary importance of the
topological information, and the secondary, but still important, role for
the geometry. It also separated the information needed for orientation
purposes from that used in building or mapping districts. The data

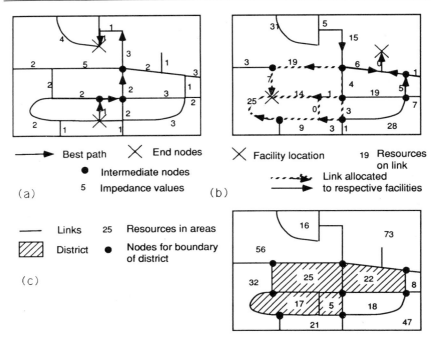

Figure 8.11 Location problem solving. (a) Routing. (b) Allocation.
(c) Districting.

encoding and structure used (Figure 8.10) therefore had different
conceptual layers for districting and orientation data, but used a
coincidence coding technique where boundaries of different types were
identical, to reduce the amount of data stored. The boundary and
landmarks were prepared spatially as in Figure 8.10a, and had data
formats as in Figure 8.10b. Special requirements included two versions of
a contiguity matrix, one allowing connections across the Chesapeake Bay,
the other not; provision for interactive adjustment of computed centroids;
routines for two compactness statistics; programs for node and polygon
edits of the boundary data, and use of bounding rectangles for the
polygons.

8.5.2 Location problem solving

It is interesting to examine a graph data structure in this regard for
comparison purposes. Suppose we wish to accomplish several kinds of
location problem solving within one information system: routing,

districting and allocation of customers to service centres (Figure 8.11). First imagine a distribution of people according to houses and street addresses, the location of schools with a fixed size, information about travel times on city streets, and a target number of people to be contained within the legislative district. If the data are structured according to city streets, population is associated with the graph edge representing the street, possibly recognizing one or the other side, the school is a node on the network, and the individual graph edges for the streets have attribute data for travel times.

An intuitively based procedure for routing the school bus would be to follow the streets, accumulating traffic time along the way, by different paths to find acceptable routes. If the number of schoolchildren are assigned to the street links, without recognizing on which side of the street they live, then the number of children along a route can be accumulated (Figure 8.11a). Indeed, the process of accumulation can begin nearest the school and work outwards to collect the 'resources', the schoolchildren, and then compare to the maximum allowed at a school (Figure 8.11b). If a district is to be created, then the city blocks must be joined – the population associated with the street segments has now to be known for each side of the street, or the aggregations have to use the data for line segments (Figure 8.11c). Indeed, the orientation of data to the graph edges makes the districting operation awkward even while facilitating the routing and allocation problem solving. This is because the district building inherently requires polygon data, not edge data.

Some location problems are not graph-based. Thus, human or natural flows can occur between specific points or between point and line features, for example, overland flows of pollutants ending up in a river. Problem solving techniques may require special path finding over surfaces with different conditions and which are represented by polygons or grid cells or they may involve the concept of a virtual link between points, that is, the dyad. For example, predicting inter-city air-traffic volumes in the past was often done by means of a gravity model, requiring data for factors such as the size of nodes and their separation (miles, air fares, travel time).

8.5.3 Map overlay modelling and analysis

Much work with polygons also uses a map overlay concept to allow association of different themes. We highlight further some of the differences in working with regular and irregular tessellation data by undertaking a map overlay model now for a topologically structured database. Tasks are very similar to the example using grid-cell data: for example, designing a predictive model, identifying logical combinations, and applying arithmetic manipulations to the attribute values to produce a new composite measure.

This time, though, many of the attributes require geometric operations, especially line intersections for creating new polygons with combined characteristics, or creating proximity zones. Where the data are encoded as thematic layers, the particular operations will come from a set like that shown in Figure 8.12a. In addition to bookkeeping or utility functions like clipping, erasing and splitting, and spatial operations like the creation of buffer zones, it may be necessary to use set operations like intersect, union and identity. The general process, illustrated first for an intersection of line and polygon units (Figure 8.12b), requires line intersections and polygon assembly operations to be performed, followed by assigning attribute codes to the new units created. However, if the

(a)

attributes are scalar quantities, there is no allocation as part of this process because the geometry and attribute data handling are separate. Figure 8.12c shows the simpler case of polygon–point overlay, requiring the determination of which polygons particular points fall into.

The comprehensive modelling task represented by Figure 2.8 requires many computational geometry operations (Figure 8.13). Three different buffer zones are established; two of these, schools and fire stations, are intersected, producing some holes. The highway buffer is used to erase part of the entire area, and a composite access layer is created by spatially joining the different buffer zones. The map layer spatial joins for union and intersection allow implementation of the elimination of certain types of land use or land cover, or for combining with buffer zones. After the final map layer, ELIGIBLE, is produced, then the modelling process using numerical computations is invoked to produce the scalar values of levels of potential for development. The conceptual and visual overlay for

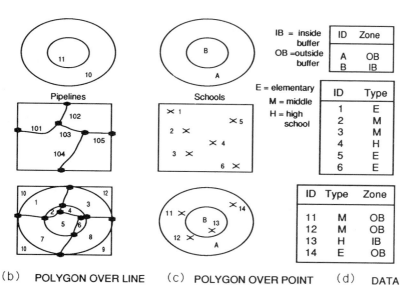

(b) **POLYGON OVER LINE**   (c) **POLYGON OVER POINT**   (d) **DATA**

Figure 8.12 Some polygon overlay operations. (a) Overlay covers, here polygons, are superimposed over the primary layers, also polygons. (b) Details for a polygon over line identity situation; lines are split where they cross the boundaries of two regions. (c) An example for a polygon over point intersect operation. (d) Data used for the polygon over point example. Of six schools, only one is in the inner buffer; three are in the outer buffer; and two are outside the study area.

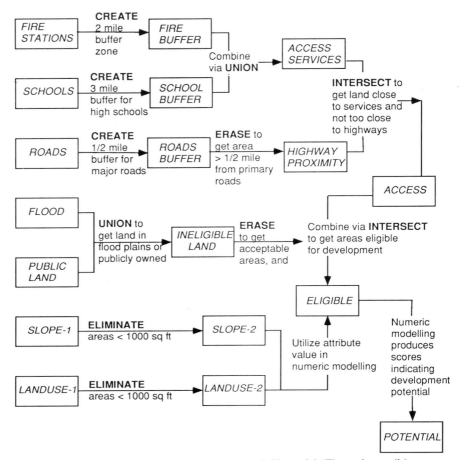

Figure 8.13 An example of overlay modelling. (a) Flow chart. (b) Resultant map – shown as Figure 2.6. A process similar to that for Figure 8.6 is here represented for information contained within a layered topologically structured database. Seven initial map themes are processed sequentially along several tracks; then combined to produce a final map with polygon units of land possibly suitable for some development. These polygons are the areas shown in Figure 2.6, where highways, high schools, and fire stations are also depicted. The impact of circle buffer zones is clearly seen in the lower part of the map. Other major areas eliminated are government owned land not available for development, and flood plains. After the hundreds of polygons have been created by the overlay operations, to produce the cover called ELIGIBLE, then factors of soil conditions, terrain slope, and land use are employed in the numeric modelling stage to create weighted scores for development potential.

such a project is not difficult, but numerical overlay is quite demanding computationally as discussed earlier in this book.

We now leave the discussion of spatial analysis to conclude Part Two. However, other location problem domains like path-finding and overland flow modelling are not forgotten; we return to many more functional needs later.

8.6 INTEGRATION AND MULTIPLE REPRESENTATION

Over time, orientation has shifted for both theoretical and applied domains from simple data sets and processing programs to more complex data organization and general purpose software systems. As this happened, interest has also grown in integrating different kinds of spatial and thematic data in one software system, or in having multiple representations within a single framework. Several aspects are pertinent:

1. The ability to store particular classes of object in different representations in a database, such as contours stored as a string of coordinate pairs, or as the parameters for a Bézier curve.
2. The provision of tools to convert the stored form of representation to another form for particular purposes.
3. The inability to deal with all situations extensionally.
4. The original capture of data in certain forms such as waveform seismic signals or raster encoded data from satellite sensors.
5. The circumstances of integration of different kinds of spatial data, whether at the level of the screen display or at the level of computing procedures called when particular spatial analysis tasks are undertaken.

This desire to be able to work with varied spatial representations partly reflects a recognition of the difficulties of undertaking some tasks with particular forms of data. In particular, there are some things done well via graph models, others by triangulated irregular networks, others by grid cells, and others by quadtrees. For example, terrain surface representation for land use analysis seems to be conceptually more appealing if triangular facets are used, while triangles and regular cells both have limitations for modelling overland water flow. For this purpose quadrilateral chunks from contour bands may be better, whereas stochastic fractals may be more effective for representing glacial landscapes, and lattices are preferred for producing interpolated values for contour mapping.

Indeed, a pertinent issue is whether users' interests may not be better served by special purpose databases and/or software systems, represented by the one for Maryland redistricting noted in section 8.5.1, or commercial software like *SALADIN* and *TRANSCAD*, geographic information systems for transportation analysis. We indicate below how some spatial databases and commercial software deal with varied spatial and attribute data.

8.6.1 Multiple representation

Multiple representation refers to the modelling of different forms for different spatial objects or instances of those. The representation may be rule based or use extensional data. Recalling an earlier discussion in this book, we note the use of splines, coordinate strings, generalized forms of original coordinate pairs, enclosing boxes, and parametric equations as possibilities for lines; perimeter boundaries, link–node structures, or a combination as representations for areal objects; and irregular triangles, grid squares, contour bands, or quadrilaterals as examples of earth surface facets. Clearly, a one-to-one relationship is unlikely between type of object and its representation within a spatial information system.

It is unlikely too that a single way explicitly to encode spatial phenomena can efficiently provide for all the diverse needs and uses of spatial databases. Different representations have distinct inherent properties in terms of:

1. Geometric computations.
2. Positional error.
3. Maintenance of topological consistency.
4. Searching for pieces of space.

Hierarchical or mixed spatial decompositions or clusterings, such as KD-, quad- or R-trees, are not well suited to maintaining topological consistency. Several topologically structured vector representations are good for this, but make it more difficult for windowing and query operations. Neither deal directly with the joining of semantic descriptions of spatial phenomena with their positional and relational components. Nor is there a one-to-one match between use or user and representation; multiple representations may be needed for different purposes, so that different user needs can be effectively addressed.

8.6.2 Integration

Integration in one sense means an ability of a software system to work directly with different spatial data types, providing tools to link and/or convert between the different forms of representation. While this may suggest only one form of representation needs to be stored, there are other considerations that might necessitate storage of multiple forms. In another sense, integration relates to the use of thematic and spatial data, and also potentially temporal and symbolic information. In yet another sense, integration encompasses the combining of data from different layers for a database with a layered architecture.

We develop in Part Three some approaches for certain types of integration of different types of data. For the moment, we consider just the question of the joint use of grid-cell and vector data, or the combination of continuous surface and entity perspectives.

For the combination of grid-cell and vector data, integration includes these possibilities:

1. The combination of both types of data in one regular cell structure.
2. The use of cells for organizing and accessing vector data.
3. The use of grid cell data as background for editing or updating vector data.
4. Having both grid-cell and vector spatial representations in one database.

The first two items are similar in implying the use of regular tessellations for spatial decomposition to facilitate access to a database. They are different in, on the one hand, including geometric data as part of the regular tessellation units, and, on the other, having spatial entities recognized as the data units. The third type provides a way to work with both image and vector data at the stage of database creation and update, not for querying or other use of information.

The fourth item (which in itself is not integration but a matter of storage of multiple representations) raises the question of at what level in an information system there should be integration. This can be at the display screen level only; or for processing operations in which both vector and image representations are included and used as appropriate for particular purposes, or at a level closer to the users, in which they see no difference between data types.

Regarding the entity or field approaches, there are these options:

1. Representation of continuously varying phenomena as point data, with tools to discretize as needed.

2. Representation of discrete phenomena with tools to convert to continuous distributions, if needed.
3. Representation in both continuous and discrete forms, for line and area data.

If uses are not known ahead of time, then an argument can be made for varied representations. The field and entity orientations can both be accommodated for a full range of contexts and data types, either by providing conversion tools or double representations. We will shortly describe the example of the *ARC/INFO* system in enabling the use of polygon and triangular irregular tessellations, graphs and regular lattices, in order to associate earth features of different spatial form, for example land conditions with terrain elevations and drainage networks.

The general organization (the architecture) of the database also influences the ability to integrate different pieces of data. First of all (as is discussed fully in subsequent chapters), attribute and spatial data may be kept separate or integrated. Recalling earlier comments, we note that the classical approach has been to keep these two types of data separate, not only when thinking of mapping systems but also in the context of database management systems. For example, the latter have limitations in assembling quickly all the large number of instances of myriad primitive level units for graphics oriented tasks like visual thematic displays or for complex queries.

Secondly, we can imagine a continuum of possibilities from a single thematic layer with varied content to a series of layers each singular in theme. Joining different themes into multiples is dependent on matters such as:

1. The nature of the users' needs for many themes at one time.
2. Practices and protocols for the creation, maintenance, integrity and updating of data.
3. The complexity of the phenomena.
4. The real world spatial coincidence of different kinds of entity.

If phenomena coexist in reality, that is, spatially covary areally or are linearly coincident, then an implementation may reasonably well take advantage of shared geometrical form at the level of spatial element primitives, thereby fostering a single layer approach. However, if phenomena do not match well, or are particularly spatially complex, then a separation of themes may be better. Users' needs, especially if they can be anticipated, may lead to different orientations; working with many attributes of a single class of spatial unit places little burden on the processing system, but undertaking analysis or modelling using different

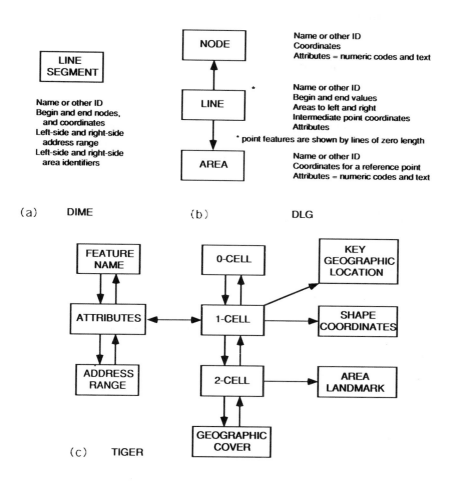

Figure 8.14 The DIME, DLG, and TIGER data encodings. (a) The Dual Independent Map Encoding concept. (b) The Digital Line Graph concept. (c) The Topologically Integrated Geographic Encoding and Referencing system concept. Features that are not part of the underlying tessellation of areal units, for example particular buildings or land reserves, are associated to either the one- or two-cells in that tessellation. (Taken from Marx, 1986 with permission.)

kinds of spatial entity may produce complexities, as demonstrated earlier in this chapter. As a practical matter, an institutional framework that places responsibility for different themes in different places or agencies will generally be better served by a layered architecture.

8.6.3 Some examples of public and commercial spatial data organization

The great variety of ways of organizing and representing spatial and attribute data, and, for many people, the resulting confusion because of a bewildering array of alternatives, is illustrated here by practical examples from public and private domains. It is the purpose of Chapter 9 to begin a theoretical structure for treating multiple representations and integration.

The data organization used by the Census Bureau of the USA for the 1990 Census enumeration and mapping has a more fully developed topological structure known as the *TIGER* (Topologically Integrated Geographic Encoding and Referencing) model (Marx, 1986). The underlying map geometry and topology are interlocked as a series of logical tabulations and links between them (Figure 8.14c). Spatial searching is facilitated by the use of Peano curve reference keys for the zero-cells spatial units.

The broad structure uses the topological one-cell as the principal element, to which are related zero- and two-cells, and the point, line and area features not part of the planar-enforced complete map of the USA. Thus, the shape points for line features or polygon boundaries and area landmarks are kept separate but connected, as are the labels for features, the address ranges for the lines and the descriptive data for the one-cells.

Points that are not part of the tessellation of polygon units representing adminstrative units, data collection areas like census tracts or physical features like blocks, are divided into two types. Key geographic locations, for example a tower, are linked to one-cells, in that it is usual for a street address to be important data for such features. The landmarks, the second type, are linked to the two-cells in which they are located.

The comprehensive *TIGER* structure is an outgrowth of the ideas of the dual independent map encoding, developed for use in census enumeration activities in the 1970 Census (Figure 5.8) suitable for matching street addresses to the lines, and indicating neighbouring-particular areal units at different levels of 'census geography' (Marx, 1986). Coordinates for nodes were integrated into the table, as could be any attribute for the line segments. Because all lines were straight there were no other coordinates to deal with, and because there was no explicit polygon encoding, there were no polygon attributes as such, although these could be associated with the polygon codes by reference to other tabulations.

In contrast to this simple concept (Figure 8.14a) the *DLG* (Digital Line Graph) format encoded the contents of the 1:24,000 topographical quadrangle map sheets as separate layers of information for the main feature classes of hydrology, transportation, political and administrative boundaries, and the public land survey system. Breaking down the geometry and topology into elemental units of nodes, lines, and areas, and coordinates, the Geological Survey of the USA provides tabulations with data as shown in Figure 8.14b. Notice that areas have a centroid carrying the polygon label and coordinates; the lines have strings of coordinates; the attributes are held in separate tables connected by means of element labels. Special conditions of unconnected polygons are shown by lines with the same beginning and ending nodes, and point features are recorded as collapsed (or degenerate) lines, having the same coordinates for beginning and ending nodes, and identical right- and left-side polygons. Well oriented to data exchange purposes, the *DLG* is less convenient to users except in mapping.

An example of a data organization for areal phenomena that are not possibly structured as a completely connected graph, that is, enclaves exist, is that of the US Geological Survey's Geographic Information Retrieval and Analysis System (*GIRAS*). Based on elemental units of nodes, arcs, polygons, islands and polygon labels, the structure was developed in the early 1970s for encoding principally land use and land cover, and associated thematic overlays such as political units and federal land ownership units (Mitchell, 1977). Tabulations are compiled for: arcs, including the label and attribute for polygons to left and right sides, begin and end node labels, and length; for polygons, including attribute, area, and centre coordinates; for coordinates for arcs, and a listing of arcs making up polygons. The question of containment was indirectly addressed by recording in the polygon record the islands within a given polygon and the polygon containing the given polygon.

The hierarchical arrangement of administrative units coded as polygons was addressed in the commercial Geographic Information Mapping and Management System (*GIMMS*) software, developed in the early 1970s, by encoding a level explicitly in the basic record for line segments. Provided nesting exists, different levels of an administrative hierarchy can be encoded by labels associated with polygon boundaries. In general the *GIMMS* software will recognize line, area and node entities, points and centroids, but is particularly oriented to thematic

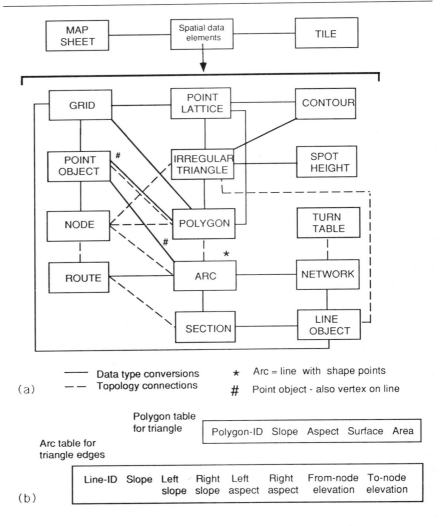

(a)

Data type conversions   ★   Arc = line with shape points

Topology connections   #   Point object - also vertex on line

(b)

Polygon table for triangle

Polygon-ID	Slope	Aspect	Surface	Area

Arc table for triangle edges

Line-ID	Slope	Left slope	Right slope	Left aspect	Right aspect	From-node elevation	To-node elevation

Figure 8.15 The ARC/INFO data elements. (a) The data elements and space subdivisions. (b) Some details for the TIN component—data items in the tables produced by the triangle to polygon transformation. (Based on material provided by The Environmental Systems Research Institute Inc., Redlands, California, about the new release, version 6.0, of the ARC/INFO system. Note that the illustration is not intended to convey the details of the ARC/INFO data structure, which has undergone substantial changes in 1990–1. The reader is referred to the Environmental Systems Research Institute for details about the ARC/INFO software.)

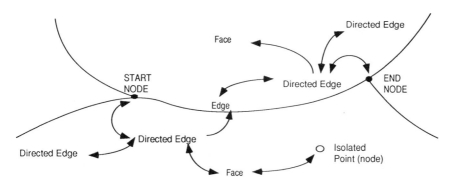

Figure 8.16 The TIGRIS approach: geometry for the topological elements. (Reprinted from Herring, 1987, with the permission of the American Society for Photogrammetry and Remote Sensing, Bethesda, Maryland, USA.)

mapping for polygons. A layer concept is not a feature of the *GIMMS*. Attribute data are tabulated separately as data items in ordinary tables, being associated with the locational data by common identifiers.

The first general purpose geographic information system commercial software package, *ARC/INFO* (Morehouse, 1989), originating about 1980 and developed by Environmental Systems Research Institute in California, has a topologically oriented structure for the spatial properties for point, line and area objects, and associated tabular data structure for attributes (Figure 8.15a). Different attribute themes are conceived and stored as layers, so that there is a separate set of attribute data for each layer. Connectivity and adjacency information are recorded for arcs representing polygon boundaries; coordinates data are kept separate from the attributes; and some spatial properties, perimeter and area, are computed and written out as attributes in feature attribute tables which are linked to the spatial data tables via the entity identifiers. Containment is not explicitly addressed in the spatial data organization, only via implicit encoding for arcs that are assembled to get polygons, by separately identifying the arcs making up the inside and exterior boundaries of enclosed polygons.

Over the years the *ARC/INFO* system has expanded to also incorporate other data types – lattice grid and triangulated irregular tessellation data for surface conditions, and a graph structure for network type spatial problems (this is described more fully in Section 12.5.1). The irregular tessellation structure produces topological encoding for triangles, as already demonstrated conceptually via Figure 6.26, and facilitates the production of proximal regions. After conversion, the triangles can be processed as polygons in conjunction with the main part of the software. Figure 8.15b indicates the data items automatically provided for the user for the polygon and line data representations produced by the conversion from the triangular irregular network. The representation for networks can be set up as a graph of line elements or as a network containing polygons within the lines as well as the line features. A turntable concept allows incorporation of data for impedances of links intersecting at junctions.

A system of more recent vintage, *TIGRIS*, developed by the Intergraph Corporation, headquartered in Huntsville, Alabama, USA has a comprehensive spatial data structuring without a need for thematic layers (Herring, 1987). Points, lines and area features are created from a basis of a topological structuring of primitive level directed edges, start and end nodes, left and right faces (Figure 8.16). Lines that are coincident for different attributes are represented only once, as shared spatial primitives. Data redundancy is reduced by encoding coincidence via shared spatial primitives. Points, lines and areas can be assembled as appropriate into higher level entities like river systems and administrative districts as discussed in section 12.5.1. Attribute data are contained in separate data tabulations connected to the spatial units by identifiers. Objects are related to each other by matched pointers.

The Prime Corporation's *SYSTEM 9* geographic information system software is similar to *TIGRIS* in its concept of only one layer encompassing multiple thematic content. *SYSTEM 9* has a topological structure built from nodes, lines, and polygons, although it does not use the directed edge concept (*PRIME*). Like *ARC/INFO* the *SYSTEM 9* separates the attribute and spatial data, dealing with them via separate management and processing environments. *SYSTEM 9* and *TIGRIS*, like *ARC/INFO*, provide analytical capabilities, often by separate program modules. For example, *SYSTEM 9* offers network analysis, polygon overlays and digital terrain modelling based on triangles.

Many other commercial software systems and publicly produced spatial databases exist (GIS World, 1990). For example, several US government agencies are supporting the development of a comprehensive grid-cell data system, *GRASS* (the Geographic Resources Analysis Support System). This system provides a range of capabilities primarily for raster data, conceptually organized into thematic layers. Grid-cell encoding is also supported by run-length data compression, but there is no hierarchical structuring. As for many regular tessellation systems, surface generation can be undertaken from point data, but triangulation is not used. Tools for vector-to-raster and raster-to-vector conversions are provided; and attribute data can be maintained separately from the cell encoding.

The Canadian company *TYDAC* markets a quadtree based system, *SPANS* (Spatial Analysis System), while the geographic information system software of the Geovision company based in Ottawa, like ESRI's *ARC/INFO*, includes a quadtree implementation to facilitate spatial searching. Professor Hanan Samet has developed a quadtree system called *QUILT* (Shaeffer *et al.*, 1990). *SPANS*, apparently the most well known quadtree geographic information system, emphasizes modelling functionality while providing for a large range of data types. Its processing capabilities are categorized into: points analysis, covering contouring, space potentials, point quadtree and spatial interaction models; network analysis, including routing and travel contours, and raster analysis, encompassing cell overlay modelling and neighbourhood functions, diffusion and visibility processing, and three-dimensional surface modelling.

8.7 SUMMARY

The material presented in Part Two has demonstrated many theoretical and practical matters that arise in working with spatial data. The first three chapters laid a conceptual foundation emphasizing semantic issues. The success in understanding and developing spatial information systems in both theoretical and practical domains must come from recognizing the complexity of the real spatial world. Relationships among phenomena are compound, comprehensive and extensive. Computers do not know about maps that are fine devices for bringing together different phenomena – they have to be given explicit information in order for particular tasks to be undertaken.

Spatial data are not straightforward compared with non-spatial data. They are special with regard to continuity in space, anisotropicity, sphericity, spatial autocorrelation, multi-dimensionality, duality (attribute and spatial) of access and intensionality. Undertaking spatial data processing may require use of one or more fundamental operations at a level of encoding rather than semantics; that is, the computational geometry operations as discussed earlier. Operators, as such, are singular in purpose, but analytical and modelling requirements may necessitate procedures incorporating several operators.

The great variety of uses of data for spatial units is summarized, recalling section 2.4 and Figure 2.20, by categories of spatial problems: flow models, optimizing models, simulation, spatial diffusion, surface models, location-allocation problem solving, map overlay mapping and inventory. Automated mapping generally lacks topological features, while some topology, and both entity and field orientations appear in thematic mapping. Map modelling puts more attention on attribute data handling, while spatial statistics are dominated by descriptive geometry, and spatial analysis is heavily graphical and topological in nature. These all contrast with the browsing and querying categories which are more concerned with access to data.

Software may be oriented to information or to analysis. That this information involves identifiers, positional data, topology and attributes is clear; how to organize the material for one or more explicit analytical purposes is the challenge, for many people. Because there are costs involved in developing and using particular structures, and unclearly articulated needs, then in a practical sense software systems may be judged inadequate. Linking data structures with requirements as represented by the particular operations to be performed, requires matching of a processing model with a data model within the context of a semantic model.

It is our view that software systems should be designed by moving from general needs to the semantics. The latter addresses functions which may need to be defined for particular classes of object. A complete data model should contain information on the phenomena types and their properties, the operations that can be applied to instances of phenomena represented by varied spatial data types, and rules which pertain to the data integrity, identifiers, geometry, topology and attributes. As a practical matter, spatial information systems are today limited to having operators defined for simple, not complex objects, although there are many developments under way to attach more meaning to current data models, generally reflecting the phenomena orientation of querying and analysis needs rather than the traditional mapping needs.

Absent a fully articulated list of processing requirements, we can but convey the flavour of the situation by referring to user tasks like path-finding and area measurements. These needs are then translated into particular processing operations that may or may not involve spatial properties; which may or may not be previously computed or explicitly stored.

Spatial data manipulation may require single-entity access, access to sets of objects, or spatial access. Or it may involve graph traversal for certain kinds of spatial analysis operation needing line entities to be associated, or it may require associations for just a single entity to any others. It should be apparent that there are many possible spatial relationships that may be encountered in the great range of spatially oriented tasks presented in Chapter 2. In the context of Chapter 7, building upon a semantic base introduced in Chapter 3, it is useful to now categorize spatial relationships as those which:

1. Can be ascertained from coordinates, for example, containment within boxes or polygons, overlaps for lines on polygons, intersections for lines, and measures of distance, length, area and form.
2. Utilize nonmetrical information, such as connecting line segments to make chains or networks, grouping lines to make areal units, and combining adjacent polygons to make districts.
3. Require attribute data to be tabulated in the database. For example, whether or not a flow is allowed for all possible combinations at intersections or the basic spatial units are to be combined to get a higher level object.

In the sense of practical implementations, recalling some general contextual matters presented in Chapter 1, we see the toolbox approach as one in which a software system provides tools, and ways to create tools, for working with spatial data held somewhere and somehow in the database. The operations described for grid-cell data are tools for users to undertake map overlay modelling. A software system oriented to vector data may have tools for creating buffer zones, solving a routing problem, and so on.

The complexities of the spatial world are seen in the many views or themes interwoven into the fabric of this book. Simplifying by using a set of bipolar dimensions, we point to the different perspectives of:

1. Spatial and attribute data.
2. Geometrical and topological spatial properties.
3. Field or entity orientation.
4. Layers or object orientation.
5. Graph or continuous surface conditions.
6. Searching or spatial analysis.
7. Query or product mode of use.
8. Inventory or reasoning.
9. Conceptual/logical or computer physical structuring.
10. Enginering or cognitive realms.
11. Intuitive approaches or mathematical formalisms.

Spatial and attribute data may be separated (georelational) or combined (regular tessellation). Drawing functions may not require topologically structured data. An entity approach is less beneficial for dealing with data errors. A field approach is good for describing uncertainty and modelling physical processes. A layer orientation is borne of, and reinforces a map model, while an object orientation is possibly more intuitively appealing but practically more difficult. It demands an *a priori* definition of objects. Users' needs may not be anticipated because they change. At the same time, confusion exists because so many fields and disciplines are involved, each with somewhat different traditions, orientations, biases, languages and requirements.

Users have better vocabularies for describing spatial relationships for entities than for continuous surfaces. A geometric orientation of people and organizations matches a geometric data structure; an image orientation matches with a grid-cell concept. Structure and vision capabilities are present to differing degrees in people as individuals; it is their views that matter the most.

8.8 BIBLIOGRAPHY

Further elaborations on spatial analysis and modelling can be found in Gaile and Willmott. A full treatment of map modelling and a map algebra oriented to data in grid-cell form are provided by Tomlin.

Aronson, Peter. 1987. Attribute handling for geographic information systems. *Proceedings of the Auto Carto 8 Conference, Baltimore*, Maryland, USA, pp. 346–355.

Benoit, David and Yann Viemont. 1990. Data structures alternatives for very large spatial databases. Paper presented at the Spatially Oriented Referencing Systems Association Conference, Freiburg, Germany.

Bracken, Ian and Chris Webster. 1989. Towards a typology of geographical information systems. *International Journal of Geographical Information Systems* 3(2): 137–152.

Broome, Frederick R. and David B. Meixler. 1990. The TIGER data base structure. *Cartography and Geographic Information Systems* 17(1): 39–47.

Carruthers, Ann W. and Thomas C. Waugh. 1988. *GIMMS Reference Manual.* Edinburgh, Scotland, UK: GIMMS Ltd.

Chrisman, Nicholas R. 1990. Deficiencies of sheets and tiles: building sheetless databases. *International Journal of Geographical Information Systems* 4(2): 157–168.

Environmental Systems Research Institute. 1988. *ARC/INFO Programmer's Manual*, 2 volumes. Redlands, California, USA: Environmental Systems Research Institute.

Environmental Systems Research Institute. 1990, *Understanding GIS: the ARC/INFO Method.* Redlands, California, USA: Environmental Systems Research Institute.

Fletcher, David and Tom Ries. 1989. Integrating network data into a transportation oriented geographic information system. Paper presented at the Annual Conference of the Association of American Geographers. Baltimore, Maryland, USA.

Gaile, Gary L. and Cort J. Willmott. 1984. *Spatial Statistics and Models.* Dordrecht, The Netherlands: D. Reidel.

GIS World. 1990. *The 1990 GIS World Software Survey.* Fort Collins, Colorado, USA: GIS World Inc.

Goodchild, Michael F. 1990. A geographic perspective on spatial data models. Paper presented at the GIS Design Models and Functionality Conference, Leicester, UK.

Herring, John R. 1980. TIGRIS: a data model for an object-oriented geographic information system. Paper presented at the GIS Design Models and Functionality Conference, Leicester, UK.

Herring, John R. 1987. TIGRIS: topologically integrated geographic information system. *Proceedings of the Auto Carto 8 Conference, Baltimore,* Maryland, USA, pp. 282–291.

Hillsman, Edward L. 1980. Heuristic solutions to location-allocation problems. Monograph Number 7, Department of Geography, University of Iowa, Iowa, USA.

Lettré, Michel and Derek Thompson. 1983. Information systems for geographic districting in Maryland. Paper presented at the Urban and Regional Information Systems Association Conference, Atlanta, Georgia, USA.

Lupien, Anthony, Willam H. Moreland and Jack Dangermond. 1987. Network analysis in geographic information systems. *Photogrammetric Engineering and Remote Sensing* 53(10): 1417–1421.

Marx, Robert W. 1986. The TIGER system: automating the geographic structure of the United States Census. *Government Publications Review* 13: 181–201.

Mitchell, William B. 1977. GIRAS: *A Geographic Information Retrieval and Analysis System for Handling Land Use and Land Cover Data.* US Geological Survey Professional Paper 1059. Washington, DC, USA: US Government Printing Office.

Morehouse, Scott. 1989. The architecture of ARC/INFO. *Proceedings of the Auto Carto 9 Conference, Baltimore,* Falls Church, Virginia, USA: American Society for Photogrammetry and Remote Sensing/American Congress for Surveying and Mapping, pp. 266–277.

Morehouse, Scott. 1987. ARC/INFO: a geo-relational model for spatial information. *Proceedings of the Auto Carto 7 Conference, Washington, DC, USA,* pp. 388–397.

Nyerges, Timothy L. 1989. Design considerations for transportation GIS. Paper presented at the Annual Conference of the Association of American Geographers, Baltimore, Maryland, USA.

Nyerges, Timothy, L. and Kenneth J. Dueker. 1988. *Geographic Information Systems in Transportation.* Report submitted to US Department of Transportation, Federal Highway Administration. Washington, DC, USA.

Peuquet, Donna J. 1988. Representations of geographic space: toward a conceptual synthesis. *Annals of the Association of American Geographers* 18(3): 375–394.

Prime Corporation. (no date). Marketing brochures and technical summaries about SYSTEM 9.

Samet, Hanan. 1990. *Applications of Spatial Data Structures: Computer Graphics, Image Processing and GIS*. Reading, Massachussetts, USA: Addison-Wesley.

Shaeffer, Clifford A., Hanan Samet and R. C. Nelson. 1990. QUILT: a geographic information system based on quadtrees. *International Journal of Geographical Information Systems* 4(2): 103–132.

Simkovitz, Howard J. 1988. Transportation applications of geographic information systems. *Computers, Environment, and Urban Systems* 12: 253–271.

Slocum, Terry A., Robert J. Hanisch and Derek Thompson. 1984. The design and applications of a cartographic data base for redistricting. *Geo-Processing* 2: 151–176.

Thompson, Derek and Terry A. Slocum. 1982. A geographic information system for political redistricting in Maryland. *Proceedings of the Applied Geography Conferences*, 5: 73–87.

Thompson, Derek *et al.* 1983. *The Maryland Reapportionment Information System*. College Park, University of Maryland.

Tomlin, C. Dana. 1990. *Geographic Information Systems and Cartographic Modelling*. Englewood Cliffs, New Jersey, USA: Prentice Hall.

Werner, Christian. 1985. *Spatial Transportation Modeling*, Scientific Geography Series, vol. 5. Beverly Hills, California, USA: Sage Publications.

Part Three

Modelling for Spatial Data

9
Design for Information Systems

Methodologies, issues

Julio Pérez wants a methodology to design the database for his town and Pedro Hernández suggests he play with a toy cadastre. For that intellectual plaything, Carmen Rodríguez sees the universe as comprising of entities and relationships, whereas for Antonia González the world is full of tables, and María Martínez sees a field planted with records and sets.

Part Three covers some foundations for procedures for designing the database for a spatial information system. The design of a spatial information system is similar to the design of any alphanumerical information system except that we have to take the geometric and the multimedia features into account. The first chapter in this part presents a commonly used conceptual modelling methodology and its mapping into different logical models. The two following chapters present many examples of the utilization of the methodology for one-, two- and three-dimensional spatial phenomena. A concluding chapter discusses different approaches and provides a basis for evaluation.

Chapter 9 presents some principles for information system design methodologies, not only for conventional information systems but also for spatial ones. After providing some background about database management systems, we give some details regarding a standard methodology, the ANSI-SPARC (American National Standards Institute, Standards' Planning and Requirements Committee), emphasizing the conceptual and logical levels, especially by means of the entity-relationship approach. Then the network and the relational models used for implementation are described and exemplified. Even though the relational algebra foundation for these tools is not be developed until Part Four, some hints about this topic are given in this chapter.

9.1 DATABASE MANAGEMENT SYSTEMS

Data may be stored in separate unrelated tables, in a single table, in connected tables, or as lists, sets or collections. A **data set** is a single collection of information, such as may be created by hand or purchased from a service bureau, without any particular requirement as to form of organization. A **database** is a collection of non-redundant data shareable among different applications (computer programs performing particular functions) representing the needs of individual or group users. Within the database, data are organized into large segments, called files. In the usual case, files contain particular records of data, the rows of a table; and the record has data for a particular place, event or entity.

A **database (management) system** is a computer based record-keeping frame (possibly better labelled as an operating system) for an integrated and shared repository of information. It allows individual data items to be (perhaps concurrently) used by different programs, and allows unification of several distinct sets of data. Most actions on data fall into categories of **query**, that is, the retrieval of certain instances according to specified conditions, or **transactions**, changing the specific values in some way, especially via updating existing entries in the database.

The database system is itself a computer program, usually very large and complex, located between the user, thinking and acting in a style customary for his needs, and the physical storage. It provides for the separation of the two, allowing the user to accomplish tasks without being knowledgeable of the physical encodings. In a practical sense this is seen for example, allowing changes to the data, perhaps updates or error corrections, without affecting other parts of the system. For instance, there could be a format change such as unpacked decimal to packed decimal; a file structure change, for example sequential to random; or a medium change, like magnetic to optical disk.

Developed largely to meet the needs of large business, government, military or educational institutions for handling large volumes of data with possibly many transactions to be recorded, the database management systems typically have a range of tools oriented towards sorting, searching, retrieval and creation of tabular reports. They work with a handful of different data types like characters, numerals or dates; have languages for describing, or manipulating the data, or for querying the database for particular pieces of information; provide programming tools, and have particular file structures. The system also includes software for establishing and maintaining relationships among tables, files and records in the database, although the nature of these associations depends on the general design (the architecture) of the database system.

TRAFFIC STUDY DATA

Record Number	Household –ID	Person –ID	Person Age	Vehicle –ID	Vehicle Type	Trip –ID	Trip Purpose	Trip Duration
1	101	A	59	10	Car	1	S	75
2	101	B	47	10	Car	1	W	38
3	102	A	24	10	Car	1	C	18
4	102	A	24	10	Car	2	R	24
5	102	A	24	10	Car	3	S	07
6	102	A	24	10	Car	4	C	18
7	102	A	24	10	Car	5	R	30
8	103	A	41	10	Car	1	S	17
9	103	B	38	11	Van	1	W	23
10	103	B	38	11	Van	2	R	54
11	103	B	38	11	Van	3	H	27
12	103	C	17	10	Car		no trips	

Figure 9.1 Conventional table format.

There are several types of data organization, known as flat file (or rectangular), hierarchical, network and relational. Suppose we have collected information for residents of urban areas as to their travel purposes, personal characteristics, details of trips made, and related information. All of these data could be entered into one large table, or **flat file**, as shown in Figure 9.1. A spreadsheet concept is one example of a tabular model; a table of data in a statistical yearbook is another. A tabular model clearly allows association of attributes to entity instances, but is not very effective for different levels of aggregation or for complex situations with many entity types. There is not much flexibility and there are risks in linking tables together and seeing that they have matching data for items occurring in more than one table.

However, using the urban travel information example, it is clear that some data pertain to households, like vehicle ownership and street address; some attributes are associated with the people, like age and number of trips made; and some features pertain to the trips, such as duration and purpose. If all the varied data are stored in one table then much has to be repeated (Figure 9.1), and it is not so easy to get at data for just the shopping trips by bus by the senior citizens in the household.

It is for situations like these that the other types of structure are appropriate. Logically, we could have separate tables for households, persons, trips or vehicles upon which we could draw for particular combinations of data (Figure 9.2). For instance, we might wish to connect the types of vehicle with the income level of the household. In this **relational** concept the user makes the linkages logically through

HOUSEHOLDS

Household – ID	Number of persons	Income category
101	2	3
102	1	5
103	3	2

PERSONS

Person – ID	Age	Gender
101A	59	male
101B	47	female
102A	24	female

VEHICLES

Vehicle –ID	Vehicle Type	Age
101–10	Car	5
102–10	Car	2
103–10	Van	7

TRIPS

Trip –ID	Trip Purpose	Trip Duration
101A–1	S	75
101B–1	W	38
102A–1	C	18

Figure 9.2 Separation of data themes.

associating data items, or by matching up the cases by their identifiers.

Sometimes a particular structure can be known in advance in the creation of the database. If we know that households can be split into persons, that persons make trips, and that particular trips have some attributes, then we can relate the three elements in a hierarchy, either in a simple tree form or in a more complex **network** with connections across branches. The former has only one-to-many associations, that is one level is split into several branches, which themselves can be split into branches (Figure 9.3). If there are many-to-many associations, paths will be converging as well as diverging for any particular level (Figure 9.4a). Simple graphs can be made for the complex case (Figure 9.4b) but they do not show the entire set of connections.

It will not be easy to relate the activities of one person to another if their records are separated. Such an arrangement is generally associated with only one viewpoint, perhaps a study of the daily time use by members of a household. But this is awkward for cross-referencing, as when we wish to retrieve data for all shopping trips performed by all members of the household. Whereas the relational model implies open-endedness for connections, for the hierarchical and network models these linkages must be known in advance. This knowledge in a practical sense means that we do not connect tables by reference to data items, but that there is a physical linkage. This, usually called a pointer, might consist of a code number in one column of the table, directing attention to another set of data for related information. In the relational model connections, joins of tables are usually temporary, providing flexibility to the user, and

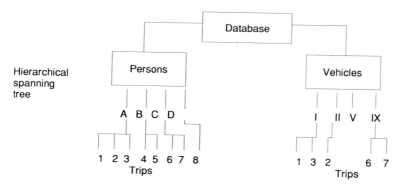

Figure 9.3 The hierarchy concept.

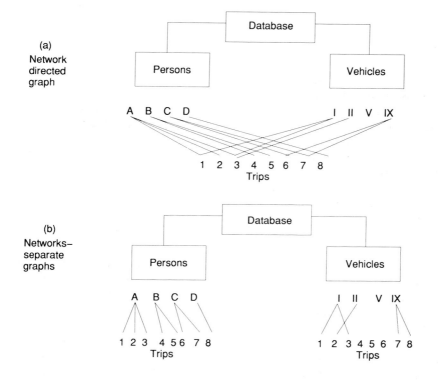

Figure 9.4 The network concept. (a) Many-to-many links are shown by a directed graph. (b) One-to-many links are shown by separate graphs.

demonstrating the decoupling of the use of data from the knowledge of how they are organized.

The comparative conditions of different forms of organization can be seen via a spatial example consisting of a study area (Figure 9.5a) made up of two regions, Blue and Green, which are land parcels, and their edges and numerous corner points (Figure 9.5b). The four tables necessary for holding the data for the four features, Territory, Parcels, Segments and Corners (Figure 9.5c), cannot be connected in any way as all items in these tables are different. A conceptual view linking the entities (Figure 9.5d) might be implemented by a flat file (Figure 9.5e) emphasizing the land parcel as the primary element, but requiring repetition of parcel and segment data.

A hierarchical representation with its one-to-many form (Figure 9.5f) will provide a path from parcels to corners, and from parcels to segments (Figures 9.5g and h). However, the common boundary line, g, and some end-points for edges are necessarily repeated, thereby causing some data duplication. The network model (Figure 9.5j) does not repeat the parcel, segment and corner elements but has a more complex web of connections (Figure 9.5i) requiring a pattern of pointers like that shown in order to link pieces together (Figure 9.5k).

On the other hand, a relational representation would be something like the next set of tables (Figure 9.5l) for the polygons, boundary edges and corner points. Connections can be made between separate tables by means of common items like the vertex ID and parcel ID.

In general, relational databases are popular because they provide simplicity of data organization for the user – a fixed structure of connections does not have to be known *a priori*. The relational model provides flexibility to the data user as it is able to join tables as needed, it provides for efficiency of storage by proper design of the tables, and it has a non-procedural style for accessing data. In other words, queries are expressed in terms of what is needed rather than how to get the data to answer the question posed. In contrast, for some purposes, the network model can deliver better performance, although it is necessary to know paths through the structure for access to particular information.

As can be seen, a database approach to handling data can lead to economical and effective sharing, enforcement of standards, maintenance of data integrity and quality, avoidance of inconsistencies, and balance of different and possibly conflicting requirements; reduction in data duplication, and the provision of a flexible environment for working with many and varied data. Rather than provide a detailed routine discussion of the merits and limitations of different types, we provide a foundation for assessing the potential value of database tools to spatial information systems purposes. We begin by examining the process of designing databases.

9.2 THE ANSI-SPARC DESIGN METHODOLOGY

For many years, and especially since 1975, four different levels of data modelling have been used as a framework in the design of an information system (ANSI/X3/SPARC). Starting from the so-called real world these are the:

External level
Conceptual level
Logical level
Internal level

Using Figure 9.6 as an overview, let us look at each in turn. Here we first describe the characteristics of each level, before going on to elaborate about only the external and conceptual levels. We see the levels as stages in the building of the database which is necessary and sufficient to accommodate the requirements of different people or organizations, or different uses by just one individual.

As a practical matter, the real world corresponds to a subset of reality that is of interest. Sometimes the notion of universe of discourse is used. In this spirit it is assumed that one can build a database of a subset of reality only if one knows how to describe it with words. However, in the multimedia world of spatial information systems, many data, such as pictures, sketches or sounds, cannot be described by words only, but they can be described by other forms of digital or analogical encoding. So we prefer to think of the **universe of modelled phenomena**.

The very beginning of the database design process is the external modelling step in which potential users define their own subset of the real word, that is, what is relevant for their needs. Commonly, we deal with as many **external models** as we have different purposes.

The **conceptual level** corresponds to a synthesis of all external models. It is called this for two reasons: firstly, because it is made of very sound concepts; secondly, because it is the basis for the conception process. Although an abstraction of the real world, the result of conceptual modelling is supposedly quite concrete in nature, consisting of schematic representations of phenomena and how they are related. The organization scheme created at this stage generally deals with only the information content of the database, not the physical storage, so that the same

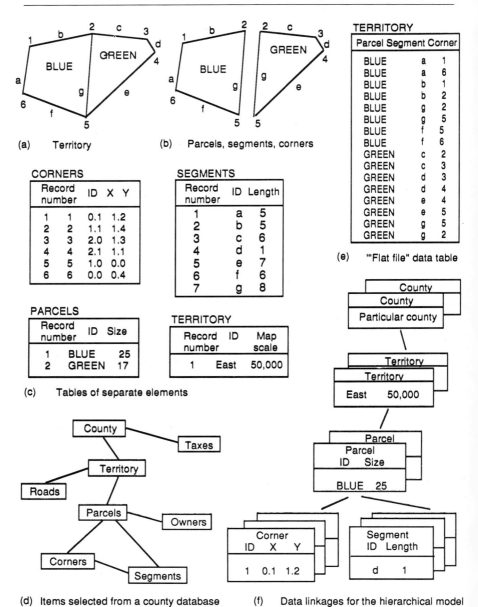

(a)   Territory

(b)   Parcels, segments, corners

CORNERS

Record number	ID	X	Y
1	1	0.1	1.2
2	2	1.1	1.4
3	3	2.0	1.3
4	4	2.1	1.1
5	5	1.0	0.0
6	6	0.0	0.4

SEGMENTS

Record number	ID	Length
1	a	5
2	b	5
3	c	6
4	d	1
5	e	7
6	f	6
7	g	8

PARCELS

Record number	ID	Size
1	BLUE	25
2	GREEN	17

TERRITORY

Record number	ID	Map scale
1	East	50,000

(c)   Tables of separate elements

TERRITORY

Parcel	Segment	Corner
BLUE	a	1
BLUE	a	6
BLUE	b	1
BLUE	b	2
BLUE	g	2
BLUE	g	5
BLUE	f	5
BLUE	f	6
GREEN	c	2
GREEN	c	3
GREEN	d	3
GREEN	d	4
GREEN	e	4
GREEN	e	5
GREEN	g	5
GREEN	g	2

(e)   '"Flat file" data table

(d)   Items selected from a county database

(f)   Data linkages for the hierarchical model

Figure 9.5 A spatial data example for hierarchical, network, and relational organization. (a) Territory. (b) Parcels, segments, and corners. (c) Tables of separate elements. (d) Selected items from a county database. (e) Flat file table. (f) Data linkages for the hierarchical model. (g) Hierarchical structure to obtain corner

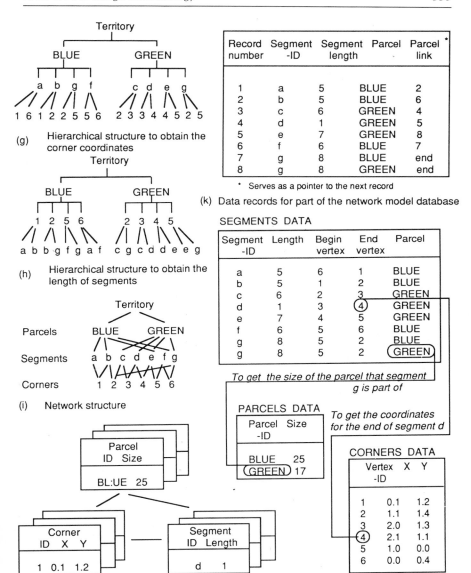

(g) Hierarchical structure to obtain the corner coordinates

(h) Hierarchical structure to obtain the length of segments

(i) Network structure

(j) Data linkages for the hierarchical model

(k) Data records for part of the network model database

Record number	Segment -ID	Segment length	Parcel	Parcel link*
1	a	5	BLUE	2
2	b	5	BLUE	6
3	c	6	GREEN	4
4	d	1	GREEN	5
5	e	7	GREEN	8
6	f	6	BLUE	7
7	g	8	BLUE	end
8	g	8	GREEN	end

* Serves as a pointer to the next record

SEGMENTS DATA

Segment -ID	Length	Begin vertex	End vertex	Parcel
a	5	6	1	BLUE
b	5	1	2	BLUE
c	6	2	3	GREEN
d	1	3	4	GREEN
e	7	4	5	GREEN
f	6	5	6	BLUE
g	8	5	2	BLUE
g	8	5	2	GREEN

To get the size of the parcel that segment g is part of

PARCELS DATA

Parcel -ID	Size
BLUE	25
GREEN	17

To get the coordinates for the end of segment d

CORNERS DATA

Vertex -ID	X	Y
1	0.1	1.2
2	1.1	1.4
3	2.0	1.3
4	2.1	1.1
5	1.0	0.0
6	0.0	0.4

(l) Data tables for the relational model

Figure 9.5 continued

coordinates. (h) Hierarchical structure to obtain length of segments. (i) Complex network structure (many-to-many). (j) Data linkages for the network model (partial). (k) Data records for (part of) the network model. (l) Data tables for the relational model.

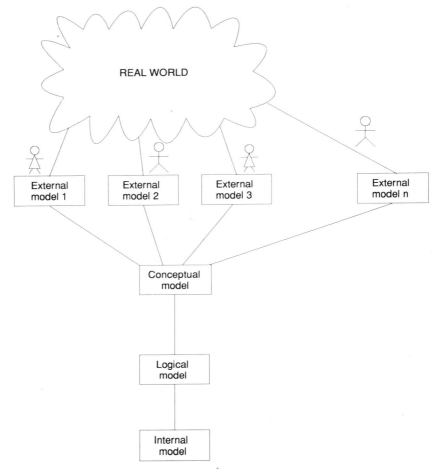

Figure 9.6 Data modelling levels.

conceptual model may be appropriate for diverse physical implementations. The conceptual model not only provides a basis for schematizing but is also a tool for discussion and, as such, a good conceptual model must be understandable.

This model sharing may be done by narrative statements, but the transfer to the next, logical stage, is easier if more formal mechanisms are used. Designers usually use the visually oriented **entity-relationship approach** in which phenomena and their associations are presented graphically as shown later. The representation of the conceptual organization substantively has varied forms. Ordinarily it can be thought of as a sort of semantic data model in which the principal types of phenomenon and their associations and constraints are laid out. **External modelling** uses similar tools but often consists simply of a narration of requirements stipulated by an individual in an interview process.

Many difficulties arise in designing the conceptual model as a synthesis of several external models, especially in dealing with the geometric data. Indeed, as illustrated later, external geomatic models can be based on different geometric representations, and the synthesis can imply either of the following:

1. Having many geometric representations and then creating procedures to transform from one state to another.
2. Keeping only one model in the database and then transforming each external model, perhaps producing some difficulties for the computer programmers.

The next step in the design of an information system is called the **logical modelling** phase. In fact, it is the first step in computing. The expression logical model has two meanings, and is not meant to convey that conceptual or external modelling are any less systematic. Firstly, it constitutes a mathematical basis or a set of mathematical concepts. Secondly, it corresponds to the transformation (mapping) of the conceptual model with the tools offered by the logical modelling. In other words, we have a transfer between the conceptual models and a new modelling level which is more computing oriented.

In the past fifteen years of so, three different approaches have usually been commercially available:

1. The hierarchical model (not examined in the treatment below), in which the only data structure is a tree, constraining the designer only to hierarchical branching relationships.
2. The network model, of which CODASYL is the most well known, in which the modelling base is a set, a unit quite appropriate for one-to-many relationships.
3. The relational model in which the basic structure is a table.

The realization of a conceptual model is achieved by mapping the conditions of the semantic data model into the definitions, constraints and procedures for one of these models, most especially today, the relational, or other newer types, extensions of the relational. In this way the permanent properties of the database are clearly specified, whatever the circumstances pertaining to a particular set of instances of phenomena. This procedure is associated, in a practical sense, in the creation of a data dictionary, a set of statements about important properties of the data items, such as name, type of data, range of values or missing values.

The **internal model** is concerned with the byte-level data structure of the database. It explains every pointer and any information necessary to access and handle correctly all data items. Whereas the logical level is concerned with tables and data records, the internal level deals with storage devices, file structures, access methods, and locations of data. For commercial database management systems, this level is not generally accessible to users.

9.3 CONCEPTUAL MODELLING: THE ENTITY-RELATIONSHIP APPROACH

Conceptual modelling in database design often makes use of a formal approach known as entity-relation modelling. First presented comprehensively in 1976 (Chen, 1976), but based on some older ideas, it is a means to organize and schematize information. We first present the formalism and then give an example which will be used often in this book.

In this formalism the basic components are:

Entities
Classes of entities
Relationships between entities or classes of entities
Attributes for both entities and relationships
Cardinalities of relationships
Integrity constraints

An entity is a person, place, thing or event. The Mr Green's parcel, the building 2314, the car numbered 34 HP89, the flight 123, are entities in the real world. But, as first illustrated in section 3.1, entities can be regrouped into classes. Relationships, definable for single entities or classes, are the **associations** between phenomena, such as the land parcel has a house, the car has an owner, and the flight goes between Greenland and Fiji. Both entities and relationships can hold attributes, their special characteristics, such as size of parcel, owner of car, or flight duration.

Generally, an association is binary, that is linking only two entities, but it may be more involved. The **degree** of relationship expresses this number of linkages. **Cardinalities** are expressed by four numbers defining the minimum and maximum number of entities occurring in a relationship, in both the forward and reverse senses. Usually the minima are 0 or 1;

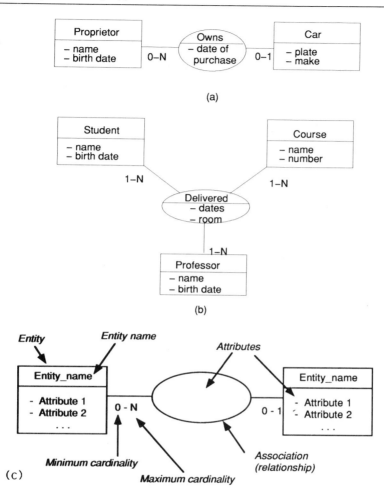

Figure 9.7 Examples of entity-relationship diagrams. (a) Example of a binary relationship. (b) Example of a ternary relationship. (c) Nomenclature for entity-relationship diagrams.

when the maximum is unknown, we again use the letter N. In some special cases, we may need to use a summary statistic, for example the average number of vehicles. An **integrity constraint** can be defined as a predicate (value or symbol) which must be matched in order to confer integrity onto the model. We can apply integrity constraints to attribute values or to attribute definitions, but the most important are cardinality constraints.

Let us take a simple example dealing with proprietors and cars. A proprietor can own several cars (minimum 0 and maximum N), and cars can be owned by zero or one or more (joint owners) people. So, the entity-relationship diagram can be designed as in Figure 9.7a. As attributes in this example, we can mention the owner's name, his or her date of birth, and so on; the make of the car, or its licence plate number. We can also consider attributes in the relationship such as purchasing dates.

In the visual convention we are using (Figure 9.7c), one of several in customary use, an entity or class of entities is depicted by a rectangle and a relationship by a circle or an ellipse linking two or more entities or entity classes. Attributes are shown as lists in the rectangle boxes, or attached to the relationship ellipses. Figure 9.7b illustrates a ternary relationship among three entity classes of student, course and professor. More generally, it is occasionally necessary to deal with n-ary relationships, where n denotes the number of classes.

A more comprehensive example, Figure 9.8, relates to a city containing land parcels, as a principal spatial phenomenon. A description of parcels and streets (referred to by us here and later as our toy cadastre) and of landowners is prepared in entity-relation form (Figure 9.9). A parcel is limited to the range from at least three up to N boundary segments; a street is limited to a minimum of two to an unknown upper limit of N segments. A segment can delimit one or two parcels and can border one street or none at all. Moreover, a segment is composed of two and only two end-points (vertices), and an end-point can be connected with one-to-many segments. Here, we can illustrate a cardinality constraint about segment cardinalities. That is, a segment can be bordered by two parcels or by one parcel and one street. In other words, the sum of cardinalities must always be equal to two.

In a typical conceptual model, it is usual to deal with scores of entities and associations. Thus, the conceptual level's statement of the complete description of a future database is very useful for discussions among people in order to check the completeness and correctness, and to bring forth modifications. During the conceptual level step it is advisable to create a data dictionary, one component of metadata, incorporating the definitions of entities, of relationships, and of their attributes. This further facilitates the discussion between all potential users, allowing a designer to detect entities having contradictory definitions or different entities having the same definition.

The entity-relation modelling design approach (and its varieties) is a general tool in the business world for designing databases. Only a handful of examples of its use for spatial information systems has been published,

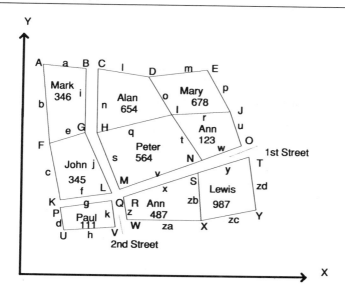

Figure 9.8 An example of land parcels: toy cadastre.

suggesting limited use in spatial information systems design. However, it would appear that the inherent complexity in spatial problems should lead to more use of design tools like these semantic data models, if only to come to grips with that complexity. The formalized data modelling can foster clear identification of entities, attributes and associations. It provides a basis for discussion and then refinement; and it is a structured foundation for designing a database by implementation techniques

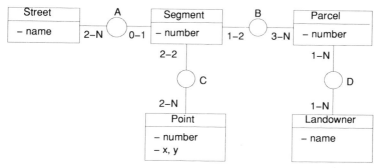

Figure 9.9 The entity-relationship diagram for land parcels. A: streets have edges (segments); B: parcels have boundaries (segments); C: line segments have two endpoints; D: parcels have owners, and people own land.

discussed below. It is most valuable for complex situations, if only to reveal how complicated reality and user perceptions are.

9.4 LOGICAL MODELLING: RELATIONAL DATABASES

Now let us turn our attention to the logical modelling of a database using the relational model first. In this section we give only a small introduction to a well known topic first formally examined in 1970 (Codd, 1970). More recent developments from this basic idea will be found in Chapter 14. A **relation** is an organized assembly of data that meets certain conditions. While all relations are tables, not all tables are, strictly speaking, relations when the formal definition of this structure is recognized. A relational database is a collection of relations represented by statements as to contents, and tables containing instances of the relations. The database has no *a priori* pattern of connections among the relations.

The basic tool is a statement corresponding to a relation. Using a conventional form of expression, a single relation for our car ownership example is:

R (Proprietor_name, Car_make, Registration_number)

where R stands for relation, and the parentheses contain the names of three data items or attributes. A relation has a collection of attributes or data items representing some property of an entity about which data are to be stored. A table containing an **instance** of this example relation is shown below.

R	Proprietor's name	Car make	Registration number
	John	Renault	1234 AB 75
	Mary	Fiat	4567 CD 76
	Mark	Volvo	8901 EF 89
	. . .	. . .	. . .

The table consists of rows, columns and cell entries. The columns, sometimes referred to as fields, contain attributes. The rows, sometimes called records or tuples, contain particular instances of an entity. The cells contain one or more values for an attribute for an entity. There does not have to be any particular order to either rows or columns. Using more technical terminology to distinguish between features of tables and characteristics of relations, the main components of the relational model are attributes, domains, keys, relations and tuples (Figure 9.10).

A **domain** D is a set of atomic values. By atomic we mean that each value in the domain is indivisible as far as the relational model is concerned, that is we do not care about digital computer bits. It is useful to specify a name for the domain to help in interpreting its values. Some examples of domains are:

Names: the set of names of persons
USA-phone-number: the set of valid ten-digit numbers
x coordinates: the set of decimal numbers
Parcel number: a set of numbers and letters
Employee age: the set of numbers between 17 and 70
Land use: the set of integers between 10 and 79

A common method of specifying a domain is to specify a **data type** from which the data values forming the domain are drawn. Historically, data types have been numerals, integers, dates, money, characters, character strings and special symbols. Different attributes can have similar data types, for example both people and towns have character names, and both land use code and street addresses of houses are usually integers. Single attributes can be represented by different data types, although not necessarily in the same database, for instance identifiers for

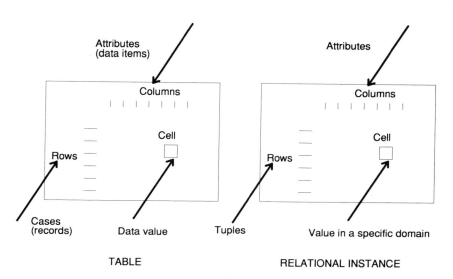

Figure 9.10 Nomenclature for relational tables.

people may be names, social security numbers or other numbers. Numerals may be used as identifiers or for computational purposes. The domain is one type of constraint; the values are also restricted to a particular format and range, for example, dd/mm/yy for dates, where d, m, y are numerical digits, or xx,xxx.xx for a coordinate, where x is a numerical digit.

For each domain, a data type and a format must be provided, and a unit of measurement must be specified. For example, for space coordinates, the metre or the foot can be selected. A domain can embrace several attributes, A_i, but an attribute corresponds to only one domain. Dom A_i is the notation for this condition for one particular attribute from the set.

One or more of the attributes is specified as a **key**, requiring that there be at least one data item column containing unique values for each row. In practice, the key is an identifier like a social security number, geocode, country name, or an arbitrarily assigned identifier like polygon number. At times the key may be a concatenation of several separate identifiers, like a household number-person number-travel trip-number combination. There may be several identifier or label attributes, not all of which may be used as keys.

A complete descriptive statement, or **schema**, for a relation looks like:

POINT (Point-ID, X, Y, Accuracy)

In this example, POINT is the name of the relation; Point-ID, X, Y, and Accuracy are attributes, each drawn from a specific domain:

Dom (Point-ID) from the set of numbers
Dom (X) = Dom (Y) from the set of decimal numbers
Dom (Accuracy) from the set of numbers

Another example of a relation schema is:

PURCHASE (<u>Plot-ID</u>, Owner_name, Purchase_date)

in which PURCHASE is the relation name, Plot-ID is the number of a plot of land (an attribute defined in an integer domain), Owner_name is the name of the landowner (an attribute defined in a domain of character string), Purchase_date is the date in which the previous landowner bought the plot of land (attribute defined in a date domain). In this case the key is identified by the underlining.

The row entries in the instance of a relation, the **tuples**, are shown in the following table.

PURCHASE	Plot-ID	Owner_name	Purchase_date
	5678	Peter	13-03-76
	4567	Mary	16-05-81
	1208	Mark	01-07-78
	3089	Peter	20-12-80
	3218	Lewis	31-01-82
	. . .	. . .	. . .

Each N-tuple T is an ordered list of N values $T = \langle V_1, V_2, \ldots, V_N \rangle$, where each value V_i is an element of $\text{Dom}(A_i)$, or is a special null value. Tuples represent the extension of the relation schema R. So, a relation schema R, denoted by $R(A_1, A_2, \ldots, A_N)$, is a set of attributes $R = \{A_1, A_2, \ldots, A_N\}$. Each attribute A_i corresponds to a domain D in the relation schema R. D, the domain of A_i, is denoted by $\text{Dom}(A_i)$. A relation, or a relation instance r of the relation schema $R(A_1, A_2, \ldots, A_N)$, also denoted by $r(R)$ is a set of N-tuples $r = \{T_1, T_2, \ldots, T_M\}$. A null value indicates a state of no knowledge of a value for a given attribute-tuple combination, or a blank space for which a datum is not yet available.

9.5 TRANSFORMING ENTITY-RELATIONSHIP MODELS INTO RELATIONAL MODELS

Starting from a conceptual model presented in the entity-relationship approach, a next very important step is to transfer or **map** it into a relational model. For doing that, let us examine some examples, starting with the general process demonstrated as Figure 9.11. The POLYLINE entity is very simple to transform into a relation:

POLYLINE (Polyline-ID, Type)

For the point entity, we have similarly:

POINT (Point-ID, X, Y)

For the binary association, though, in this example between line and point, several situations can occur, such as the polyline may have several points, or the polyline may have a sequence of points. Here we mention only the order of points.

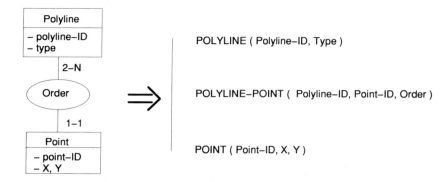

Figure 9.11 Mapping a conceptual model into a relational model.

Generally speaking, an association is mapped into a relation for which either keys are the proper identifiers of the entities associated by the relationship, or attributes are data items which are not identifiers, that is, non-key attributes, for the relationship, if they exist at all. Consequently, for our example, we have:

POLYLINE_POINT (<u>Polyline-ID, Point-ID</u>, Order)

where the attribute Order has data values indicating the sequence of points on the polyline in order to draw the shape correctly. This relation has two keys, necessary for linking to the two other relations (Figure 9.11). (Subsequently, our descriptions do not identify keys, under the assumption that the ID variable is the key.)

In the case of associations among several entities, a corresponding relation will be created having keys as numerous as the number of connected entities. Returning to the example of the toy cadastre, the transformation is as follows.

First of all, five entities are translated into five relations:

STREET	(Street-ID, Street_name, etc.)
OWNER	(Owner-ID, Owner_name, etc.)
COORDINATES	(Point-ID, X, Y)
SEGMENT	(Segment-ID)
PARCEL	(Parcel-ID, Address)

Then four relations are created corresponding to four associations each with two keys and with zero attributes:

STREET-SEGMENT	(Street-ID, Segment-ID)
PARCEL-SEGMENT	(Parcel-ID, Segment-ID)
SEGMENT-POINT	(Segment-ID, Point-ID)

OWNER-PARCEL   (Owner-ID, Parcel-ID)

However, rearrangement is possible. Since a segment has always and only two end points, the SEGMENT and the SEGMENT–POINT relations can be combined to give:

SEGMENT-ENDPOINTS (Segment-ID, Point1-ID, Point2-ID)

Data tables for our relations are shown in the tables below.

STREET	Street-ID	Street_name
	101	1st street
	102	2nd street

STREET-SEGMENT	Street-ID	Segment-ID
	1	f
	1	g
	1	v
	1	w
	1	x
	1	y
	2	i
	2	j
	2	k
	2	n
	2	s
	2	z

PARCEL-SEGMENT	Parcel-ID	Segment-ID
	111	d
	111	g
	111	h
	111	k
	123	r
	123	t
	123	u
	123	w
	. . .	. . .

SEGMENT-ENDPOINTS	Segment-ID	Point1-ID	Point2-ID
	a	A	B
	b	A	F
	c	F	K
	d	P	U
	e	F	G
	f	K	L
	g	P	Q
	. . .	. . .	. . .

COORDINATES	Point-ID	X	Y
	A	209.1	488.4
	B	224.5	487.0
	C	230.7	487.1
	D	251.3	484.7
	E	275.5	486.9
	F	211.0	455.5
	G	216.8	459.4
	H	230.7	459.9
	I	261.6	467.7
	J	284.0	469.2
	K	216.7	431.7
	L	236.5	434.9
	M	240.3	436.1
	N	273.2	448.0
	O	289.0	454.6
	P	216.7	427.9
	Q	241.9	432.3
	R	242.5	432.2
	S	272.3	443.9
	T	283.1	450.5
	U	217.0	418.1
	V	238.7	420.7
	W	232.5	423.8
	X	273.2	424.0
	Y	295.8	428.4

OWNER	Owner-name	Owner-ID	Address
	Alan	5667	6 Churchill Street
	Ann	3445	1678 Shakespeare Avenue
	John	1888	3456 Chaucer Road
	Lewis	3459	24 Elizabeth Street
	Mark	5210	4567 Bridge Street
	Mary	6089	8734 King Street
	Paul	4529	St Peter Road
	Peter	1007	5690 St Paul Street

PARCEL	Parcel-ID	Address
	345	2nd Street
	346	2nd Street
	654	2nd Street
	564	1st Street
	678	3rd Street
	123	1st Street
	487	2nd Street
	987	1st Street
	111	1st Street

OWNER–PARCEL	Owner-name	Parcel-ID
	Alan	654
	Ann	123
	Ann	487
	John	345
	Lewis	987
	Mark	346
	Mary	678
	Paul	111
	Peter	564

Rearrangement, a very important aspect to relational modelling, may be necessitated if relations created by entity-relation mapping are not acceptable according to certain criteria. That is to say, they may be necessary for consistency checking, to better track errors. Some of these conditions are elaborated in section 9.7; in Chapter 13 there is discussion of the technicalities of this procedure, known as **normalization.**

The commercial database systems that are built using the relational model, for example *INGRES* and *ORACLE*, provide users with a language for data retrieval. Recently approved as a standard in the USA by the American National Standards Institute (1986), this most commonly encountered language, the **Structured Query Language** (SQL), provides a quite natural way to access data and to perform some operations upon them. Initially developed by IBM, the SQL provides a means to define and manipulate data, shielding the user from needing to know how the data are structured internally, that is, the user undertakes referencing by table and item names.

Part of the language is for data definition, used for setting up tables and data, or establishing views, that is subsets of the entire suite of tables and items, by means of instruction keywords like **CREATE** or **VIEW**. The data manipulation language provides a way to select data from rows and columns, via basic commands like **MODIFY**, **DELETE**, **INSERT**, and various relational operators discussed in Chapter 13. The database systems also provide some capabilities for arithmetical and other mathematical operations like averages or exponentiation, and operations appropriate to special data types like time data, money data or date data. But, without special extensions, they do not address operations needed for geometric entities, as in a request to a database to connect all water pipes in order to track the flow of a dye through a water system.

At this point, though, we leave relational modelling temporarily to present a different approach to building a database.

9.6 LOGICAL MODELLING: CODASYL DATABASES

CODASYL is another logical database model, originated in the late 1960s by the Data Base Task Group of a USA committee named CODASYL (Conference on Data Systems Languages). Even though the CODASYL model, a classic **network model**, is not much used in spatial information systems, it seems important to present an idea of another model for describing spatial databases, if only in order to emphasize the advantages of the relational model.

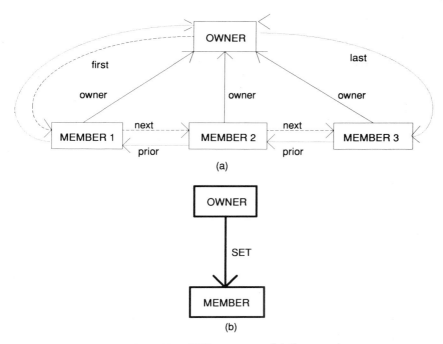

Figure 9.12 The CODASYL SET structure. (a) Structural representation of a Set with all pointers. (b) Schematized representation of a
Set.

9.6.1 The modelling

The aim of this section is not to give a complete and detailed presentation
of the CODASYL network system but rather some hints for modelling
spatial information systems. The **SET** structure is the basic tool of a
CODASYL database system built to support one-to-many relationships.
A set corresponds to a one-to-many relationship. It is essentially a ring
data structure with two kinds of record linked by several pointers (Figure
9.12a):

1. An 'owner' record.
2. Several 'member' records.
3. Two pointers in an owner record, pointing, respectively, to the first
   and the last members of the set.
4. Three pointers per member record, pointing, respectively, to the prior
   member, the next member and the owner.

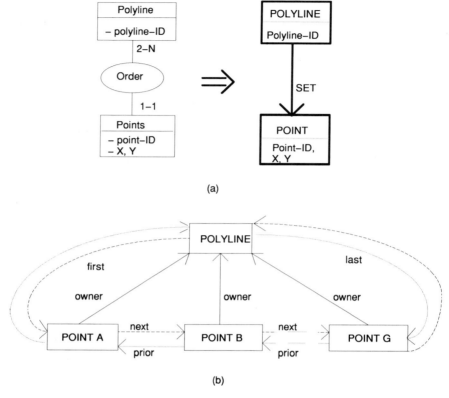

Figure 9.13 Mapping a one-to-many relationship into a CODASYL SET. (a) Example of a one-to-many relationship. (b) CODASYL representation of the polyline-point Set.

A ring, then, links the Member1, Member2, Member3 records in turn to the owner record, going in each of two directions to facilitate access to the neighbouring records in the set. Each Member record is also directly connected by a pointer to the Owner record. The schematic diagram (Figure 9.12b), known as a Bachman diagram, represents the logical model of the database, and is formulated from the starting point of an entity-relationship model. The example of polylines and points (Figure 9.13) not only shows the correspondence, mapping from the conceptual to logical model, but also demonstrates the network model in contrast to the relational presented earlier (Figure 9.11).

Since the CODASYL basic element is a set allowing only the modelling of one-to-many relationships, it should be apparent that there is a problem for representing many-to-many relationships. However, this is possible by introducing a new element, a **pseudo-record** (Figure 9.14a), one without information, and by using two sets which have it as a common member: as Figure 9.14b when no attributes occur in the relationship, and as Figure 9.14c in the case of attributes.

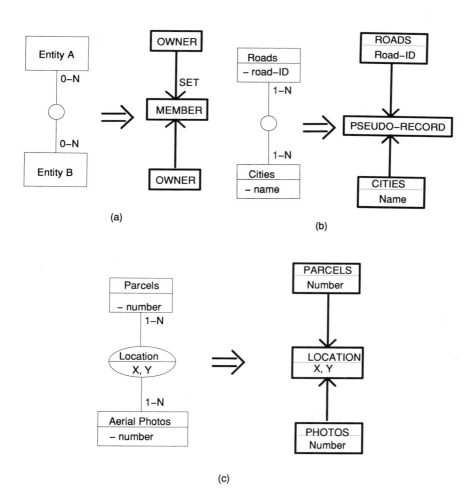

(a)

(b)

(c)

Figure 9.14 Mapping a many-to-many relationship into CODASYL Sets. (a) Transformation of a many-to-many relationship into two sets with a void extra record. (b) Example of mapping a many-to-many relationship. (c) Example of mapping a many-to-many relationship with attributes.

Now we have all the tools to map the toy cadastre conceptual model into the CODASYL logical model. For the example given in Figure 9.9, we have the following transformations (Figure 9.15a). Each of the A and D relationships is mapped into a set; and the B relationship, a many-to-many association, requires introduction of an extra record and two sets, the Par-seg record respectively with the Segment ID-number and the Parcel ID-number. The C relationship mapping implies the creation of an extra record, Seg-points, and two sets, as shown.

In this case, though, considering that the two-to-two relationship has to be regarded as a one-to-many association, the representation is ameliorated by modifying the Segments record and removing the Seg-point record, as illustrated in Figure 9.15b.

9.6.2 The Data Definition and Manipulation Languages

In a CODASYL approach, a database description is given in three parts:

1. A description of records and sets for the global database, the schema.
2. A description of the physical structure.
3. For each application, a subschema describing only the appropriate part of the database required.

For describing the database, a special language, called the **Data Definition Language** (DDL), must be used. For the logical aspects, we have two things to describe:

1. The records, with their characteristics and their fields.
2. The sets, with their characteristics.

Here is the CODASYL description corresponding to Figure 9.15. The role of this example is not to present all data definition language components but only to give a hint of what is possible. Nor is it important to consider the details of the syntax of the coding. (Note though, that the spatial terms AREA and LOCATION do not refer to spatial semantics but to physical storage.)

```
SCHEMA NAME IS TOY-CADASTRE.
AREA NAME IS AREA1.

RECORD NAME IS STREET
LOCATION MODE IS CALC USING STREET-NAME
    DUPLICATES ARE NOT ALLOWED WITHIN AREA1.
02 STREET-NAME TYPE IS CHARACTER 30.
```

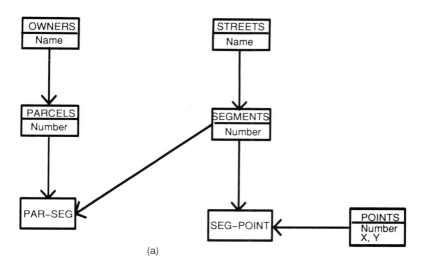

(a)

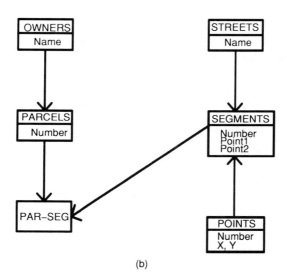

(b)

Figure 9.15 The Bachman diagram for the toy cadastre. (a) The Bachman diagram without ameliorations. (b) After some modifications.

RECORD NAME IS LANDOWNER
LOCATION MODE IS CALC USING OWNER-NAME
    DUPLICATES ARE NOT ALLOWED WITHIN AREA1.
02 OWNER-NAME TYPE IS CHARACTER 30.

RECORD NAME IS PARCEL
LOCATION MODE IS CALC USING PARCEL-NUMBER
    DUPLICATES ARE NOT ALLOWED WITHIN AREA1.
02 PARCEL-NUMBER TYPE IS DECIMAL 10.

RECORD NAME IS SEGMENT
LOCATION MODE IS CALC USING SEGMENT-NUMBER
    DUPLICATES ARE NOT ALLOWED WITHIN AREA1.
02 SEGMENT-NUMBER TYPE IS DECIMAL 6.
02 POINT1 TYPE IS DECIMAL 6.
02 POINT2 TYPE IS DECIMAL 6.

RECORD NAME IS PAR-SEG
LOCATION MODE IS VIA PARCEL-PAR-SEG
    WITHIN AREA1.

RECORD NAME IS POINT
LOCATION MODE IS CALC USING POINT-NUMBER
    DUPLICATES ARE NOT ALLOWED WITHIN AREA1.
02 POINT-NUMBER TYPE IS DECIMAL 6.
02 X TYPE IS DECIMAL 10.
02 Y TYPE IS DECIMAL 10.

SET NAME IS LANDOWNER-PARCEL
OWNER IS LANDOWNER
ORDER IS PERMANENT INSERTION LAST.

MEMBER IS PARCEL
INSERTION MANUAL RETENTION OPTIONAL
SET SELECTION THRU LANDOWNER-PARCEL SET
OWNER IDENTIFIED BY APPLICATION.

SET NAME IS PARCEL-PAR-SEG
OWNER IS PARCEL
ORDER IS PERMANENT INSERTION LAST.
MEMBER IS PAR-SEG
INSERTION MANUAL RETENTION OPTIONAL
SET SELECTION THRU PARCEL-PAR-SEG SET
OWNER IDENTIFIED BY APPLICATION.

SET NAME IS STREET-SEGMENT

OWNER IS STREET
ORDER IS PERMANENT INSERTION LAST.
MEMBER IS SEGMENT
INSERTION MANUAL RETENTION OPTIONAL
SET SELECTION THRU STREET-SEGMENT SET
OWNER IDENTIFIED BY APPLICATION.

SET NAME IS SEGMENT-PAR-SEG
OWNER IS SEGMENT
ORDER IS PERMANENT INSERTION LAST.
MEMBER IS PAR-SEG
INSERTION MANUAL RETENTION OPTIONAL
SET SELECTION THRU SEGMENT-PAR-SEG SET
OWNER IDENTIFIED BY APPLICATION.

SET NAME IS POINT-SEGMENT
OWNER IS POINT
ORDER IS PERMANENT INSERTION LAST.
MEMBER IS SEGMENT
INSERTION MANUAL RETENTION OPTIONAL
SET SELECTION THRU POINT-SEGMENT
OWNER IDENTIFIED BY APPLICATION.
END_SCHEMA.

The navigation in the database is made through pointers by a language called the **Data Manipulation Language** (DML), hosted by the programming language COBOL. A selection of the main statements is:

1. Opening or closing the database.
2. Retrieving records directly or following certain pointers.
3. Storing records and making connections to sets.
4. Modifying and deleting records.

It is necessary to mention that with CODASYL only **feature based queries** can be solved, that is retrieving descriptive information about entities, like size of lot or owner's name. Regarding location queries, we need to implement special procedures, the algorithms for which are given in Chapter 14. For relational databases, we have exactly the same difficulty, but here we will examine only some CODASYL examples in order to illustrate the possibility of employing this network system.

QUERY 1

Let us retrieve all parcels belonging to Peter. We will first find Peter and then retrieve all his parcels by following the links, giving:

> MOVE "PETER" TO OWNER-NAME.
> FIND ANY LANDOWNER WITHIN AREA1.
> GET. DISPLAY OWNER-NAME
> FIND FIRST PARCEL WITHIN LANDOWNER-PARCEL.
> PERFORM A UNTIL END-OF-SET.

  A.

> GET. DISPLAY PARCEL-NUMBER.
> FIND NEXT PARCEL WITHIN LANDOWNER-PARCEL.

QUERY 2

Suppose we want to draw Churchill Street and have a subprogram called DRAW (X1, Y1, X2, Y2). First we have to get the street record, then the segments and their coordinates in this manner:

> MOVE "CHURCHILL STREET" TO STREET-NAME.
> FIND ANY STREET WITHIN AREA1.
> FIND FIRST SEGMENT WITHIN STREET-SEGMENT.
> PERFORM B UNTIL END-OF-SET.

  B.

> GET. MOVE POINT1 TO POINT-NUMBER.
> FIND ANY POINT.
> GET. MOVE X TO X1. MOVE Y TO Y1.
> MOVE POINT2 TO POINT-NUMBER.
> FIND ANY POINT.
> GET. MOVE X TO X2. MOVE Y TO Y2.
> CALL DRAW (X1, Y1, X2, Y2).

QUERY 3

Suppose we now wish to draw Peter's parcel, we have to follow the same schema, except that we have to take the owner link of Par-seg via the Segment-par-seg set in order to reach the segment that has to be drawn.

```
    MOVE 'PETER' TO OWNER-NAME.
    FIND ANY LANDOWNER WITHIN AREA1.
    GET. DISPLAY OWNER-NAME
    FIND FIRST PARCEL WITHIN LANDOWNER-PARCEL.
    PERFORM C UNTIL END-OF-SET.
C.
    GET. DISPLAY PARCEL-NUMBER.
    FIND FIRST PAR-SEG WITHIN PARCEL-PAR-SEG.
    PERFORM D UNTIL END-OF-SET.
    FIND NEXT PARCEL WITHIN LANDOWNER-PARCEL.
D.
    FIND OWNER WITHIN SEGMENT-PAR-SEG.
    GET. MOVE POINT1 TO POINT-NUMBER.
    FIND ANY POINT.
    GET. MOVE X TO X1. MOVE Y TO Y1.
    MOVE POINT2 TO POINT-NUMBER.
    FIND ANY POINT.
    GET. MOVE X TO X2. MOVE Y TO Y2.
    CALL DRAW (X1, Y1, X2, Y2).
```

These examples of CODASYL uses were given to demonstrate that it is possible to solve certain spatial problems with a so-called network modelling approach. However, as it is quite impossible to implement computational algorithms in COBOL, it is not feasible to solve location based queries like those dealt with in Chapter 13. Consequently CODASYL examples will not appear again in this book. We have merely chosen to illustrate that, while combining business data processing tasks with some spatial data is feasible, it is not very rewarding in general.

9.7 SOME ISSUES IN ENTITY-RELATIONSHIP AND LOGICAL MODELLING

Effective modelling for spatial or other databases requires attention to detail for various elements. Not the least of these are the degree and cardinalities of associations, and the interdependencies of attributes. To start with, it is valuable to determine the existence of many-to-many associations. For example, expanding on the toy cadastre, it is necessary to know if parcels have several owners and are located in particular cities or counties. Assessment and taxation conditions may be different

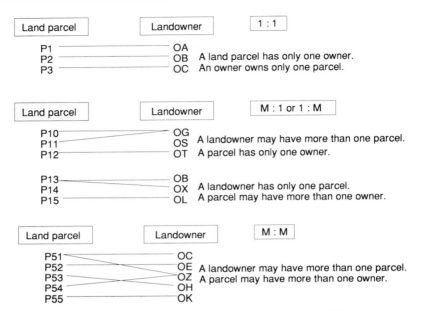

Figure 9.16 Possible real associations for two entities.

between the different local government units, or some parcels may be split by administrative boundaries.

9.7.1 Implied relationships

Various circumstances may arise, as shown in Figure 9.16, where M indicates many. First of all, a land parcel may have only one owner, and each owner may have only one piece of land, although this is a situation that does not arise in many economies. Or, landowners may have several lots, or land parcels may have several owners and different owners have more than one parcel. The different situations are clearly indicated by the narrative statements in the diagrams. One-to-many associations are straightforward but many-to-many relationships must be carefully decomposed into one-to-many connections as will be demonstrated. Both the cardinality and degree must be related to the real world conditions, which themselves provide a basis for establishing integrity constraints.

Let us consider the case of a land parcel which may be in a city which is located in a county. If all land space is not incorporated to the city or town government, then it is important to establish a direct parcel–county

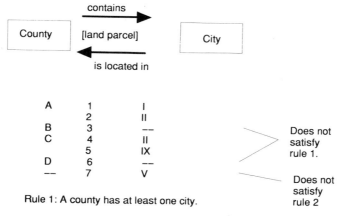

Rule 1: A county has at least one city.

Rule 2: A city must be entirely in a county.

Figure 9.17 Hidden relationships for two entities.

association. These various situations, shown in Figure 9.17, can be dangerous if not known in advance; for instance, **implied relationships** of the kind that say if a county then also a city. First of all, a connection between county and city may not be sufficient to understand why there is an assignment of city to county and vice versa. In the example, counties B and D do not meet the first rule; and city V is at odds with the second rule. However, while rule 2 is appropriate for the state of Maryland, it is not correct for the state of Virginia, in which cities are independent entities at the same level of local government as counties. Different possibilities of this condition, termed **obligation of membership**, are demonstrated in Figure 9.18.

Now, adding the land parcel element, it may arise that a lot may not be in incorporated territory (Figure 9.19a), but our representation will not allow us to know that. There is not a one-to-one match of parcel-to-city. Next the land parcel may be in a county but does not belong to a city (Figure 9.19b). County-includes-city-includes-land parcel is implied, but this may not always work. In such a case a ternary association is necessary, either a set of three binary relationships or a triadic association represented by a three-pronged diagram containing a pseudo-relation, with three-way linkages (Figure 9.20).

It may also arise that a boundary splits a land parcel into two or more parts. Putting to one side the probable, although not necessary, condition of all land being administered via one government unit only, it will be necessary to recognize that there is not a nesting of units. Consequently,

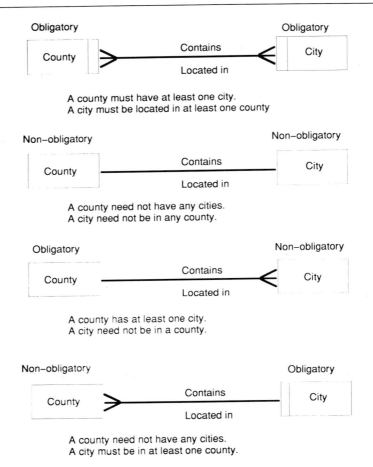

Figure 9.18 Obligatory relationships.

there is not a one-to-many association of, for example, state, county, city, parcel; instead there are several many-to-many associations.

9.7.2 Person-made and natural rules

In the field of information management for business organizations, it is customary to talk about **enterprise rules** for situations like these (Howe, 1983). The rules represent the real conditions or the practices of governments and corporations. In our case we also deal with **natural rules**

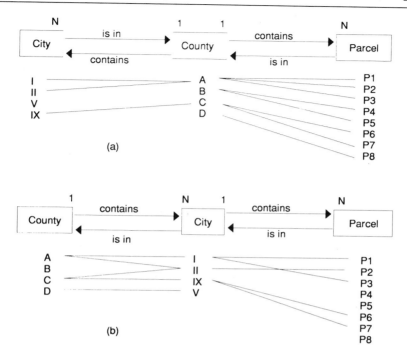

Figure 9.19 Relationships for three entities. (a) Parcel-city links are uncertain. (b) Parcel-county links are uncertain.

of the world – the empirical facts. In spatial information systems it can be important to know situations such as:

Land parcels, or even worse, buildings, can be split between different government units
Not all territory is incorporated
All highways do not intersect (there are flyovers, etc.)
There are political exclaves and enclaves
Streets and highways have more than one name
Some waterways cross each other without intersection
Mineral deposits are found in only certain geological formations
All rivers are not acyclic graphs
Barrier islands tend to be elongated

However, in the geographical world, take away comprehensive inventories and it may arise that there is not enough information to establish a limit to the maximum cardinality so that using the indefinite *N* cannot be avoided. For example, if we can know there are exactly fifty

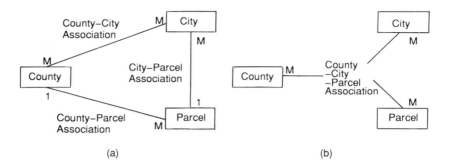

Figure 9.20 Ternary associations. (a) Three two-way linkages. (b) One three-way association.

states in the USA then a cardinality constraint is very apparent. It would be most welcome to know that polygons in an irregular tessellation have no more than four edges so as to implement a cardinality constraint as an integrity check rather than using the more time consuming topological consistency check for irregular polygons with an unknown number of edges. We develop this topic of spatial integrity constraints in Chapter 15.

While enterprise rules do change in space and time, and are context dependent, it is still important to look for determinancies like cardinality constraints and obligatory memberships. Public utility companies with established inventories, and governments with good census or survey data collections and administrative records can be expected to make a good attempt at that, but, alas, the empirical knowledge about natural features is not yet widely enough established for many phenomena with the level of certainty necessary to produce the equivalent of enterprise rules.

9.7.3 Table organization

Relational tables can be better designed if relationships and cardinalities can be clearly known, but the construction of the tables as a process itself implies certain conditions must exist for effective database operations of certain kinds. That is, at the elementary level some restrictions can be applied in the interest of logical simplicity. Row order is not important, but each row must be distinct. A row–column intersection should have only one attribute value. In other words, with reference to Figure 9.21a, the double entry of Grass and Forest for polygon 31 is not good. Changing to the extra line in Figure 9.21b is not good either because the key data item now has two entries the same, P31.

(a)

Polygon–ID	Area in hectares	Land use type
P27	502	GRASS
P36	28	FOREST
P31	1234	GRASS, FOREST
P82	297	WETLAND
P30	28	FOREST

(b)

Polygon–ID	Area in hectares	Land use type
P27	502	GRASS
P36	28	FOREST
P31	1234	GRASS
P82	297	WETLAND
P30	28	FOREST
P31	1234	FOREST

Figure 9.21 Some elementary considerations in table construction. (a) More than one attribute value. (b) Repetition of a record.

Tables satisfying these conditions may still have some limitations, for they may contain many redundant data. While not all duplicate data are unnecessary, the existence of redundant data can produce problems of lack of data consistency and integrity. One record may have a value changed, but other records with the same values may be forgotten. It is customary to decompose tables with redundant data into several smaller tables, following certain rules. For example, the initial unsuitable table of Figure 9.22a with repetition of the grass code, could be split into two tables, as shown, in which the type of land use for polygons is separated from the description of the land use codes (Figure 9.22b). Thus, if these descriptions or codes change, then only one data entry needs to be updated. Flat files and spreadsheets often contain repetitions for data items, or worse still, the equivalent of the ditto (``) symbol.

Also, the presence of the repeated information may suggest that the irrelevance of row order is violated. With reference to the example in Figure 9.23a, it is not desirable to have rows of different lengths to contain the lists of census tracts. Nor is it desirable to add to the rows, for there will be much duplication, or worse still, null values, for the additional rows, and the sequence of rows is now important for associating the ID and name information with each of the census tracts (Figure 9.23b). The preferred solution is to create more tables, linking

(a)

Polygon–ID	Polygon type	Polygon description
P2	1	GRASS
P7	6	FOREST
P2	4	WETLAND
P5	1	GRASS

(b)   LAND USE   LAND USE CODE

Polygon–ID	Polygon type
P2	1
P7	6
P2	4
P5	1

Polygon type	Polygon description
1	GRASS
4	WETLAND
6	FOREST

Figure 9.22 Elementary data reorganization. (a) Repetition of attribute values. (b) Splitting into two tables.

them by appropriate identifier items, the keys (Figure 9.23c).

In addition to these general difficulties, which can be systematically treated by a process known as **normalization**, the relational model has some particular special drawbacks for spatial data contexts. Leaving a fuller explanation until later, we here refer simply to conditions like row ordering. For some purposes in spatial databases the order of data in rows is important, for example the sequence of vertices for a polyline (assuming that the order is not indicated by an attribute). However, it is often interesting to use **unnormalized relations**, as now demonstrated.

We consider the frequent case in working with geometric or topological encodings, that we have sets of data rather than individual values. For example, we may use a string of coordinate pairs for a line entity, or we may have a matrix of grid cells for an image, or a dyadic matrix of contiguity relations. In these, and other situations, we have a **nested table** of data, for example:

SEGMENT_COORDINATES (Segment-ID, (X, Y)*)

in which the * indicates the nesting. This implies that the X, Y coordinate data are in a separate table; however, they may be stored as a long list of coordinate pairs in this one table SEGMENT_COORDINATES, potentially causing other problems inherent in having a variable length field of data.

County–ID	County–name	Census–tracts
A	Montgomery	C1, C5, C22
B	Eisenhower	C3, C5
C	Churchill	————
D	De Gaulle	C8, C9

(a)

County–ID	County–name	Census–tracts
A	Montgomery	C1
		C5
		C22
B	Eisenhower	C3
		C5
C	Churchill	————
D	De Gaulle	C8
		C9

(b)

(c)

County–ID	County–Name
A	Montgomery
B	Eisenhower
C	Churchill
D	De Gaulle

County–ID	Census–tracts
A	C1
	C5
	C22
B	C3
	C5
C	————
D	C8
	C9

Figure 9.23 Variable length records. (a) Multiple attribute values. (b) Extra records with duplication of some data. (c) Splitting into two tables.

Other examples are:

STREET_SEGMENT (Street-ID, Street_name, (Segment-ID)*)
OWNER_PARCEL_SEGMENT   (Owner-ID,   (Parcel-ID, (Segment-ID)*)*)

the latter implies a double nesting of segments making up parcels, and several parcels for an owner.

The need for subdivision of entities to get like entities is foreign to the basic concepts of relational systems used for standard data types of fixed size. Recent developments in the use of long or bulk data types suggest some flexibility may be possible for handling some spatial data, as studied at the University of Edinburgh in the GEOVIEW implementation (Waugh and Healey, 1986) and used by the commercial SYSTEM 9. But more promise seems to lie in the area of object-oriented databases, a topic for Chapter 17.

9.8 THE PROCESS FOR THE DESIGN OF SPATIAL INFORMATION SYSTEMS

The design of the database contents is not the only component of the more encompassing design of an information system, but it is a most important part. The conceptual and logical modelling approaches presented in this chapter provide a good basis for organizing the substance of the database. However, they are in a sense the tools used in a process that has many parts and can be most difficult for complex situations. A long discussion of the overall methodology is beyond the intent of this book, but we present here a general picture of what is involved.

9.8.1 The information discovery

Assuming that an organization's requirements have been ascertained, we enter the process at a first stage of data modelling, treating the data requirements and the creation of the conceptual model. We may think of this stage as one of information discovery, that is clearly ascertaining the entities, attributes and associations in the context of the processing requirements established by a functional analysis. Part of this process is learning the enterprise rules about spatial units, extremes of cardinality and other possible integrity constraints. A simplified procedure is to:

1. Ascertain processing tasks which the data model must support.
2. Determine the spatial data representation model and rules.
3. Plan dialogues with users to determine external models.
4. Prepare lists of entities and attributes.
5. Make preliminary entity-relationship diagrams for the known associations among entities and the classes into which they may be grouped.
6. Prepare a synthesis of the various separate models.
7. Create skeleton tables showing attributes and identifiers.
8. Review these tables for omissions and conflicts.
9. Revise as necessary.

The nature of the entity-relationships is an important determinant of the tables. If there is a one-to-one relationship then a separate table for one of those two entities is probably not necessary: it can be carried as an attribute of the first entity. For example, in associating attribute data to polygons via polygon seeds (generally a centroid in the polygon), if it is established that there can be only one such seed point per polygon, then the centroid will be an attribute of a polygon, not a separate entity. If membership relations are non-obligatory, as in the examples given in the previous section for cities and counties, then separate tables are necessary. Moreover, it may be that different users have different needs, necessitating a combination of external models into the higher order comprehensive conceptual model. This process, called **view integration** will be treated in Chapter 12.

Entities and attributes should be examined as to their properties and roles. Generally speaking, entities have descriptive information; attributes which are identifiers do not. Those entity descriptors will be turned into the list of attributes, including labels or other forms of identifier, at least one of which becomes a candidate to be a key. Simple identifiers are usually preferred to composites, but the latter are created for the relationship tables. Identifiers cannot have null values, nor must they have several instances with the same name or number so, for example, geographic names are often ruled out because not all spatial units have unambiguous names. There may be dependencies among attributes. For example, population density may have to be computed from number of persons and land area, so that it is necessary to check attribute lists against requirements and the spatial data model, since the spatial units may include water as well as land.

9.8.2 Information system design

The conceptual design is, though, but a part of a larger task of information system design. Using Figure 9.24 as a model of the process, the main steps concern needs assessment and conceptual modelling. Needs assessment, often involving painstaking surveys and interviews of staff in an organization, establishes the general categories of needs in the context of policies, goals, objectives or missions. Details of particular data, transactions, products and the like, ascertained via the requirements analysis, are the basis for the conceptual modelling and design, involving numerous components as shown. The conceptual design is followed by physical design, piloting, testing and final implementation.

The effectiveness of the procedure rests upon adequate identification of needs. But, as a long history of failures in spatial information systems in the 'early days' of the 1970s has shown, the requirements analysis is at times not done well. This state of affairs may occur for several reasons:

1. Users do not know clearly what they want.
2. Users cannot articulate their needs effectively.
3. Users' needs change over time.
4. Users may not be approaching the design task in the right way.

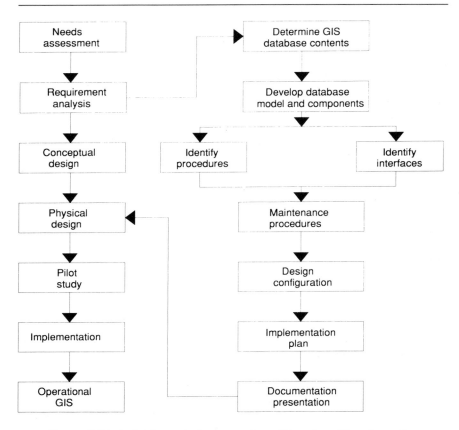

Figure 9.24 A database design procedure. (Based on Chambers; the diagram is a modified version of a diagram that appeared in *ARC News*, Volume 11, Number 2, 1989, and is used, slightly modified, with the permission of the Environmental Systems Research Institute, Inc.)

In the environment of uncertainty and change, it is not necessary to have a full, complete, single design. Consequently, to improve the chances of success, several design methodologies can be used.

Different methodologies may be used according to varied scenarios. That is, for example, requirements may already be known, a vendor may already be identified, needs are expected to change dramatically, and so on. In fact, two main cases can be distinguished:

1. Where the spatial information system vendor and system are known, perhaps because another section of an agency already uses the software.
2. When it is necessary to make the specification list, to launch a call to vendors for proposals, and to select the optimal system taking into account specificities, constraints, and prices.

Some of the design methodologies are now presented in summary form only, as it is not in the scope of this book to present the practicalities of implementation. First of all, there is a **top-down** approach which assumes that one can begin from the organizational and management aspects, decomposing the needed works into successive steps and finishing by programming. This approach assumes also that, starting from any specification list, it is possible to construct a system matching these specifications. However, it may be that some specifications are contradictory or not possible by means of actual technology. In addition, it is difficult to establish clearly all entities, attributes and associations if the enterprise rules are not well known.

If the entire specification list cannot be met, an idea is to offer only possible systems. This **bottom-up** approach then assumes that one must begin by low-level programs and build a system gradually by successive integration from low levels to high levels, trying to meet the users' needs as well as possible. Disadvantages of this methodology include the difficulty in knowing in complex situations the entire set of attributes or the dependencies among entities. An intermediate way is the **walk-through** approach which supposes that one begins at both ends, the low level and the management level, and tries to develop the system gradually and to match the intermediate results.

A **prototyping** approach assumes that previous approaches do not match the users' needs, so instead offers simple systems with reduced specifications, and, by successive corrections and integrations, tries to construct a satisfactory system. Another approach is called **technology driven** which supposes that the starting point is to know what kind of technology there is, and what type of characteristics to use, and only afterwards is it possible to build a definitive system. Some other methodologies exist but the prototyping approach seems to be the most popular, perhaps because it does not require comprehensive knowledge of all requirements at the beginning of the development process. It provides flexibility and gives users oportunities to try out and suggest changes. A cyclic process is a practical alternative to a linear/sequential design and implementation.

9.9 SUMMARY

The special natures of spatial data, the multiple and complex types of use of such data, and the uncertainties about enterprise rules, lead to the conclusion that it is not straightforward to design the databases for spatial information systems. We have to deal with many many-to-many associations, entities may be accessed spatially or via attributes, multiple keys may be needed, and many entities have several possible represent-ations.

The associations of entities may not be very clear at all at the outset. Indeed, the simple entity-relationship modelling concepts presented in this chapter must be expanded as comprehensive semantic data models, as we do later in this book, in order to deal with the many types of association. Moreover, the integration of multiple views requires the use of additional semantic concepts like generalization, aggregation and hierarchy. For example, several water features may be aggregated into one group based on the common property of water, but a different set may arise if they are features used for recreation purposes. Compound and complex objects may need to be created from different entities, for example a drainage basin made up of rivers, underground aquifers, lakes, terrain and boundary. Groupings into higher-level objects may arise by different criteria.

More generally, semantic data modelling, in a loose sense used as an intellectual aid in scientific conceptualization for a long time, attempts through systematic procedures to organize data and the meaning associated with phenomena and relationships. Databases themselves contain little material as to what their contents really mean. The challenges to the spatial information systems community appear to be as much to know more about reality, as the development of new techniques and methodologies for implementing logical designs. Relational modelling and their counterparts in practice, the relational database management systems, are established features of the information handling world. Tools are being developed for the creation of intelligent databases. But effort must be directed to the special needs for unravelling and modelling of the complexities of the real spatial world.

9.10 BIBLIOGRAPHY

The reader may obtain extensive treatment of entity-relation modelling from the works of Barker and Howe. Database management systems concepts are covered thoroughly by Date and Elmasri and Navathe, and the reader is referred to Codd (1990) for elaboration on relational database management systems. Howe provides a treatment of the practicalities of employing relational modelling concepts.

American National Standard Institute. 1975. *A Study Group on Data Base Management Systems Report.* New York: American Association for Computing Machinery.

ANSI/X3/SPARC. 1978. *Framework Report on Database Management Systems.* American National Standards Institute, Standards, Planning and Requirements Commitee, Database System Study Group; Montvale, New Jersey: AFIPS Press.

Barker, Richard. 1990. *CASE Method: ER Modelling.* Reading, Massachusetts, USA: Addison-Wesley.

Batini, Carlo, Stefano Ceri S. and Shamkant B. Navathe, 1990. *Conceptual Database Design, an Entity-Relationship Approach.* New York: Benjamin Cummings.

Bedard Yvan and F. Paquette. 1989. Extending entity/relationship formalism for spatial information systems. *Proceedings of the Auto Carto 9 Conference, Baltimore, Maryland, USA.* Falls Church, Virginia, USA: American Society for Photogrammetry and Remote Sensing/American Congress for Surveying and Mapping, pp. 818–827.

Bracken, Ian and Chris Webster. 1989. Towards a topology of geographical informations systems. *International Journal of Geographical Information Systems* 3(2): 137–152.

Calkins, Hugh W. and Duane Marble. 1987. The transition to automated production cartography: design of the master cartographic database. *The American Cartographer* 14(2).

Chambers, D. 1989. Overview of GIS database design. *ARC News.* Redlands, California, USA: Environmental Systems Research Institute, 11(2).

Chen, Peter P. 1976. The entity-relationship model: toward a unified view of data. *ACM Transactions on Database Systems* 1(1): 9–35.

Codd, Edgar F. 1970. A relational model for large shared data banks. *Communications of the ACM* 13(6): 377–387.

Codd, Edgar F. 1990. *The Relational Model for Database Management, Version 2.* Reading, Massachusetts, USA: Addison Wesley.

Date, Chris J. 1985. *An Introduction to Database Systems*, 4th edn. Reading, Massachusetts, USA: Addison-Wesley.

Elmasri, Ramez and Shamkant B. Navathe. 1989. *Fundamentals of Database Systems.* New York: Benjamin Cummings.

Feuchtwanger, Martin. 1989. Geographic logical database model requirements. *Proceedings of the Auto Carto 9 Conference, Baltimore.* Falls Church, VA: American Society for Photogrammetry and Remote Sensing/American Congress for Surveying and Mapping, pp. 599–609.

Howe, David R. 1983. *Data Analysis for Data Base Design.* London: Edward Arnold.

Kennedy, Hubert C. (editor and translator) 1980. *Peano: Life and Works of Giuseppe Peano*, Studies in the History of Modern Science, No. 4. Dordrecht and Boston: Kluwer Academic.

Laurini, Robert. 1990. Introduction to entity-relationship modelling for urban data management. In S. Spaccapietra (ed.), *The 9th International Conference on the Entity Relationship Approach*. Amsterdam, Netherlands: North-Holland.

Lu, Wei and Han Jiawei. 1990. Decomposition of spatial database queries by deduction and compilation. *Proceedings of the Fourth International Symposium on Spatial Data Handling, Zurich*, Switzerland, pp. 579–588.

Olle, T. W., H. G. Sol and Stuart A. A. Verrijn. 1982. *Information Systems Design Methodologies: A Comparative Review*. Amsterdam: North-Holland/Elsevier.

Waugh, Thomas C. and Richard G. Healey. 1986. The GEOVIEW design: a relational database approach to geographical data handling. *Proceedings of the Second International Symposium on Spatial Data Handling, Seattle*. Williamsville, New York, USA: International Geographical Union Commission on Geographical Data Sensing and Processing, pp. 193–212.

10
Spaghetti

Conceptual modelling of line oriented objects

Franco Maragoni glues a spaghetti noodle on a contour line and says, 'Oh! Oh! It works.' So Antonella Baraldo uses her spaghetti to model the polygonal terrain she has just bought. Massimo Rumori generalizes this way of modelling to city blocks, parcels, streets, rivers and engineering networks. In contrast, Maria-Elena Jacotto tries to use spaghetti to model her house.

Recalling that in Euclidean geometry we essentially focus on points and line segments, so spatial objects can be defined in a certain way called **vector modelling**, otherwise known in the computer graphics field as the wireframe approach. In this chapter, we examine the entity classes of polylines and isolated simple polygons, sets of polygons, terrains, other two-dimensional geomatic representations, and then polyhedra. The chapter concentrates on demonstrating the utility of the conceptual and relational modelling, and the variety of situations that might exist. It does not discuss the particulars of the spatial problems: the necessary background has been covered earlier in this book.

10.1 REPRESENTATION OF SEGMENTS, POLYLINES AND MIXTILINES

Spatial objects encompass an infinite number of points. Since it is impossible to store an infinite number of data records to match these points, extensional models are not possible, so we need to develop intensional models. In other words, conceptual models of spatial objects will be based on some privileged data, together with intensional–extensional rules and spatial integrity constraints. Figure 10.1 sets out the general case: we have several ways to select the so-called privileged points and so multiple representations of the same object can exist. In another role, we must argue from those privileged points when undertaking a task like the point-in-polygon query. For this we use the

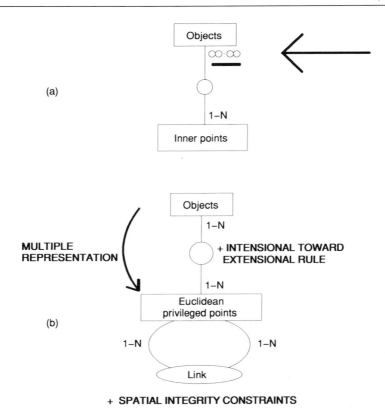

Figure 10.1 Ideal schema of a spatial object and its practical equivalent. (a) Theoretical model–impossible to implement. (b) Practical equivalent model.

concept of spatial indexing, that is the mapping between space and objects.

We begin with the development of conceptual models for one dimensional objects, such as segments, polylines and mixtilines. The entity-relation diagrams that reveal these models incorporate the rule for the intensional representations by reference to the symbol denoting the association between entities. We refer to two types of rule: **generating** and **membership**. The rule for generation refers to mechanisms like equations which are used to create more spatial objects from stored entities. The membership rules are those for testing if a particular instance meets specified conditions.

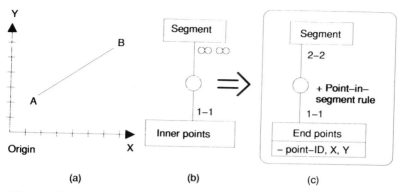

Figure 10.2 Storing a segment by end points and rules. (a) Line segment. (b) Entity-relationship model. (c) Model including point-in-line rule.

10.1.1 Segments

For storing a segment, we do not have to store only the end-points, but also the intensional–extensional rule needed to represent fully the entity, the point-in-segment rule (Figure 10.2). Recall that the generating rule for the point in segment case is given by an equation generating all points within the two end-points:

$$x = x_A + t(x_B - x_A)$$
$$y = y_A + t(y_B - y_A) \qquad (0 \leqslant t \leqslant 1)$$

The extensional versus intensional situation, which employs a membership rule, uses the same equations in a process in which, instead of generating points, we have to check whether there exists a parameter T meeting the specified constraints. Sometimes, due to spatial resolution, a candidate point can approximately meet the constraints, and in this case, the membership can be accepted. As an example, given a segment limited by A $(x_A = 1, y_A = 2)$ and B $(x_A = 3, y_B = 5)$, consider a candidate point C $(x_C = 1.5, y_C = 2.4999)$ the true point T in the segment being $(x_t = 1.5, y_t = 2.5)$. The point C can be declared as belonging to the AB segment depending on the level of resolution or an acceptable threshold of discrepancy tolerance.

In the relational representation of this context, a segment can be encoded:

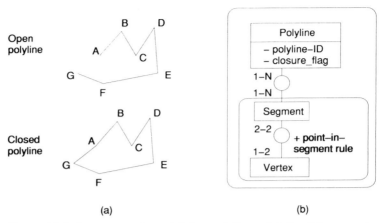

Figure 10.3 A conceptual model for polylines. (a) Spatial concept. (b) Entity-relationship model.

SEGMENT (Segment-ID, Point1-ID, Point2-ID)
POINT (Point-ID, X, Y)
+ RULE: point-in-segment rule

In order to accelerate the use of the point-in-segment rule, if the segment equation is given by $aX + bY + c = 0$, then the numerical coefficients can be stored in the SEGMENT relation:

SEGMENT (Segment-ID, Point1-ID, Point2-ID, a, b, c)

10.1.2 Polylines

To represent polylines, we have to store a set of vertices with their coordinates. The only ancillary information is to know whether the polyline is closed or not. In this case, the rules are the same as before for segments, but there are as many rules as there are segments in the polyline.

Initially, the polyline can be encoded as segments:

POLYLINE0 (Polyline-ID, Closure-flag)
POLYLINE1 (Polyline-ID, Segment-ID)
SEGMENT (Segment-ID, Point1-ID, Point2-ID)
POINT (Point-ID, X, Y)
+ RULE: point-in-polyline by iteration of the point-in-segment rule

On the other hand, if the polyline is represented by a set of sorted points, it is necessary to mention also the possibility of closure by a flag (yes/no), as well as an order (Figure 10.3):

POLYLINE2 (Polyline-ID, Closure_flag)
POLYLINE3 (Polyline-ID, Point-ID, Order_of_point)
POINT (Point-ID, X, Y)
+ RULE: same as aforementioned

A third way of expressing polylines is by using:

POLYLINE (Polyline-ID, (Point-ID)*)

to replace the relations POLYLINE2 and POLYLINE3. In this representation, a set of vertices is encoded, and a polyline is created by listing the sequence of points by ID-number. For closing the polyline, a possibility is to repeat the first point at the end of the sequence.

10.1.3 Representation of a mixtiline

By mixti- or mixedform-lines we mean arbitrary lines consisting not only of linear segments but also of portions of other curves such as circles and parabolas (Figure 10.4). We can define a mixed form line by a set of successive objects, for example, a portion of a circle (AB), then a set of segments (BC, CD and DE), and ending with a portion of a circle (EF).

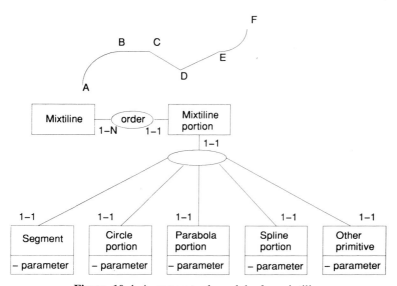

Figure 10.4 A conceptual model of a mixtiline.

In such a case we must represent the object by a set of vertices characterized by intensional–extensional rules, not only associated with segments, but also attached to circles and parabolas whose parameters must be stored. Depending on resolution, those portions of mixtiline objects can be replaced either by portions of rectilinear lines, so giving polylines, or by a succession of pixels.

A common method is by the use of spline curves (Figure 10.5) as discussed in section 4.5.2. The example depicts three spline pieces, their control points B and C, and their influences illustrated as springs. The complete conceptual model has the spline parameters as attributes in the SPLINE-SEGMENT table, and the entire mixed form line has a closure flag as an attribute.

10.2 ONE-DIMENSIONAL REPRESENTATION OF POLYGONS AND AREAS

Recall that polygons (or rings) may have special conditions giving rise to several situations: isolated simple polygons, isolated complex polygons, irregular tessellations of polygons and tessellations of areas limited by mixtilines. In all these cases, for retrieving the inner part of a polygon, we will use the point-in-polygon rule.

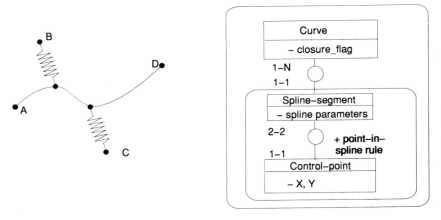

Figure 10.5 A conceptual model of a curve described by a spline.

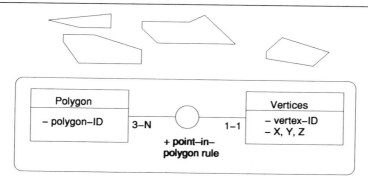

Figure 10.6 A conceptual model of simple isolated polygons.

10.2.1 Isolated polygons

For simple polygons, that is non-connected polygons without holes and islands, the relational model can use an encoding like a polyline with a succession of points, for which closure must not be forgotten (Figure 10.6):

POLYGON (Polygon-ID, Point-ID, Order_of_point)
POINT (Point-ID, X, Y)
+ RULE: point-in-polygon

It is important to realize that in the pure relational model nothing distinguishes a closed polyline from a polygon. Taking rules into account, though, it is clear that a closed polyline is a boundary and the rule governing it is the point-in-polyline rule; whereas, for a polygon, we are interested in the inner part, the area entity and we need a point-in-polygon rule.

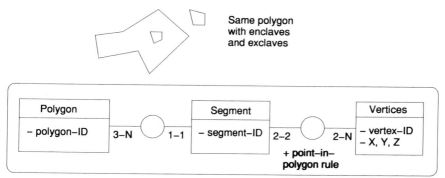

Figure 10.7 A conceptual model for complex isolated polygons.

Unfortunately, real world polygons are not simple. As an example, Italy has many islands plus two holes (the Vatican City and the principality of San Marino). In order to deal with a complex polygon of enclaves and exclaves, we have to take the non-connectivity features into account. In this situation, the next representation (Figure 10.7) is more appropriate, and the relational model is:

> POLYGON (Polygon-ID, Segment-ID)
> SEGMENT (Segment-ID, Point1-ID, Point2-ID)
> POINT (Point-ID, X, Y)
> + RULE: point-in-polygon

10.2.2 Sets of polygons

An irregular tessellation can be seen as a set of polygons, possibly non-connected, for example countries in Europe. In this case, a polygonal segment is shared (except at the outside boundary). So the prior conceptual model yields to one which stresses the segment entity (Figure 10.8). Here a polygon can have from three to N line segments for its boundary, and a segment is associated with one or two polygons, assuming the outside area is not an entity in the database.

For many purposes, counties and states as an example, polygons must be considered as bordered by polylines, not by straight lines. In this case (Figure 10.9) the representation recognizes that a polygon may have just one polyline and two categories of point (the end points, and the intermediate points, here called inner points). Rules are necessary for constructing polylines from the points, and associating a line to a

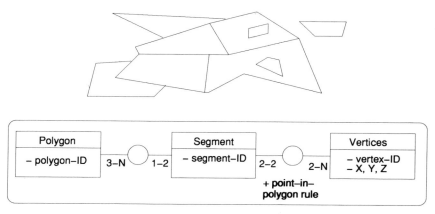

Figure 10.8 A conceptual model of an irregular tessellation.

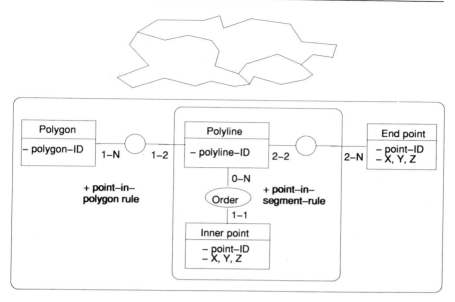

Figure 10.9 Polygons delimited by polylines.

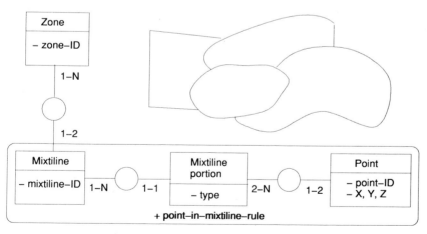

Figure 10.10 A conceptual model of a tessellation delimited by mixtilines.

polygon. In some cases, areas are bounded by mixtilines instead of being limited by segments. Such a tessellation is given in Figure 10.10 together with its conceptual model.

For some applications, it is desirable to orient polygons, as when creating polygonal areas from spaghetti data. For orienting a polygon, we select a direction, for instance clockwise or counterclockwise. When the polygon is punctured the orientation sense will change at the hole, reversing direction, in order always to have the polygon interior at the left when clockwise (or right when counterclockwise). Figure 10.11 provides an example showing the double vertex representation for the inner and outer boundary, respectively.

In the relational representation, we have:

> ORIENTED_POLYGON (Polygon-ID, Segment-ID, Orientation)
> SEGMENT (Segment-ID, Point1-ID, Point2-ID, Left_polygon-ID,
>   Right_polygon-ID)
> POINT (Point-ID, X, Y)
> + RULE: point-in-polygon

We need the orientation ($+1$ or -1) as an attribute in the first relation, because the SEGMENT relation favours one orientation ($+1$ means good orientation; -1 reverse orientation). Moreover, at the tessellation boundary, Left_polygon or Right_polygon attribute values might be null.

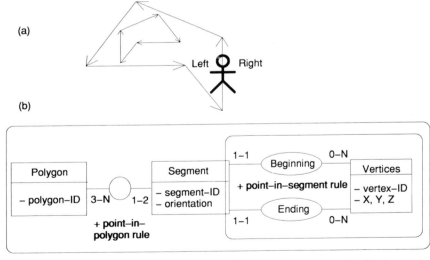

Figure 10.11 Polygon orientation. (a) Spatial concept. (b) Entity-relation model.

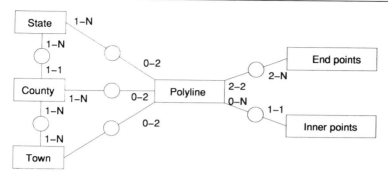

Figure 10.12 Geometric description of hierarchical territories.

It is often necessary to store in the same database geometric characteristics of several levels of hierarchically organized territories; for example, administrative units from parcels, to towns, to counties, to states, and so on. In this case a single segment can demarcate several levels, for instance, building, land parcel and town limits. The problem is how to store this kind of information yet limiting the redundancy. Using the USA as the example (Figure 10.12), the conterminous territories can be broken down into states, counties, cities and towns. Here the polyline approach is taken, emphasizing that one polyline can belong to several territory levels. Moreover, the relationship cardinalities between territories also can be indicated as an example of a many-to-many association, assuming nesting of county and state, but no nesting of town and county. We show a one-to-many relationship between town and county, because there exist some towns belonging to several counties, but a county can be in only one state.

10.3 MODELLING FOR GRAPHS

Recalling that in spatial information systems we represent water or road networks as graphs, two representations can arise depending on whether the graph is oriented or not. When the graph is not oriented, for an edge we have to store only the extremity nodes and the linked edges (Figure 10.13). The recursive association for edges represents the connections necessary to make chains and allow for paths. Any edge can be linked to two others, except if it is isolated or a termination like a railway branch line or street cul-de-sac. The example assumes there are at least two links for an edge.

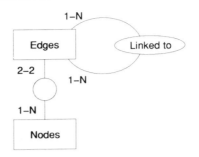

Figure 10.13 A model for a graph without orientation.

For an oriented graph (Figure 10.14), like a street system with one-way streets or a river network, we need to store the origin and destination nodes for each arc, together with previous and next arcs in the graph, giving a sort of small sequence. So the additional topological property of sequence is added to the connectivity feature by means of the types of association among entities.

The conceptual model, given as Figure 10.13 for **graphs without orientation**, implies several relations. First of all, there is a relation for the edges giving the extremity nodes:

EXTREMITY (Edge-ID, Node1-ID, Node2-ID)

then a relation giving the links between nodes and edges:

NODE (Node-ID, Edge-ID)

and lastly a relation for the links between edges:

LINKS (Edge-ID, Neighbour_edge-ID)

In the case of the **oriented graph** (Figure 10.14, in which arc is used as synonym for edge), the relations are a little different:

EXTREMITY (ARC-ID, To_node-ID, From_node-ID)

and, in order to show the direction of the arc:

TO_ARC (Node-ID, To_arc-ID)
FROM_ARC (Node-ID, From_arc-ID)

and, for depicting a succession of links:

TO_ARC (Arc-ID, To_arc-ID)
FROM_ARC (Arc-ID, From_arc-ID)

10.4 CONCEPTUAL MODELLING OF TERRAINS

Landscape terrains require that we store a two-dimensional surface in an area unit. Since it is not possible to store the infinite number of points of such areas, intensional models are necessary; so we encounter digital terrain or digital elevation models. The more common forms of representation are gradients, orthogonal grids, contours and triangulated irregular networks, as discussed in earlier parts of this book.

Instead of speaking about the z coordinate as landscape elevation or depth relative to sea level, let us recall that these models can also be used to store any kind of continuous phenomenon deriving from a function $f(x, y)$. For instance, we refer to examples in meteorology where isarithmic maps are used for barometric pressure (isobars) or temperature (isotherms) or precipitation (isohyets), all of which are similar to contours in concept. Other examples are the surfaces derived from point data for social, economic or demographic characteristics, like distributions of average income or education levels.

10.4.1 Gradients, grids and contours

For some digital terrain studies or analyses, it is worthwhile not only to store the point location coordinates but also their gradients. Recalling that the gradient is a unit vector pointing towards the maximum slope, it is then defined by three vector components: g_x, g_y, g_z. This approach implies having some more attributes in the point entity table than are usually present (Figure 10.15). Other slope angles can also be computed of course, but storage will require identifying the other end of a dyad too.

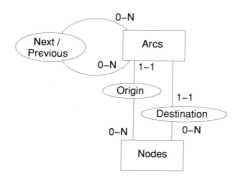

Figure 10.14 A conceptual model of an oriented graph.

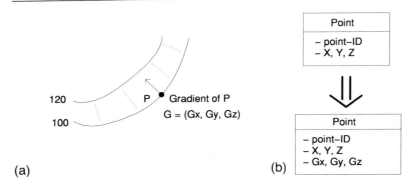

(a)   (b)

Figure 10.15 Taking gradients into account. (a) The spatial situation.
(b) Entity-relationship model.

In an **orthogonal grid**, we try to store only points at the intersection of
two orthogonal lines in x and y. Should it be necessary, perhaps in case of
irregular terrains with much roughness, we can have tighter grid lines in
selected parts of the study area. Figure 10.16 has an example of a grid
and two conceptual models, the first using a set of rectangle meshes, and
the second having sets of orthogonal lines (X_line and Y_line) for which
the beginning and ending coordinates are specified as well as the
particular ordinate or abscissa value.

With these two intensional representations, we need a rule to model all
points in the terrain. This rule is called the **point-in-grid rule**, which,
generally speaking, is approximated by the following procedure. Suppose
we have an (x, y) point and need to retrieve the corresponding elevation.
Should the point be in the database, the result is immediate. More likely,
though, (x, y) is not stored in the database. In this case, since it is not
possible to discover the right value of this elevation, an approximation
can be desirable. For this purpose we look for neighbouring lines, such as

$$x_i < x < x_i + 1$$

$$y_j < y < y_j + 1$$

We delineate a rectangle whose corner elevations are known $(z_{i,j},$
$z_{i,j+1},$ $z_{i+1,j}$ and $z_{i+1, j+1})$. Starting from these four points, we can
estimate the elevation, z, by interpolation, using the **elevation rule**
(Figure 10.17), which most commonly is a bilinear interpolator, using the
ruled surface $z = Axy + Bx + Cy + D$, where A, B, C, D depend on the
four points in the mesh.

In the relational representation we have the following. Firstly, for
rectangular or **squared meshes**:

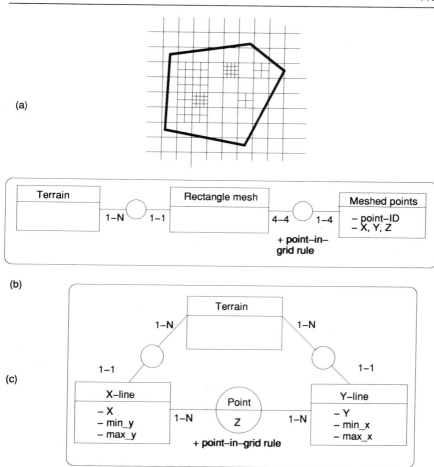

Figure 10.16 Examples of a grid for terrain and its two conceptual models. (a) Mesh of grid lines. (b) Mesh representation. (c) Grid-line model.

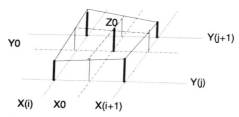

Figure 10.17 Bilinear interpolation to approximate elevation using the point-in-grid rule.

TERRAIN (Terrain-ID, Mesh-ID)
MESH (Mesh-ID, Point1-ID, Point2-ID, Point3-ID, Point4-ID)
POINT (Point-ID, X, Y, Z)
+ RULE: point-in-grid

Secondly, using lines:

X_LINE (Terrain-ID, X_line-ID, Min_Y, Max_Y)
Y_LINE (Terrain-ID, Y_line-ID, Min_X, Max_X)
POINT (Point-ID, X_line, Y_line, Z)
+ RULE: point-in-grid

Another way to store elevations is by **contours**, that is lines of equal elevation. In this approach, privileged points are taken along a contour curve which may, for convenience, be split into several portions. The model is given in Figure 10.18, assuming contour portions. A closure flag is necessary for single portions that have a common starting and ending point. The elevation value is taken care of by a separate entity – the level of curve – which can have many pieces but for which a curve portion can be associated with only one level.

In this case, should we want to retrieve the elevation of any point (x, y) we try to retrieve neighbouring contour curves and make an interpolation using a **point-in-contour rule**. The relational model is:

TERRAIN (Terrain-ID, Z_level)
CURVE (Z_level, Curve_portion-ID, Closure_flag)
PORTION (Curve_portion-ID, Point-ID, Order)
POINT (Point-ID, X, Y)
+ RULE: point-in-contour

In the previous conceptual model, contours were approximated by linear segments, but spline approximations are common in this domain.

10.4.2 Triangulated irregular networks

As discussed in section 6.6, triangulated irregular networks can represent a nice way to store terrains. The basic entity, the triangle, can be stored as a set of segments or directly as a set of vertices. Both are shown in Figure 10.19. The intensional rule corresponds to the point-in-polygon method that for this present purpose can be simplified as a **point-in-triangle rule**.

The relational representation, first for the segment oriented representation is:

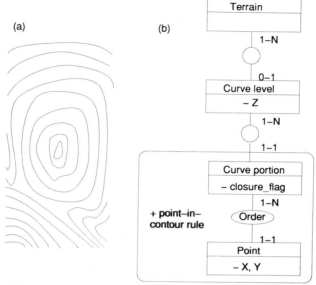

(a) (b)

Figure 10.18 A conceptual model for contours. (a) Spatial model. (b) Entity-relationship diagram.

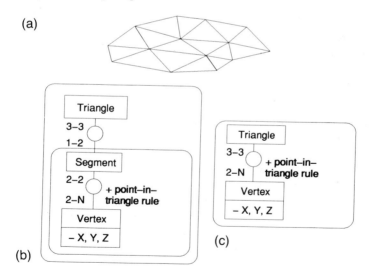

(a)

(b)

(c)

Figure 10.19 Modelling triangulated irregular networks. (a) Spatial concept. (b) Model for a topological representation. (c) Model for a geometric representation.

TRIANGLE (Triangle-ID, Segment1-ID, Segment2-ID,
Segment3-ID)
SEGMENT (Segment-ID, Point1-ID, Point2-ID)
VERTEX (Vertex-ID, X, Y, Z)
+ RULE: point-in-triangle

In order to accelerate some algorithms, it can be worthwhile to store the left and the right triangles for each segment. The second relation given above is then changed into:

SEGMENT (Segment-ID, Point1-ID, Point2-ID, Left_triangle-ID,
Right_triangle-ID).

Since the border of the triangular network has only one neighbouring triangle, the second value can be put to null, denoting an inappropriate item.

The direct representation relations are:

TRIANGLE (Triangle-ID, Vertex1-ID, Vertex2-ID, Vertex3-ID)
VERTEX (Vertex-ID, X, Y, Z)
+ RULE: point-in-triangle

In this second, simpler, representation a portion of the complete topology is not included. In so doing, if the main advantage is to save some memory space, then some drawbacks can appear in some algorithms.

10.5 REPRESENTATION OF POLYHEDRA

Even though polyhedra are not yet much used in spatial information systems, it is interesting to present them very rapidly in order to give a flavour of what could be done. A polyhedron, a solid limited by planar facets, can be represented by a set of segments. In the case of simple polyhedra, the entities are facets, edges and vertices. In this section, several cases of polyhedra will be introduced: an isolated simple polyhedron, a tessellation of simple polyhedra, then complex polyhedra and their tessellation.

10.5.1 Simple polyhedra

By an isolated simple polyhedron we mean a connex figure without holes; for instance, a kind of diamond (Figure 10.20). Concerning the

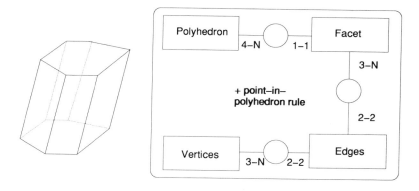

Figure 10.20 A model of a simple polyhedron.

intensional rule, we have to apply a **point-in-polyhedron rule** which is a three-dimensional extension of the Jordan half-line theorem.

A tessellation of simple polyhedra is easy to imagine from the case of one isolated polyhedron, and the corresponding conceptual model (Figure 10.21), is different only with respect to certain cardinalities. That is, edges are part of two or more facets, rather than just one, and a facet can be part of two or more polyhedra, rather than just one. With the relational model, we get:

R1 (Polyhedron-ID, Facet-ID)
R2 (Facet-ID, Edge-ID)
R3 (Edge-ID, Vertex1-ID, Vertex2-ID)
R4 (Vertex-ID, X, Y, Z)
+ RULE: point-in-polyhedron

10.5.2 Complex polyhedra

In the general case, a polyhedron can be non-connex with holes and separate parts. In this case, we need to add entities such as shells and loops. A shell is a singly connected surface, and a hole is an empty part of a shell, like a mine shaft bored through a part of the earth underground. If we assume geological strata as being limited by planar facets, the subsoil can be considered as a tessellation of complex polyhedra. Such a model is given in Figure 10.22, differing from the isolated complex polyhedron in some cardinalities. For the relational model we obtain:

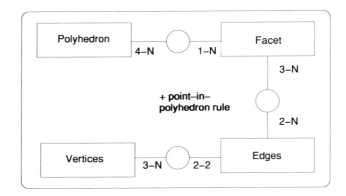

Figure 10.21 A model of a tessellation of simple polyhedra.

R1 (Polyhedron-ID, Shell-ID)
R2 (Shell-ID, Facet-ID)
R3 (Facet-ID, Loop-ID)
R4 (Loop-ID, Edge-ID)
R5 (Edge-ID, Vertex1-ID, Vertex2-ID)
R6 (Vertex-ID, X, Y)
+ RULE: point-in-polyhedron

Instead of limiting solids by planar facets, we can limit them by different kinds of surface. A possibility is to use spline surfaces. A **patch** is a portion of a surface and the solid is supposed to be defined by a set of contiguous patches. This approach provides a more realistic model for terrain and for geological layers (Figure 10.23). A **point-in-patch rule** must be defined in order to retrieve or to test all points lying on the patch surface or inside the solids.

10.6 SOME EXAMPLES OF VECTOR ORIENTED GEOMATIC MODELS

For storing city blocks and land parcels, we have principally a set of polygons. Usually blocks are separated by streets and so they have no polygon boundary segment in common. So, blocks can be seen as isolated polygons (Figure 10.24a), whereas parcels sharing boundaries must be modelled as an irregular tessellation (Figure 10.24b). (In order to focus the presentation the intensional rules are not indicated.)

Care is needed, though: the city blocks model given in Figure 10.24 is

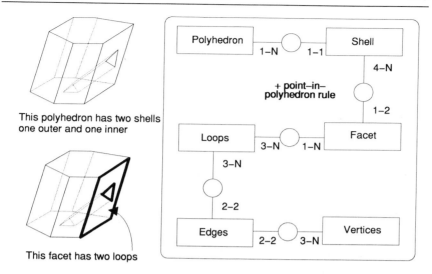

This polyhedron has two shells one outer and one inner

This facet has two loops

Figure 10.22 A complex polyhedron model. (a) Shells and loops. (b) Entity-relationship model.

valid only when we are using this particular geometric information. Should we have a database incorporating both blocks and parcels, we have to change the model saying that blocks are defined by segments and not by vertices (Figure 10.25). In this case we have two external models with different spatial representations and, for obtaining the new conceptual model, we have to harmonize the two geometric representations, and an integrity constraint must hold if we want a consistent result. In this case, for example, the Euler equality checking procedure implies

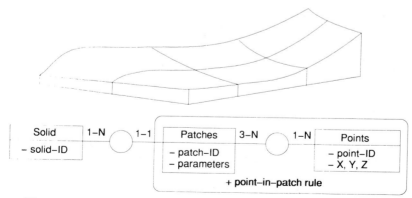

Figure 10.23 A conceptual model of a solid limited by patches.

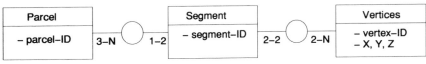

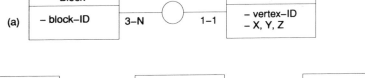

(b)

Figure 10.24 Conceptual models for city-blocks and land parcels.
(a) City-blocks. (b) Land parcels.

that the number of vertices plus the number of parcels equals the number of city blocks plus the number of segments.

Depending on applications and on the desired scales, roads and rivers may be modelled by polylines or by a set of bordering segments. We present first (Figure 10.26) an example of streets, for which, continuing the previous example, we want to model the parcel boundaries or the city block boundaries with a segment oriented representation.

We have similar problems for rivers except that, firstly, we often need to represent some islands. Secondly, in contrast to the case of the boundaries of a road being generally well defined, for a riverbed, especially for rivers with very different flows, it is possible to define several limits, for example for rivers in spates, braided streams or conventional flooding limits. Figure 10.27 gives the entity-relation diagram for a river defined by sets of open polylines. Let us mention that

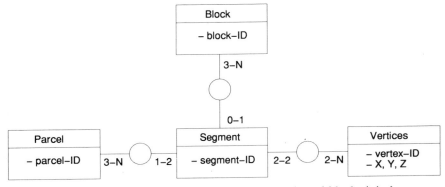

Figure 10.25 A conceptual model for land parcels and blocks jointly.

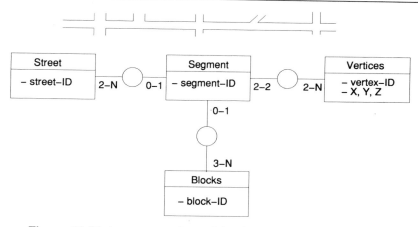

Figure 10.26 A conceptual model of a street with city-block boundaries.

rivers and roads can also be used as boundary lines. In this case, we have to be very precise as to what exactly is the boundary, perhaps the medial axis or one of the rims.

Pipe modelling for urban utilities is very common, covering phenomena such as water, gas, electricity, sewerage, cable television and traffic flows. For modelling these situations we need a good system for graph representation. Generally speaking, an unoriented graph is accepted for water and gas; for electricity, we prefer an oriented one; for sewerage, depending on the water pressure, the flow can change. Anyway, the model given in Figure 10.28 is robust enough to handle several kinds of network and types of node, tap, manhole, junction box and the like.

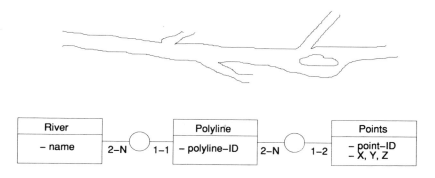

Figure 10.27 A conceptual model of a river.

The situation for transportation networks can be quite complex because along roads we have several reference systems ranging from coordinates to mile-posts and section controls (Nyerges, 1989). This aspect is illustrated by Figure 10.29 in which we have:

1. A chain and node locational reference scheme.
2. Road surface conditions based on control sections.
3. The number of lanes based on offset distances, here mile-posts.

For this need the corresponding conceptual model is given as Figure 10.30, where the main entities are:

1. The portions of routes.
2. Network chains with coordinate lists for the spatial representation of the polyline.
3. Nodes for the graph structure.

In this case it can be possible to store information intensionally, that is to store the segments of roads having the same characteristics concerning either pavement conditions or the number of lanes. The corresponding relations are:

R1 (From_route_mile, To_route_mile, Control_section-ID,
    Pavement_conditions)
R2 (From_route_mile, To_route_mile, Chain-ID, Number_of_lanes)
R3 (Chain-ID, Polyline-ID)
R4 (Polyline-ID, From_node, To_node, From_route_mile,
    To_route_mile, (X, Y)*)
R5 (Node-ID, X, Y)

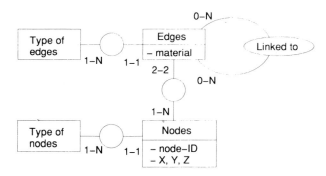

Figure 10.28 A conceptual model for pipes.

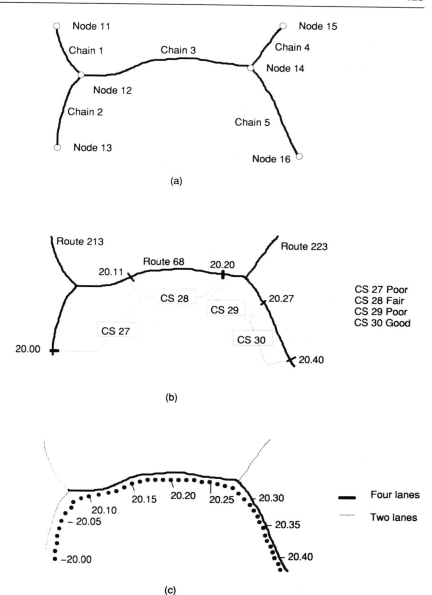

Figure 10.29 Data elements for roads maintenance. (a) Chain and node locational reference scheme. (b) Pavement conditions based on control sections. (c) Number of lanes based on run-length attribute coding. (From Nyerges, 1989.)

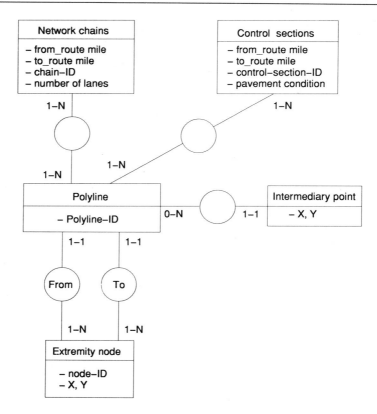

Figure 10.30 A conceptual model for roads maintenance.

10.7 SUMMARY

This chapter has demonstrated the vector or spaghetti representation approach by means of various spatial cases. The main characteristics are that models are simple but manipulation can be quite difficult. Leaving general discussion until Chapter 12, let us conclude simply by saying that the Euclidean vector approach needs little storage, whereas the algorithms for processing and manipulating are time consuming.

10.8 BIBLIOGRAPHY

The reader is referred to several works for more extensive reviews of geometric modelling, especially Günther, and Kemper and Wallrath.

Günther, Oliver. 1988. *Efficient Structures for Geometric Data Management*, Lecture Notes in Computer Sciences. Netherlands: Springer-Verlag.

Kemper, Alfons and Mechtild Wallrath. 1987. An analysis of geometric modelling in database systems. *Association for Computing Machinery Computing Surveys* 19(1): 47–91.

Laurini, Robert and Françoise Milleret-Raffort. 1989. A primer of multimedia database concepts. In Robert Laurini (ed.). *Multi-media Urban Information Systems*. Urban and Regional Spatial Analysis Network for Education and Training, Computers in Planning Series (N.D. Polydorides, ed.), vol. 1, pp. 7–75.

Laurini, Robert and Françoise Milleret-Raffort. 1989. *L'ingenierie des Connaissances Spatiales*. Paris, France: Hermès.

Nagy, George and Sharad Wagle. 1979. Geographic data processing. *Association of Computing Machinery Computing Surveys* 11(2): 139–181.

Nyerges, Timothy L. 1989. Design considerations for transportation GIS. Paper presented at the Annual Conference of the Association of American Geographers, Baltimore, Maryland, USA.

Peuquet, Donna J. 1984. A conceptual framework and comparison of spatial data models. *Cartographica* 21(4): 66–113.

11
Pizza

Conceptual modelling for areas and volumes

> After eating spaghetti, Franco Maragoni tastes a pizza and the marvellous colours on it suggest to him the various colours of a landscape and he says, 'Oh! Oh! I can use my pizza piece to model territories.' Antonella Baraldo cuts her pizza into little squares with different colours but Sanan Hamet prefers sometimes to regroup various small squares into bigger ones. Mohamed Faroodi would rather climb up a pyramid to see whether he can spot some petrol in cubic barrels.

In this chapter we present some area and volume based models useful for geomatic objects. We present the quadtree family, the pyramids and the octtrees, and then provide examples from the field of geology.

In the previous chapter we introduced the conceptual modelling of zero- and one-dimensional objects especially by means of segment-oriented models where geometric objects are defined by their boundaries. In a sense, they are always intensive models since even if we store their limits we have to take into account that we deal with objects of other dimensions. With the segment-oriented models we have to remind ourselves that we are dealing not only with the boundaries but, overall, also with the inner parts. That is, we need areal representations for areal objects and volumic representations for volumic objects; we also need the generation and membership rules to derive the extensional data.

The question, then, is whether it is possible to store areal and volumic objects with an extensive aspect, as treated in Chapter 6. It is feasible by using very small two- or three-dimensional cells; and in order to avoid dealing with an infinite number of cells, some approximations can be used. In fractal geometry, a point can be defined as a small area (or a small volume) for which the size is tending towards zero. Consequently, this provides a possibility for defining a solid by a set of fractal points with a certain level of resolution, and those fractal points can be subdivided into other smaller points (with bigger resolution). Thus, Figure 11.1 shows another way to define solids. In this case, the intensive–extensive rule becomes an aggregation–disaggregation rule. The hierarchical data structure approach of the quadtree and octtree is very nice in this domain.

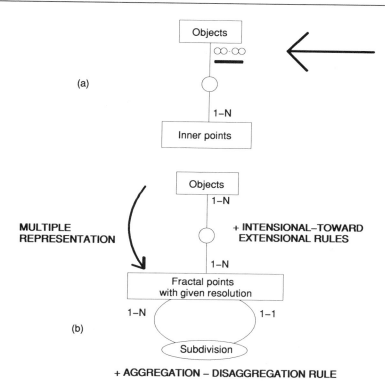

Figure 11.1 Ideal schema of a spatial object and its practical equivalent. (a) Theoretical model–impossible to implement. (b) Practical equivalent via fractal geometry.

11.1 REGULAR CELL GRID REPRESENTATION

As explained in Chapter 6, it is possible to decompose space into tessellations based on regular cells. Such a tessellation is based on regular squared tiles which can be organized as either *opus quadratum* or *opus lateritum* (Figure 6.5). Here we will examine only the former type of representation. An example, covered in more detail in section 16.2, could be image encoding in which pixels are the basic cells.

The first alternative is to store each square cell with a locator, the (x, y) coordinates, and its attribute:

GRID_CELL (X, Y, Attribute)

This constitutes a matrix in which x and y respectively correspond to the

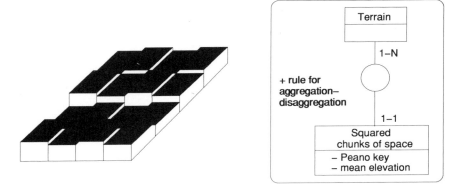

Figure 11.2 Representation of a terrain as a devil's staircase.

row and column numbers. In this model, generally speaking, x and y are integer coordinates. Supposing we want to know the value of the attribute in the case of a decimal coordinate value, say $x = 2.427$ and $y = 9.083$, an intensional-towards-extensional rule can be employed, meaning that the coordinates can be truncated. For terrain modelling the cell matrix representation will imply a devil's staircase approximation, as illustrated in Figure 11.2.

When the spatial distribution is such that some cells have the same value as one or more neighbours, a row or a column regrouping can be done in order to demarcate a sort of range, usually called **run-length encoding** in the case of images. With row-wise regrouping we can have:

ROW_REGROUPING (Y, X_begin, X_end, Attribute)

in which X_begin and X_end are the coordinates of the start- and end-points of the regrouping. Similarly a column-wise grouping may be performed.

11.2 QUADTREES

In this section we present the classical hierarchical representation for quadtrees and their linear forms. Afterwards, some extensions will be presented.

11.2.1 Review of the concept of quadtrees

To store a quadtree we generally use a hierarchical data structure, with four links, as discussed in section 6.4 and depicted in Figure 6.13.

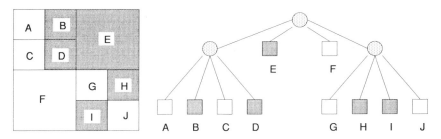

Figure 11.3 Storing a quadtree as a tree.

Conventionally, the tree leaves – the quadrant blocks – are encoded using B for a black to indicate presence of an attribute condition, W for the white (absence), and G for a grey node, the intermediate steps before the branch terminates (Figure 11.3). The initial square, the complete large square, is the neutral colour, grey. Then we have a grey square, mixing A, B, C and D, implying WBWB, followed by E with only the presence attribute. At this same, second, level, the quadrant F has only the absence state, encoded white, and finally there is a grey square with G, H, I and J, giving the code GWBBW. So the quadtree is encoded as a chain of digits GGWBWBBWGWBBW. Since we have only three

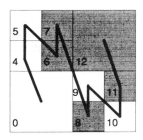

Object	Peano key	Side-length
A	6	1
A	7	1
A	8	1
A	11	1
A	12	2

(a)

Object	Hilbert key	Side-length
A	6	1
A	7	1
A	8	2
A	12	1
A	14	1

(b)

Figure 11.4 A linear quadtree ordered by Peano and Hilbert keys. (a) Example of a black object encoded in a linear quadtree ordered by Peano keys. (b) Example of a black object encoded in a linear quadtree ordered by Hilbert keys.

different symbols, two computer bits are sufficient, for example, using 00 for black, 01 for white, and 10 for grey (11 can possibly be used as a terminator), meaning that the chain, and therefore the quadtree, can be encoded with only 26 binary digits.

In contrast to this conventional hierarchical structure, Hilbert and Peano or other space-filling curves can be used to order the square cells; we call such a quadtree a **linear quadtree**. In Figures 11.4a and b, the Peano and the Hilbert keys, respectively, are used to organize linear quadtrees for the same spatial data. In the entity-relation modelling examples that follow, we use the space-filling curve ordered quadtrees.

We emphasize, too, that the quadtree representation may be used either as an object description or as a spatial index. A great advantage of the quadtree concept for indexing is that we can remove the white and grey nodes, using only the black, thereby conserving storage space. For the moment we are concerned with only object description; the spatial index domain is treated in section 15.2. We use these spatial orderings extensively in Chapter 13, at which place their properties will be examined by means of Peano relations and Peano tuple algebra.

11.2.2 Modelling polygons and terrains by quadtrees

Returning to the use of quadtrees for spatial modelling, let us consider the representation of areal and volumic objects. Instead of employing segment oriented models (nodes and edges) for zones, we can also use quadtree models. When modelling a polygon by a quadtree, we have two possibilities:

1. To model the polygon exactly in the vicinity of edges and vertices, getting squares and quadrants smaller and smaller, and so creating an infinite number of blocks, as demonstrated in section 6.4.
2. To approximate the polygon by modelling edges by staircases limited by the resolution level.

The conceptual model for the quadtree is shown in Figure 11.5, and in this case, the aggregation–disaggegation rule is given by the decomposition of quadrants into smaller quadrants. In the alternative case, we have a region with three polygonal zones that will be modelled by quadrants by a staircase approximation of the main parts of the polygons. For allocating remaining edge quadrants to zones, we can follow common practice and use a majority rule for assigning cut squares based on their relative proportions, as illustrated later in Figure 11.8. The relational model becomes:

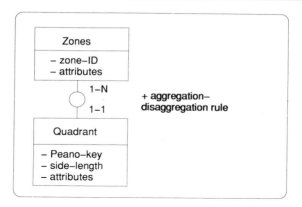

Figure 11.5 A conceptual model of zones described by quadtrees.

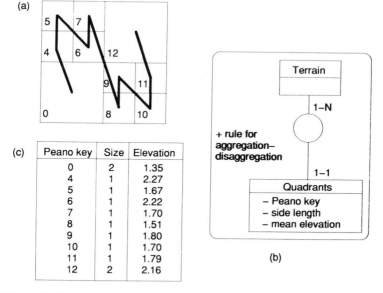

Figure 11.6 An example of terrain encoding. (a) Peano key encoding. (b) Entity-relation model for quadrants. (c) Example of tabulated data.

QUADRANT (Zone-ID, Peano_key, Quadrant_size, Quadrant_
attributes)
ZONES (Zone-ID, Zone_attributes)

where Zone-ID stands for the zone number, Peano_key is the key for each quadrant, Quadrant_size is the side length of the quadrant located via the Peano_key, and Attributes, one or several non-spatial attributes.

A linear quadtree (Figure 11.6) may be an effective encoding for terrains, requiring three pieces of information in the table (the space-filling curve index, the quadrant size, and the elevation data, recorded as a mean elevation across the block).

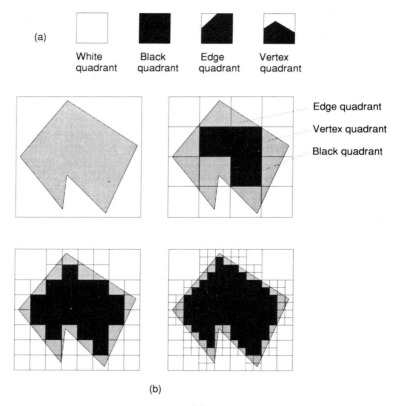

Figure 11.7 Extended quadtrees. (a) Different types of quadrants in extended quadtrees. (b) Example of an extended quadtree for three levels.

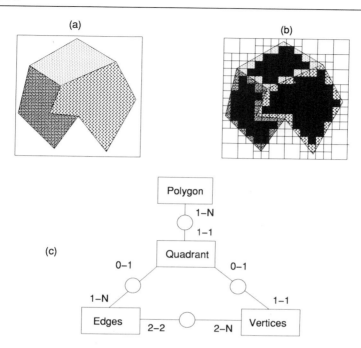

Figure 11.8 A conceptual model combining quadtrees and segment-oriented representation. (a) Three polygons. (b) Quadtree representation. (c) Conceptual model (rules are not mentioned.)

11.2.3 Extended quadtrees

To avoid multiplying excessively the number of small squares, and in order to store also the edges, a nice representation is to use an extended quadtree in which we have not only full squares but also squares cut by one edge or two edges. So, for this purpose, we can distinguish several kinds of quadrants (Figure 11.7): full squares (black quadrants), white squares, edge quadrants and vertex quadrants. In this way we are able to represent geometric objects taking both the interior and the boundaries into account.

An example of a combined representation, based partly on segments and partly on quadrants, is given in Figure 11.8. A ternary association links vertices and edges to the quadrants and to each other. The vertex to quadrant cardinality is one-to-one, since we have defined a vertex quadrant as having only one vertex. Edges can cross several quadrants, but a quadrant can have only one edge for a polygon tessellation model.

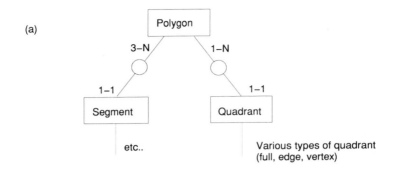

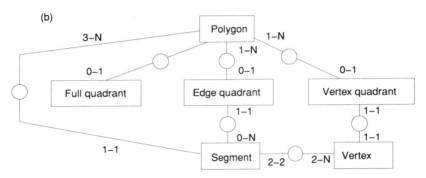

Figure 11.9 A conceptual model for an extended quadtree. (a) Concept of the extended quadtree. (b) Completed conceptual model.

A more complete model (Figure 11.9) depicts the linkage of edge quadrants to segments, and the linkage of vertex quadrants to vertices. For such a spatial representation, we have the following relations for the relational model. First of all, for the quadtree:

QUADRANT (Polygon-ID, Peano_key, Side_length, Type)

in which Peano_key refers to a key for full, edge or vertex quadtree blocks, and Type is the type of quadrant. The example has all three types.

Next, there are three relations giving the polygon geometry, using the segment-oriented approach:

POLY1 (Polygon-ID, Edge-ID)
POLY2 (Edge-ID, Vertex1-ID, Vertex2-ID)
POINT (Vertex-ID, X, Y)

in which Edge-ID means a polygon segment, and Vertex-ID, Vertex1-ID, and Vertex2-ID are some point numbers. If Peano keys are used instead of the vertex numbers, then the relation POINT is no longer necessary, and POLY2 becomes:

POLY2-BIS (Edge-ID, Peano_key1, Peano_key2)

Thirdly, there are two relations describing the edge and the vertex quadrants:

EDGE_QUADRANT (EQ_Peano_key, Edge-ID), and
VERTEX_QUADRANT (VQ_Peano_key, Edge1-ID, Edge2-ID,
Vertex-ID)

in which EQ_Peano_key and VQ_Peano_key are some Peano keys for edge and vertex quadrants respectively, Edge-ID, Edge1-ID and Edge2-ID are some polygon segments, and Vertex-ID is a point number which also can be replaced by its Peano_key.

11.3 PYRAMID MODELS

An efficient way to store images and raster data is to use the pyramid data structure first presented in Chapter 6. It consists of storing not only the image itself (usually in the bitmap form), but also the same image at a lower resolution. In other words, with a 512×512 cell image, we store also the 256×256 corresponding image by regrouping four pixels together. And we continue similarly, using 128×128, 64×64, and so on. It can be shown that we only increase the space occupancy by one-third by this process of storing several levels.

Figure 11.10 gives the conventional structure of a pyramid, and the conceptual model of a pyramidal object. These model forms can be used for any object modelled by grid squares. There exist several possibilities for transforming this pyramid structure into a relational model. The more immediate way is:

PYRAMID (Pyramid-ID, (Level-ID, (Locator, Colour)*)*)

in which Locator may be either a coordinate pair (x, y) or a Peano key (P). For accessing the four subdivided cells at the succeeding level, we have two possibilities. With the X-Y locator, the four subdivided pixels can be accessed by the locators $(2X, 2Y)$, $(2X+1, 2Y)$, $(2X, 2Y+1)$, $(2X+1, 2Y+1)$. In the case of the Peano keys the corresponding cells are located by $(4P)$, $(4P+1)$, $(4P+2)$, $(4P+3)$.

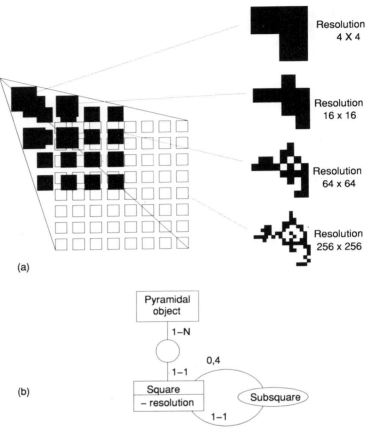

(a)

(b)

Figure 11.10 A model for a pyramid. (a) Example of a pyramid at four resolution levels. (b) Conceptual model of a pyramidal object.

11.4 MODELLING VIA OCTTREES

Recalling from section 6.4 that at the three-dimensional level we have octtrees instead of quadtrees, we can split the original cube octant into eight smaller cubes and subdivide each of them recursively until they are homogeneous or until a certain level of resolution is achieved. Since they are a three-dimensional extension of quadtrees, all quadtree considerations can be extended to octtrees. So, it is possible to define hierarchical trees, with eight branches, linear octtrees, and extended octtrees. In this section, we will examine successively the hierarchical octtrees, the linear

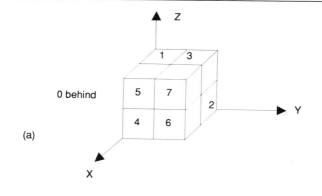

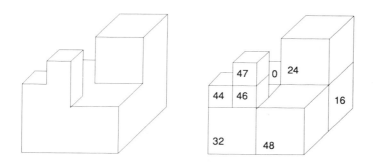

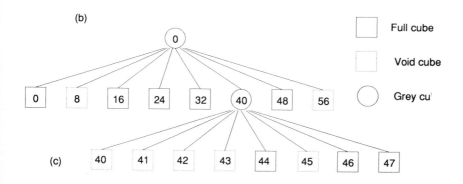

Figure 11.11 Hierarchical octtrees. (a) Peano order for an octtree. (b) Example of a solid described by an octtree. (c) Hierarchical structure of an octtree.

octtrees ordered by Peano keys, extended octtrees limited by planes, and finally, extended octtrees limited by ruled surfaces.

11.4.1 Hierarchical and linear octtrees

The first representation for octtrees is the hierarchical model with eight octants per node. Figure 11.11 shows such an octtree with three kinds of node: the full octant which has the subject of interest, for example a presence of mineral deposits, a void octant, and the so-called grey cubes which must be more decomposed into eight other octants. Similarly, as for linear quadtrees, employing Peano keys to convey an efficient order, we have the linear octtrees.

The entity-relationship model is provided as Figure 11.12 for a solid described by linear octtrees. For this example the relational schema is:

OCTANTS (Solid-ID, Peano_key, Side_length, Matter)
SOLID (Solid-ID, Attributes)

However, this kind of relation must be handled by means of the Peano tuple algebra, a topic which is not discussed until Chapter 13.

11.4.2 Extended octtrees

By extended octtrees, we mean not only pure cubes, full or void, but also cubes cut by planes. Consequently, it is possible to define a plane octant

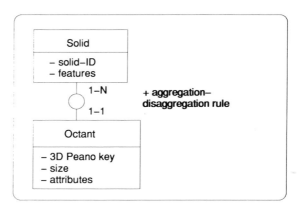

Figure 11.12 A conceptual model of a solid represented by linear octtrees ordered by 3D Peano keys.

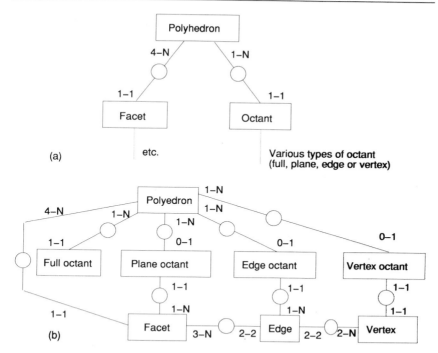

Figure 11.13 A conceptual model of an extended octtree. (a) Concept of the extended octtree. (b) Completed conceptual model.

having one intersecting plane, an edge octant, with two intersecting planes, and a vertex octant having three intersecting planes. Figure 11.13 gives the conceptual model of an extended octtree, and we have the following relations.

Firstly, the relation describing the octtrees:

OCTANT (Solid-ID, Peano_key, Side_length, Matter, Type)

in which Type refers to full, facet, edge or vertex type of octant.

Secondly, we have three relations giving the solid geometry via the segment approach:

POLYHEDRON1 (Polyhedron-ID, Facet-ID)
POLYHEDRON2 (Polyhedron-ID, Edge-ID)
POLYHEDRON3 (Edge-ID, Vertex1-ID, Vertex2-ID)
POINT (Vertex-ID , X, Y)

Instead of the vertex numbers and their coordinates, Peano keys can also

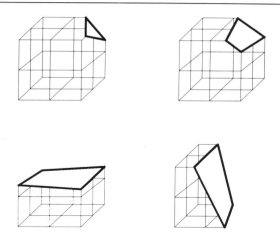

Figure 11.14 Different types of cubes cut by ruled surfaces.

be used. In this case, the relation POINT is no longer necessary and POLYHEDRON3 becomes:

POLYHEDRON3-BIS (Edge-ID, Peano_key1, Peano_key2)

Thirdly, there are three relations describing the facet, edge, and vertex octants:

FACET_OCTANT (FQ_Peano_key, Facet-ID)
EDGE_OCTANT (EQ_Peano_key, Edge-ID)
VERTEX_OCTANT (VQ_Peano_key, Edge1-ID, Edge2-ID,
Vertex-ID)

in which FQ_Peano_key, EQ_Peano_key and VQ_Peano_key are some Peano keys. FQ, EQ and VQ stand for full-, edge- and vertex-quadrants. Edge-ID, Edge1-ID, Edge2-ID refer to polygon segments, and Vertex-ID is a point number which also can be replaced by its Peano_key.

The main drawback of the previous representation is that it is suitable only to model polyhedra or, in better words, to approximate solids by polyhedra. Another possibility is to cut the cubes by **ruled surfaces**, thereby allowing more flexibility in the modelling. Note that ruled surfaces are defined by a rigid rule sliding on two other rigid rules, that is, a ruled surface is a surface consisting of straight lines, the straight segments, glued together. Concerning cubes, a ruled surface cuts a cube only with straight lines. Figure 11.14 shows some examples; the entity-relationship model is in Figure 11.15.

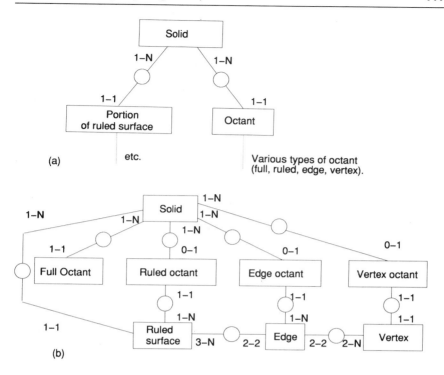

Figure 11.15 A conceptual model of an octtree cut by ruled surfaces. (a) Idea of an octtree cut by ruled surfaces. (b) Completed conceptual model.

11.5 EXAMPLE: MODELLING OF GEOLOGICAL OBJECTS

An important use of three-dimensional models in geomatics is in modelling geological or hydrological objects, like underground ore bodies or aquifers. In this chapter the case of geological strata as represented by octtrees is presented. A spatial representation of layers using octants is illustrated schematically in Figure 11.16; and the entity relation model is shown schematically as Figure 11.17.

The geological strata may have zero or a very large number of borings, and every boring passes through at least one layer. The details of the layers are interpolated from the borings as described in Chapter 7. Octants represent the strata, and contain from zero to a large number of borings. Subsoil characteristics are described by the Terrain subsoil

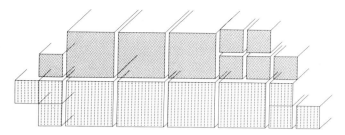

Figure 11.16 Modelling geological strata by octtrees.

entity, which is associated with the individual quadrants at the surface. A surface layer quadrant matches only the upper surface of the uppermost cube, representing the subsurface.

11.6 SUMMARY

For geomatics, spaghetti modelling seems to be the first idea which comes to people who know something about Euclidean geometry. Yet for pizza modelling, based as it is on fractal geometry, the basic assumptions are at

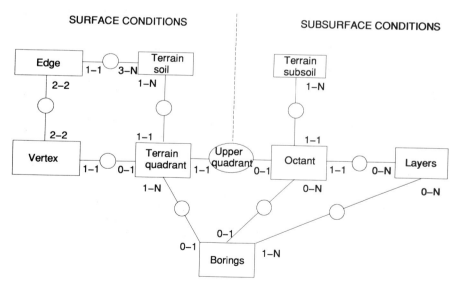

Figure 11.17 Conceptual modelling of terrain and its geological layers.

first somewhat disconcerting. This may be because people either are not familiar with the idea of space comprising 'squared points whose size tends towards zero', or they may have some trouble with the notion of space-filling curves, even though they accept the idea of paths through space.

However, for a computer, fractal representations imply simplicity in spatial data manipulations, whereas vector representations mean difficulties in manipulations even though storage requirements are low. It seems, then, that spatial information systems might offer users several varieties of geometric or topological representations for cognitive comfort or practical precision, but they must use different internal representations to speed up manipulations. There could be overall utility in allowing mixed vector and raster encodings of the same entities. Partly as a matter of philosophy, but also a matter of good practice, it seems that we need to reorient our thinking about spatial information systems away from a dominance by Euclidean geometry to more use of fractional geometry.

11.7 BIBLIOGRAPHY

A detailed treatment of quad-, oct-, R-tree, and other hierarchical data structures is provided by Samet.

Gargantini, Irene. 1982. An effective way to represent quadtrees. *Communications of the ACM* 25(12): 905–910.

Jones, Christopher B. 1989. Data structures for three-dimensional spatial information systems in geology. *International Journal of Geographical Information Systems* 3(1): 15–32.

Laurini, Robert. 1985. Graphics databases built on Peano space-filling curves. *Proceedings of the Eurographics Conference, Nice.* Amsterdam: North-Holland, pp. 327–338.

Mantyla, Martti. 1988. *An Introduction to Solid Modeling.* San Francisco: Computer Science Press/W. H. Freeman.

Requicha, Aristides A. G. 1980. Representations of solid objects: theory, methods and systems. *ACM Computing Surveys* 12(4): 437–464.

Samet, Hanan. 1990. Applications of Spatial Data Structure: Computer Graphics, Image Processing and GIS. Reading, Massachusetts, USA: Addison Wesley.

12
Spatial Object Modelling
Views, integration, complexities

> Eventually, Mr Cheese wanted to know what was the most adequate description of his city for pizza deliveries. He was told it was the spaghetti modelling of streets, exactly the oriented graph version. But Red Pepper disagreed, saying he was concerned about the analysis of sales in different districts; while Willie Makeit thought the speed of delivery was not good enough.

As the concluding element in Part Three, this chapter presents some topics of utility in the choice of data models adequate to users' requirements and orientations. It reviews criteria for selection of representations and provides a basis for the synthesis of several external models. To illustrate the alternatives available in the marketplace, one section provides brief descriptions of some current commercial geographic information systems. Some of the characteristics and limitations of the entity-relation and relational modelling approaches are then summarized. In particular, there is a discussion of semantic data models, the trends in working with complex features, and a review of the role of relational database management systems for dealing with spatial problems.

12.1 SELECTION CRITERIA FOR A GOOD REPRESENTATION

Out of the numerous criteria for selecting a good representation, we mention only three. First of all, the model must be **appropriate**. This implies that there is a matching between the user's vision of the real world and the vision offered by the spatial information system. This aspect is generally not adequately met today by off-the-shelf systems because they impose a special representation form not convenient for users. For example, this happens when someone needs graph models but has available only a quadtree based spatial information system. Moreover, the multiplicity of users, each with a different vision, makes it difficult for one system adequately to address all needs. So-called general purpose systems even may be inappropriate if they have limited flexibility of data representations.

Secondly, the **generalized costs** regarding time and space consumption are important factors in an evaluation. Some spatial information systems can represent spatial items with a very low level of computer memory but they are time-consuming for manipulation, and vice versa. To the user, a generalized cost function translating time and space consumption into monetary terms can reveal comparative costs of different theoretical representations or commercial systems.

Thirdly, some new spatial information systems can have a very attractive spatial representation structure, but they are not totally mature in the sense of **maintenance**. That is, a commitment to a particular framework may make it practically impossible to produce marginal changes to a data structure or any other increase in flexibility.

Starting from these considerations, some performance indices can be built into the software, in order to compare the different possibilities. One of the options is to design several data structures for the same purpose and to compare them by simulation, either for a commercial system or a theoretical data structure. In this case, we generate randomly some queries with given probabilities and compute the mean response time and the values of the other performance indices. Formalized procurement practices may require competing companies to run various tasks known as **benchmarks** to provide information to potential customers as to whether a system can perform specific functions, and how well it may undertake a set of related tasks.

12.2 EXTERNAL MODELS: SYNTHESIS WITH DIFFERENT REPRESENTATIONS

One of the special characteristics of spatial information is the difficulty in synthesizing external models in order to design the conceptual one, as explained in Chapter 9. Indeed, the synthesis must take into account that some users may need different geometric representations. Generally speaking, two ways coexist for accommodating different views:

1. The imposition of one model.
2. The employment of several geometric models.

In the first place, one model is selected for a particular, presumably most important need, forcing the other users (by default) to modify their

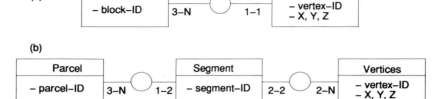

Figure 12.1 Conceptual models for city-blocks and land parcels. (a) Conceptual model for city-blocks. (b) Conceptual model for land parcels.

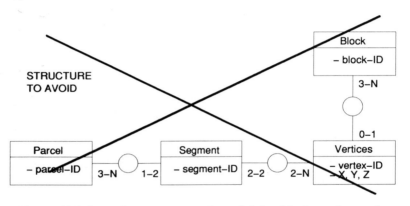

Figure 12.2 Inconsistent conceptual model for blocks and parcels mixing different geometric representations.

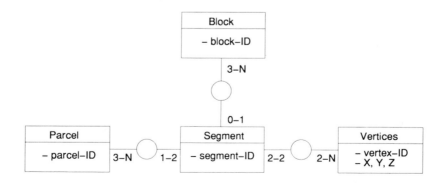

Figure 12.3 A consistent conceptual model for blocks and parcels with unification of geometric representation.

real world perception. One consequence is that the disadvantaged users may have difficulties or be reluctant to use a standardized model. On the other hand, the coexistence of two or more geometric models for the same entity can give rise to difficulty in answering several questions while keeping the database internally logically consistent after some data manipulations. Let us illustrate this dilemma by two models for the case of land parcels and city blocks. The latter can be represented as isolated simple polygons (Figure 12.1a) and the parcels by a polygonal tessellation (Figure 12.1b).

12.2.1 Standardization of geometric representation

The first approach is to say that both models include a 'vertex' entity and so the synthesis can imply Figure 12.2. This model is not acceptable because different geometric rules must be used, leading to some inconsistencies. The preferred solution will be to ask the users in charge of city blocks to modify their model in order to represent the blocks as tessellations. In this case, city blocks and parcels can have the same representations as given in Figure 12.3.

12.2.2 Coexistence of several geometric representations

There are various circumstances in which it is difficult to unify geometric representations. As an instance, let us examine the case of streets with four different viewers (Figure 12.4).

1. The office in charge of traffic management sees the street network as a directed graph because some streets are oriented, that is there are one-way streets. For this user, the geometric representation is an oriented graph, like that given in Figure 10.14.
2. The office in charge of the city cadastre is interested in knowing the landowners, possibly the city itself. For this office a street is seen as a set of open polygons limited by segments, as given in Figure 10.27.
3. The office in charge of pavement revetments considers streets as surficial and as area objects; for instance, a quadtree representation can be nice for this department because one of this user's main needs is to determine the material for the 'hardtop' applied to the streets (Figure 11.4).
4. The office in charge of the city engineering network sees the street as volumic objects including pipes. This user's model will include octtrees (Figure 11.11) and pipes (Figure 10.28).

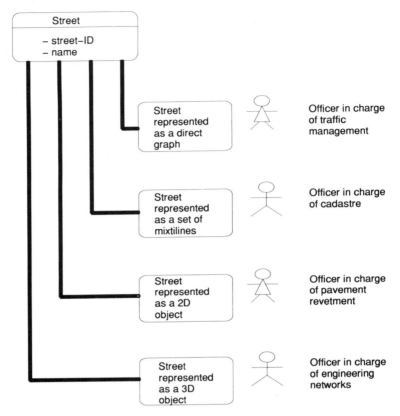

Figure 12.4 Different user views.

In these cases it seems very difficult to select one representation in order to build the conceptual model. The first two are poor for the third and the fourth office viewers, whereas the last model is too rich for the first pair of city offices. The only truly harmonious possibility is to have the four representations coexist, as illustrated in Figure 12.5.

12.2.3 An additional step in conceptual modelling

This synthesis corresponds to a design step in database systems development known as **view integration**. This integration requires a modification to the design process outlined in Chapter 9, where we identified these steps as necessary in order to model a portion of the real world:

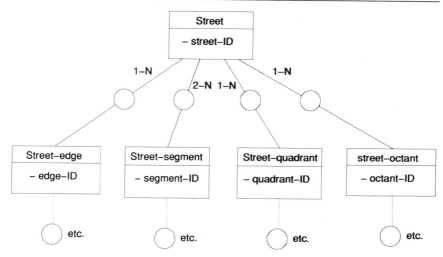

Figure 12.5 Different co-existing geometric representations for streets.

1. The external levels, one per user.
2. The conceptual level, the synthesis of the external models.
3. The logical model, taking account of database management system types like network, or relational.
4. The internal model (byte level).

The discussion in this section has emphasized the necessity of either accepting the coexistence of several spatial representations for the same objects or of agreeing on a standardized spatial representation. Indeed, the coexistence of several geometric representations implies many difficulties when updating, and several inconsistencies can occur. On the other hand, to impose a common spatial model will create some difficulties for some specific users.

Notwithstanding, it seems that the majority of people or organizations using spatial information systems are more in favour of standardizing the spatial representation models. The consequence is the necessity of splitting the so-called conceptual level into two other levels. The first one corresponds to the pure synthesis of the various spatial representations, as asked for by the users, which we call the **conceptual level without spatial standardization**. The other corresponds to the selection of a standard spatial model and its implications everywhere, referred to as the **conceptual level with spatial standardization**. Now some form of feedback will be necessary for those users for whom the representation has been changed. The new flow of conceptual modelling is shown in Figure 12.6.

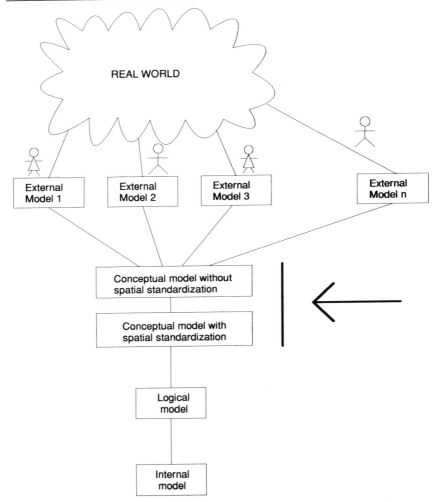

Figure 12.6 New modelling steps taking spatial standardization into account.

Recently, Armstrong and Densham (1990) have provided an example of spatial data modelling integration, undertaking a process close to that summarized by Figure 12.6, except that they use a variety of the entity-relationship modelling called the entity-category-relationship (ECR) approach. They contrast a cartographic view based on the topological arc–node model, for which the entities are point coordinates, nodes and chains, with a spatial analysis view (Figure 12.7). For this the spatial units

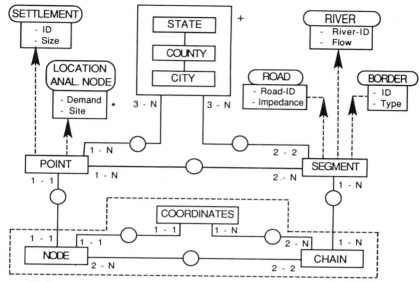

Chain: a sequence of connected line segments
Location analysis node: a node identified especially for purposes of location-
  allocation modelling
Node: a topological junction where chains meet
* For location analysis nodes demand = demand for services, and site =
  code indicating presence of existing facility
+ Nested hierarchy of area units

┌─ ─ ─ Cartographic view. The rest is the spatial analysis view

◯ Categories for point and segment entities

Figure 12.7 Integration of cartography and spatial analysis views for a mixture of point, line, and area spatial units. (After Armstrong and Densham, 1990; reprinted, in slightly modified form, with the permission of the authors and the publisher, Taylor and Francis.)

are nodes of a certain type, lines of a certain type, and a hierarchical organization of administrative areas.

The analysis view, required for establishing a decision support computer system for location-allocation analysis, places more attention on certain attributes of the lines and nodes. Nodes are either topological junctions or places which are candidates as sites for providing services. For the latter, the important attributes are service demand and if there is an existing facility. The nodes, therefore, have attributes reflecting their role or purpose. The line features making up the transportation network have data for impedances to travel, necessary for undertaking route-finding. Areas are represented by an entity set, encompassing states, counties and cities.

In this case the two views are reconciled, as shown, with overlap occurring for the chains and nodes, rather than maintaining different spatial data representations for different purposes. The logical level implementation uses an extended network model, a sort of combination of the network and the relational. Flexibility is retained via the designation of different sets of entities and attribute-coding techniques. Performance in the sense of facilitating retrieval, is improved by categorizing spatial units of a particular type into subsets, and speeding up search and access by working on a reduced number of records. In addition, all principal entities are declared as sets in the network implementation, thereby improving retrieval by providing direct access, without passing through subsidiary record types.

12.3 WORKING WITH COMPLEX FEATURES

Recalling our earlier discussion of complex spatial phenomena, let us now extend this discussion of conceptual modelling and practical realities to include objects made up of several basic spatial primitive units. The conceptual models and many data tables for the commercial systems described earlier relate to the spatial primitives of edge, node, triangle or polygon. However, for a user, the concept of a set of entities is important, for example a chain made out of connected links in a sewer network, or a combination of points, links, nodes and facets making up a surface for a watershed. Moreover, there are many ways in which entities may be connected or combined.

For our purposes, let us refer to a collection of spatial units as a **feature**. Firstly we distinguish between complex and compound features. A complex feature is a set of phenomena of different spatial form; a compound feature is a set of phenomena of like spatial form. The latter includes a network made up of links; the former includes an improved real estate lot made up of polygons (the garden), points (water connection), lines (sewer) and volumes (the house). Complex and compound features may vary substantially in their range of heterogeneity, for example the units of an educational institution campus, and watersheds, demonstrated as Figure 12.8.

Features may be produced in three ways:

1. By grouping entities spatially.

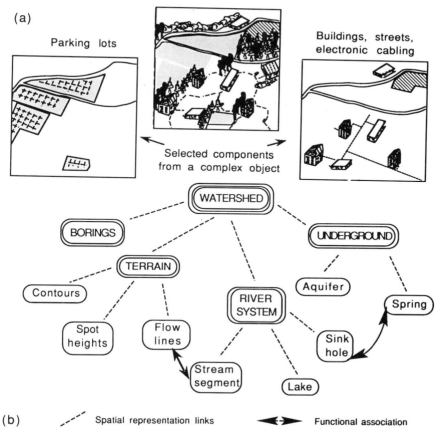

(a)

Figure 12.8 Heterogeneous features. (a) Educational campus – map model. (b) Watershed – entity model based on Figure 2.14.

2. By clustering based on attributes or entity types.
3. A combination of both.

For the first of these we have already noted topologically based associations like the neighbours of land parcels. Other examples are linking graph edges, joining pieces of polylines together, combining nodes into dyads, creating tessellations from triangles and joining polyhedral units. Connecting may be done at the level of the geometric encoding, for example, by matching nodes by identifiers or coordinates, by pointers, as in putting together a string of line segments, or through recursive linkages for database relational tables.

Higher order features may be created from components by reference to named objects or specific attributes. Thus, a watershed may be assembled by collecting the rivers, underground aquifers, terrain facets and pipelines. Or, for an example of counties, groups can be formed for low, middle and high degrees of urbanization based on the precomputed attribute of population density. Any spatial data combination is incidental to the primary objective of attribute clustering.

The case of creating groups of polygons is a good example of the combination of attributes and spatial properties. Geographers have traditionally recognized three types of region: formal, where different pieces of territory are grouped on the basis of homogeneity of one or more attributes into larger units, not necessarily contiguous; functional, where larger units exist by reason of some interdependence among smaller pieces of land, such as for newspaper circulation areas; or administrative, in which smaller units are combined to larger units, usually in a nested hierarchy, for governance purposes. Creating new features like these requires an ability to process both kinds of data in one environment, an undertaking that may not be easy if the data model is inappropriate.

We develop this discussion of more complicated situations by going further into the spatial semantic content of the information than is contained in the basic entity-relation model. While distinct spatial elements may be assembled in different ways, it is important to know how they might be connected. We identify a few ways:

1. By a functional association, for example water flow.
2. By a computational association, for example population density.
3. By grouping, for example human settlements.

For the first category, two spatial units are associated by a physical or virtual **flow**, covering many phenomena, like tourists, telephone calls, sewerage or air masses. Figure 12.8b shows functional associations for water flows. Recall the concept of the dyad as a necessary spatial unit representing this situation. While many software systems recognize the edge of a graph as a connection, they do not usually provide for virtual connections among point positions.

While attributes may be combined computationally in many ways, some derived statistics have implications for spatial representation. Thus population density requires access not only to the number of people, which may be associated with one set of polygons, but also to land area, which is a different set and may very well not be a complete tessellation of space. Obtaining land area may require special processing such as having to erase the water area from administratively defined polygon units.

Elements of one kind may be grouped into larger sets on the basis of some criterion of similarity or function. For example, hamlets, villages, towns and urban areas can be grouped as human settlements. Railways, stations, airports, streets, etc., may be assembled as a transportation system.

As one example of the recognition of the need to treat higher level objects, we can look at the proposal of the United States Geological Survey (Guptill, 1990) for an enhanced digital line graph (DLG-E). Defining a feature as a set of phenomena with common attributes and relationships, the idea is to orient towards models of geographic reality. This approach adds concepts of features and views to the interpretive side of the spatial data found on the topographical maps produced by that agency. Accordingly, the proposal, not yet formally adopted, provides for recognizing different views of real world phenomena. These mutually exclusive external models, covering about two hundred features, are: cover, division, ecosystem, geoposition and morphology. In the words of the Geological Survey, each view reflects a self-contained analytic approach to world features:

1. **Cover** reflects physical or material features at a location on or near the surface of the earth. While this view is based on form, at the lowest level, features may be differentiated by function.
2. **Division** reflects cultural demarcation of the earth's surface for a particular purpose, or for separations resulting from human activity.
3. **Ecosystem** is based on climate, vegetation, soils and other controlling environmental factors that result in unique ecological units.
4. **Geoposition** reflects measurement data about the earth's surface and contains points or lines on the earth or its representation for which the location, relative to a particular datum, is well known.
5. **Morphology** is based on the form of the land. Morphological features are those landform features that are named, labelled or symbolized as distinct entities on current map products.

About two hundred landscape features are recognized, ranging from barren land to graves to reservoirs to pipelines. The barren land subview, itself falling under the earth features view, comprises several entities of similar character based on surface material: barren land, firebreaks, ice mass and mud pots.

We have presented this classification not as a definitive way to view the mapped or mappable features, but as an indication of the difficulties of dealing with reality when, in a sense, each person on the earth has a view. It seems that there are two approaches to the semantic level organization for complex features:

1. Categorization of entities and classes of entities by the providers of data, perhaps most likely by attribute codes specifying membership of classes to represent particular views.
2. To let users of spatial data define their own views, facilitating this process by providing necessary tools.

The practical aspects of working with complex and compound entities is that the 'things' in a database may be handled in two ways. On the one hand, elements may be placed in sets identified as higher order features by assigning a code value to an attribute in the data table for a lower order feature. On the other hand, groups of spatial elements can be defined as distinct entities in a database, perhaps after some initial processing of the lower level spatial units. Indeed, in the spirit of the first of these, the grouping of entities associated with the entity-category-relation approach can be handled by attribute encoding, rather than by establishing special entity categories.

We develop this theme of semantic organization a little more below, and return to it in Chapter 17 in the discussion of object oriented databases.

12.4 SEMANTIC DATA MODELS

As discusssed in Chapter 9, conceptual modelling tools developed in cognitive and information science fields can provide for effective procedures for sorting out complex situations. Our concern now is to present some matters that extend the entity-relation model to cover more semantic content. Using the term semantic data models as a general category for extended entity-relation models, let us look first at one particular model used for a spatial data context, and then present some organizing principles.

Developed in the mid 1970s by François Bouillé at the University of Paris, the **hypergraph based data structure** (HBDS) formalism allows the identification and representation of classes of entities and their relation-ships (Bouillé, 1978). For the visual convention used in Figure 12.9, an oval embraces both entity (object) and particular instances; a class is indicated by an oval with a black disk, and all the attributes by small squares. Solid lines show the links between entities. Thus, roads, linked to themselves, may have tunnels, crossroads and sections. Crossroads may have traffic lights; tunnels may have sections which have lights. These two types of light, and possibly others, are combined to a higher

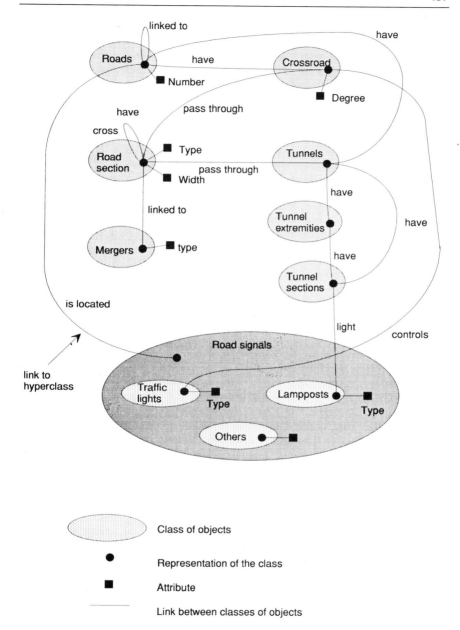

Figure 12.9 An example of the hypergraph-based data modelling.

level object, a hyperclass, road signals, to which the road object is linked by the concept of signals being located on roads.

The organization uses notions of sets and hypergraphs. The latter allows for graph linkages at different levels. The idea of sets is used to encompass objects or instances of objects into sets. So we have a **class**, a set of elements with one or more common properties, for example, road signals. Within the class are specific objects, like the traffic light and lamppost. Attributes are associated with the particular objects within the class (for instance, type of traffic light), or with the class. Objects may be linked among themselves (for example, the tunnels, tunnel extremities and tunnel sections); singly or in a group associated with an object in a class (for example, lampposts or roads), or an object (for example, roads) may be associated with a higher level entity, a superclass like road signals. This model will be used in Chapter 17.

More generally, we deal with a situation such as that simplified as Figure 12.10. We have an object type of a square which has certain properties, for which an object instance A fits because among its attributes it has those of four equal sides and four right angles required by the definition of a square. Two other regular geometric figures (the rhomb and the rectangle) are like the square, in having four sides, but only the square is also a rectangle and a rhomb when the respective properties of equality of side lengths and ninety degrees for each of the four angles are considered. So we say that the square inherits the properties of the rectangle and the rhomb. Both types of association are referred to as **is-a** links. The instance A *is a* square, the inheritance of properties of an object type by an instance. The object square *is a* rectangle, the inheritance of a property by an object subclass from an object class, the rectangle. Consequently, the grouping follows two dimensions, possibly virtually simultaneously:

1. Generalization of specialization on the basis of some properties to get classes of object types.
2. Aggregation of instances of objects to form a higher level compound or complex feature.

As with the square, **generalization** is oriented to properties, the attributes of objects, which may be static, dynamic or functional. A class known as human settlements is created out of phenomena called villages, hamlets, cities, etc., on the basis of form and function of clusters of residences for humankind, but ignoring the size dimension. The settlements class inherits the properties of the city, village or other type of residential cluster. Differences among the entities are ignored in order to produce the higher order entity.

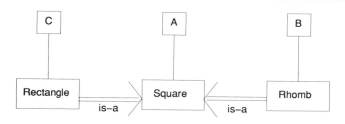

A is a square   (property: 4 equal side–lengths and 4 right angles)
B is a rhomb   (property: 4 equal side–lengths)
C is a rectangle (property: 4 right angles)

Figure 12.10 Example of an object-oriented model.

Similarly oriented to specified properties, the opposite orientation of the same process, **specialization**, can produce new object types out of others. For instance, a lake may be described by its use for recreation or water supply, producing new objects by subdivision, the recreational lake and reservoir, respectively. The lake characteristics of water polygon surrounded by a closed ring of polygon edges are inherited by the recreational area and reservoir, but the use function difference serves to distinguish the two types of waterbody.

Particular cases of objects can be **aggregated** on the basis of some attributes to produce new types. Just as we could identify a particular geometric instance A is-a square, then taking the instances for the x and y coordinate attributes, whether nodes, label points or vertices, produces the point spatial unit. All such instances could be joined in a group called points. Yet the new type does not exist as a higher level object as in the case of square-relative-to-rectangle. New types are created by combining instances that are similar for specified attributes. For example, if we have the characteristics of date, author and publisher of a book we have a publication, whether this is a book, audio tape or map.

Aggregation can also produce a new named object of a different kind, examples of compound as opposed to complex objects. Thus, the points mentioned in the previous paragraph could be taken together as an irregular lattice unit, for which a statistic indicating average spacing distance could be computed. Or all edges could be combined to form a graph of a transportation network, whether road, rail or water mode.

New objects can also be created by **association of entities**. In the entity-relation concept, the traffic lights, hazard warnings or turn restriction signs may be different objects but they can be grouped into a class called road signs. If we have structures codified as residential and vehicular storage, we have the house with garage, but if we have a house and a garage and a garden encoded as entities, we have an improved real estate lot, a higher order entity produced by an association.

The structuring of new objects is partly influenced by how entities are defined in the first place and by the processes used. What is more important than labels like association is the recognition that a structure is not independent of the concept of a view. The difficulties in fusing several external views into one conceptual model may result from conflicting definitions of entities or failure to distinguish between grouping on the basis of properties or instances. Indeed, the notion of classes of entities is, in itself, not necessary for entity-relation modelling, for categorization can be done using optional attributes.

Although semantic models have been constructed and used as intellectual aids in scientific fields for much time, it is only recently that database oriented formalisms and working systems have been developed. We present more details of the realization of two main types of semantic model in following chapters. The more loosely semantically connected pieces of data (the associative networks) exemplified by hypertext systems, are treated in Chapter 16. The more structured approaches (building upon the concepts of inheritance, abstraction, aggregation and so on) are covered in Chapter 17. It is here that we use the hypergraph based data structure (HBDS) for the object oriented modelling.

12.5 MODELS USED IN SOME SPATIAL INFORMATION SYSTEMS AND DATABASES

The majority of commercial spatial information systems employ the concepts of the geometric and topological models discussed in prior chapters, at times with some special variants. To demonstrate some alternatives available in the marketplace, we will very rapidly cover the topological data structures used in four general purpose commercial spatial information systems. Let us introduce the *SICAD* model, which is a sort of *CODASYL* structure; *TIGRIS*, which appears to have a relational structure for spatial data; *SYSTEM 9*, which, like *TIGRIS*, has a non-layered feature oriented approach; and *ARC/INFO*, perhaps the more widely known spatial information system, which is called geo-relational. Then we will look briefly at some digital cartographic databases being produced by two national organizations.

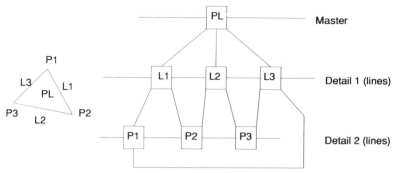

Figure 12.11 Example of the SICAD data structure for storing polygons. (Based on a paper presented at the Conference on GIS Design Models and Functionality, Leicester University, UK, March 1990.)

12.5.1 Commercial spatial information systems software examples

The *SICAD* system, created by Siemens AG, a German company, is a very commonly encountered system in Europe. It has its topological structure governed by a Master–Detail system; internal address mechanisms generate topological relations between the Master and Detail elements. This is complemented by a second set of addresses that links all the elements of the same specification to prevent hierarchical structures and to allow flexible data access. In a sense, *SICAD*, illustrated in Figure 12.11 for a polygon entity, can be seen as having a sort of network structure.

This system, among others, can also integrate vector and raster graphics. Apparently, the hybrid processing is done at the level of graphics display by superimposing the two kinds of data, or at the level of interpretation by vector-to-raster and raster-to-vector conversion.

Originally designed for thematic mapping and map analysis, especially for dealing with natural resources themes, the *ARC/INFO* general purpose geographic information system lends itself to tabular data processing applications as well as to automated cartography. It is a combination of spatial representation, to represent feature locations and topology, and the relational model, for representing nonspatial attributes, referred to by its authors as a **georelational model** (Morehouse, 1987). The previous description of this system's architecture, in Section 8.7.2 and Figure 8.19 is expanded here.

Using the *ARC/INFO* terminology, an important object is the **coverage** which is the basic unit of data storage, for a thematic layered database. A coverage is defined as a set of spatial units, called features, for a theme, where each feature has a location defined by coordinates, and topological pointers to other features, and possibly some non-spatial attributes. Some spatial data and all attributes are stored in feature attribute tables, oriented to either point, node, arc or polygon spatial units, the tables directly used by the users. In the following rapid description, we do not give the complete list of tables included in the *ARC/INFO* system (ESRI), but only some that are appropriate for understanding the geometric representations used.

The BND file stores the coverage's spatial extent in minimum and maximum x and y coordinates corresponding to a coverage:

BND (Coverage-ID, X_minimum, Y_minimum, X_maximum,
     Y_ maximum)

and map sheet registration points are stored in a TIC (tic point) table.

Attribute data are available via several principal tables, although one structure is used for both point and polygon spatial units. The polygon attribute table, PAT (polygon attribute table) has standard items consisting of identifier, polygon perimeter and polygon area, and any other attributes desired by the user:

PAT (Coverage-ID, Polygon-ID, Polygon_area, Polygon_perimeter,
     Attributes)

Point feature attributes are handled like polygons via the polygon attribute table, except that for points, the area and perimeter computed measures are set to zero. Data for nodes are tabulated via a Node Attributable Table (NAT):

NAT (Coverage-ID, Node-ID, Arc-ID, Attributes)

The AAT (Arc Attribute Table) contains the attributes for the arcs in a coverage:

AAT (Coverage-ID, Arc-ID, From_node-ID, To_node-ID,
     Left_polygon-ID, Right_polygon-ID, Length_of_arc,
     Attributes)

while the arc coordinates and connectivity and contiguity topology are stored in a separate ARC table. The coordinates and label for a centroid of each polygon in the coverage and links between the centroid and polygon are stored in other tables. The polygon topology and coordinates of bounding rectangles are also stored.

The *ARC/INFO* system also provides for representations of surfaces and linear features. The latter may be seen as either a system of connected linear entities through which resources flow, that is, a network, or as independent phenomena. The links representing phenomena such as highways or rivers may have barriers or pick-up points, both point entities. Turning limitations are handled via special turn-tables, the concept of which was presented in Chapter 8. The link data are stored in the Arc Attribute Table which contains special attributes for link impedance and resource demand.

AAT (Coverage-ID, Arc-ID, From_node-ID, To_node-ID,
    Left_polygon-ID, Right_polygon-ID, Arc_length,
    Arc_impedance, Arc_demand)

and the turn conditions in a turntable (TRN):

TRN (Coverage-ID, Node-ID, (From_arc-ID, To_arc-ID,
    Angle_of_turn, Azimuth_degree, Turn_impedance)*)

Linear features can be dealt with as separate entities, standing alone, or in combination to create networks; or can be produced by splitting arcs. A user of the system may define routes, a particular kind of linear feature with an associated measuring system, and/or sections. A section is a spatial unit of particular attributable character, not necessarily also defined topologically. An arc may have several sections, akin to line segments, and a section is associated with only one arc. A route has several sections and/or arcs, but a section can be in only one route. The data are prepared via a Route Attribute Table (RAT), Section Attribute Table (SAT), and an events table, a record of attributes associated with a route. Linking across these tables uses the section or route items, and the locator item:

SEC (Coverage-ID, Section-ID, Arc-ID, From_position, To_position)
RAT (Coverage-ID, Section-ID, Route-ID), and
EVENTS (Coverage-ID, Route-ID, From_position, To_position,
    Attributes).

The section, and events, like a traffic accident site, or change from one road surface type to another, are spatially referenced by a linear addressing system, using relative or absolute distances like offset mileages or house numbers. The locations at which the attribute value of a linear feature changes are recorded by such distances rather than by splitting arcs by pseudonodes.

Spatially continuous phenomena can be handled by grid or lattice spatial data types, with conversions possible between these and triangles, and between them and polygons (Figure 8.15). There is a raster counterpart to the layered vector data model, using the square cell grid. Descriptive information for grid cells is taken care of via a value attribute table (VAT), which has the attribute codes and cell frequency counts for each attribute value. Surface representations for terrain or other attributes may also employ a triangulated irregular network within the *ARC/INFO* system, utilizing a different set of tables, as shown already in Figure 6.26. The surface facets can be converted to the polygon form by a user initated procedure. The relational equivalents for the data for the nodes and edges comprising triangles are:

NODE (Triangle-ID, Node1-ID, Node2-ID, Node3-ID)
EDGE (Triangle-ID, (Adjacent-triangle-ID)*)

The positional information is recorded in two tables:

ELEVATION (Node-ID, Height)
COORDINATES (Node-ID, X, Y)

There are also ancillary data tables for coverage boundary coordinates, registration points, the nodes in the triangular convex hull, and general information about the triangulated irregular network.

Originally developed in Switzerland and now a product distributed by the American computer manufacturer, Prime Inc., the SYSTEM 9 commercial product has a non-layered topological structuring, integrates a relational database system, and can treat higher order objects building up from lower level elements (Prime, 1989). These basic components are the spatial primitives of nodes, which are discrete points, not necessarily topological junctions; lines, which may be straight, curved, splines or circles; and surfaces, essentially polygons. Nodes, lines and surfaces are interrelated by pointers; and the use of coincidence encoding via shared primitives, either nodes or lines, reduces storage volumes by avoiding duplication of the geometric information. In this way it is not necessary to have a physical layering to the database, although it is still useful to conceive of different themes as layers.

The SYSTEM 9 concept identifies complex and simple feature classes, like bridges or service zones for residences, using a two part identifier for each feature in the database. That is, the thematic relational tables, for example soils and land use, each contains a class identifier and a class membership for every tuple. Both pairs are written to a new table for the result of a spatial overlap of soils and land use. Feature classes, which can overlap, follow a logic of similarity of topology and attributes.

The TIGRIS system, first mentioned in section 8.6.3, rather like SYSTEM 9, does not use the concept of layers. It too employs shared primitives, although the particular data structure is different. Both systems, having an object orientation, have a different structure from the layer-oriented ARC/INFO. Expanding on the presentation of the TIGRIS model in section 8.6.3 and Figure 8.16, we now describe the model via relational tables (Herring, 1990). The basic relations are the following:

NODE (Node-ID, Class-ID, X, Y, Z)
DE (DE-ID, Class-ID, Orient, Face-ID, Next-DE, Node-ID)
EDGE (Edge-ID, Class-ID, Positive-ID, Negative-ID, Coord_count,
      Coordinates)
FACE (Face-ID, Class-ID)
NODE_TO_FACE (Node-ID, Node_class-ID, Face-ID,
               Face_class-ID)
AREA_TO_FACE (Area-ID, Area_class-ID, Face-ID,
              Face_class-ID)
LINE_TO_DE (Line-ID, Line_class-ID, DE-ID, DE_class-ID)
POINT_TO_NODE (Point-ID, Point_class-ID, Node-ID,
               Node_class-ID)

Distinctive point entities are recognized as such, and associated with the face (polygon) in which they are located. Directed edges (DE) are used so that the left face as well as the next-directed edge can be stored. The system does not have a layered database architecture; instead it recognizes coincidences of features, and encodes only one representation of such objects. Encoding of shared linear features facilitates dealing with nested boundaries; and the assembly of linear features into networks is simplified by the use of the directed edge primitive element. In order to distinguish several kinds of object, and thereby facilitate dealing with complex entities, class identifiers are used, and spatial primitives assembled (Figure 12.12). Extensional–intensional rules are treated separately by programs.

At the topological level, an object corresponds to one of three geometric entities, node, edge or face. The next level collects topological objects into features components, for example road segment or forest stand. Higher levels collect features and components into more complex entities (Herring, 1989).

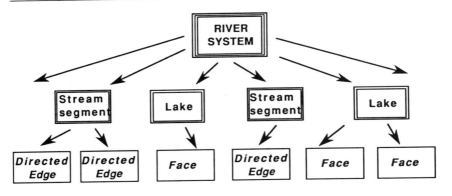

(a)

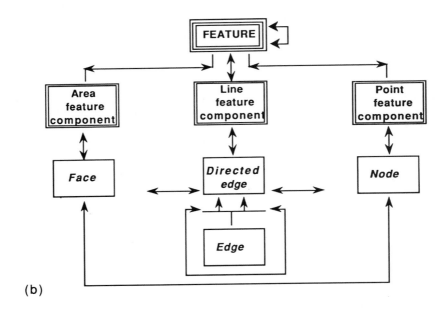

(b)

Figure 12.12 The TIGRIS approach. (a) Example of a complex feature; the hierarchy implies primitive units are combined to get specific features, the lake or stream, which are then combined to make up the river system. (b) Class relationships. (Based on Herring, 1987, and reproduced with the permission of The American Society for Photogrammetry and Remote Sensing.)

12.5.2 National cartographic databases

Like companies producing software for spatial information, so national government organizations, frequently major builders of digital spatial databases, are faced with addressing the issue of varied needs. In a practical sense, the agencies may provide alternative representations as products for different categories of user, or they may provide one structuring of data which others can import to their own systems with or without remodelling.

The Ordnance Survey of Great Britain, the major producer in the United Kingdom of topographical data, produces digital spatial data from its large and medium scale topographical map series. It is evolving a concept of three distinct data sets developed from the underlying digital encoding of the geometry of map phenomena, OSMAP, OSLAND and OSCAR (Figure 12.13) (Smith, 1989). The first, OSMAP, is essentially a topographical view, a set of links and nodes topologically structured as a digital map useable as a basis for geographical information systems and graphic plot production. A network set of data, OSCAR, would be focused on street and highway centrelines, to provide a resource for navigation and road management and maintenance. A land parcel model, OSLAND, covering land and highway parcel polygons and street addresses, would be oriented towards administrative functions. By this group of data sets, the identified major external views can be accommodated, and complex objects, definable for a particular purpose, can be handled in a practical sense by simple tables containing ordered

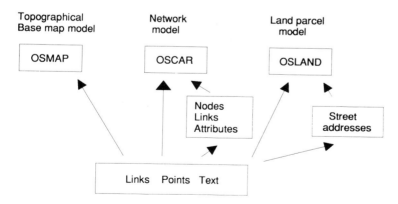

Figure 12.13 The Ordnance Survey of Great Britain topographic database views. (Based on Smith, 1989.)

lists of spatial units that make up particular complex or compound entities.

The United States Census Bureau has created a computer database storing geometric representations of natural and man-made landscape features and administrative and statistical reporting zones. The database known as the TIGER (Topologically Integrated Geographic Encoding and Referencing) database, was built primarily to accomplish mapping tasks in conjunction with the decennial census of 1990. Consisting of several directories and lists linked by an elaborate set of pointers, as shown in summary form in Figure 12.14, the 'file' is structured around the zero-, one- and two-cell topological units, matched by the major files C0RALS, C1RALS and C2RALS. It is a tightly integrated, planar enforced, single layer database of a very large volume of spatial and attribute information. It is more like a network model concept than a relational one; this is not surprising given the mapping, rather than semantic query, orientation, and the need to establish clearly the detailed requirements before implementation.

Several database views have been implemented, primarily for the internal purposes of the US Census Bureau, for example, the geometric view, the linear attributes view, the geographical view, and the 1990 census geography view (Figure 12.14). For example, the linear attributes view encompasses the one-cell data (C1RALS) and associated linear features (FIDDIR, FIDRALS, FIDCONT), and street addresses and postal information (ARRALS and ZIPRALS). The 1990 census geography view, aimed at assembling the information for polygonal units encompasses the primitive units of two-cells (C2RALS), the GTUBs (geographic tabulation unit base) (GTUB90, GTUBAN, GT90DIR, GTANDIR), and the block units (BKARA, BKARADIR) as well as the geographic (statistical, political, or other) entities made up of one or more two-cells, for example, regions, and zones. This allows processing akin to that presented in Section 5.4.2. Several data extracts are created from the master database, for mapping purposes, or for release to the public for use in conjunction with their own spatial information systems or mapping software.

12.6 ISSUES IN REPRESENTATIONS AND CONCEPTUAL MODELLING

The dynamism in the field of spatial information systems, on the one hand, makes it difficult to make statements that will remain valid over an extended period of time; but on the other, reveals the excitement over the development of even better ways to handle spatial information. At the moment, as new semantic modelling concepts and their implementations are being developed or devised, the practical questions relate to how useful are classical types of database management system. We try here to distil some of the main issues.

12.6.1 Metadata

A complete design will incorporate supplementary information about the database. These metadata include matters of encoding practices, decision rules regarding spatial data representation, definitions of attributes and attribute values, and so on. In a sense, there needs to be a physical counterpart of view integration, setting out details of the encodings; in practice this component is often known as the data dictionary. The important matters are:

1. Guidance as to semantic integrity.
2. Encoding integrity.
3. Supplementary information, providing context.

Database creation for spatial data generally will require some statements about how entities are represented, even if it is not possible to establish cardinalities. If nothing else is provided in the way of metadata, the **representation rules** should be stated, covering the geometry and composition of entities. Composition description includes language like, 'a river is composed of a sequence of connected directed line segments, some of which meet at nodes that have more than two lines (that is, a confluence)'. Or, for a reservoir, representation consists of an area object, with the attribute of water, and completely bounded by line elements and the shoreline, at least one of which has an attribute which is person-made and probably straight: the dam. Such matters of semantic integrity are akin to the enterprise rules discussed earlier. They treat phenomena at the level of the real world.

The quality of data encoding and storage can be addressed via the incorporation of material on the particulars of recording attributes, like the ranges of values possible, identification of different kinds of null value, what topology properties are explicitly encoded, methods for spatial indexing, resolution of the data, tolerances used when building the graphic database, colour encoding for image data, and map projections used. Spatial integrity constraints are an important part of this endeavour, establishing the parameters by which internal consistency and

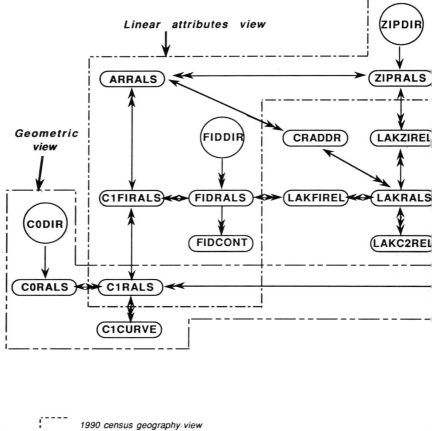

Linear attributes view

Geometric view

ZIPDIR

ARRALS ⟷ ZIPRALS

FIDDIR CRADDR LAKZIREL

CODIR C1FIRALS ⟷ FIDRALS ⟷ LAKFIREL ⟷ LAKRALS

FIDCONT LAKC2REL

CORALS ⟷ C1RALS

C1CURVE

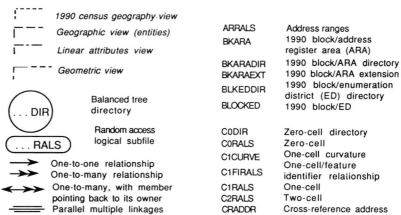

⌐ - - - ¦ 1990 census geography view	
⌐ - - - ¦ Geographic view (entities)	
⌐ - · - ¦ Linear attributes view	
⌐ - - - ¦ Geometric view	

(... DIR) Balanced tree directory

(... RALS) Random access logical subfile

→ One-to-one relationship
↠ One-to-many relationship
↞↠ One-to-many, with member pointing back to its owner
≡ Parallel multiple linkages

ARRALS	Address ranges
BKARA	1990 block/address register area (ARA)
BKARADIR	1990 block/ARA directory
BKARAEXT	1990 block/ARA extension
BLKEDDIR	1990 block/enumeration district (ED) directory
BLOCKED	1990 block/ED
CODIR	Zero-cell directory
CORALS	Zero-cell
C1CURVE	One-cell curvature
C1FIRALS	One-cell/feature identifier relationship
C1RALS	One-cell
C2RALS	Two-cell
CRADDR	Cross-reference address

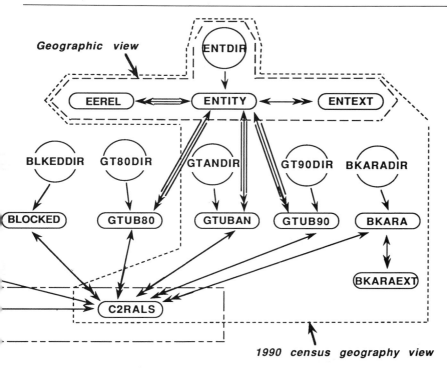

Geographic view

1990 census geography view

EEREL	Entity to entity relationship	GT80DIR	1980 geographic tabulation unit
ENTDIR	Geographic entities directory		(GTUB) base directory
ENTEXT	Geographic entities extensions	GT90DIR	1990 GTUB base directory
ENTITY	geographic entities	GTANDIR	1990 ancillary GTUB base directory
		GTUB80	1980 GTUB base
FIDCONT	Feature name continuation	GTUB90	1990 GTUB base
FIDDIR	Feature identifier directory	GTUBAN	1990 ancillary GTUB base
FIDRALS	Feature identifier	LAKC2REL	L:andmark, area, key geographic
			location two-cell relationship
ZIPDIR	Zip code directory	LAKRALS	Landmarks, areas, key geographic
ZIPRALS	Zip code		locations
LAKZIREL	Landmark, area, key geographic	LAKFIREL	Landmark, area, key geographic
	location/ zip code relationship		location feature identifier relationship

Figure 12.14 Different views for the TIGER database. (This diagram, slightly modified, is reproduced with the permission of the US Bureau of the Census, Washington, DC, USA.)

some external validation might be judged.

Otherwise, there could be statements about geometric data quality, principally the data for location. As treated extensively in Goodchild and Gopal (1989) data quality is not a simple matter, and may be overlooked in the process of building spatial databases. Indeed, while there are many delightful benefits to processing data via a spatial information system, the software may be frighteningly revealing of weaknesses in spatial data. A good understanding requires an appreciation for the accuracy and precision of source documents, the precision of the machines storing the data and the effects of processing steps on the data, topics not treated in this book.

Supplementary information includes material as to sources of data, and comments about the quality of original materials from which the database is developed. As a practical matter, virtually all the supplementary information, and details about representation rules and integrity constraints could themselves be stored in a database management system and spatial variations in data reliability could be represented as map data.

12.6.2 Database concepts and practical matters

Database management systems, particularly relational but to some extent also network, have been used by software vendors in their spatial information system products. We have already mentioned the use of the tabular data structure by ARC/INFO for attribute data; and various other systems like SYSTEM 9, GEO/SQL and TIGRIS also use relational database systems, embedded or external, for some or most of their data handling needs. Most vendors of spatial information systems software have chosen to use existing database systems rather than develop their own because of the tremendous development costs.

It is clear that relational databases offer some controls over data inconsistencies by normalization processes and cardinality constraints. Even if spatial data may be treated with some awkwardness it is possible to implement checks like the Euler number in the relational model, thus avoiding, or as an alternative to, the topological editing checks like node chaining that require algorithms and a data structure facilitating linking topological elements. Moreover, some conditions can be checked by using the attribute data for ascertaining **feature integrity**. For example, the characteristics of line elements forming the boundary of a body of water should show a water attribute on only one side.

Clearly, though, topological and geometric data representations do not mesh well with any of the classical database system types. These software systems do not recognize the concept of Euclidean distance, and themselves cannot handle computational geometry so necessary for so many purposes. They are also, generally, quite restrictive in the data types they offer to the database builder. Some systems now allow for user definitions via the concept of the **abstract data type**, providing the means to establish entity forms like point, line, box, region or image, to go alongside the conventional numeral, character string, symbol or date. There are still uncertainties, though, as to how many types of object can be provided for effectively in this manner.

We have already noted the awkwardness in handling some data characteristic of spatial information contexts. The sequence of objects can matter, and is, therefore, in conflict with the tabular model principle of row order not being important. Variable length fields, for example coordinate pairs for lines, produce embedded relations. The inclusion of complex spatial definitions and the multiplicity of relationships create complexity in implementations in the network and hierarchical models, and even with the relational concept. The complexity of spatial information system requirements results from geometric operations, the combinatorial explosion when associating multiple components in many-to-many linkages, and the embedding of modelled objects in Euclidean space.

As a whole, the relational model places a great deal of onus on the user to know how to use a database system effectively, possibly diverting attention from the knowledge of a problem. The user must know the database software well in order to do things effectively, for example how to use indexing. While the relational model provides a great deal of flexibility and broad applicability, it may not be the best for all purposes. Long sequences of commands may be necessary for undertaking particular operations, and performance may be degraded by creating the many tables that normalization usually entails.

In addition, the classical database management system is oriented more to high volumes of short transactions, for example, updating bank balances, not to the generally less frequent longer transactions associated with much spatial data processing. For instance, the update to a map feature in a cartographic database, such as a property subdivision or a new road alignment, consists of a series of operations involving topology recreation, inconsistency verification, etc. In this regard, it may be more prudent, for a variety of reasons, to take care of map updates separately from attribute data changes, suggesting merit in the georelational model for a spatial information system database.

Generally, the relational model and relational algebra are very interesting for structuring a database and querying against it. However, because the spatial data have an intensional component, and the relational model is valid only for extensional data, problems occur in structuring and querying. The main advantage of the relational approach might be simply in using it to produce a smart and clear model for representing phenomena.

We have also indicated the difficulties produced by the need to accommodate different views, which produce practical problems of multiple representations, groupings, or decompositions. Selecting a representation is rather difficult. For the moment, it seems that we can safely draw two conclusions. Firstly, that the Euclidean representations (vector, spaghetti, etc.) are very close to personal intellectual habits because we are used to thinking with Euclidean geometry. However, if geometric transformations are reputed to be simple, Boolean logic operations and spatial queries are thought to be more difficult to perform for this kind of data.

Secondly, even though the fractal representations (raster, pizza and so on) are not familiar to everybody, geometric transformations are reputed to be more difficult whereas Boolean operations and the majority of spatial queries are simpler to perform. In the next two chapters we clarify this dilemma by examining the manipulation and retrieval of spatial data.

Moreover, if for a moment we can be a champion for the database users, we emphasize that relation schemata for a polygon, consisting of a handful of relational statements for polygon, ring, chain, node and point, or other examples from previous chapters, require them to work at a low level. Implementations, oriented to higher semantic order, will be most welcome. Putting more meaning into spatial information systems might be undertaken in two contrasted ways. First of all, extra material could be included as metadata, that is descriptive information about definitions, grouping of entities, associations of phenomena and forms of representation. Or, in the spirit of encoding meaning directly, via implementations of newer semantic modelling forms.

12.7 BIBLIOGRAPHY

The papers by Hull and King, Kemper and Wallrath, and Peckham and Maryanski provide fairly recent reviews of semantic and geometric modelling methods and tools.

Armstrong, Marc and Paul J. Densham. 1990. Database organization strategies for spatial decision support systems. *International Journal of Geographical Information Systems* 4(1): 3–20.

Blaha, Michael R., Wilhan J. Premerlani and James E. Rumbaugh. 1988. Relational database design using an object-oriented methodology. *Communications of the Association for Computing Machinery* 31(4): 414–427.

Bouillé, François. 1978. Structuring cartographic data and spatial processes with the hypergraph-based data structure. In Geoffrey Dutton (ed.), *First International Symposium on Topological Data Structures for GIS.* Cambridge, Massachusetts: Laboratory for Computer Graphics and Spatial Analysis, Harvard University.

Bouillé, François. 1984. Architecture of a geographic structured expert system. *Proceedings of the International Symposium on Spatial Data Handling, Zurich,* Switzerland. International Geographical Union Commission on Spatial Data Handling, pp. 520–543.

Broome, Frederick R. and David B. Meixler. 1990. The TIGER database structure. *Cartography and Geographical Information Systems* 17(1): 39–47.

Calkins, Hugh W. and Duane F. Marble. 1987. The transition to automated production cartography: design of the master cartographic database. *The American Cartographer* 14(2): 105–119.

Domaratz, Michael and Harold Moellering. 1986. The encoding of cartographic objects using HBDS concepts. *Proceedings of Auto Carto London Conference.* London, UK: International Cartographic Association, pp. 96–105.

Egenhofer, Max J. 1987. Appropriate conceptual database schema designs for two-dimensional spatial structures. *Proceedings of the 1987 American Society for Photogrammetry and Remote Sensing/American Congress on Surveying and Mapping Convention, Baltimore, Maryland, USA,* vol. 5, pp. 167–179.

Environmental Systems Research Institute. 1988. *ARC/INFO Programmers Manual* (2 volumes). Redlands, California, USA: Environmental Systems Research Institute, Inc.

Frank, Andrew, F. 1989. Spatial database query language. Unpublished remarks, Symposium on Geographical Data Structures, Virginia.

Goodchild, Michael, F. and S. Gopal. 1989. Accuracy of Spatial Databases. London, UK: Taylor and Francis.

Guptill, Stephen C. (ed.). 1990. *An Enhanced Digital Line Graph Design.* US Geological Survey Circular 1048. Washington, DC, USA: US Printing Office.

Harel, David. 1988. On visual formalisms. *Communications of the ACM* 31(5): 514–530.

Herring, John R. 1990. TIGRIS: a data model for an object-oriented geographic information system. Paper presented at the GIS Design Models and Functionality Conference, Leicester University, UK.

Howe, David R. 1983. *Data Analysis for Data Base Design.* London, UK: Edward Arnold.

Hull, Richard and Roger King. 1987. Semantic database modeling: survey, applications, and research issues. *ACM Computing Surveys* 19(3): 201–260.

Kemper, Alfons and Mechtild Wallrath. 1987. An analysis of geometric modelling in database systems. *ACM Computing Surveys* 19(1): 47–91.

Morehouse, Scott. 1987. ARC/INFO: a georelational model for spatial information. *Proceedings of the Auto Carto 7 Conference, Washington, DC,* pp. 388–397.

Morehouse, Scott. 1989. The architecture of ARC/INFO. *Proceedings of the Auto Carto 9 Conference, Baltimore, Maryland, USA.* Falls Church, Virginia, USA: American Society for Photogrammetry and Remote Sensing/American Congress for Surveying and Mapping, pp. 266–277.

Nyerges, Timothy L. 1989. Schema integration analysis for the development of GIS databases. *International Journal of Geographical Information Systems* 3(2): 153–183.

Peckham, Joan and Fred Maryanski. 1988. Semantic data models. *ACM Computing Surveys* 20(3): 153–189.

Price, Stephen. 1989. Modelling the temporal element in land information systems. *International Journal of Geographical Information Systems* 3(3): 233–243.

Prime Computer. *SYSTEM 9 Geographic Information System Product Description.* Natick, Massachusetts: Prime Computer.

Salgé, François and Marie Noëlle Sclafer. 1989. A geographic data model based on HBDS concepts: the IGN Cartographic Data Base Model. *Proceedings of the Auto Carto 9 Conference, Baltimore, Maryland, USA.* Falls Church, Virginia, USA: American Society for Photogrammetry and Remote Sensing/ American Congress for Surveying and Mapping, pp. 110–147.

Smith, Neil. 1989. The Ordnance Survey pilot topographic database since the [SORSA] Durham Symposium. *Proceedings of the Spatially Oriented Referencing Systems Association Colloquium, University of Maryland*; March 29–31, 1989; College Park, Maryland, USA.

Stonebraker, Michael (ed.). 1986. *The INGRES Papers: Anatomy of a Relational Database System.* Reading, Massachusetts: Addison-Wesley.

Stonebraker, Michael and L. Rowe. 1986. The design of POSTGRES. *Proceedings of the ACM – Special Interest Group for Management of Data International Conference on Management of Data, Washington, DC.* New York: Association for Computing Machinery, pp. 340–355.

Teorey, Toby J. *et al.* 1989. ER model clustering as an aid for user communication and documentation in database design. *Communications of the ACM* 32(8): 975–987.

Tuori, M. and G.C. Moon. 1984. A topographic map conceptual data model. *Proceedings of the First International Symposium on Spatial Data Handling, Zurich, Switzerland, pp. 33–50.*

Van Roessel, Jan W. 1987. Design of a spatial data structure using relational normal forms. *International Journal of Geographical Information Systems* 1(1): 33–50.

Waugh, Thomas C. and Richard G. Healey. 1986. The GEOVIEW design: a relational database approach to geographical data handling. *Proceedings of the Second International Symposium on Spatial Data Handling, Seattle.* Williamsville, NY: International Geographical Union Commission on Geographical Data Sensing and Processing, pp. 193–212.

Waugh, Thomas C. and Richard G. Healey. 1987. The GEOVIEW design: a relational database approach to geographic data handling. *International Journal of Geographical Information Systems* 1(1): 101–112.

Part Four

Spatial Data Retrieval and Reasoning

13
Algebras

Relational and Peano tuple

Antonia González sees the world full of tables with colums and rows. Petrov Pavlov looks for keys, not for opening doors but for accessing the tables. Jacques Charpentier tries to find how to operate on tables and to normalize them. And Geo Matix is always looking for his neighbours in his toy cadastre. One day, Fran Miller and Bob McLaren discover that they can use their thin ribbon to fill a spatial object and to cut it accordingly. But when the object is moving, the ribbon moves totally differently and they have to cut it diversely in order to avoid getting nodes. This ribbon can be manipulated nicely to fill other objects.

In Part Four we treat first the topic of retrieval of spatial data held in spatial databases of the relational kind, and then, in Chapters 16 and 17, we present some principles about newer different approaches to spatial information: the hypermedia and object oriented.

Chapter 14 provides examples of several categories of queries, after the foundation principles are laid in Chapter 13. Here we present mainly the tools for manipulating relations. However, in order to do that, first some elaboration about the relational model must be made, essentially concerning the properties of relations. After that, the relational operators will be presented, then the normalization of databases, followed by some geomatic examples using the structured query language (SQL). The chapter then develops a framework for the use of Peano relations and algebra for manipulating spatial objects, considering three topics: the definition and usage of Peano relations, the presentation of the operators, and the methodology for solving some queries.

13.1 FEATURES OF RELATIONS

The definition of relations given in Chapter 9 revealed certain characteristics that make a relation different from a table. We now discuss some of these properties in detail.

13.1.1 Some properties of tuples

According to Codd (1970), the ordering of tuples is 'immaterial', meaning that the sequence of rows has no importance. Yet it often seems necessary to order the tuples by their identifiers. This may be simply for user convenience, for historical reasons, or as a device to speed up some algorithms, especially if they involve sorting. The ordering of values within tuples also is immaterial, according to Codd; that is, the attributes can be in any column sequence. However, when an order is initially given, indeed it must be followed afterwards. We have already given examples in section 9.7 of some situations in which row ordering arises if only a flat table concept is used rather than relations. And, for certain situations of spatial data, there is a real need to be able to put tuple elements in sequence; for example, a chain or arcs for a polygon boundary.

The values of the tuples are drawn from the specified domain. Indeed, often we need what is called a **null value**. In databases the null designation is used for several reasons for recording values for the instances recorded in the table rows:

1. When a value is unknown, even temporarily.
2. When a piece of data is not yet loaded (at the time of tuple creation).
3. When a value is not applicable.
4. In some other cases.

A null value is not the same as a zero or blank; for practical purposes it also seems that there should be two distinct types of null value, one kind representing missing data, the other referring to inappropriate data. Social scientists and statisticians generally recognize missing values in their data sets, in this way acknowledging the limitations of data collection. In spatial information systems some attributes may be inappropriate for different circumstances. For example, a polygon or triangle cell does not have a real entity on its outside if it lies on the perimeter of a study area, or a cul-de-sac street segment will not have two connecting line elements. In practice, it may be preferable to use descriptive metadata to distinguish the kinds of null value.

Often, database tuples can be used as **facts**. As an example, recall the PURCHASE relation (section 9.4) in which each tuple can be considered as a fact, the combined data for buyer of a lot and the purchase date. In another sense, tuples are acceptable only if they satisfy one or more conditions that relate to their semantic or geometric validity (see Chapter 15).

13.1.2 The Cartesian product for relations

Another important characteristic of relations is the derivation of a new relation called a Cartesian product. Consider two relations R(A, B) and S(C, D) where A, B, C, D are attribute domains. We can consider creating a new relation T(A, B, C, D) in which the attributes will be A, B, C, D, and each tuple will be the concatenation of every tuple of R paired with every tuple of S. Let the tuples of R(A, B) be $R(a_1, b_1)$, $R(a_2, b_2)$, $R(a_3, b_3)$. For S(C, D), we have $S(c_1, d_1)$, $S(c_2, d_2)$, $S(c_3, d_3)$ and $S(c_4, d_4)$. The new relation T(A, B, C, D) will have twelve tuples:

A	B	C	D
a_1	b_1	c_1	d_1
a_1	b_1	c_2	d_2
a_1	b_1	c_3	d_3
a_1	b_1	c_4	d_4
a_2	b_2	c_1	d_1
a_2	b_2	c_2	d_2
a_2	b_2	c_3	d_3
a_2	b_2	c_4	d_4
a_3	b_3	c_1	d_1
a_3	b_3	c_2	d_2
a_3	b_3	c_3	d_3
a_3	b_3	c_4	d_4

containing the values for the attributes A through D. This new relation *T* is called the Cartesian product of R and S. This definition can be extended to the case of several relations.

Another way of defining relations is to say that a relation $r(R)$ is a subset of the Cartesian product of the domains that define R. As an example, let us define a relation LAT_LONG, containing the latitude and longitude for cities:

LAT_LONG (City_name, Latitude, Longitude)

with tuples as shown in the table below (values are approximate).

CAR_PROD	City_name1	Latitude1	Longitude1	City_name2	Latitude2	Longitude2
	Paris	48° 50' N	2° 20' E	Paris	48° 50' N	2° 20' E
	London	51° 31' N	0° 06' W	Paris	48° 50' N	2° 20' E
	Rome	41° 53' N	12° 33' E	Paris	48° 50' N	2° 20' E
	Lisbon	38° 44' N	9° 05' W	Paris	48° 50' N	2° 20' E
	Paris	48° 50' N	2° 20' E	London	51° 31' N	0° 06' W
	London	51° 31' N	0° 06' W	London	51° 31' N	0° 06' W
	Rome	41° 53' N	12° 33' E	London	51° 31' N	0° 06' W
	Lisbon	38° 44' N	9° 05' W	London	51° 31' N	0° 06' W
	Paris	48° 50' N	2° 20' E	Rome	41° 53' N	12° 33' E
	London	51° 31' N	0° 06' W	Rome	41° 53' N	12° 33' E
	Rome	41° 53' N	12° 33' E	Rome	41° 53' N	12° 33' E
	Lisbon	38° 44' N	9° 05' W	Rome	41° 53' N	12° 33' E
	Paris	48° 50' N	2° 20' E	Lisbon	38° 44' N	9° 05' W
	London	51° 31' N	0° 06' W	Lisbon	38° 44' N	9° 05' W
	Rome	41° 53' N	12° 33' E	Lisbon	38° 44' N	9° 05' W
	Lisbon	38° 44' N	9° 05' W	Lisbon	38° 44' N	9° 05' W

LAT_LONG	City_name	Latitude	Longitude
	Paris	48°50′ North	2°20′ East
	London	51°31′ North	0°06′ West
	Rome	41°53′ North	12°33′ East
	Lisbon	38°44′ North	9°05′ West

Suppose we are interested in obtaining a table with the direct air flight distance between those cities. Firstly, we perform a Cartesian product with the same relation of itself giving:

CAR_PROD (City_name1, Latitude1, Longitude1, City_name2, Latitude2, Longitude2)

The tuples are shown opposite.

For this table we suppress the city paired with itself (the diagonal entries of a spatial interaction matrix), and the duplication of each dyad because the flight distance is symmetrical for the two directions between any pair of cities. So we reduce from the original sixteen dyads to just six, that is half of $N(N - 1)$ where N is the number of cities. We also add another attribute, Fly_distance to contain the new values based on the latitude and longitude coordinates. Thus, we have the relation:

FLY_DIST (City_name1, Latitude1, Longitude1, City_name2, Latitude2, Longitude2, Fly_dist)

Its instance is shown on p. 484.

13.2 RELATIONAL OPERATORS AND RELATIONAL ALGEBRA

To manipulate relations we need some operators organized in an algebra. Let us examine the operators intersection, union, difference, join, relational projection, restriction and division. We have already discussed the most basic operator, the Cartesian product.

13.2.1 Intersection

Intersection produces the subset of tuples present in both relations of a pair. As an example let us consider two relations, the first containing English speaking countries and their capitals, and the second European countries and their capitals:

FLY_DIST	City_name1	Latitude1	Longit1	City_name2	Latitude2	Longit2	Fly_dist
	London	51°31' N	0°06' W	Paris	48°50' N	2°20' E	325
	Rome	41°53' N	12°33' E	Paris	48°50' N	2°20' E	1020
	Lisbon	38°44' N	9°05' W	Paris	48°50' N	2°20' E	1350
	Rome	41°53' N	12°33' E	London	51°31' N	0°06' W	1325
	Lisbon	38°44' N	9°05' W	London	51°31' N	0°06' W	1220
	Lisbon	38°44' N	9°05' W	Rome	41°53' N	12°33' E	1420

ENGL_SPEAK_COUNTRY (Country_name, Capital)
EUROPE_COUNTRY (Country_name, Capital)

The tuples are:

ENGL_SPEAK_COUNTRY	Country_name	Capital
	USA	Washington
	Eire	Dublin
	Canada	Ottawa
	Australia	Canberra
	Nigeria	Lagos
	United Kingdom	London

EUROPE_COUNTRY	Country_name	Capital
	Eire	Dublin
	France	Paris
	Italy	Rome
	Spain	Madrid
	Portugal	Lisbon
	Norway	Oslo
	United Kingdom	London

Obtaining the subset of English-speaking countries that are European is accomplished by an intersection of the two relations, producing:

Eire	Dublin
United Kingdom	London

So, if we consider two relations R_1 and R_2 whose attributes are defined in the same domains, the tuples in the intersection are common to both.

13.2.2 Union

Next imagine we need only one list of countries. Such a table will be produced by the union of the relations ENGL_SPEAK_COUNTRY and EUROPE_COUNTRY by finding all countries mentioned in either table:

USA	Washington
United Kingdom	London
Eire	Dublin
Canada	Ottawa
Australia	Canberra
France	Paris
Italy	Rome
Spain	Madrid
Portugal	Lisbon
Norway	Oslo
Nigeria	Lagos

The operation also carries the other attributes in the tables: in this case just the capital city item.

The union, then, of relations R_1 and R_2 whose attributes are defined in the same domain is defined by all tuples which are present in both R_1 and R_2.

13.2.3 Difference

Finding the English-speaking countries which are not in Europe is accomplished by taking the list of English speaking countries and removing countries which appear in the other list. We get:

USA	Washington
Canada	Ottawa
Australia	Canberra
Nigeria	Lagos

This result is obtained through a difference between two relations. So, the difference of two relations R_1 and R_2 whose attributes are defined on the same domains corresponds to the list of tuples which are present in R_1 and not present in R_2.

13.2.4 Join

Suppose we have another relation giving country population, COUNTRY_POP (Country_name, Population), with the following tuples:

COUNTRY_POP	Country_name	Population
	USA	250
	Eire	4
	United Kingdom	57
	USSR	284
	Italy	57
	Nigeria	108
	Chile	10

We can build a new relation COUNTRY_POP_CAPITAL (Country_name, Population, Capital) by using Country_name as a sort of bridge between both relations, producing:

COUNTRY_POP_CAPITAL	Country_name	Population	Capital
	USA	250	Washington
	United Kingdom	57	London
	Eire	4	Dublin
	Nigeria	108	Lagos
	Italy	57	Rome

The USSR and Chile are not mentioned because we have no corresponding tuple in the other relation. This operation is known as a join: it takes all tuples in the first relation and looks in the second relation for tuples having the same element in common. Then the operation produces a new relation for which the tuples are formed with data from all the attributes of the specified relations, except that the bridge attribute is cited only once.

A join, then, between two relations R_1 and R_2 which have at least one attribute defined in the same domain is defined by a list of tuples for which the attributes belong to both the tuples of R_1 and of R_2. The join does not require union compatibility; it does need a domain match. It is important to note that null values of the linking attributes can imply some inconsistencies; we have to treat this case very cautiously.

13.2.5 Relational projection

Suppose we want only European capitals. This can be obtained by considering only one column of the European countries relation, giving:

| Dublin |
| Paris |
| Rome |
| Madrid |
| Lisbon |
| Oslo |
| London |

This operation of selecting one or more attributes is known as a relational projection. Thus, the projection of a relation onto a list of attributes is composed of tuples whose only attributes belong to this list. Simply put, relational projection allows selection of a vertical subset of a table.

13.2.6 Restriction

Suppose we want to retrieve only countries with more than 200 million inhabitants. This selection by a particular specification for one or more attributes can be obtained by keeping only the tuples in the COUNTRY_ POP relation for which the value for the attribute population is greater than 200, giving:

| USA | 250 |
| USSR | 284 |

This operation selects particular rows meeting some specified condition, otherwise called a **predicate**. So a restriction of a relation R_1 is a relation R_2 for which the predicate P is composed only of tuples satisfying this predicate. In other words, we take a horizontal subset. Sometimes also known as the **selection** operator, the restriction takes tuples according to stated conditions, expressed as a Boolean combination of terms, each invoking a simple comparison of two states (true or false) for each individual row in the table.

13.2.7 Division

Now we have additional data, a list of countries and their official languages stored in the relation OFF_LANG (Country_name, Language) with tuples like:

OFF_LANG	Country_name	Language
	Italy	Italian
	Chile	Spanish
	Canada	English
	Canada	French
	Nigeria	English
	Belgium	French
	Belgium	Flemish
	Switzerland	German
	Switzerland	French
	Switzerland	Italian
	Cameroon	French
	Cameroon	English
	Argentina	Spanish

Consider that now we are looking for countries for which English and French are the official languages. For that, we have to create a relation ENG_FRENCH (Language) with only two tuples:

ENG_FRENCH	Language
	French
	English

The desired table is given by the division of OFF_LANG by ENG_FRENCH on the item Language:

Country_name
Canada
Cameroon

In summary, the different operators may be classified as to whether they are oriented to attributes or tuples, and how many relations they involve at one time. Of those mentioned, all deal with the selection of rows, that is, they are **query tools**, except for projection which selects attributes. Conditional selection of rows, the restriction operation, and projection operate on single relations; joins can be accomplished for two

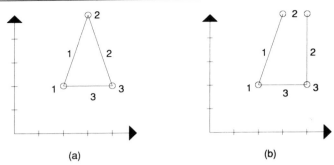

Figure 13.1 Consequence of absence of normalization on a polygon.
(a) Data for a triangle. (b) Repositioning of one point.

or more relations, while other operations are binary. Intersection and division are likely to be least used, although, as demonstrated in the material above, there are interesting needs in spatial data handling where even they are useful.

13.3 NORMALIZATION

In order to work with only well-formed relational databases, a process known as normalization can be implemented. The goal of normalization is to certify that the database holds no anomalies when certain changes (updating, deleting or insertion) are undertaken. Section 9.7 introduced this topic; now we treat it more formally.

13.3.1 Necessity for normalization

Taking the case of updating, suppose we have duplicate information in one table, or the same data present in more than one relation. It is possible to modify only one tuple forgetting that the same information is also present in other tuples, thereby creating an inconsistency in the database. In order to avoid this kind of problem, and also for deletion or insertion inconsistencies, relations must be split into very small tables so that we have only one modification to perform, and this single modification will imply always dealing with an internally consistent database.

Suppose we have data for polygons (Figure 13.1) with the following descriptions:

POLYGON (Polygon-ID, Segment-ID)
COORD (Segment-ID, X_1, Y_1, X_2, Y_2)

and, for the polygon of Figure 13.1a:

COORD	Segment-ID	X_1	Y_1	X_2	Y_2
	1	2	2	3	5
	2	3	5	4	2
	3	4	2	2	2

Alas, suppose we want to move point 2 from (3,5) to (4,5) and we change only the first line segment. In this case, the polygon is no longer closed; the correction requires that the coordinates in the second segment must also be changed. Consequently, via this small example, we can see that the solution to give an identifier to points and then use it for the coordinate description is worthwhile. So the previously inconsistent COORD relation must be split into two other consistent relations COORD1 and COORD2:

POLYGON (Polygon-ID, Segment-ID)
COORD1 (Segment-ID, Point1-ID, Point2-ID)
COORD2 (Point-ID, X, Y)

For this schema the positional data instance for Figure 13.1 is:

COORD1	Segment-ID	Point1-ID	Point2-ID
	1	1	2
	2	2	3
	3	1	3

COORD2	Point-ID	X	Y
	1	2	2
	2	4	5
	3	4	2

Normalization has numerous facets, of which the situation described above is just a part. Initially three normal forms were proposed by Codd, called first, second and third normal forms, denoted respectively as 1NF, 2NF and 3NF. Subsequently, other definitions and normalization rules were proposed. For instance, there is another definition of the 3NF known as the Boyce–Codd normal form. All these forms are based upon the functional dependencies among the attributes of a relation.

13.3.2 Functional dependencies

In order to avoid the kind of anomalies resulting from inconsistent updating, some constraints between attributes must be defined. This topic is approached by looking at some special linkages among attributes, and the associated normalization procedures.

A functional dependency is a constraint between two sets of attributes in the database. Suppose our relational database schema has N attributes $\{A_1, A_2, \ldots, A_N\}$. Instead of thinking about the set of tables which undoubtedly exist for the database, let us think for the moment that the whole database is described by simply a single universal relation schema $R = \{A_1, A_2, A_3\}$ encompassing all attributes concerned. This concept does not mean that we are attempting to store the database as a single relation; it is necessary only to explain the idea of functional dependencies.

A functional dependency, denoted $X \rightarrow Y$, between two sets of attributes X and Y that are subsets of R specifies a constraint on the possible tuples that can form a relation instance r of R. This constraint states that for any two tuples t_1 and t_2 in r, such that $t_1(X) = t_2(X)$, we must also have similar parts in Y, that is: $t_1(Y) = t_2(Y)$. This means that the values of the Y component of a tuple in r depend on the values of the X component, or, alternatively, the values of the X component of a tuple uniquely determine the values of the Y component. For instance, a Social Security Number (SSN) determines the name of a person, or a point longitude and latitude determines a city name:

SSN $\rightarrow$ Person_name
(LON, LAT) $\rightarrow$ City_name

Another way to define functional dependencies is to say that in a relation R, X functionally determines Y if and only if whenever two tuples of $r(R)$ agree on their X-value, they must necessarily agree on their Y-value. For example, if the supposed two persons with social security number 123 45 6789 each have the same name, then the condition of dependency of the name (Y) on the number (X) is met. However, if the names are different, the condition is not met and we might suspect we have some bad data because there is supposed to be a one-to-one relation between name and number. However, the situation of identical names does not presuppose the same person, so that the social security number could be different. Moreover, if R states that there cannot be more than one tuple with a given X-value in any instance $r(R)$, X is called a candidate key for R.

13.3.3 First normal form

The first normal form (1NF) was introduced to disallow multivalued attributes, composite attributes and their combinations, as discussed in section 9.7. While it is considered to be more or less a part of the formal description of relations, meeting the properties of first normal form ensures a solid basis to design the database. The domains of attributes must include only **atomic values** (simple, indivisible) and the value of any attribute in a tuple must be a single value taken from the domain of that attribute (described in Section 9.7.3). In other words, 1NF allows dealing only with atomic or indivisible values.

It is also desirable to avoid having nested relations, that is, relations within a relation as attributes. Suppose we have a relation giving the capital city population of a country:

R (Country_name, CAP_POP)

in which CAP_POP is another relation so that:

CAP_POP (Capital_name, Population)

Then, in order to a have a 1NF valid relation, we must integrate attributes of the second relation into the first one, producing relation R':

R' (Country_name, Capital_name, Population)

Formally, suppose we have a relation R, where S is another relation and A_1, A_2, A_3, A_4 are atomic attributes:

R (A_1, S, A_2)
S (A_3, A_4)

In this case, we must write:

R' (A_1, A_2, A_3, A_4)

and so we obtain a valid 1NF relation.

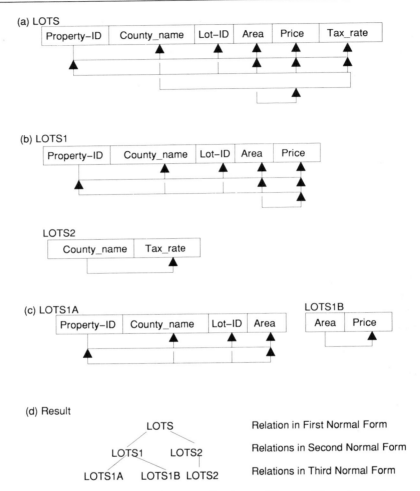

Figure 13.2 Example of normalization. (From Elmasri and Navathe 1989, with permission.)

A second example (Elmasri and Navathe, 1989), depicted in Figure 13.2, is:

LOTS (Property-ID, County_name, Lot-ID, Area, Price, Tax_rate)

This describes parcels for sale in various counties of a country. Suppose that we have two candidate keys: Property-ID and {County_name, Lot-ID}; that is, the Lot-ID code numbers are unique only within each county, but the Property-ID values are unique across counties for the entire country. The property-ID is the suitable key, and the combination of country-name and Lot-ID is not necessary as an ID, leaving each attribute as a separate item. In this second example, no composite attributes occur, so this relation is a valid 1NF relation.

The avoidance of nesting is very basic to the effective operation of routines applied to the data; that is, to guarantee that the database system software will work. We are not talking about external procedures regarding the data, but refer rather to the relational queries. Relational algebra cannot be applied to vectors, or matrices, only to individual tuples; the latter are not decomposable into other units.

13.3.4 Second normal form

The second normal form (2NF) is based on the concept of a **full functional dependency**: $X \rightarrow Y$ is said to be a full functional dependency if removal of any attribute A from set X means that the dependency does not hold any more. We must have only a partial dependency if there is some attribute which can be removed from X and the dependency will still hold for some A: $X \rightarrow \{A\} \rightarrow Y$. Suppose that, for the LOTS relation, we have functional dependencies as given in Figure 13.2a, we can see that we must split it into two relations (Figure 13.2b):

LOTS1 (Property-ID, County_name, Lot-ID, Area, Price)
LOTS2 (County_name, Tax_rate)

in order to meet the property of second normal form, because the tax rate is a constant for a particular county. That is, all properties in one county have an identical tax rate. So county A has $4.15 per assessed value unit, and county B has $3.85 per unit.

A relation schema R is in 2NF if every non-key attribute in R is not partially dependent on any key of R. This condition is met by the removal of Tax_rate to a separate table. For a polygon spatial data model:

POLYGONS (Chain-ID, Point-ID, X, Y)

there is a joint key of chain and point, but X and Y coordinates are themselves dependent on only the point element. Because a point may be shared by two or more chains, it is necessary to have two relations:

CHAIN (Chain-ID, Point-ID)
POINT (Point-ID, X, Y)

13.3.5 Third normal form

The third normal form (3NF) is based on the concept of a **transitive dependency**. A functional dependency $X \rightarrow Y$ in a relation R is a transitive dependency if there is a set of attributes Z that is not a subset of any key of R, and both $X \rightarrow Z$ and $Z \rightarrow Y$ hold. In this case we must split this relation into two new relations and a join between both will recover the original relation. In the case of our example, for LOTS2 there is no problem, but for LOTS1 there is a transitive dependency: Property-ID $\rightarrow$ Area, and Area $\rightarrow$ Price. So we need to change LOTS1 into (Figure 13.2c):

LOTS1A (Property-ID, County_name, Lot-ID, Area)
LOTS1B (Area, Price)

The resulting rearrangement to satisfy the three normal forms is a set of three relations as shown in Figure 13.2d.

13.3.6 Other normal forms and implications for spatial data

In some special cases, some other normal forms can be defined – the Boyce–Codd normal form which is similar to 3NF, and fourth and fifth normal forms that take multi-valued dependencies into account, but we do not discuss them here. Instead, we examine briefly some implications for spatial data, although the implications in the design of geomatic databases are more or less the same as for other databases.

We have already noted in Chapter 9 some circumstances in which non first normal form may arise for spatial data. Another aspect is the necessity of dealing with positional information, or **locators**. Even while identifiers are well treated in the relational model, locators are absent except if they can be written by alphanumerical strings, which is not always the case.

Normalization can be seen, then, to have several benefits even for spatial data. The process reduces unnecessary duplication, fosters integrity and encourages thinking about well-formed databases, if only to produce a healthy respect for the difficulties inherent in building large databases. At the same time, normalization has costs like the extra effort required to use or keep track of many tables, the cases in which non-normalization may be preferred for spatial data, and the difficulties in knowing enough about enterprise rules to establish functional dependencies or other kinds of association. The relative benefits and costs also depend on the ways in which data will be used. Database querying may place a premium on rapid access to valid and correct numbers that are spread across several tables, but visual display uses may be better served by not normalizing.

13.4 STRUCTURED QUERY LANGUAGE EXAMPLES IN GEOMATICS

The role of the Structured Query Language, SQL, first mentioned in section 9.5, is to allow the writing of query instructions for a relational database. It allows users to pose queries to data according to the model previously described. In this section we will give only a short formal treatment about the SQL, and then give some examples in geomatics.

Other than the general utility actions for table or database creation, deletion, and the like, the most important feature of the SQL is the **SELECT** command. The basic form of this operation consists of three clauses represented by the keywords **SELECT**, **FROM** and **WHERE**, having the following form:

SELECT ⟨ attribute list ⟩
FROM ⟨ table list ⟩
WHERE ⟨ condition ⟩

in which the attribute list is a list of attribute names for which values are to be retrieved by the query; table list is a list of the names of relations containing the data required to process the query, and condition is a Boolean expression that identifies the tuples to be selected by the query.

QUERY 1

As a first illustration, let us take the COUNTRY_POP relation with the schema:

COUNTRY_POP (Country_name, Population)

Looking for the names of the countries having more than 100 million inhabitants, we formulate this query as follows:

SELECT Country_name
FROM COUNTRY_POP
WHERE Population > 100

QUERY 2

Boolean expressions can be combined for any query by means of OR and AND **logical connectors**. For instance, to retrieve countries with population ranging from 10 to 50 the instructions are:

SELECT Country_name
FROM COUNTRY_POP
WHERE Population > 10
   AND Population < 50

QUERY 3

Suppose we have also a relation CITY_POP giving the population of some cities:

CITY_POP (City_name, City_population)

Suppose also that we are looking for any European country capital city with more than 5 million inhabitants. For such a query we need to use also the relation EC (Country_name, Capital), specifying that the Capital is also a City_name. Technically, we must join the relations. For that we need to have a combination of relations as follows:

SELECT Capital, City_population
FROM CITY_POP, EUROP_COUNT
WHERE City_name = Capital
   AND City_population > 5

QUERY 4

To retrieve Spanish speaking countries having more than 10 million inhabitants we also use the OFF_LANG (Country_name, Language) relation. The query will deal with ambiguous attributes names; indeed, Country_name is an attribute of both OFF-LANG and of COUNTRY_POP. In order to overcome this ambiguity, the relation names are included as prefixes to the attributes:

SELECT OFF_LANG.Country_name, Population, Language
FROM OFF_LANG, COUNTRY_POP
WHERE Language = "Spanish"
   AND Population > 10

AND OFF_LANG.Country_name = COUNTRY_POP.Country_ name

QUERY 5

Sometimes we need to work with several instances of the same relation. In this case, the prefix is also ambiguous and thus, aliases, declared in the **FROM** clause, are used as prefixes. Suppose we have a set of lines and we are looking for segments in which only one end-point must lie in a specific spatial range, say $x > 5$ (Figure 13.3a). The tables to be utilized are:

SEGMENT (Segment-ID, Point1-ID, Point2-ID)
POINT (Point-ID, X, Y)

which are fully illustrated in Figure 13.3a.

Letting A and B stand for two tuples of the POINT relation, we have

SELECT Segment-ID
FROM SEGMENT, POINT A, POINT B
WHERE (A.Point-ID = Point1-ID **AND** B.Point-ID = Point2-ID)
    AND ((A.X > 5 AND B.X < 5) **OR** (B.X > 5 AND A.X < 5))

In a sense, double access to the tables is required to retrieve beginning and ending points, as shown in Figure 13.3a. For this example, only one line segment out of the three matches the double condition.

QUERY 6

Turning back to the Spanish speaking countries, this type of query can be solved also by nesting the **SELECT** statements. In other words, in the criterion we can compare something with the result of another **SELECT** operation:

SELECT Country_name, Population
FROM COUNTRY_POP
WHERE Country_name = **SELECT** Country_name
                FROM OFF_LANG
                WHERE Language = "Spanish"

By nesting **SELECT** instructions, we can often avoid using prefixes.

QUERY 7

Now let us again utilize the toy cadastre example given in Chapter 9, including the following relations:

STREET	(Street-ID, Street_name)
OWNER	(Owner-ID, Owner_name, Address)
PARCEL	(Parcel-ID, Address)
BLOCK-PARCEL	(Block-ID, Parcel-ID)
STREET-SEGMENT	(Street-ID, Segment-ID)
PARCEL-SEGMENT	(Parcel-ID, Segment-ID)
SEGMENT-ENDPOINTS	(Segment-ID, Point1-ID, Point2-ID)
COORDINATES	(Point-ID, X, Y)
OWNER-PARCEL	(Owner-ID, Parcel-ID)

Against these relations, we will pose some more complex queries. As a first example (Figure 13.3f), let us retrieve the addresses of the neighbours of the landowner, Ann. This is a topological query taking account of the adjacency property. In this case, we must:

1. First retrieve tuples with the name Ann: A.Owner_name = 'Ann', so: A.Parcel-ID; via OWNER-PARCEL to get the Owner-ID before identifying and retrieving the parcels, for which in this example there are two, as shown, numbers 123 and 487 (Figure 13.3b).
2. Secondly, join to PARCEL-SEGMENT: A.Parcel-ID = C.Parcel-ID.
3. Then find the corresponding segments: C.Segment-ID. Here there are four for each of the two parcels (Figure 13.3c).
4. Look for the parcels B sharing this segment: B.Segment-ID = C.Segment-ID, but we must eliminate the case B.Parcel-ID = C.Parcel-ID, the situation of adjacent parcels also owned by Ann (if there are any) (Figure 13.3d).

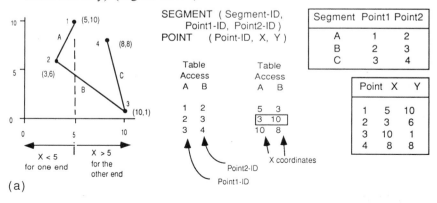

(a)

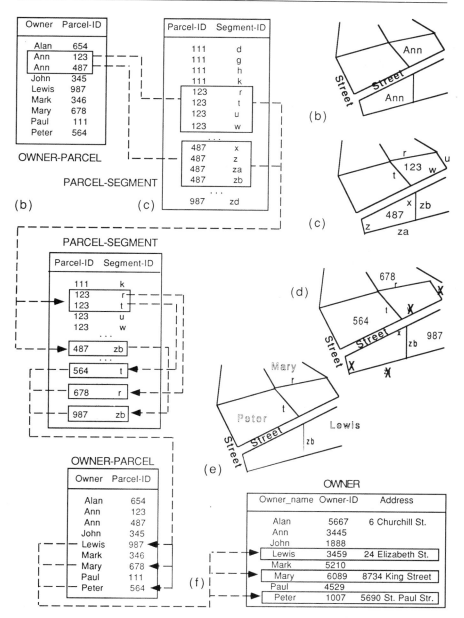

Figure 13.3 Two queries using spatial data. (a) Coordinate value for one end of a line segment. (b) Parcels owned by specific owner. (c) Boundary segments. (d) Adjacent parcels. (e) Neighbouring landowner. (f) Address of landowner.

5. Followed by finding the corresponding D owner with a join so that C.Parcel-ID = D.Parcel-ID, as shown in Figure 13.3e.
6. Finally, by performing a join to OWNER, the address of Ann's neighbours can be found (Figure 13.3f).

This query can be schematized as follows:

Ann → A: (Owner-ID, Parcel-ID) → B: (Parcel-ID, Segment-ID) → C: (Parcel-ID, Segment-ID) → D: (Owner-name, Parcel-ID) → E.owner-name → answer E. Address

It requires the following instructions:

SELECT E.Address
FROM OWNER-PARCEL A, PARCEL-SEGMENT B, PARCEL-SEGMENT C, OWNER-PARCEL D, OWNER E
WHERE A.Owner_name = 'Ann'
  AND A.Parcel-ID = C.Parcel-ID
  AND B.Segment-ID = C.Segment-ID
  AND B.Parcel-ID ≠ C.Parcel-ID
  AND C.Parcel-ID = D.Parcel-ID

QUERY 8

Now we need to know the perimeter of Ann's parcel. In this case, we have to compute the length of each segment by means of Pythagoras' theorem, and for that it is necessary to employ a square root function which can be applied to the coordinates of the two end-points D and E:

SQR((D.X − E.X) * (D.X − E.X) + (D.Y − E.Y) * (D.Y − E.Y))

This value must be summed, using a **SUM** function, in order to get the result. For our specific case, in order to compute the perimeter of a polygon, this **SUM** function must be applied to the length of all parcel segments, giving:

SELECT A.Parcel-ID,
  SUM (**SQR** ((D.X − E.X) * (D.X − E.X) + (D.Y − E.Y) * (D.Y − E.Y)))
FROM OWNER-PARCEL A, PARCEL-SEGMENT B, SEGMENT ENDPOINTS C, COORDINATES D, COORDINATES E
WHERE A.Owner_name = 'Ann'
  AND A.Parcel-ID = B.Parcel-ID
  AND B.Segment-ID = C.Segment-ID

    AND C.Point1-ID = D.Point-ID
    AND C.Point2-ID = E.Point-ID

QUERY 9

As another example, let us retrieve the address of the parcel 345. This can be considered as a topological query using a segment as a boundary between a parcel and a street.

    SELECT Street-ID
    FROM PARCEL-SEGMENT, STREET-SEGMENT
    WHERE Parcel-ID = 345
        AND PARCEL-SEGMENT.Segment-ID = STREET-SEGMENT.
        Segment-ID

TRAFFIC FLOW EXAMPLE

Other geomatic queries can be posed against a spatial database. Let us mention the determination of areas of zones based on the cross-products of coordinate segments, using the formula given in section 7.3.3. However, this computation of area is quite difficult to do via the SQL. One other example is given here for a more complex situation in more complete form, covering data, relations, and queries. A toy traffic town is depicted in Figure 13.4 as an expression of the common occurrence of one-way streets and turning limitations. In order to study or simulate traffic conditions, especially the traffic regulations, it is necessary to distinguish several entities: roads, road sections, roadways and crossroads, with the following particularities:

1. A road must be constituted of several road sections, each of them limited by two crossroads.
2. Roads cross at some crossroads.
3. Road sections can be one- or two-way roads.
4. Traffic vehicles in a roadway can go to other roadways (to-roadways) and can come through different other roadways (from-roadways).

In some applications, it is also necessary to split roadways into several lanes for some needs, and in this case, vehicles come from and go to different lanes.

The conceptual model of Figure 13.4 can be mapped into the following relations:

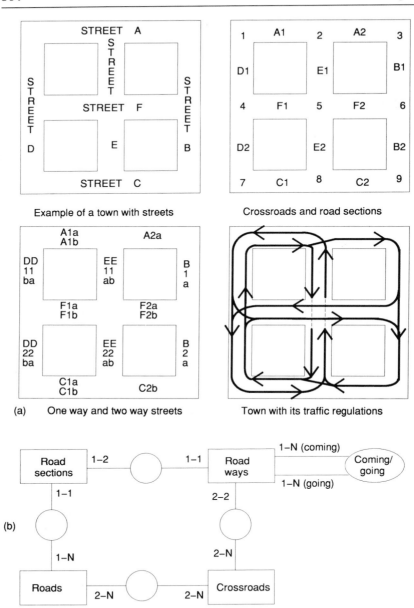

(a)   One way and two way streets   Town with its traffic regulations

(b)

Figure 13.4 A toy town and its traffic regulations. (a) Streets for the toy town. (b) The conceptual model corresponding to the toy town.

ROADS (Road-ID, Crossroad-ID)
SECTIONS (Road-ID, Road_section-ID)
ROADWAYS (Road_section-ID, Roadway1-ID, Roadway2-ID)
CROSSROADS (Roadway-ID, From_crossroad-ID, To_crossroad-ID)
PRIOR_ROAD (Roadway-ID, From_roadway-ID)
NEXT_ROAD (Roadway-ID, To_roadway-ID)

So we have the following tables:

ROADS	Road-ID	Crossroad-ID
	A	1
	A	2
	A	3
	B	3
	. . .	. . .

SECTIONS	Road-ID	Road_section-ID
	A	A1
	A	A2
	A	B1
	B	B2
	. . .	. . .

ROADWAYS	Road_section-ID	Roadway1-ID	Roadway2-ID
	A1	A1a	A1b
	A2	A2a	Null
	B1	B1a	Null
	B2	B2a	Null
	C1	C1a	C1b
	. . .	. . .	. . .

CROSS ROADS	Roadway-ID	From_crossroad-ID	To_crossroad-ID
	A1a	2	1
	A2a	2	3
	B1a	3	6
	B1a	3	6
	B2a	6	9
	C1a	8	7
	. . .	. . .	. . .

PRIOR_ROAD	Roadway-ID	From_roadway-ID
	A1a	E1b
	A2a	A1a
	A2a	E1b
	B1a	A2a
	B2a	B1a
	B2a	F2b
	C1a	C2b
	C1a	E2a

NEXT_ROAD	Roadway-ID	To_roadway-ID
	A1a	D1b
	A1b	A2a
	A2b	E1a
	B1a	F2a
	B1a	B2a
	. . .	. . .

As an example, in order to retrieve the roadways in a crossroads into which it is forbidden to enter, the solution is to make two **SELECT**s in the SQL and to perform a difference (**MINUS**) between them:

1. The first **SELECT** in order to retrieve all existing arriving roadways in a crossroad.
2. The second **SELECT** in order to retrieve To_roadways:

> **SELECT** A.Roadway-ID, B.Roadway-ID
> **FROM** CROSSROADS A B

WHERE A.From_crossroad-ID = B.To_crossroad-ID
MINUS
SELECT Roadway-ID, To_roadway-ID
FROM NEXT_ROAD

So, at this point, we have really reached the limit of exploitation of the SQL. Recalling that we cannot solve problems requiring computational geometry, other approaches are needed.

13.5 PEANO RELATIONS

Because the relational model is valid only for extensional data, and in essence spatial data are intensional as well as extensional, sometimes problems occur not only in structuring but also in querying. One aspect is the difficulty in solving spatial queries that involve computational geometry such as the point-in-polygon query. However, thanks to Peano relations and Peano tuple algebra, some of these difficulties can be overcome. The main interest of the process is to allow the modelling of geometric objects by an intensive–extensive method, allowing the solution of spatial queries by relational algebra. In a sense, then, the attribute and some spatial queries can be undertaken without using any computational geometry algorithms.

13.5.1 Peano relations concept

In Chapter 11 we presented a solution to storing quadtrees with a linear model based on Peano space-filling curves and Peano keys. Using the relational notation, a quadtree can be represented by the relation schema:

QUAD (Object-ID, Peano_key, Side_length)

in which a tuple represents a quadrant, and Object-ID represents the object number. Peano_key (sometimes denoted by P) is the Peano key of the quadrant foot, and Side_length (sometimes represented by S or Size) is the length of the side of the quadrant square. Figures 13.5a and 13.5b show, respectively, the data encoded for an (area) object and the quadtree space itself, where the presence or absence of an attribute is denoted by the colour.

In geomatic applications, a tuple representing a quadrant can model different situations. First of all, for an image the quadrant can be a pixel

(a)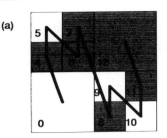

Object	Peano key	Side length
X	4	1
X	6	1
X	7	1
X	8	1
X	11	1
X	12	2

(b)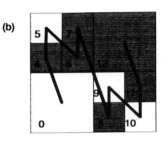

Peano key	Side length	Colour
0	2	White
4	1	Black
5	1	White
6	1	Black
7	1	Black
8	1	Black
9	1	White
10	1	White
11	1	Black
12	2	Black

Figure 13.5 Linear quadtrees. (a) Example of a quadtree object ordered by Peano keys. (b) Quadtree space ordered by Peano keys.

or a group of pixels. In this case, the minimum value of the size is 1, that is the resolution; and we are dealing with **pictorial objects**. Secondly, for example in many uses in geography, we may employ a squared zone whose minimum size depends on the requirements of the task; this minimum resolution can be noted ε (epsilon). Here we deal with **geographical objects**. Even though the minimum square is not an image pixel, in what follows we will often refer to it as a pixel.

Partly as review, and partly as a visual guide for understanding Peano keys, we present Figure 13.6 at this point. The diagram shows the minimum square block denoted epsilon, with a resolution of side length, S, of 1. For quadrant decomposition, where the integer n takes on values of 1, 2, 3 . . . , the side length is 2^n, a geometric series 1, 2, 4, 8 . . . . The range of coordinates in X and Y are nicely given by the expression $2^r - 1$, where r is an integer in the arithmetic series: 1, 2, 3, ... The (-1) is necessary because we begin the coordinates with the value 0. The maximum value for the Peano key is obtained from $4^r - 1$, as shown (Figure 13.6a) for the block of sixteen squares. So, when $\varepsilon = 1$, and the coordinates X and Y vary from 0 to 2^r, then $0 \leq P \leq 4 - 1$. For instance, in a 1024×1024 image matrix, $r = 10$.

Because each tuple represents a quadrant and since, according to the

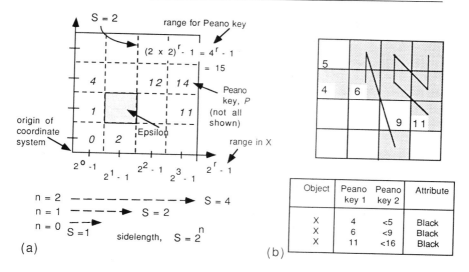

Figure 13.6 Some components of linear quadtree space. (a) Basic components. (b) Run-length encoding.

definition of quadtree, a quadrant can be split into four other quadrants except at the limit set by ε in practical terms, a tuple is also equivalent to four other tuples. As an example, the tuple R(X, 12, 2) from Figure 13.5a can be split into R(X, 12, 1), R(X, 13, 1), R(X, 14, 1) and R(X, 15, 1). Conversely, if we have four tuples with some good characteristics, they can be aggregated to form only one tuple. However, in the conventional relational model, a tuple cannot be equivalent to several other tuples; in other words, the Codd relations are extensional. This fact is a characteristic of the relations as previously given. It means that a tuple as defined here is both extensional and intensional. That is the reason why we must define a new kind of relation, called a **Peano relation**.

13.5.2 Definition of a Peano relation

Let us define a Peano relation schema as a relation schema in which each tuple describes a quadrant in a linear quadtree, ordered by Peano keys. Let us work with:

PR (Object-ID, P, S, Attributes)

in which:

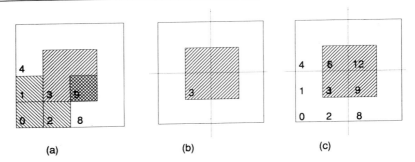

Figure 13.7 Example of a first conformance level. (a) Set of tuples. (b) Tuple with key 3 is not conformant. (c) Split into four tuples.

PR denotes a Peano relation
Object-ID represents the object number
P is the Peano key of the quadrant foot
S is the side length of the quadrant square
Attributes represent a set of non-spatial attributes

The non-spatial attributes can be removed; in this case, we have only a spatial description of an object. Another possibility is to mention only interesting Peano keys. For instance, to encode a black/white quadtree, we can either mention all quadrants, and a colour for each of them, or mention only black quadrants, or conversely white quadrants. Another notation can be in non-1NF:

PR1 (Object, (P, S, Attributes)*)

where, as before, * indicates the nesting. A third possibility is to regroup, according to the runlength encoding scheme, all successive tuples with the same attribute value. The relational schema will be:

PR2 (Object-ID, P_1, P_2, Attributes)

or

PR3 (Object-ID, (P_1, P_2, Attributes)*)

where P_1 and P_2 are the beginning and ending Peano keys. In such a manner, we obtain another intensional aspect because a PR2 or PR3 tuple represents also a PR tuple whose Peano keys range from P_1 up to, but excluding P_2 (Figure 13.6b). So, PR3 (Object-ID, (P_1, P_2)*) is equivalent to:

PR (Object-ID, P_1, ε)
PR (Object-ID, $P_1 + \varepsilon^2$, ε)
PR (Object-ID, $P_1 + 2\varepsilon^2$, ε)
.
.
.
PR (Object-ID, $P_2 - \varepsilon^2$, ε)

In this notation, ε, as before, means the minimum size. A run-length encoding clearly facilitates identification of consecutive tuples. With this notation, the quadtree given in Figure 13.5b will be represented by the following tuples (P_1 standing for beginning keys and P_2 for ending keys).

Object-ID	P_1	P_2
X	4	5
	6	9
	11	16

Instead of ordering by the Object, we can organize by an increasing order:

PR4 (Peano_key, (Side_length, Object-ID)*)

With this form, we can allow several objects to lie at the same place.

Notwithstanding our continued use of black–white colour coding for quadtree blocks, we assure the reader that any kind of attribute may be modelled. For example,

PR (Peano_key, Side_length, Land_use_category)

and, in demography,

PR (Peano_key, Side_length, Number_of_people)

In case of overlapping zones, the object number must be mentioned. When the Peano_key is the same, several quadrants must receive different attributes. For instance, taking the example of newspaper circulation we can write:

PR (Newspaper-ID, Peano_key, Side_length, Circulation)

It is necessary to give a reminder that a relation with a Peano key is not necessarily a Peano relation. For instance, suppose a relation named SEGMENT, in which all end point numbers are identified by their Peano keys, is a good standing Codd relation:

SEGMENT (Segment-ID, Endpoint_Peano_Key_1, Endpoint_Peano_Key_2)

in which Endpoint_Peano_Key_1 and Endpoint_Peano_Key_2 are the Peano keys of the segment end points.

13.6 CONFORMANCE LEVELS AND EXTENSIONS

Suppose we have a set of tuples as candidates for describing an object. The problem is to know whether this object is correctly defined in order to deal with consistent Peano relations. As an example, let us examine the following instance:

Object-ID	Peano_key	Side_length
A	0	1
A	1	1
A	2	1
A	3	2
A	9	1

In order to obtain a consistent description, three steps are necessary, each of them consisting of checking what are known as **consistent conformance levels** (CL):

1. Good positioning of quadrants: first conformance level (1CL).
2. Absence of overlapping quadrants: second conformance level (2CL).
3. Maximal quadrant compaction: third conformance level (3CL).

With regard to the example (Figure 13.7a), we find that there is an overlap of cell 9 recorded as a single tuple (A, 9, 1) and as part of tuple 3 (A, 3, 2). In addition, tuple (A, 3, 2) is misaligned (Figure 13.7b); that is, it crosses over the axes used for quadrant decomposition, assuming this begins at the middle. The decomposition of tuple (A, 3, 2), as shown in Figure 13.7c, will be necessary, and, of course, now clearly shows duplicaton of the cell with Peano key = 9.

In general, a conformance level is a consistent state in a database, in the same sense as normal forms and integrity constraints.

13.6.1 First conformance level: well-positioned object

The first thing to check is whether each tuple corresponds to a real quadtree quadrant. According to the quadtree definition, not any quadrant is possible; these must be arranged in a special hierarchical structure even though they are linearly encoded. It can be established that a set of tuples denoted R(P, A) correspond to a **well-positioned object** if and only if:

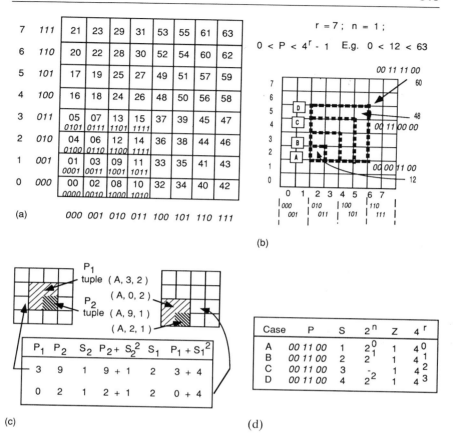

(a)

(b)

(c)

(d)

Figure 13.8 Well-formed Peano quadtrees. (a) Examples of possible well-positioned quadrants. (b) Binary encoding for Peano keys for a block of sixty-four squares. (c) An example of the square located at position 12. (d) An illustration of the alternative test.

1. The side length attribute is a power of 2, that is $S = 2^n$ up to the size of the study area set by the range r.
2. Quadrants are well-aligned; in other words, there exists an integrity constraint between the key and the side length.

In order to verify this constraint, it is necessary to compare the number of successive zeroes at the rightmost side of the Peano key represented in binary form (least significant bits) and the square side. Due to the bit interleaving of coordinates to obtain Peano keys, in order to deal with a well positioned object, the number of zeroes of the side length value must be less than half the number of rightmost zeroes of the Peano keys.

With reference to the example of Figure 13.8 for Peano key 8 (1000 in binary numbers, Figure 13.8a), the only possible square size-values are 2 (10 in binary) or 1 (1 in binary). It is the same for 12 (1100 in binary). For 16 (10000 in binary) we have four zeroes at the rightmost side, so the possible square side values can be 1, 2 and 4, corresponding respectively to the power 0, 1 and 2 of the number 4.

Taking into account this aspect, we can see that the tuple R(A, 3, 2) is not consistent (Figure 13.7b) so the offending tuple must be changed into four other tuples.

Object-ID	Peano_key	Side_length
A	3	1
A	6	1
A	9	1
A	12	1

The side length of 2 with a binary value of 10, is not less than half the number of zeroes at the right of the Peano key for which the binary equivalent of key 3 is 0011. Building upon this, we obtain the following relation:

Object-ID	Peano_key	Side_length	
A	0	1	
A	1	1	
A	2	1	
A	3	1	Set of new
A	6	1	tuples
A	9	1	replacing
A	12	1	(A, 3, 2)
A	9	1	

Knowing that Z(P) is the number of rightmost zero pairs of a given Peano key, to have well-positioned objects (1CL), we can establish that the following integrity constraints must hold for each tuple:

$$S = 2^n$$

$$0 < P < 4^r - 1$$

$$0 \leqslant n \leqslant Z(P) \leqslant r$$

These properties, illustrated in Figure 13.8b for four sizes of a block with P = 12 (00 11 00), clearly indicate, first, if the size of a square is a power of 2 (which a set of nine squares cannot be); secondly, if the Peano key lies in the total space; and thirdly, and most especially, if the block of squares is correctly aligned.

All four cases, A, B, C, D lie within the total space, that is, $0 \leqslant 12 \leqslant 63$; but case C does not meet the first test, for 2 to the power n cannot produce a set of nine pixels. For the third test, only blocks of size 1 and 2 based on square 12 can be correctly aligned. This is revealed for case D by the $0 \leqslant 2 \leqslant 1 \leqslant 3$ result as shown in the small table.

A single quadrant covering the entire area has a Peano key of 00 00 00, and the inequality $0 \leqslant n \leqslant Z (P) \leqslant r$ holds, that is $0 \leqslant 3 \leqslant 3 \leqslant 7$. For Figure 13.7b the square with P = 3 has a correct side length but is not correctly aligned with the two axes used for decomposition and, therefore, cannot have the correct number of rightmost zero pairs: P = 00 11, and Z (P) = 0.

If these constraints are not met the tuple(s) must be disaggregated into four other tuples, possibly recursively if the conditions continue to be violated.

13.6.2 Second conformance level: removal of overlaps

After having replaced inconsistent tuples by good ones, a reorganization by sorting tuples according to their Peano keys is needed. In this process, it can happen that some tuples can be mentioned several times. In the previous example, the tuple R(A, 9, 1) is mentioned twice; one of the two occurrences should be removed.

In other cases, suppose we have R(B, 16, 2) and R(B, 18, 1). In this case, since the first one is equivalent to four other tuples, for example, R(B, 16, 1), R(B, 17, 1), R(B, 18, 1) and R(B, 19, 1), we can see that the quadrant represented by the tuple R(B, 18, 1) is overlapped by the quadrant represented by R(B, 16, 2), which should be removed.

Continuing with the previous examples, that is, selecting the identical tuples (A, 9, 1) and (A, 9, 1), we produce:

Object-ID	Peano_key	Side_length
A	0	1
A	1	1
A	2	1
A	3	1
A	6	1
A	9	1
A	12	1

In other words, a set of tuples already in 1CL must be in 2CL when, if considering two arbitrary tuples $R(P_1, S_1)$ and $R(P_2, S_2)$ in any object descriptions so that P_2 is only after P_1 (not necessarily an ending key), we never have the following inequality:

$$P_1 \leq P_2 < P_2 + S_2^2 \leq P_1 + S_1^2$$

With reference to Figure 13.8c, we have, at the left, two tuples $(A, 3, 2)$ and $(A, 9, 1)$. For these the computed values in the first line of the small table show that the inequality does not hold. So the second tuple $R(P_2, S_2)$ must be eliminated because it corresponds to a quadrant overlapped by $R(P_1, S_1)$.

Similarly, if the overlaps occur in the bottom-left block of four squares (second row of the table in Figure 13.8c), that is, the tuples now are $(A, 0, 2)$ and $(A, 2, 1)$ perhaps resulting from a spatial translation, the misalignment no longer exists, but the overlap is detected by the inequality.

13.6.3 Third conformance level: compact objects

By considering Figure 13.7 again, we can see that the first four tuples can be regrouped to form a new tuple $R(A, 0, 2)$, giving:

Object-ID	Peano_key	Side_length
A	0	2
A	6	1
A	9	1
A	12	1

With this form, we have a compact object, that is, an object with its **minimal valid description**. Generally speaking, suppose we have four tuples such as:

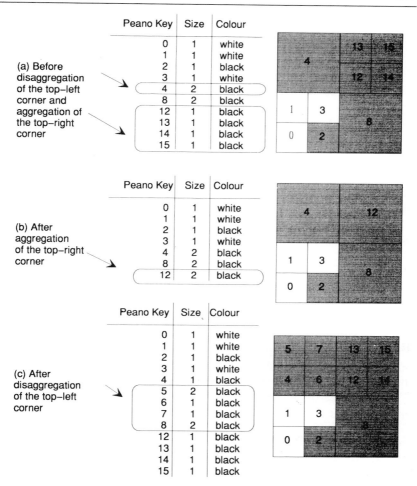

Figure 13.9 An example of aggregation and disaggregation of quadrants. (a) Before aggregation. (b) After aggregation. (c) After disaggregation.

$R(P + 0 * 2^i, 2^i)$
$R(P + 1 * 2^i, 2^i)$
$R(P + 2 * 2^i, 2^i)$
$R(P + 3 * 2^i, 2^i)$

They can be aggregated to get: $R(P, 2^{i+1})$, where P is the Peano key, and i is any integer.

After each operation, as presented subsequently in this discussion, a conformance procedure will be run in order to always get a consistent and compact object. In summary, this entails the processes of:

1. Disaggregation, the fact that one tuple (quadrant) can be disaggregated into its four equivalent tuples (subquadrants), except when the minimum resolution is reached.
2. Aggregation, the fact that four consecutive tuples (quadrants) can be aggregated to a single one provided that some equalities hold.

The performance procedure is like a spatial integrity constraint checking process, appropriate for linear quadtrees ordered by Peano keys.

The processes of aggregation and disaggregation are illustrated for a simple quadtree in Figure 13.9. The homogeneous block of four tuples at the top-right fits into the regular decomposition and can be amalgamated to produce a higher level quadrant of Peano key 12, and side length of 2 Figure 13.9b). The block at the top-left similarly occupies a space fitting into the Peano key multiples of 1 or 2 and can be split as shown (Figure 13.9c).

13.6.4 Extension beyond two dimensions

It is possible to extend the previous definitions in several ways. On the one hand, they can be used for higher dimensions; on the other, space-filling curves and keys in addition to the Peano type can be employed.

Now, a three-dimensional Peano relation tuple will represent an octant and the recursive splitting will imply eight suboctants or eight tuples:

PR3D (Object-ID, 3DPeano_key, Side_length, Attributes)

For instance, PR3D (A, 16, 2) is equivalent to the eight tuples by disaggregation:

PR3D (A, 16, 1) PR3D (A, 17, 1)
PR3D (A, 18, 1) PR3D (A, 19, 1)
PR3D (A, 20, 1) PR3D (A, 21, 1)
PR3D (A, 22, 1) PR3D (A, 23, 1)

Similarly, other forms of three-dimensional Peano relations can be defined:

PR3D1 (Object-ID, (3D_Peano_key, Side_length, Attributes)*)

or by using the runlength encoding scheme:

PR3D2 (Object-ID, $3DP_1$, $3DP_2$, Side_length, Attributes) or

PR3D3 (Object-ID, (3DP$_1$, 3DP$_2$, Side_length, Attributes)*)

As an example, for encoding a solid made of several materials, we can describe it by:

SOLID (Object-ID, (3D_Peano_key, Side_length, Material)*)

In geology, strata can be modelled by octtrees, for which the corresponding Peano relation is:

STRATA (Layer-ID, (3D_Peano_Key, Side_length, Layer_type)*)

Similarly, the three conformance levels must hold:

1. Badly positioned octants must be split into eight other octants, possibly recursively.
2. Overlap must be removed: the integrity constraints must take three-dimensional aspects into account.
3. Maximal compaction must be examined, with possibly eight octants being regrouped to be replaced by a new octant.

All these definitions and procedures can be extended to higher dimensions. At four dimensions the Peano keys can be generated by interleaving x, y, z and t (time) – in this way we can encode moving objects.

13.6.5 Hilbert keys

Hilbert or other keys may be used instead of Peano keys. As noted earlier in Chapter 4, the Hilbert key is also a good candidate. So we can have:

PRHK (Object-ID, Hilbert_key, S, Attributes)

or

PRHK1 (Object-ID, (Hilbert_key, S, Attributes)*)

or, with H$_1$ and H$_2$ standing for two Hilbert keys such that H$_1$ < H$_2$:

PRHK2 (Object-ID, H$_1$, H$_2$, S, Attributes)

or

PRHK3 (Object-ID, (H$_1$, H$_2$, Attributes)*)

The same properties hold: a tuple is equivalent to four tuples, conformance levels can be defined similarly, and three-dimensional Hilbert keys can be used to define three-dimensional objects.

For this context Peano and Hilbert keys have the same characteristics. However, recall from section 4.8.2 that Hilbert keys are more difficult to compute and are not stable when the occupied space is extended. These properties suggest avoidance of the Hilbert keys, especially when union operations have to be considered. It should be possible similarly to define Hilbert relations, but the characteristics are practically the same as Peano relations. We propose to call them Peano relations with Hilbert keys, knowing that the algorithm to generate Hilbert keys is different from that for Peano keys.

13.7 THE PEANO-TUPLE ALGEBRA

The role of the Peano-tuple algebra is to offer some tools to manipulate Peano relations in the same vein as relational algebra operators. Three general kinds of operator can be envisioned, including some that deal explicitly with spatial properties:

1. Boolean operators: union, intersection, difference.
2. Geometric operators: translation, rotation, scaling, symmetry, window extraction, replication, simplification.
3. Relational operators: geometric projection and Peano join.

Even though in this chapter we speak about quadtree-based Peano relations, all the ideas can be extended to octtree-based Peano relations. In the ensuing discussion we employ the example, Figure 13.10, containing two areal objects A and B.

13.7.1 Boolean operators

By performing the **union** of two objects A and B described by quadtrees or octtrees, we obtain another object, C, for which the second conformance level is not satisfied due to overlaps (Figure 13.11). This operation can easily be done by merging the two sets of tuples, and while it is performed, conformance normalization can also be done. In other words, the union of two spatial objects is realized by tuple union, and not by geometric considerations.

Figure 13.11a has the union:

C = UNION (A, B)

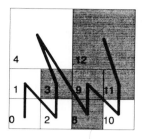

Object	Peano key	Side length	(a)
A	3	1	
A	8	1	
A	9	1	
A	11	1	
A	12	2	

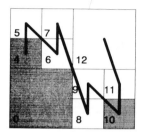

Object	Peano key	Side length	(b)
B	0	2	
B	4	1	
B	10	1	

Figure 13.10 Two objects described by Peano-based quadtrees. (a) Object A, map and tabulated data. (b) Object B, map and tabulated data.

of the objects depicted in Figure 13.9. Since R(3, 1) is overlapped by R(0, 2), it must be cancelled (2CL). Moreover, R(8, 1), R(9, 1), R(10, 1) and R(11, 1) must be aggregated to give R(8, 2), to satisfy the third conformance level condition.

The **intersection** of two spatial objects is performed by a tuple intersection. If an object tuple is overlapped by another one, the second object tuple needs to be disaggregated until intersection is achieved. The result meets the second conformance level specification. The intersection of objects A and B, producing D via:

D = INTERSECTION (A, B)

is shown as Figure 13.11b. In this case, R(0, 2) must be disaggregated and only R(3, 1) is kept.

The **geometric** intersection of two quadtrees, then, is performed by a tuple intersection without using any geometric algorithm. We take the two lists of tuples and we compare them, keeping only tuples in common, possibly after disaggregation.

The **difference** is performed similarly. In this case, we have to remove parts of object B which are also present in A (Figure 13.11c):

E = DIFFERENCE (A, B)

UNION BEFORE CONFORMANCE

Object	Peano key	Side length
C	0	2
C	3	1
C	4	1
C	8	1
C	9	1
C	10	1
C	11	1
C	12	1

UNION AFTER CONFORMANCE

Object	Peano key	Side length
C	0	2
C	4	1
C	8	2
C	12	2

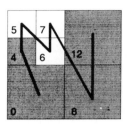

(a) Result of the union of objects A and B

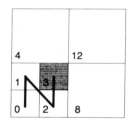

INTERSECTION

Object	Peano key	Side length
D	3	1

(b) Result of the intersection of objects A and B

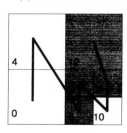

DIFFERENCE

Object	Peano key	Side length
E	8	1
E	9	1
E	11	1
E	12	2

(c) Result of the difference of objects A and B

Figure 13.11 Result of the union, intersection, and difference of two objects. (a) Result of the union of objects A and B. (b) Result of the intersection of objects A and B. (c) Result of the difference of objects A and B.

13.7.2 Geometric operators

Any object A can be **translated** from a point $(x_0, y_0: P_0 = P(x_0, y_0))$, where x_0, y_0 are coordinates of any point, to give another object F:

$$F = \text{TRANSLATION } (A, P_0)$$

where P_0 is the Peano key of the derived point $P_0 = P(x_0, y_0)$. Denoting by t the number of pairs of rightmost zeroes of P_0, $t = Z(P_0)$ when performing a translation, we can see that:

1. All tuples change their Peano keys accordingly.
2. Quadrants whose size is less or equal to 2^t are unchanged.
3. The others must be disaggregated into 2^t-sized quadrants.

The result obtained satisfies the first conformance level, as it does the second, too, because there are no overlaps, but must be transformed into the third conformance level. As an example, let us translate the object A in Figure 13.10 to the point $x_0 = 1$, $y_0 = 1$. The Peano keys become:

$$3 + (1, 1) \rightarrow 12$$
$$8 + (1, 1) \rightarrow 11$$
$$9 + (1, 1) \rightarrow 14$$
$$11 + (1, 1) \rightarrow 36$$
$$12 + (1, 1) \rightarrow 15$$

This gives, after sorting (Figure 13.12):

Peano_key	Side_length
11	1
12	1
14	1
15	2
36	1

but the tuple R(15, 2) is not acceptable *vis-à-vis* the first conformance level. It needs to be disaggregated, giving:

$$R (15, 2) \rightarrow R (15, 1), R (26, 1), R (37, 1), R (48, 1)$$

So the result, after sorting is (Figure 13.12):

Peano_key	Side_length
11	1
12	1
14	1
15	1
26	1
36	1
37	1
48	1

In some cases with negative coordinate values, Peano keys are not defined. In this case, a change of axes must be performed.

Generally speaking, in a **rotation**, quadrants need to be disaggregated into the smallest squares. All squares must be rotated and, after this operation, the result needs to be reordered and normalized directly from the first to third conformance level, as previously explained for the translation. Denoting the rotation angle by α, we have:

G = ROTATION (A, α)

For the **scaling** operation, where SC is the scaling factor, except when

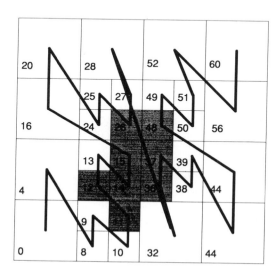

Figure 13.12 Translation of object A.

$SC = 2^i$, the quadtree object is totally disturbed and needs to be normalized.

$H = SCALING (A, SC)$

When $SC = 2^i$, the quadtree is not changed and all Peano keys are obtained by multiplying them by 4^i and the size by 2^i. The result automatically meets the requirements of the 3CL. A possibility for avoiding the quadtree disorganization is to allow quadrant sizes to be not exclusively powers of 2. In this case, all sizes are multiplied by SC, but this solution leads to difficulties with the conformance procedure.

Another geometric operation, **symmetry**, can be defined in several ways, including symmetries to a point, to an axis or to any line. Denoting the core of this symmetry by SYM, we have:

$I = SYMMETRY (A, SYM)$

In some cases, in order to avoid negative Peano keys, some additional translation must be performed.

Extraction of a subobject consists of keeping only quadrants belonging to a special window W either entirely or partially after a disaggregation. After this initial step, the algorithm is similar to translation:

$J = EXTRACTION (A, W)$

Replication consists of constructing a new object K in which a previous object A acts as a pattern to be replicated. Replication can be described as a sort of spatial multiplication including translations, rotations and scalings. Generally speaking, replication must also be defined through a **shape grammar**, a layout pattern, denoted by G, applied to object A:

$K = REPLICATION (G(A, PAT))$

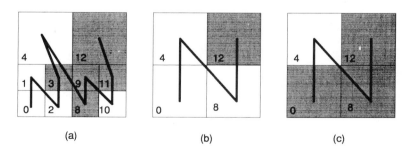

(a) (b) (c)

Figure 13.13 Different types of quadtree simplifications. (a) Original object A. (b) Simplification by default. (c) Simplification by excess.

where PAT represents the pattern. In some cases, a shape grammar may incude several patterns: for example, wall papers or building façades, some of them deriving from other shape grammars.

Simplifying a quadtree can mean one of two operations. The first is to keep only quadrants for which the size is bigger than a given threshold, THRES, thereby removing smaller sized quadrants; the second is to replace those smaller sized quadrants by quadrants having this threshold as their size. Let us call these respectively, simplification by default and simplification by excess (Figure 13.13). So

L = SIMPLIFY (A, THRES)

This particular operation is pertinent when transforming a quadtree into a pyramid.

13.7.3 Relational operators

Two relational operators are of particular interest in the Peano tuple algebra domain: **geometric projection** and **Peano join**. The former is taken here to be the geometric meaning of the term projection. If a three-dimensional object described by octtrees is projected to a plane, a two-dimensional object represented by a quadtree results, provided that the plane is perpendicular to the axes. Otherwise, for any plane, the spatial object produced is a tessellation which has to be transformed into a quadtree, where in this case, owing to overlaps in the projection, the second conformance level is not satisfied.

Taking the octtree given in Figure 13.14 as an example, a projection

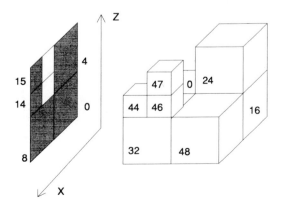

Decimal numbers	Binary numbers
00	00 00 00
04	00 01 00
08	00 10 00
14	00 11 10
15	00 11 11
16	01 00 00
24	01 10 00
32	10 00 00
44	10 11 00
46	10 11 10
47	10 11 11
48	11 00 00

Figure 13.14 Projection of an octtree to give a quadtree.

onto the plane XZ transforms the three-dimensional Peano keys by cancelling the Y-bit components from the sequence XYZXYZ. Three examples for Figure 13.13 are:

$$44_{(10)} = 10\mathbf{1}100_{(2)} \rightarrow 1110_{(2)} = 14_{(10)}$$
$$46_{(10)} = 10\mathbf{1}1\mathbf{1}0_{(2)} \rightarrow 1110_{(2)} = 14_{(10)}$$
$$47_{(10)} = 10\mathbf{1}1\mathbf{1}1_{(2)} \rightarrow 1111_{(2)} = 15_{(10)}$$

where the decimal and binary numbers are indicated by the base (10) and base (2), and the digits that disappear are marked in **bold**. The resulting quadtree, for which the side-length sizes are, of course, not altered when going from cubes to squares, is:

3D Peano_key	$\rightarrow$	2D Peano_key
0	$\rightarrow$	0
16	$\rightarrow$	0
24	$\rightarrow$	4
32	$\rightarrow$	8
44	$\rightarrow$	14
46	$\rightarrow$	14
47	$\rightarrow$	15
48	$\rightarrow$	8

We then subject it to the conformance tests. First we sort the derived two-dimensional Peano keys, and then we have some overlaps to cancel, but no aggregation to perform. So, the result is:

Peano_key	Side_length
0	2
4	2
8	2
14	1
15	1

The geometric projection is obtained, not by a geometric algorithm, but by a relational operation. Denoting the projection plane by P, we can define:

M = PROJECTION (A, P)

When dealing with a projection plane not perpendicular to one of the axes, the result is totally different; we obtain a tessellation not based on squares.

A **relational join** for Peano relations means joining the Peano keys of the descriptions of two objects:

N = JOIN (A, B)

This operation is extremely interesting for solving the point-in-polygon test and the region query (as demonstrated in sections 14.2 and 14.3). In this procedure, we keep tuples which have matching Peano keys in both quadtree objects. As an example, suppose we have three objects making up a portion of some territory (Figure 13.15) and an arbitrary query region. We wish to retrieve the objects belonging totally or partially to the query zone. That is, the operation must correctly reveal that the query area (made up of six squares as shown) overlaps with A and B (no matter how small the overlap) but not C. We have the following relations:

<table>
<tr><th colspan="3" align="center">OBJECT_LOCATION</th><th colspan="2" align="center">REGION_QUERY</th></tr>
<tr><th>Object-ID</th><th>Peano_key</th><th>Side_length</th><th>Peano_key</th><th>Side_length</th></tr>
<tr><td>A</td><td>16</td><td>2</td><td>30</td><td>1</td></tr>
<tr><td>A</td><td>24</td><td>2</td><td>31</td><td>1</td></tr>
<tr><td>A</td><td>28</td><td>1</td><td>52</td><td>2</td></tr>
<tr><td>A</td><td>30</td><td>1</td><td>53</td><td>1</td></tr>
<tr><td>B</td><td>29</td><td>1</td><td>55</td><td>1</td></tr>
<tr><td>B</td><td>31</td><td>1</td><td>62</td><td>1</td></tr>
<tr><td>B</td><td>48</td><td>2</td><td></td><td></td></tr>
<tr><td>B</td><td>52</td><td>2</td><td></td><td></td></tr>
<tr><td>C</td><td>0</td><td>4</td><td></td><td></td></tr>
</table>

The result is obtained by joining the Peano relations, Object_location and Region_query, followed by a relational projection to get the object data. Because the tuple R(B, 52, 2) can be disaggregated into four other tuples R(B, 52, 1), R(B, 53, 1), R(B, 54, 1) and R(B, 55, 1), the result of the join contains five tuples, three from the block number 52 of side length of two units, as well as blocks 30 and 31. The entire procedure ends with a relational projection into the object identifier, revealing that there is overlap with areas A and B:

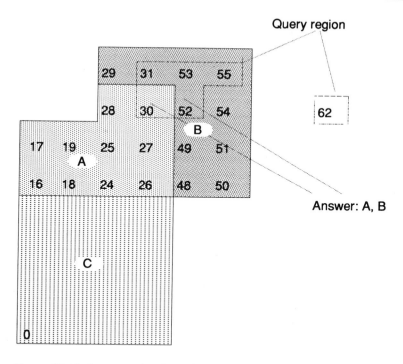

Figure 13.15 An example of spatial query solving by a Peano join.

Object-ID	Peano_key	Side_length
A	30	1
B	31	1
B	52	1
B	53	1
B	55	1

RESULT OF THE JOIN

OBJECTS

Object-ID
A
B

PROJECTION →

Let us mention that the point-in-polygon query is solved similarly; in this case, the size is reduced to the minimum allowed by the desired resolution.

13.7.4 Examples of Peano-tuple algebra queries

Two examples of the use of the Peano-tuple algebra will be given in order to give a flavour of the way it may be used for solving spatial queries, for, respectively, a cadastre, a two-dimensional case, a geological problem, and a three-dimensional situation.

CADASTRE EXAMPLE

Suppose we have a cadastre whose geometric representation is given by Peano relations:

LAND_OWNER (Parcel-ID, Landowner-ID)
PR-PARCEL (Parcel-ID, Peano_key, Side_length)

and a relation for a planned road passing through the area of the cadastre:

PR-ROAD (Peano_key, Side_length)

One task is to ascertain the name of the landowner affected by this proposed highway; a second is to determine the affected land area. In order to obtain the answers several steps must be considered.

First of all, we make a Peano join between PR-ROAD and PR-PARCEL giving an intermediate piece of information, the Peano description of only the parcels affected:

A1 (Parcel-ID, Peano_key, Side_length)

Then it is necessary to make a relational join between A1 and LAND_OWNER on Parcel-ID in order to get:

A2 (Landowner-ID, Parcel-ID, Peano_key, Side_length)

Now in order to ascertain the landowner, a projection must be performed, resulting in:

A3 (Landowner-ID)

In order to estimate land price, the corresponding area will be easily computed from the sum of the relevant square areas, producing:

Affected_area = Σ (Side_length * Side_length)
A4 (Landowner-ID, Affected_area)

TUNNEL EXAMPLE

This three-dimensional example is similar in procedure to the previous case. Geological layers are described by the following octtree Peano relations:

PR-LAYERS (3DPeano_key, Side_length, Layer_type)

and a tunnel is projected with the same description:

PR-TUNNEL (3DPeano_key, Side_length)

By a Peano join between PR-LAYERS and PR-TUNNEL, we will get the layer type to drill in order to construct this tunnel:

B1 (3DPeano_key, Side_length, Layer_type)

Now, similarly, if we want to know the volume of each layer material to drill, we have:

Volume = Σ (Side_length * Side_length * Side_length)
B2 (Layer_type, Volume)

13.8 SUMMARY

In general, considering spatial information systems, the relational model and the relational algebra are very worthwhile for structuring a database and querying against it. Taking advantage of the relational modelling approach in providing a smart and clear model for representing data, the relational algebra allows the efficient and elegant solving of many kinds of query, especially by means of the SQL. However, even if spatial data can be stored in the relational model, nothing at all can be done concerning the generating and membership rules for point-in-polygon, and other tests, even at the retrieval level. The only possibility is to integrate these via some specially prepared modules at the SQL level in order to tackle this problem correctly.

Recalling prior discussion of integration of thematic and spatial data, the classical approach, seen in mapping software, and in spatial information systems using relational database management systems software, has been to logically separate the two. As a practical matter, recently some attempts have been made to use relational systems more fully, not so much in the sense of the Peano-tuple approach, but in encompassing a variety of vector data. An experimental project, GEOVIEW, developed at the University of Edinburgh, has shown that a comprehensive range of different types of spatial entity can be stored and retrieved efficiently in a relational database provided certain conditions are met. Principally, the GEOVIEW design uses bulk data storage (the long data type of the ORACLE system used for implementation) for variable length coordinate data items; and it stores all spatial entities and attributes in a unified structure in the single relational database management system.

An alternative approach is represented by the algebra for pieces of space. Peano relations and algebra are good tools to represent and manipulate spatial data, especially allowing easy performance of common queries such as point-in-polygon and zones. The main drawback is the complexity of geometric transformations such as translations and rotations. Yet it appears that rotation and translation of spatial objects in a database are needed only rarely. Moreover, the conformance and normal form conditions provide some means for checking data integrity in a manner similar to the devices described in Chapter 15 for topologically structured spatial data.

Even though the approach of pizza modelling is not very familiar *vis-à-vis* spaghetti modelling, it provides advantages for manipulation. A commonly cited limitation is the need for big computer memories, inherently a feature of regular cell encodings, but currently some simulations and other experiments show that this need not be a drawback.

13.9 BIBLIOGRAPHY

Further reading on relational and Peano tuple algebras for spatial data is provided by several papers by Laurini or Laurini and Milleret. Concepts of relational modelling and database management systems are covered extensively by Codd and Date.

Codd, Edgar F. 1970. A relational model for large shared data banks. *Communications of the Association for Computing Machinery* 13(6): 377–387.

Codd, Edgar F. 1990. The Relational Model for Database Management, Version 2: Reading, Massachusetts, USA: Addison-Wesley.

Date, Chris J. 1985. *An Introduction to Database Systems*, 4th edn. Reading, Massachusetts, USA: Addison-Wesley.

Date, Chris J. and Colin White. 1988. *A Guide to SQL/DS*. Reading, Massachusetts, USA: Addison-Wesley.

Diaz, Bernard and Sarah B.M. Bell (eds). 1986. *Spatial Data Processing Using Tesseral Methods*. Swindon, UK: Natural Environment Reseach Council.

Elmasri, Ramez and Shamkant B. Navathe. 1989. *Fundamentals of Database Systems*. New York, New York, USA: Benjamin Cummings.

Gargantini, Irene. 1983. Translation, rotation and superposition of linear quadtrees. *International Journal of Man–Machines Studies* 18(3): 253–263.

Laurini, Robert. 1987. *Manipulation of spatial objects with a Peano tuple algebra.* Report CAR TR 311, Center for Automation Research, University of Maryland, College Park, Maryland, USA.

Laurini, Robert and Françoise Milleret. 1986. Les clefs de Peano: un nouveau modele pour les bases de données multidimensionnelles et les bases d'images. *2emes journées. Conference Bases de Données Avancées, Giens, France.* Rocquencourt, France: INRIA, pp. 211–230.

Laurini, Robert and Françoise Milleret. 1988. Spatial data base queries: relational algebra versus computational geometry. *Proceedings of the Fourth International Conference on Statistical and Scientific Database Management, Rome, Italy,* M. Rafamelli *et al.*, (eds). Berlin; Germany: Springer Verlag. pp. 291–313.

Laurini, Robert and Françoise Milleret-Raffort. 1989. Solving spatial queries by relational algebra. *Proceedings of the Auto Carto 9 Conference, Baltimore.* Falls Church, Virginia, USA: American Society for Photogrammetry and Remote Sensing/American Congress for Surveying and Mapping, pp. 426–435.

Meixler, David. 1983. Peano keys. Paper presented at the Spatially Oriented Referencing Systems Association Forum, University of Maryland, College Park, Maryland, USA.

Mitchell, Williams J. 1979. Synthesis with style. *International Conference on the Application of Computers in Architecture, Building Design and Urban Planning, Berlin.* London, UK: Online Conferences, pp. 119–134.

Peano, Giuseppe. 1890. Sur une courbe qui remplit toute une aire plane. *Mathematische Annalen* 36(A): 157–160.

Peano, Giuseppe. 1957 edn. La curva di Peano nel 'formulario mathematico'. In *Giuseppe Peano. Opere Scelte di Giuseppe Peano.* Rome, Italy: Edizioni Cremonesi, vol. 1, pp. 115–116.

Peuquet, Donna J. 1984. Data structures for a knowledge-based geographic information system. *Proceedings of the Fourth International Symposium on Spatial Data Handling, Zurich, Switzerland,* pp. 372–391.

Samet, Hanan. 1984. The quadtree and related hierarchical data structures. *ACM Computing Surveys* 16: 187–260.

Samet, Hanan. 1990. *Applications of Spatial Data Structures: Computer Graphics, Image Processing and GIS.* Reading, Massachusetts, USA: Addison-Wesley.

Stiny, George. 1980. Introduction to shape and shape grammars. *Environment and Planning,* Series B, vol. 7, pp. 343–351.

Van Roessel, Jan W. 1987. Design of a spatial data structure using the relational normal forms. *International Journal of Geographical Information Systems* 1(1): 33–50.

Waugh, Thomas C. and Richard G. Healey. 1987. The GEOVIEW design: a relational database approach to geographic data handling. *International Journal of Geographical Information Systems* 1(1): 101–112.

14
Spatial Queries

Types, algorithms

Pedro da Silva wants to know how many trees are located less than 100 metres from him. Mr Point wants to know whether he is inside or outside a zone, while Mrs Region looks for the intersections. Ten year old Mary is interested in searching what is in a trench newly dug in the street outside her home. A king explores where to find his lost crown and he finds it in a buffer. Joao da Costa tries to roll a ball along the sea coast. A salesman, Willie Makeit, is looking for a travel itinerary made of economical routes and paths through terrain. Mohamed El Arab is looking for petroleum and for a customer to whom to sell land. Paulo do Campo takes an aircraft in order to be indemnified after a flooding.

This chapter embellishes the previous discussions of relational and Peano tuple algebras by means of numerous examples of spatial queries. After describing the general querying process, the discussion treats point-in-polygon, region, vacant place, distance, path and multimedia queries, ending with a general discussion. The role of this chapter is to define the specificities of spatial queries and to suggest algorithms to solve them, stressing that those algorithms must be different due to the various geometric representations used.

14.1 THE PROCESS FOR SPATIAL QUERIES

Suppose we have a wood which is split into arbitrary zones, for each of which we have the number of trees as given in Figure 14.1a. Using the SQL to retrieve the number of trees in a zone, as in this relation:

WOOD (Zone-ID, Number_of_trees)

for example, 3, causes no problem:

SELECT Number_of_trees
FROM Wood
WHERE Zone-ID = 3

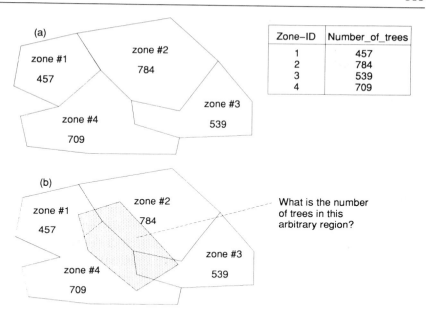

Zone–ID	Number_of_trees
1	457
2	784
3	539
4	709

Figure 14.1 Example of a spatial query. (a) Example of a wood. (b) Example of a query in the wood.

Similarly for two zones, for instance 3 and 4, we can write the SQL instruction:

SELECT SUM (Number_of_trees)
FROM Wood
WHERE Zone-ID = 3
    OR Zone-ID = 4

However, since the wood subdivision is totally arbitrary, suppose now we want to ascertain the number of trees in an arbitrary region cutting across several zones. No solution is possible within the SQL world. For this task, relational algebra is not sufficient and some geometric algorithms must be invoked. Indeed, we need to deal with a spatial description of the woods area together with the query zone.

In this case, the database view will be:

WOOD (Zone-ID, Number_of_trees)
WOOD_ZONE (Zone-ID, Segment-ID)
WOOD_SEGMENT (Segment-ID, Point1-ID, Point2-ID)
WOOD_POINT (Point-ID, X, Y)

And the query requires:

QUERY_ZONE (Zone-ID, Segment-ID)
QUERY_SEGMENT (Segment-ID, Point1-ID, Point2-ID)
QUERY_POINT (Point-ID, X, Y)

The answer, in this case, will be a single tuple relation as a solution of an area intersection algorithm and an areal interpolation technique assuming, as an example, that the trees are uniformly distributed within zones:

ANSWER (Number_of_trees)

The algorithm to solve this query will be:

Begin
       Set Number_of_trees = 0
    For each WOOD_ZONE
      compute the intersection with QUERY_ZONE
      If any THEN compute its area
          compute the number of trees as an areal interpolation and
          add it to Number_of_trees
      Endif
    Endfor
      Print Number_of_trees
End

So, this kind of query cannot be solved with relational algebra, because even though it is good for attribute based queries it is not efficient for location queries. And generally, there exist several sorts of spatial queries which cannot be solved by relational algebra and for which geometric algorithms must be invoked.

Consequently, the role of spatial queries is to provide algorithms to solve questions in which the spatial coordinates are involved. Although one can define several kinds of spatial query, the more important seem to be:

Point-in-polygon queries
Zone queries
Vacant place queries
Distance and buffer zone queries
Path queries

Let us examine these queries and ways to solve them in connection with various geometric representations. As a continuing example, let us again take the toy cadastre. We have the following relations for the segment oriented representation:

R0 (Landowner-ID, Plot-ID)
R11 (Plot-ID, Segment-ID)
R12 (Segment-ID, Point1-ID, Point2-ID)
R13 (Point-ID, X, Y)

With Peano relations, the geometry is represented by:

PR1 (Plot-ID, Peano_key, Size)

It would have been interesting to compare other geometric representations in order to have a full evaluation of performances of the target spatial queries *vis-à-vis* geometric representations, but this chapter seeks only to provide an introduction to the ideas.

Of course, these are not the only kinds of query in spatial information systems. We may also be interested in non-spatial queries, addressing the retrieval questions to only one or more attributes. These may include name based queries, such as obtaining descriptive information about places, and they may incorporate spatial data which are tabulated as attributes, such as the areal size of land parcels.

14.2 POINT-IN-POLYGON QUERIES

Starting from a point with particular (x, y) coordinates, perhaps entered as a set of numbers or by mouse pointing at a screen, the task is to find into which objects it falls. An example is to determine who is the landowner of a specific point of space in a cadastre. Let us suppose we have N plots of land (Figure 14.2).

Assuming a segment-oriented representation, considering an (x_0, y_0) point, that is any particular point for which we provide the coordinates, we need to know the name of the landowner on whose property it falls.

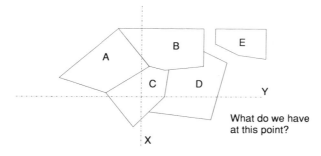

Figure 14.2 Example of a point-in-polygon query.

So we build a single tuple relation, $Q1(x, y)$, with only one tuple $Q1(x_0, y_0)$.

To solve this problem by relational algebra, the only thing we can do is a join between R13 and Q1. But, since *a priori* the (x_C, y_C) point is not stored in the database, we have no direct solution to this query. However, by using the half-line algorithm as described in section 7.23 we can find to which polygon the point belongs.

So, the complete algorithm is:

1. Use the half-line algorithm applied to the relations R11, R12 and R13 to obtain A111 (Plot-ID), the parcel of interest.
2. Join between A111 and R0 to obtain A112 (Plot-ID, Landowner-ID), that is the parcel concerned and its landowner.
3. Make a relation projection of A112 to get the desired answer A113 (Landowner-ID).

With Peano relations, considering the (x_0, y_0) point, we can determine its corresponding Peano key by bit interleaving P_0, and forming a Peano relation:

PQ11 (Peano_key, Side_length)

with only one tuple (P_0, ε) where ε is the smallest square size in the database. By a Peano join between PQ11 and PR1 (Plot-ID, Peano_key, Side_length), we obtain A211 (Plot-ID, Peano_key, Side_length) with only one tuple. Then a natural join and a relational projection will produce the answer. That is, in sequence:

1. By a Peano join between PR1 (Plot-ID, Peano_key, Side_length) and PQ11, we get A211 (Plot-ID, Peano_key, Side_length) with only one tuple having the plot of interest.
2. By a relational join between A211 and R0 we obtain A212 (Plot-ID, Peano_key, Side_length, Landowner-ID).
3. By projecting A212, we get the desired answer A213 (Landowner-ID).

In this case, an alternative is that the Peano join can be replaced by a restriction for a faster operation, since there is only one tuple.

14.3 REGION QUERIES

Here, starting from a zone which we call a region to distinguish it from area objects to which it is applied, we need to determine which objects belong to it. It is the same problem as the point-in-polygon query except that the point element is replaced by the query region. As an example, in town planning, we can try to retrieve all the landowners affected by the creation of a new motorway, or the list of entities in a zone defined by its boundary (Figure 14.3).

In this case, the query consists of two relations if we use the segment oriented representation. For the query region the geometry is:

Q21 (Segment-ID, Point1-ID, Point2-ID)
Q22 (Point-ID, X, Y)

For the pizza approach, we have a Peano relation with several tuples:

PQ2 (Peano_key, Side_length)

Now, for the segment oriented representation, using computational geometry we can determine a polygon, possibly non-connected, representing the geometric intersection with the cadastre. So for the algorithm we have the following steps:

1. By geometric intersection we obtain A121 (Plot-ID) in which we have all plots of concern.
2. A relational join between A121 and R0 produces A122 (Plot-ID, Landowner-ID).
3. Projecting A122, we get the desired answer A123 (Landowner-ID), giving the list of all landowners affected by the motorway.

For Peano relations, the algorithm is similar to the point-in-polygon case except that the query is formed of several tuples:

1. By a Peano join between PR1 and PQ2, we get A221 (Plot-ID, Peano_key, Side_length) giving all plots of interest.
2. A relational join between A221 and PR21 produces A222 (Landowner-ID, Plot-ID, Peano_key, Side_length).
3. A relational projection of A222 gives the desired answer A223 (Landowner-ID).

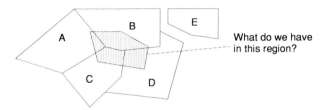

Figure 14.3 An example of a region query.

In spatial information systems, out of all spatial queries, the two-dimensional region queries are perhaps the most used. We mention typical tasks like the following:

1. The retrieval of point encoded entities within a region.
2. The retrieval of lines within a region.
3. The retrieval of areas within a region.

In addition, the nature of the query zone may itself be quite varied, possibly a circle, other regular figure, or any kind of irregular object. Using the example of Figure 14.4a, which shows a study area containing a variety of entities, consider first that we want to retrieve all information contained within a circle (Figure 14.4b). For solving this query, we have to find the punctual objects, linear objects and areal objects lying completely or partially in this circle. Suppose that we have (x, y) coordinates for candidate objects, and that x_c, y_c are the centre coordinates of a circle of radius r; then, for inside objects, the following inequality must hold:

$$(x - x_c)^2 + (y - y_c)^2 < r^2$$

Often, when we may be astride objects, as in this case, we can possibly agree that elements are inside only if their centroids are inside. For linear objects, we can choose only the inner portion. Another case is to retrieve all information within a range alongside a road or a river. So, for example as depicted in Figure 14.4c, we are trying to retrieve all objects within a certain distance from river K.

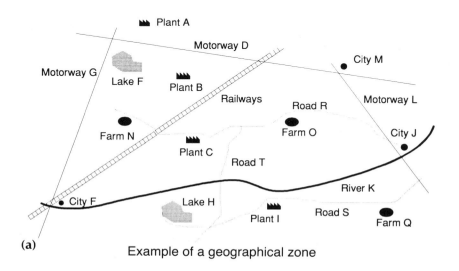

(a)   Example of a geographical zone

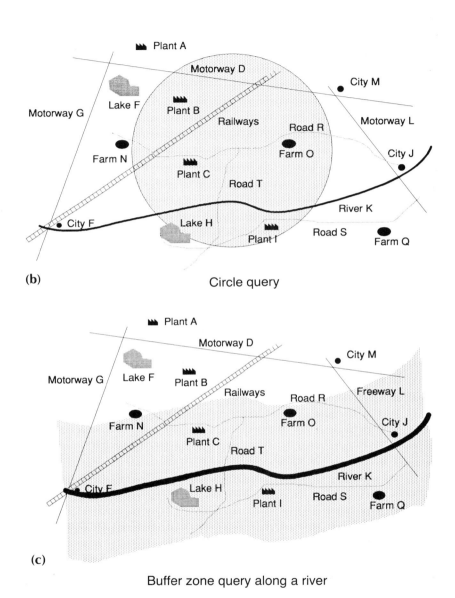

(b)   Circle query

(c)

Buffer zone query along a river

Figure 14.4 Examples of different geographical queries.

For a three-dimensional example of the range concept, consider a common query in engineering utility management, the trench query: what kind of utility networks and objects are in a future trench? For this type of query we have to find the intersections of points, lines, and areas with a three-dimensional volume (Figure 14.5).

14.4 VACANT PLACE QUERIES

Here, the problem is to retrieve vacant places within a predefined zone. For example, we have a query defining a window in which we must find the empty places (Figure 14.6). Vacant spots may be of interest in town planning or factory siting; we imagine we look for space, not for particular attributes of space. With the segment oriented representation:

Q131 (Segment-ID, Point1-ID, Point2-ID)
Q132 (Point-ID, X, Y)

For the Peano relation, we have a relation with several tuples:

PQ3 (Peano_key, Side_length)

For the segment oriented representation, instead of an intersection, a geometric difference algorithm must be executed for this query. It will give nonconnected polygons, and proceeds as follows. Initially, the geometric difference algorithm applied to the cadastre (R11, R12, R13), on the one hand, and the query window (Q131 and Q132) on the other, will produce two relations describing the vacant places:

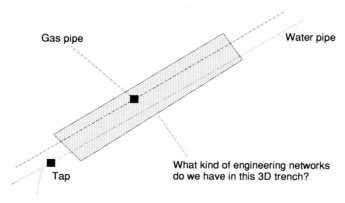

Gas pipe   Water pipe

Tap   What kind of engineering networks
             do we have in this 3D trench?

Figure 14.5 The trench query.

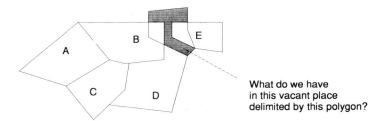

Figure 14.6 An example of a vacant place query.

A131 (Segment-ID, Point1-ID, Point2-ID)
A132 (Point-ID, X, Y)

These relations may have no tuples, that is a void difference polygon exists.

For this same task, using Peano relations, the PR1 relation can be reduced by projection to a spatial index which can be combined with PQ3 in a difference to get the desired result:

1. A projection of PR1 (Plot-ID, Peano_key, Side_length) gives A211 (Peano_key, Side_length).
2. Applying a Peano difference PQ3 minus A211 gives the desired result A232 (Peano_key, Side_length).

For the vacant place queries let us note that this Peano algorithm is faster than the one for the segment-oriented representation, and that the difference query implies non-connected resulting polygons. However, the segment-oriented representation provides more compact vacant places than the Peano approach which furnishes disseminated quadrants which often must be regrouped into connected polygons.

14.5 DISTANCE AND BUFFER ZONE QUERIES

In the distance query the task is to delineate a new geometric object defined at a specified distance from a first object and then to retrieve all objects lying at a certain distance from a given zone. An example is the delineation of maritime territories along sea coasts. Let us call this object a **crown**; in several domains it is known as a **buffer**. If the distance is a reduction then the resulting object is a skeleton.

For a zone example, suppose we have a polygon ABC for which we intend to define a new outer polygon A′B′C′ at a ten metre distance

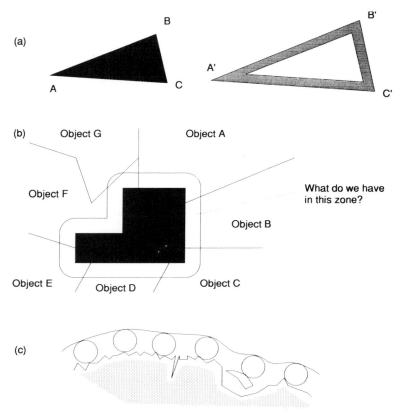

Figure 14.7 Crown determination and distance queries. (a) Buffer zones defined by parallels to another polygon. (b) Example of buffer zone query. (c) Buffer zone definition for a jagged polygon.

(Figure 14.7a). The first step is to draw parallel lines to the polygon sides and then to look for their intersections A', B' and C'. But when an angle, for example A', is very acute, the corresponding A' point is projected too far, longer than the ten metres. Therefore, we have to 'round' the angle in order that the chosen distance be obeyed. In this case, the transformed polygon, the crown, is no longer a polygon, but instead is a mixed geometric object formed of segments connected with portions of circles as depicted in Figure 14.7b.

Some awkward cases arise, such as, when we have a very jagged or indented polygon, especially with islands, the crown is not very easily described. In this case, the offset distance must be demarcated by a ball rolling along the polygon and so smoothing the jagged angles (Figure 14.7c).

Now, having defined the buffer zones, the distance query can readily be solved. For an example, consider that we need to retrieve the neighbours of a parcel within a 200 metre distance in the toy cadastre defined by R0, R11, R12 and R13, and we have a region defined by:

Q141 (Segment-ID, Point1-ID, Point2-ID)
Q142 (Point-ID, X, Y)

For the segment-oriented representation there is a four-step process:

1. Transform Q141 and Q142 to get the corresponding crown defined by segments and circle portions and then remove the inner part of Q141, Q142:

   A411 (Segment-ID, Point1-ID, Point2-ID)
   A412 (Point-ID, X, Y)
   A413 (Circle-ID, Centre_point-ID, Radius, Begin_point-ID, End_point-ID)
2. Compute the intersection of the crown and the cadastre to get a relation A414 (Plot-ID).
3. Perform a relational join with R0 to get A415 (Landowner-ID, Plot-ID).
4. Perform a relational projection to get the buffer zone neighbours A416 (Landowner-ID).

In the case of a Peano relation model, the crown, represented by PQ4 (Peano_key, Side_length), can be defined in two ways: either by Euclidean distance, so the circle portion must be approximated by quadrants given the resolution level; or by Manhattan distance, which is simpler to handle in the quadtree domain. Anyway, the crown will be defined by Peano relations and afterwards the algorithm is similar:

1. Transform the region PQ4 to get the crown A421 (Peano_key, Side_length).
2. Perform a Peano join between A421 and PR2 (Plot-ID, Peano_key, Side_length) to get A422 (Plot-ID, Peano_key, Side_length).
3. Make a relational join with R0 to get A423 (Landowner-ID, Plot-ID, Peano_key, Side_length).
4. Lastly, make a relational projection to get the buffer zone neighbours A424 (Landowner-ID).

14.6 PATH QUERIES

Path queries can be defined very differently, according to the geometric representation. Let us examine successively, paths within a simple graph, paths within a hierarchized graph, the travelling salesman path, and paths in polygons and across terrains.

In the case of paths within a graph, all possible paths are represented by a graph as is the case when studying transportation networks. There are different possibilities as to the shortest route from an origin, say A, to a destination, say B, depending on the weights, or impedances on the edges of the graph, and depending on whether we are looking for the shortest in distance, the shortest in time, the lowest in price, or using other criteria (Figure 14.8a).

Sometimes we have to deal with networks of different levels, an example of a hierarchized graph. This is the case, for instance, when somebody wants to go from one village to another village far away, possibly hundreds or thousands of miles distant. In this case, it is necessary to take a secondary road to reach a national turnpike, then to

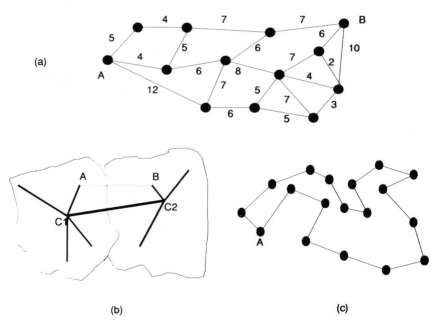

Figure 14.8 Different kinds of path query (topological). (a) Search for the shortest path in a graph. (b) Selection of a path in a hierarchized graph. (c) Example of the travelling salesman problem.

continue with provincial or national roads, and finally, to use some other secondary roads to reach the destination. The case of a hierarchical network is depicted in Figure 14.8b, assuming someone has to go from A to B, passing by central places C1 and C2.

Another example is provided by aerial traffic. For going from A to B, it is first necessary to fly to the airport at C1 with a small aircraft, then take a large aircraft from C1 to C2 and then finish with a small aircraft. Of course, in these cases, the selected path could be longer in kilometres than the distance as the crow flies, and we need to measure distances from the graph not in Euclidean space.

A very common problem of path selection is known as the travelling salesman route, in which the salesman wants to visit a large number of places without passing any place twice and then return home, as shown in Figure 14.8c. This and other graph-oriented problems are dealt with by numerous algorithms developed in the field of operation research and related areas.

Path selection may also involve polygons instead of graphs (Toussaint, 1989). This is the case, first of all, when going from from one point to another separated by a barrier such as a lake. Using the example of Figure 14.9 in the case of going from A to B, there is no problem at all since the crow distance path is totally included in the polygon. However, for going from C to D the directed path is not totally included in the polygon and the best way will be to pass via I and G to reach D.

Secondly, paths may be seen as passing through polygon segments for terrains (Barrera and Vazquez-Gomez, 1989) of different qualities, perhaps with different modes of transportation, or different costs. Here the path query is to find the best way for going from A to B over a terrain represented by a grid or by countour lines (Figure 14.10). Several kinds

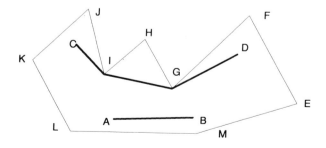

Figure 14.9 Paths within a polygon.

of path can be defined: the shortest in the on-ground distance, that is, taking the three-dimensional features of the terrain into account; the shortest time distance, which, for a special case like skiing may require different speeds for climbing and descending; or the path with the most regular gradient.

14.7 EXAMPLES OF MULTIMEDIA QUERIES

Now let us study two examples of multimedia spatial queries, the first from the domain of geology (ore ownership) and the second in town planning (flooding) to illustrate more comprehensive needs.

EXAMPLE OF MINERAL OWNERSHIP

For the first query, the mineral ownership, we have the landowners and their parcels decribed in segments and we have the geological description of the subsurface. The task is to retrieve the list of landowners whose subsoil has ore. For simplicity we assume that all coordinates are based on the same reference system. Figure 14.11 has the complete process.

Let us consider a relation giving a parcel and its landowner, and three other relations for parcel boundary description:

R1 (Parcel-ID, Owner-ID)
R2 (Parcel-ID, Segment-ID)

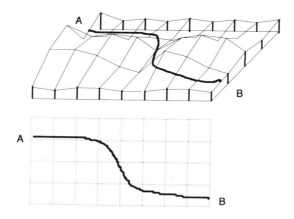

Figure 14.10 Selection of a path in a mountain region.

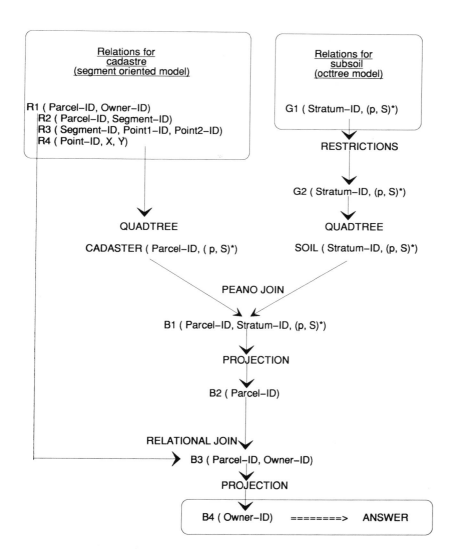

Figure 14.11 The structure of the algorithm for solving the ore query.

R3 (Segment-ID, Point1-ID, Point2-ID)
R4 (Point-ID, X, Y)

Here the subsoil can have a segment-oriented description which is not very handy, so we prefer a cell representation based on octtrees governed by three-dimensional Peano keys to describe the geometry of the strata:

G1 (Stratum-ID, (Peano_key, Side_length)*)

The process for obtaining the desired results requires, first of all, the key step of transforming the cadastre into a quadtree giving:

CADASTRE (Parcel-ID, (Peano_key, Side_length)*)

The second step is to perform a geometric projection of the geological octtree into the soil to give a quadtree. So we get a two-dimensional Peano relation by removing the z bits of the 3D Peano keys, and then complete this operation by an aggregation to obtain a well-formed quadtree. Let us use Peano_key [X, Y] for this two-dimensional Peano key reduced to x and y coordinates:

SOIL (Stratum-ID, (Peano_key [X, Y], Side_length)*)

Now we can perform a Peano join between CADASTRE and SOIL to give B1:

B1 (Parcel-ID, Stratum-ID, (Peano_key, Side_length)*)

which is followed by a relational projection:

B2 (Stratum-ID)

Now, a relational join with R1 to give B3 is undertaken:

B3 (Stratum-ID, Owner-ID)

followed finally, by a relational projection to get the answer as to which landowners own land above the mineral deposits:

B4 (Owner-ID)

EXAMPLE OF FLOODING

Suppose after a flooding one wants to locate all farmers affected by this event in order to indemnify them. For this, let us start from a cadastre and aerial photographs of flooded fields. After giving the structure of the land data and pixel-based photographs, we present the solution process (Figure 14.12).

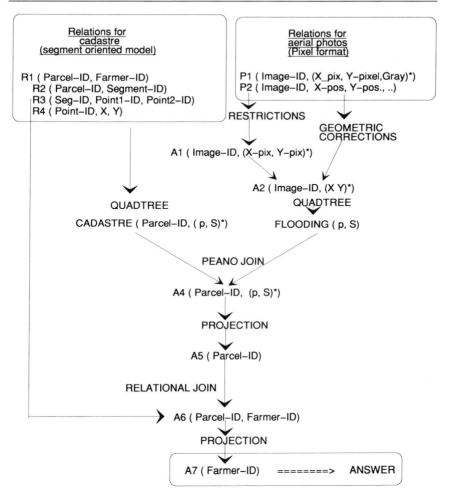

Figure 14.12 The structure of the algorithm for solving the flooding query.

As in the previous example, let us have a relation giving a parcel and its landowner and three other relations for parcel boundary description:

R1 (Parcel-ID, Farmer-ID)
R2 (Parcel-ID, Segment-ID)
R3 (Segment-ID, Point1-ID, Point2-ID)
R4 (Point-ID, X, Y)

Suppose our evidence of flood damage comes from aircraft based digital photographs of flooded areas, with several photographic grey levels. Moreover, suppose that the exact position of each photograph in terms of coordinates X_pos and Y_pos, given by the coordinates for one point (and length and width if the photograph is not square) and orientation is also known. Then we have relations:

P1 (Image-ID, (X_pixel, Y_pixel, Grey_level)*)
P2 (Image-ID, X_pos, Y_pos, Length, Width, Orientation)

To solve this spatial query implying spatial objects described by various geometric representations, it seems interesting to map into the linear quadtree representation (Peano relations) by using relational algebra. Firstly, let us deal with aerial photographs for which a relational restriction must be applied to remove all pixels not corresponding to water. By examining the grey levels of pixels, this operation will be performed, so giving a relation A1 corresponding to P1 reduced to water data:

A1 (Image-ID, (X_pixel, Y_pixel)*)

Secondly, all these pixels have to be positioned in the coordinate system and geometrically corrected due to photographic distortions transforming some pixels into rectangles:

A2 (Image-ID, (X_corrected, Y_corrected, Length, Width)*)

The next step will be to regroup all these relations in order to cover the whole territory by a quadtree governed by Peano keys:

FLOODING (Peano_key, Side_length)

In the same manner, we have to transform land information into Peano relations. So, starting from R2, R3 and R4, we can construct the following relation:

CADASTRE (Parcel-ID, (Peano_key, Side_length)*)

To obtain the result, a Peano join has to be performed between FLOODING and CADASTRE to give A4 corresponding to only flooded parcels:

A4 (Parcel-ID, (Peano_key, Side_length)*)

Now, that we know the list of flooded farmers, in a first step we will perform a projection on A4 to give only the parcels (A5); then a relational join of A5 with R1 will give A6 which is the answer (Figure 14.12):

A5 (Parcel-ID)

A6 (Parcel-ID, Farmer-ID)
A7 (Farmer-ID)

14.8 IMPLICATIONS FOR SPATIAL INFORMATION SYSTEMS

Spatial queries are perhaps the most important and specific part of a spatial information system. Should we have a conventional database system, a substantial portion of probable queries cannot be posed against it, especially queries dealing with geometric features. However, in practical situations, queries are a mixture of relational and spatial queries. Even at this early stage of understanding of how best to design spatial information systems it appears sensible to have spatial data processors perform operations on spatial data, and relational systems respond to attribute data queries. The two domains are, though, fused through the concepts and implementations of the Peano-tuple kind.

The topic of spatial queries also causes us to return to our earlier distinction between analysis and information, or between operations on data and querying. A toolbox approach provides an environment of particular processing functions, simple or complex, an approach growing out of mapping systems. In contrast, a database management system perspective, as seen in relational modelling and its modern extensions towards object orientation, offers an extension of a query and transaction processing mode to a spatial data context.

In relational database management systems analytical operations are performed by separate computer programs not within the query environment. For spatial contexts this situation applies not only to the computational geometry based rules, but also to other requirements. For example, following recursively a relation of segments of highways for a routing problem cannot be done within the query language. However, the algorithmic view of a relational system is not tenable from a fractional geometry perspective. Peano-tuple algebra, as seen in the examples given in this chapter, provides a way to incorporate spatial data operations directly.

Progress in using spatial information systems is also being facilitated by the extension of relational database management systems to encompass more data types than the classical ones oriented to non-spatial contexts. Thus, several of the commercial database management systems now offer tools for the creation of units like point or polygon, and the definition of operators that are appropriate for these new types (Guptill, 1990). For example, without using the specific syntax, POSTGRES, a theoretical model that is a basis for the recent developments of the commercial product, INGRES, provides for creating a point data type:

DEFINE type **POINT** (internal length = 32)

a relation for such data:

CREATE Locations (ID = integer, description = point)

and an operator useful for a point-in-polygon test:

DEFINE operator **WITHIN** (procedure = point-in-polygon)

However, querying is not only giving instructions to an information system using the Structured Query Language; it is also a matter of the specifics of a language that might be most relevant to tasks involving objects located in space. The query process involves stages of compilation of one or a set of questions, obtaining the data needed to begin responding to the query, undertaking computations or data synthesis where there are different geometric representations, and providing the answer in an acceptable and usable form. We have oriented the main thrust of our discussion to the topics of spatial data representations and computations; but the beginning and ending stages, to which the user most directly relates, are nonetheless, important.

Recalling earlier comments about personal frames of reference and spatial misperceptions, we now mention some cognitive aspects of querying, particularly the language for giving instructions, and the spatial concepts that individuals may have and use. Ideally, the language for spatial-data oriented processing should:

1. Have a terminology for cartographic or other spatial situations.
2. Use a language that is cognitively appropriate.
3. Provide flexibility and efficiency.

For example, the map overlay modelling process would be more natural to users if they could give their instructions in the form of statements like: find the difference between this map and that other map, or show me the lakes that have summer cottages on their shores. Non map oriented queries might be more appealing if couched in familiar terms, such as: do large towns have a high incidence of toxic chemical waste dumps?, rather than, say, retrieve the names of the cities with more than ten waste dumps. Grid cell data systems like MAP, IDRISI and GRASS that are oriented towards map analysis have a language incorporating some spatial terminology like 'distance', or 'overlay one map on top of another'; but generally speaking, spatial information systems, for reasons reflecting their roots and data structures, make less use of spatial terminology than we may wish.

This lack is especially true for the employment of the relational database systems which generally brings with it the use of the Structured Query Language (SQL). Notwithstanding that this language is somewhat inflexible anyway, it has special limitations for spatial data contexts. Its orientation to non-spatial data naturally means it does not inherently recognize spatial objects. Some spatial information systems now commercially available, as well as research laboratory experimental software, have implemented **spatial extensions** to SQL. The extensions, an example of which is the GEO/SQL system of Generation-5 Technology, in Denver, Colorado, consist of the ability to refer to spatial objects, like polygon or network, and the use of spatial operators like distance or intersection, as in:

SELECT* FROM (the set of polygons called) COUNTRIES
    WHERE (the) Population is more than 10,000,000
    AND (the country) Chile is a neighbour

although this is not necessarily the actual syntax used.

Of course, the operations and retrievals that are possible in an extended language depends on the data structure, possibly undertaken by storing some descriptive data that will cause the assembly of all the primitive pieces necessary to make up the polygon for Chile and ascertain the neighbouring countries. Generally, we see the development of interfaces that have a query compilation layer separate from the query execution layer. The data types and task specifications associated with the language of the user will be turned into the relational and spatial operators needed for the execution of the query.

In this regard we think of requests for specific data not only as the retrieval of records in a database meeting the specified conditions, but, ideally, the return of information that facilitates the intellectual accomplishment of the user's task. For example, revealing the prettiest part of the city of Paris might be undertaken by an interpretation of pretty, the completion of a multivariate statistical analysis, the comparison of different candidate neighbourhoods, and the presentation of the answer in the form of map, picture and words for a visually non-impaired person. More simply, we may think of having an answer as to where something is located given in the form of 'near to . . .' instead of as a pair of coordinates.

While we may see this development as valuable, numerous challenges remain in creating spatial information systems that are cognitively appropriate. Difficulties here are, firstly, that there is no standard terminology for spatial concepts. Recall that the terms arc, line, edge, chain all mean the same thing, yet are also defined as different types of object. Not only are there such variations among the technical fields that deal with space, but there are differences among cultures in the use of words for describing space or actions in space. Secondly, people have different and changing spatial frames of reference. These may be egocentric, local or global; they may be physical, experimental, perceptual or conceptual; they may be geometric, topological, raster or linguistic. The manner of posing questions and the effectiveness of maps as an output form depend in large measure on the nature of personal frames of reference.

14.9 BIBLIOGRAPHY

Barrera, Renata and J. Vazquez-Gomez. 1989. A shortest path method for hierarchical terrain models. *Proceedings of the Auto Carto 9 Conference, Baltimore*. Falls Church, Virginia, USA: American Society for Photogrammetry and Remote Sensing/American Congress for Surveying and Mapping, pp. 156–163.

Guptill, Stephen C. 1990. Multiple representations of geographic entities through space and time. *Proceedings of the Ninth International Symposium on Spatial Data Handling, Zurich, Switzerland*, pp. 901–910.

Laurini, Robert and Françoise Milleret. 1988. Spatial data base queries: relational algebra versus computational geometry. *Proceedings of the Fourth International Conference on Statistical and Scientific Database Management, Rome*, M. Rafamelli *et al.*, (eds), Berlin, Germany: Springer Verlag. pp. 291–313.

Laurini, Robert and Françoise Milleret-Raffort. 1989. *L'ingenierie des Connaissances Spatiales*. Paris, France: Hermès.

Laurini, Robert and Françoise Milleret-Raffort. 1989. Solving spatial queries by relational algebra. *Proceedings of the Auto Carto 9 Conference, Baltimore*. Falls Church, Virginia, USA: American Society for Photogrammetry and Remote Sensing/American Congress for Surveying and Mapping, pp. 426–435.

Samet, Hanan. 1989. *Applications of Spatial Data Structures: Computer Graphics, Image Processing and GIS*. Reading, Massachusetts, USA: Addison-Wesley.

Sinha, Anil K. and Thomas C. Waugh. 1988. Aspects of the implementation of the GEOVIEW design. *International Journal of Geographical Information Systems* 2(21): 91–99.

Toussaint, G. 1989. Computing geodesic properties inside a simple polygon. *Revue d'Intelligence Artificielle* 3(2): 8–42.

Waugh, Thomas C. and Richard G. Healey. 1987. The GEOVIEW design: a relational database approach to geographic data handling. *International Journal of Geographical Information Systems* 1(1): 101–112.

15
Access and Quality
Spatial indices and integrity constraints

Peter Irish searches in his database for everything that could cover a single identified place. Pierre Français proposes to split the space into squared grids, but Alexander Deutschmann suggests rectangular grids. Antonio Italiano prefers to play music with Peano keys. Karima Bennis prefers collections of rectangles, Round Billy likes circles; and Ivan Kitaigorodski wants any kind of polygon for encompassing the Amour River between Russia and China. Beelzebub is judging a triangle to see whether it can enter the tessellation inferno. But can they be sure about what data they get in return?

It seems in this context there are two substantial challenges ahead for builders of spatial databases. The one is to improve the speed in data retrieval; the other is to provide checks on the integrity of the contents of the database.

Generally speaking, for alphanumerical databases data retrievals are only entity or attribute based, whereas in spatial databases we also have to deal with obtaining data based on location. The role of spatial indexing is to accelerate the retrieval of any object based on location. This chapter, building upon general ideas of access to spatial phenomena presented earlier, presents some general concepts for indexing using space-filling curves, quadtrees and R-trees. The background for these three categories of data structures was covered in sections 4.7, 6.4 and 8.6. The first two sections of this chapter provide some background about indexing in conventional file management systems and relational databases.

Then, the second major portion of this chapter treats the topic of consistency in spatial information in order that we may feel assured that we are dealing only with valid spatial objects. So we present some principles regarding the rules for checking spatial consistencies, drawing upon some background material covered earlier in sections on topology and geometry, especially section 5.3, and also in section 9.7 on some issues in modelling.

15.1 INDEXING

The subject of spatial indexing in a database context is perhaps the most difficult problem in spatial information systems. The majority of books and commercial products on spatial information systems say little about the topic of indexing, even though data retrieval performances are directly linked to spatial indices. The same omission is apparent in writings about non-spatial database systems.

Indexing is a widely used concept in manual retrieval systems such as dictionaries or card catalogue systems for libraries. It goes beyond the basic concept of a primary key, the identifier or label associated with a tuple of information, for retrieving data. As personal experience will testify, an index provides a way to speed up the process of retrieving a particular piece of information and may indeed be required if the individual items have no unique label. Indexes like the telephone 'Yellow Pages' or the subject index for a library collection, represent the general function of classification which can be used in setting up database access paths based on descriptive information rather than identifiers.

An indexed file provides an alternative way to organize data as opposed to a 'simple' list, that is an unordered (except for chronology of creation) arrangement of records. Ordering using alphabetical sequence or numerical identifier sequence, for example post code number, is an improvement over an unordered list for searching quickly using hierarchical binary search techniques. Even so the indexed files may provide more effective access to data because information other than the key identifier is used. However, indexed files are generally more awkward to deal with if a database is changed frequently, for the index itself as well as the data file must be modified.

The issues involved in accessing particular pieces of information in a computerized data world are:

1. The way in which the computer physical encoding is undertaken.
2. The design of data structures to facilitate index building.
3. The provision of tools by the software systems used for storing and managing data.
4. The attributes that are used for building an index.

Before moving to the special topic of spatial indexing, we look first at some of the background technical issues to better appreciate the nature of the topic. Then, by reviewing the process of indexing in database management systems, we demonstrate the importance of choice of attributes for indexing.

15.1.1 Indexing in file management systems

Indexing has important implications for computerized data that must be efficiently stored and retrieved because it suggests an alternative approach to direct access for random processing. An index is a file in which each entry has a data value and one or more pointers to data elsewhere. The pointer identifies a record in the indexed file having that particular value for the data item. A full index is a file organization in which an index entry is provided for each individual record occurrence in the file or subfile. An index entry consists of the value of a primary key, for instance an identifier, and a pointer to the record containing that value (Figure 15.1). There is usually some ordering to the index so as to facilitate a fast search, but no order or physical contiguity is necessarily associated with stored records.

Sequentially stored physical records are a special case of this type of organization (Figure 15.2). Recall that storage devices such as disks are organized by fixed length blocks, for instance, 4,096 bytes, obliging us to read a full block even when wishing to read just a record. Arbitrary ordering of data records has the advantages of physical data independence at the stored record level and also a low update cost. The major disadvantage of arbitrary ordering of data records occurs when the user specifies a statement requiring retrieval of all data in a specified order.

Arbitrary ordering for index entries (each entry can be considered as an 'index record') has the same advantages and disavantages as any

Figure 15.1 Full index access method.

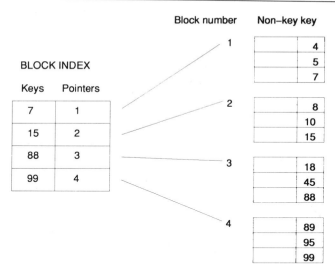

Figure 15.2 Access by block index.

physical sequential file. Firstly, unsuccessful searches for a key match often require that the entire index be scanned. Secondly, only sequential searches are possible. On the other hand, ordered indices greatly increase the potential for fast access for commands for retrieving unique records and for batch retrieval operations. Unfortunately, full indices are dense; and update operations, especially insertions of new keys and key

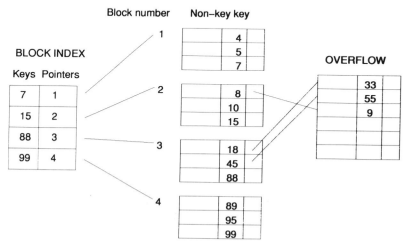

Figure 15.3 Access by block index with overflow zone.

changes, require retrieval and rewriting of the entire index, although deletions can be flagged to avoid rewriting each time. When inserting a new key, the rewritten file contains one more record than the original file, but in order to avoid the rewrite, generally speaking, this is done by means of an overflow block (see Figure 15.3).

Attributing disk addresses to keys can be performed according to several possibilities. Firstly, they may arise sequentially, or rather as indexed sequential, in which the key order is also conferred to address order, increasing or decreasing; but this method requires having all entries at the index creation. Secondly, in an hierarchical–sequential fashion, an extension of the indexed sequential methods by splitting large indices into several smaller indices, so necessitating the creation of a primary index (the master index) pointing towards several secondary indices. For huge indices, more than two hierarchical levels are necessary (Figure 15.4).

Thirdly, direct access is possible only when keys have consecutive numbers without gaps, for example the complete set of numbers for the days in a month. Major administrative regions may meet this condition at certain points in time, for instance the French administrative departements, in the range of 1 to 97. However, the method requires reservation of space for each entry. Sometimes voids are created, such as when personal identification numbers are not reissued for another person upon the death of the identity holder.

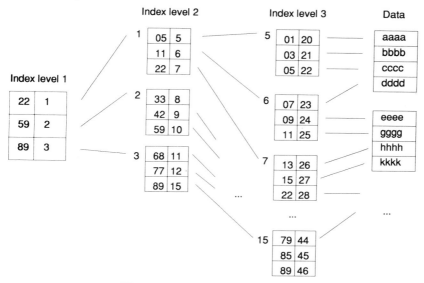

Figure 15.4 Hierarchy of indices.

Fourthly, random access, which is a direct access via a special key-to-address transformation, called a **hashing function**, may be employed. The term hashing function indicates that a class of randomizing functions is used to transform a possibly non-uniform distribution of key values into a uniform set of physical addresses. The mapping for direct access is one-to-one, but for hashing it is many-to-one; that is, several keys can be transformed into the same physical addresses, called a home address. If this occurs, those keys are referred to as synonyms and the transformation to an already occupied home address results in a collision. Several ways exist to solve collisions, especially by means of overflow areas.

15.1.2 Indexing in relational databases

In relational database commercial software products, the user has two options for indexing:

1. To do nothing, in which case a product-specific tuple-to-address procedure is provided in order to place the tuples arriving at the physical storage.
2. To create special indices based not only on identifiers (keys) but also on some carefully chosen attributes.

Unless some action is taken, then, the tuples of data are located in a sequence in the database in order of creation. In an analogous manner, records that might be created via software for digitizing will initially be ordered in the sequence of their creation. While some control can be exercised at the time of data creation that might help to order the data tuples sensibly (be the tuples spatial or non-spatial, for example digitizing in one area of a map before moving to another), it is unlikely that all future needs for data retrievals, other than by sequential record number can be anticipated.

In the SQL type databases, the index creation using attributes is realized by a **CREATE INDEX** statement. Suppose we have a relation

POP (Country_name, Population, Capital_city)

We can create an index for Country_name with the following statement:

CREATE INDEX COUNTRY_INDEX **ON** POP (Country_name)

A priori, this is an ascending order (**ASC**). However if we want also to classify population in a descending order (**DESC**), we have to write:

CREATE INDEX INDEX_POP **ON** POP (Population **DESC**)

Such indices will provide an improved way to get at the data records if population size is a commonly used item, just as accessing books in libraries might overall be done most often by major category of subject. We can create combined indices such as:

CREATE INDEX DOUBLE_INDEX **ON** POP (Population, Capital_city **DESC**)

Notice that in the previous statement we create a single index with concatenated keys. Whenever we want two different indices, we must write two statements such as:

CREATE INDEX FIRST_INDEX **ON** POP (Population)
CREATE INDEX SECOND_INDEX **ON** POP (Capital_city **DESC**)

This process provides means to accelerate retrieval although prior knowledge is needed in choosing the best attributes. There is no reason for building an index using Capital_city if we never need the information for capital city. For another example, if we search for certain types of land, perhaps those with average slope of over thirty degrees, dry soils, and elevation under 3,750 metres, the process will proceed faster if the attribute that discriminates sufficiently well to pick out the smallest number of cases is used as the first index level. For example, if there are very few cases of land terrain units with slope over thirty degrees, then searching on the other features will be facilitated if at the first level a large number of cases (records) are put to one side. Data retrieval performance depends not only on the technical aspects of the indexing process, but upon the choice of the attribute(s) for making the index. How well this choice is made clearly depends, as the examples have indicated, on knowledge of the phenomena.

Moreover, the attributes chosen for building indices may need to vary depending on purpose, and will almost certainly vary with differing views of a database contents by different users. In this regard it is an important technique to separate the index from the items being indexed. So, returning to the discussion of the indexing in file management systems, we must have a sensitivity to both selection of data items to be used for building an index and to the physical organization of them.

15.2 SPATIAL INDEXING

Recalling the discussion of space partitioning by tessellations (section 6.1) and the spatial referencing by continuous or discrete methods (section

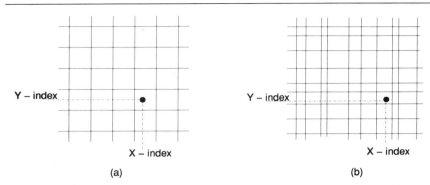

Figure 15.5 Double indexation with square or rectangular grid. (a) Fixed grid indexing. (b) Variable grid indexing.

4.2), it is appropriate in considering various issues of indexing spatial data also to consider the most important matter of access to spatial data in a database. The role of spatial indexing is to accelerate the retrieval of information based on location, especially for large databases. An access mechanism can be as basic as a geographic name or as involved as a concatenated numerical code made up of several parts. However, a spatial index should provide an access path to a location or a block of earth space, not necessarily directly to a particular object.

Using the term cover to refer to the territory we want to spatially index, the index is essentially a numerical key associated with a piece of space within that cover. A first possibility is to superimpose a grid, having squares or rectangles as shown in Figure 15.5, and to use the coordinates for an object as a spatial access for that object. But the necessity for creating two distinct indices, for the x and y orientations, is a substantial disadvantage. That is to say, for the retrieval of data recorded in two space dimensions we need to search into those two indices and then combine the results. Similarly, at three dimensions, there are disadvantages in working with the three necessary indices.

If we adopt the classical indexing, previously defined, in its focus on access via identifiers, then locators like mileposts, postal addresses or post codes, not having the problem of multidimensionality, can be treated as classical keys for access. That is, we can retrieve an object with a street address like One Front Street or an event that occurred at milestone 245. However, locators like (x, y) or (x, y, z) coordinates, expressed with two or three dimensions, cannot be treated in this way directly. Indeed, even if the problem of multidimensionality can be solved by composite indices, there still remains the most important problem that we must deal with an infinite number of keys.

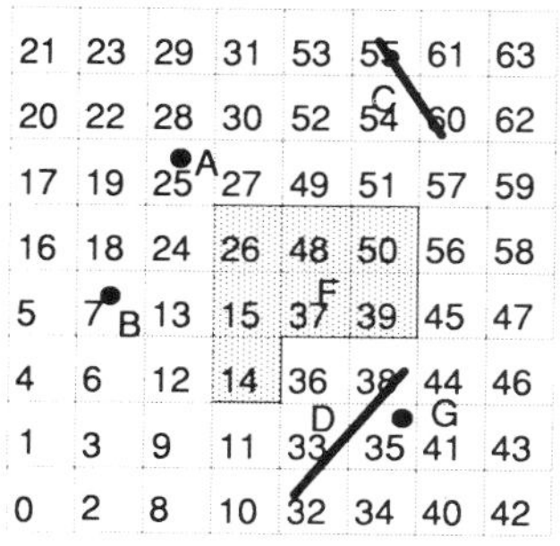

SPATIAL INDEX

OBJECT INDEX

```
21  23  29  31  53  55  61  63
                    C
20  22  28  30  52  54  60  62
17  19  25  27  49  51  57  59
         •A
16  18  24  26  48  50  56  58
 5   7  13  15  37  39  45  47
    •B          F
 4   6  12  14  36  38  44  46
                 D    • G
 1   3   9  11  33  35  41  43
 0   2   8  10  32  34  40  42
```

Peano key	Objects
7	B
14	F
15	F
25	A
26	F
32	D
33	D
35	D,G
37	F
38	D
39	F
48	F
50	F
54	C
55	C
60	C

Objects	Peano keys
A	25–25
B	7–7
C	54–55
C	60–60
D	32–33
D	35–35
D	38–38
F	14–15
F	26–26
F	37–37
F	39–39
F	48–48
F	50–50
G	35–35

(a) (b) (c)

Figure 15.6 Example of spatial indexing. (a) Objects in space – map. (b) Spatial index. (c) Object index.

The need to spatially order objects with a single index is demonstrated by a second example. Consider the territory represented in Figure 15.6, encompassing three point objects A, B and G, two line objects C and D, and one area F. Recalling earlier discussions about space-filling curve orderings, we can imagine a spatial index built with Peano keys with the runlength encoding scheme in the object index, that is giving the beginning and ending Peano keys as shown in Figure 15.6. Entry to a particular piece of space by a single dimension leads us to the objects in the cover. Notice that a single Peano key, 35 in this example, can locate several objects, here line D and point G. For completeness we also show the object keys, simply alphanumerical identifiers. These serve a purpose in revealing the location of objects or their descriptive properties (not shown in the diagram).

Several ways are possible for producing the desirable spatial indices (Figure 15.7):

1. To consider points as fractal and to order them by a space-filling curve, determining a specific level of resolution with fractal geometry.
2. To construct extents, like minimum bounding rectangles, circles and so on, and classify them into a hierarchy via a valid splitting rule, selecting privileged points in the Euclidean spirit.
3. To use a quadtree mixing Euclidean and fractal geometry.

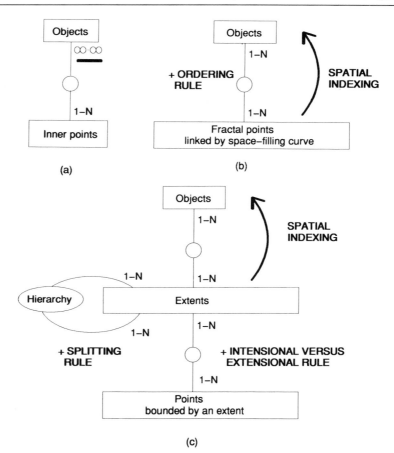

Figure 15.7 Ideal schema of spatial objects and their practical indexing. (a) Theoretical model, impossible to implement. (b) Practical equivalent model based on fractal geometry. (c) Practical model based on hierarchy of extents.

So, as for spatial object representation, we regard the indexing situation as a search for a practical alternative for the impossible situation of dealing with an infinite number of points. In the case of the fractal geometry alternative, points need to have an ordering rule; in the other case, the subdivision process requires a splitting rule; that is, the aggregation/disaggregation rule described in Chapter 11. Fractal geometry leads us to build indices using space-filling curves and quadtrees, whereas Euclidean geometry suggests the use of R-trees and the like.

15.2.1 Indexing by space-filling curves

Space-filling curves, such as the Peano (N order) or Hilbert (Π order), can order all points within the cover by means of the one-dimensional keys, but may not be the preference for all situations. Moreover, as discussed in section 4.7, different curves have different properties of ordering, stability or computational simplicity. Selection can thus be made on the basis of several factors.

A first criterion for evaluating the different types of curve is the ability to provide a spatial reference for every entity. There is no problem here for both curves, although for the Hilbert order, we need to know *a priori* the cover in order to prevent instability. A second criterion is the facility for passing from one point to its neighbours. Since two neighbouring points in the Hilbert curves are adjacent in the space, this implies that the Π order is a good candidate; this aspect is not always guaranteed in the N order.

A third criterion is the rapidity of computing keys from coordinates and vice versa. Due to the bit-interleaving procedure, the N order is the quicker ordering out of the Peano and Hilbert curves, and, as we have demonstrated in section 4.7.2, it is much easier to create keys for the Peano curve then for the Hilbert.

Another criterion is the utility of spatial indexing in conjunction with quadtrees in order to get hierarchical spatial indexing, as discussed later. In addition, a spatial index must be able to organize punctual, areal and possibly volumic objects. For points, Peano keys are fine and they can easily be extended to areas in connection with quadtrees. However, for long lines or curve portions, this space-filling curve kind of spatial indexing is not sufficient.

As an illustration of the general process, Figure 15.8 has first and second level indices for point objects using Peano keys. These objects A–H are simple and discrete for purposes of illustration. The index is a hierarchical directory allowing more efficient retrieval than a sequential index, especially when a large number of objects are dealt with. The keys are shown here in decimal form for the purpose of simplifying their presentation. In practice they are, of course, binary digits.

A spatial index for areal objects can be similarly constructed, although in this case, it is necessary to mention the low and high values for the range of space covered because areal objects are likely to be located in several portions of the curves (Figure 15.9). So the secondary index shows the low and high value Peano keys. Generally speaking, when we need to know what is in or at an (x, y) position, we convert the Cartesian coordinates into a Peano key and we search using that index.

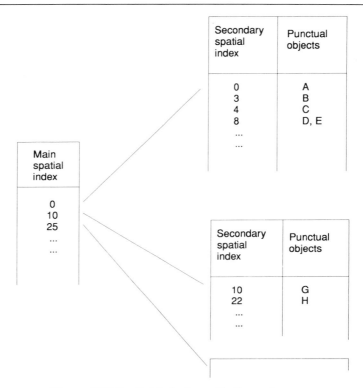

Figure 15.8 Spatial indexing with Peano keys.

However, the Peano ordering is not necessarily the most efficient. Recently a comparative study of orderings based on space-filling curves has been undertaken by Abel and Mark (1990), and Faloutsos and Roseman (1989) undertook a simulation study including a comparison of Peano keys and Hilbert keys. The numerical results of this latter study show that Hilbert indexing is the best for the rapidity of spatial retrieval. Recall, though, that the creation of the Hilbert keys is difficult, and that they are not stable when the space has to be extended. However, in the following pages we will continue to speak about Peano keys even though the Hilbert variety are more efficient in some cases. In this case, Peano keys refer to any space-filling curves rather than just the N-curve.

The main drawbacks of this space-filling curve approach is that the keys are sensitive to orientation and to the position of the Cartesian space origin. Suppose we want to merge two covers with different origins and orientations, one of them must be totally reconstructed before the merger. Indeed, since Peano keys must be recalculated after a

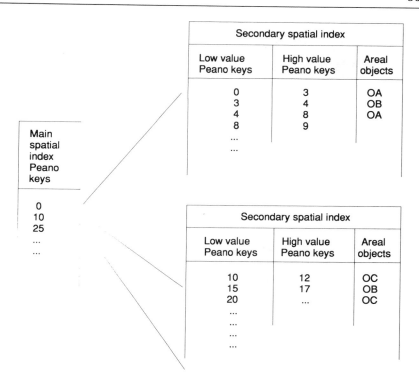

Figure 15.9 Spatial indexing with Peano keys for areal objects.

translation, as demonstrated in Chapter 13, the structure of the index is altered. For instance, an object along the Peano curve can be split into several sub-objects after the translation and, vice versa, several sub-objects before the translation can be regrouped into a bigger object. The consequence is the necessity of reorganizing indices after merging different covers.

15.2.2 Indexing by quadtrees

The use of quadtrees is an interesting possibility for spatially indexing objects. We illustrate the concept by means of Figure 15.10, using, respectively, Peano and Hilbert keys. The data tabulated here consist of the key for the quadtree block, the size of the block, that is the level in the tree, and the attribute state, here referred to as the colour of the tree node for the block.

(a) Peano keys

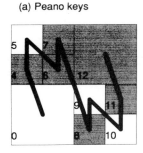

Peano key	Side Length	Colour
0	2	White
4	1	Black
5	1	White
6	1	Black
7	1	Black
8	1	Black
9	1	White
10	1	White
11	1	Black
12	2	Black

(b) Hilbert keys

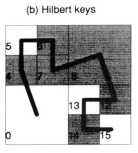

Hilbert key	Side Length	Colour
0	2	White
4	1	Black
5	1	White
6	1	Black
7	1	Black
8	2	Black
12	1	Black
13	1	White
14	1	Black
15	1	White

Figure 15.10 Spatial index encoding with Peano and Hilbert keys. (a) Peano curve. (b) Hilbert curve.

In order to organize several objects in a quadtree, we take for each of them its minimum quadrant, that is the smallest entire square bounding the object. Figures 15.11 and 15.12 depict an example of such a spatial index for a cover containing several different objects, using alternatively a linear quadtree and a hierarchical quadtree. Line entity E stretches across three of the blocks of side length of 2, so it must be referenced by the entire larger block of side length 4, which has its key as zero in the bottom-left corner. The hierarchical representation of Figure 15.12 clearly reveals the different entities retrievable at a given spatial resolution. Within each unit of two pieces of data, the left value indicates the location, and the right is the alphanumerical identifier.

Because the use of space-filling curves for indexing will imply a large number of indices, in contrast a nice possibility is to regroup fractal points into quadrants in order to use quadtrees. The latter are also valuable because they provide the ability to store objects with different sizes. Consequently, geographical objects of large areal extents will be located near the root of the tree and small objects in the terminal leaves.

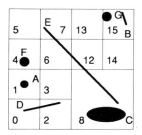

Peano keys	Side length	Objects
0	4	E
0	2	D
1	1	A
4	1	F
8	2	C
15	1	B,G

Figure 15.11 Example of a spatial index organized with linear quadtrees.

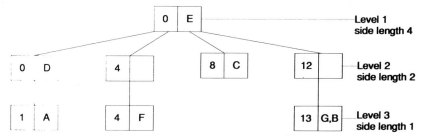

Figure 15.12 A spatial index organized with hierarchical quadtrees. The diagram shows the labels of objects located at different levels. The blocks, Peano keys, and objects are shown in the map form as part of Figure 15.11.

15.2.3 Indexing by R- and R$^+$-trees

The other possibility for spatial indexing is to use extents bounding spatial objects. One alternative is to use minimum-bounding rectangles, organized either in R-trees or in R$^+$-trees. First of all, for R-trees, consider we have N objects bounded by N rectangles and we are looking for a specific rectangle or we wish to check the membership of a specific point within a rectangle. Without any special organization, the number of tests is directly proportional to the number of rectangles. However, when we deal with, say, millions of rectangles, this time consuming process is not acceptable and we need to find a faster procedure.

This can be achieved, in one way, by regrouping adjacent rectangles within a bigger pseudo-rectangle. By repeating this operation, we construct a hierarchy of rectangles with the result that the number of tests varies with the logarithm of the number of objects instead of varying linearly. For instance, 10,000 rectangles implies an average of 5,000 tests without organization, and only 100 tests with the R-tree structure. This

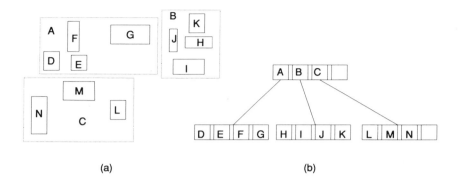

Figure 15.13 Example of an R-tree. (a) Nonoverlapping rectangles at two levels. (b) Hierarchical structure.

hierarchy is illustrated via Figure 15.13. Simply, one pseudorectangle at the higher level, say B, encompasses a record which has data for the enclosed smaller rectangles H, I, J, K.

The basic intent behind range trees is to create rectangles aligned with the orthogonal axes of the coordinate spaces, in order to:

1. Embrace as many objects as possible; and
2. Have as little overlap as possible between rectangles, but
3. Allow for subdivision to get smaller boxes within each existing rectangle.

The spatial index is determined as the rectangle in which the object is contained, with a level in a tree conveying information about resolution. Each object is associated with an R-tree node, just as for a quadtree. Precision of location may be determined for coordinate data contained in the relation.

With the relational model, it is very easy to encode an R-tree. In relational format, the rectangle description is:

RECT (Rectangle-ID, Type, Min_X, Max_X, Min_Y, Max_Y)

for which Rectangle-ID means any rectangle number so that Min_X, Min_Y, Max_X and Max_Y correspond to the coordinates of its vertices, and Type is the rectangle type, whether real or pseudo. And, for overlappings, pseudo-rectangles can have the same kinds of numbering as the real bounding rectangles.

There is also a need for a relation for the assignment of rectangles to higher order units:

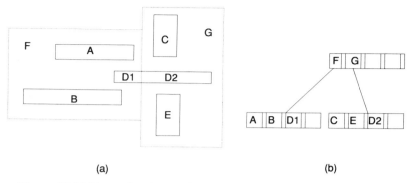

(a)

(b)

Figure 15.14 Example of an R$^+$-tree. (a) Rectangle split by higher level rectangle. (b) Hierarchical structure.

PS (Higher_Level_ps-rectangle-ID, Lower_level_ps-rectangle-ID)

Due to the spatial distribution of rectangles, though, we sometimes need to split some of them and also to balance the R-tree, producing, in this case, R$^+$-trees (Faloutsos *et al.* 1987). Balancing a tree, for instance, all levels having four rectangles and/or pseudo-rectangles, is necessary to allow the same access time to any object (Figure 15.14). An R$^+$-tree encoding with the relational model of data will give the same relations as for the R-trees. For split rectangles, an extra relation:

DECOMP (Initial_rectangle-ID, Rectangle1-ID, Rectangle2-ID)

can be useful for identifying the assignment of the pieces.

15.2.4 Indexing by other kinds of trees

The main drawbacks of minimum-bounding rectangles is that this way of spatial indexing is very sensitive to orientation. Recently some other methods have been proposed based on spheres and polygons. Instead of using a bounding rectangle, Van Oosterom and Classen (1990) have proposed enclosing objects by circles (or spheres at three dimensions). Even though it is often not easy to compute the circle, it is obvious that the extent of this geometric figure is not orientation sensitive (Figure 15.15). Moreover, this kind of spatial indexing is insensitive to orientation if the axes are rotated. Perhaps the main challenge is to find a method to determine automatically the bounding circle or sphere for any object; afterwards the addition or deletion of objects is not a problem.

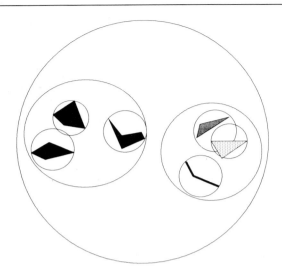

Figure 15.15 Indexing with sphere trees.

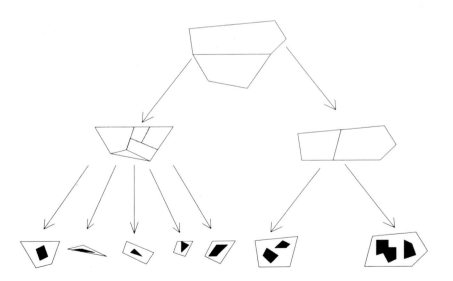

Figure 15.16 Example of indexing with a cell tree.

Another possibility is to index using polygons, called **cell trees** (Günther, 1990). In this case, each object is bounded by a convex polygon as illustrated by Figure 15.16. The main challenge is to determine rapidly the convex polygon for bounding the objects, especially the number of sides.

Also quite recently, Faloutsos and Rong (1989) have combined the R-tree and fractals by a so-called **double transformation** (Figure 15.17). Rectangles, defined by minimum and maximum x and y, can be represented by a point in a four-dimensional space (the min_X, min_Y, max_X and max_Y); this represents the first transformation. Then, all 4D points representing rectangles are ordered by four-dimensional Hilbert or Peano keys, being the second transformation. Their results show that 4D Hilbert keys give the better performance for their criteria.

15.2.5 Some practical aspects of spatial indexing

As a practical matter, only a few commercial spatial information systems today provide spatial indexing capabilities. Some systems allow access to database objects via mouse or other graphic cursor input for points or boxes or other shapes. Otherwise there is access via names or numerical identifiers in the attribute data tables. Sometimes topological neighbour-hoods provide a means of access, by following line segment or graph links for a specified polygon or line. Indexing capabilities are much rarer. For one commercial system in which indexing tools are made available, the user manuals for the ARC/INFO system (ESRI) indicate that indices for both attributes and the spatial domain can be created. The latter indexing

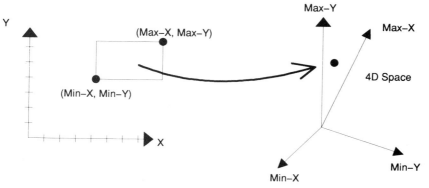

Figure 15.17 Representation of a rectangle by a point in four-dimensional space.

process uses adaptive grid-cells, the former use a binary searching mechanism, operating on data stored as modified binary (B-) trees.

The task of spatial indexing is very challenging. At present there are several techniques but none emerges as the best; although some form of hierarchical organization is generally advantageous. Moreover, two main secondary issues must also be solved: multi-layer indexing, and taking the physical disk structure into account. In several practical situations, spatial databases are split into several layers, each of them concerning a particular theme, for instance, a layer for streets, a layer for gas networks, a layer for sewerage, and so on. For this type of database it is interesting to create as many indices as there are layers and, for practical reasons, different indices may be established for different types of spatial unit. But when it is desirable to work with several thematic layers within one cover area, then the layers must be combined adequately. With a structure such as Peano keys it is simple to merge two indices, but for R-trees and cell trees, the tree branches must be redetermined, a time consuming task.

15.3 INTEGRITY CONSTRAINTS

The quality of the data obtained from a database, whether in direct form by retrieval of stored records, or resulting from some data manipulation procedure, reflects many considerations. For our purposes, we isolate here only the one aspect of spatial data consistency for some detailed treatment. Inconsistencies in data perhaps are logical defects within the spatial structure, or perhaps reflect a conflict with external validation rules.

15.3.1 Basic integrity constraints

The idea of integrity constraints, mentioned earlier in discussions of topology (section 5.3) and entity-relation modelling (section 9.3) can be invoked practically for different purposes in ascertaining consistencies, as in:

1. Checking whether an attribute value is consistent with its domain.
2. Checking whether an attribute value falls within an appropriate range, for example, a month number must be in the interval 1–12.
3. Checking the existence of another tuple in the same or in different

relations, such as the case of: if a point is mentioned, we must have either its coordinates or a way to deduce them.
4. Checking the cardinalities, for instance a segment must have two end points.

Some inconsistencies may relate to attribute data; others may arise for spatial elements. Spatial information typically is characterized by the existence of many **consistency rules** in order always to deal with valid geographical objects; for example, triangular cells in a tessellation must have three sides. Here we shall first define the set of rules necessary for spatial consistency, and then examine how to utilize them *vis-à-vis* different geometric representations. We treat this topic at this point only for spatial databases that are implemented via the relational database model.

15.3.2 Spatial data checking

In the spatial information systems world, some commercial and public domain products have error-checking procedures undertaken by special purpose computer programs. For instance, the database creation stage of the *TIGER* project of the United States Census Bureau has some editing programs for quality control of the digital version of the topographical and other maps used (Marx and Saalfeld, 1988). This checking is achieved by several tests for tracking topological inconsistencies and the like, as discussed in section 5.3. We mention, too, one software system, *SYSTEM 9*, which advertises that automatic topological checks are executed either during data capture or editing. The product description for this software indicates that there are topological checks for geometric primitives, range checks for attributes, as well as semantic topographical checks, such as a house cannot be in a river.

However, it seems that nothing has been done regarding checks on the declarative statements, that is using explicit statements via the database system, except perhaps the work of Pizano *et al.* (1989). They examine some spatial constraints, for example 'automobiles and people cannot be in a crosswalk at the same time' or 'state highways must not cross state boundaries'. Even so, it appears there is nothing quite so valuable as the topological nature in order to confer consistency on the geometric representations of spatial objects.

In a spatial database, integrity constraint checking can be performed at two times:

1. At database creation, especially when incorporating huge line segment files.

2. Incrementally at each spatial object insertion, deletion and update.

Whenever we have the possibility of creating each object completely at one point in time, we can check it carefully, which is simple but time consuming if there are many objects to deal with. The majority of cases seems to be the re-utilization of existing geographic data files, implying executing the consistency procedure when the corresponding file is integrated into the target database. For instance, points have to be examined after the point data are loaded, or line segments must be verified after segment files are dealt with. Moreover, the integrity checking is particularly important for layered databases, for new objects are created by operations like polygon overlays and buffering.

Moreover, the multiplicity of representations, a consequence of the intensional nature of spatial information, and the strong spatial relationships inherent in spatial data structuring imply that for each representation there exists a group of integrity constraints which are more readily identified with one representation than another, although at the same time meaning that several rules will be necessary.

15.3.3 Example of a cadastre

As an introductory example, let us examine the geometric description of a cadastre in order to demonstrate the consequences of consistency checking in a given representation. Consider a city (Figure 15.18) with the entity-relationship modelling for our toy cadastre of parcels as given in Figure 15.19. Land parcels are grouped into several city blocks. A

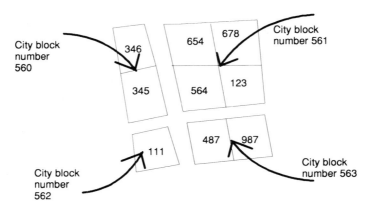

Figure 15.18 Parcels and blocks in a toy cadastre.

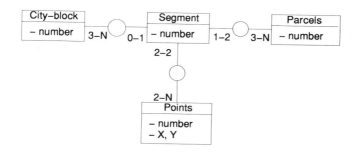

Figure 15.19 The entity-relationship diagram for the toy cadastre.

parcel represented as a polygon is limited to three or more line segments. A segment generally limits two parcels, or at times only one when there is a city block boundary for a street. Moreover, a segment is composed of two end-points and can limit two parcels. For the relational model, we have:

PARCEL (Parcel-ID, Owner-ID)
PARCEL-SEGMENT (Parcel-ID, Segment-ID)
SEGMENT-VERTEX (Segment-ID, Vertex-ID)
SEGMENT-PARCEL (Segment-ID, Parcel1-ID, Parcel2-ID)
COORDINATES (Vertex-ID, X, Y)
BLOCKS (Block-ID, Segment-ID)

One of the integrity constraints we have to check is that, for a segment (say 345) we have only two tuples in the relation SEGMENT-VERTEX. If there are three tuples with this characteristic, then an inconsistency occurs. To avoid this, we can replace the SEGMENT-VERTEX relation scheme by ENDPOINTS (Segment-ID, Vertex1-ID, Vertex2-ID). With this relation, the previous integrity constraint will be automatically matched except if one vertex identifier is null. But in the SEGMENT-PARCEL relation, sometimes Parcel1-ID or Parcel2-ID do not exist when the particular segment corresponds to a street boundary. In this case, having a null identifier is consistent with the proposed description.

Similarly, at the minimum, a polygonal parcel must have three segments, so implying that, for a given parcel, there are always three or more tuples of the PARCEL-SEGMENT relation. When a vertex exists, it will be mentioned in two tuples of the SEGMENT-ENDPOINTS relations, and when a segment exists, it must be mentioned but once or twice.

Since such integrity constraints were already revealed in the conceptual model, it is easy to verify them. But in this example, as in general, there will usually be others (Figure 15.20), such as:

1. The closure of a polygon (for example, parcel 321).
2. The complete coverage of the space (except streets).
3. A segment vertex must not belong to another parcel or be outside the space (segment D'E').
4. No vertices are located inside another polygon (for instance H).
5. There are no free-standing points or vertices.

In addition, we can work with mathematically derived data. Suppose we have information about the length of each segment, and the perimeter and the area of each parcel:

PARCEL (Parcel-ID, Perimeter, Area)
SEGMENT (Segment-ID, Point1-ID, Point2-ID, Length)

It is easy to see that some integrity constraints must hold since length, perimeter and area can be computed from coordinates, using methods described in sections 7.2.5 and 7.3.3. In this case, we can have two possibilities:

1. To have both numerical values and formulae.
2. To store only formulae.

In the first case, the stored data can be compared to the result of applying the formulae as for length or area. The formulae are used as integrity constraints for seeing if the recorded values are equal or

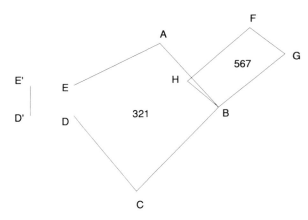

Figure 15.20 Examples of spatial inconsistencies.

approximately the same. In the second case, only the formulae are stored and they must be invoked if the result values are asked for by users, that is, they receive only mathematically derived information.

Now we list a set of integrity constraints for a cadastre:

SIC1 – all parcels will have three or more segments.

SIC2 – all segments will have two different extremities.

SIC3 – all segments border two parcels except when there is a street.

SIC4 – all parcels must be closed (all segments must have a successor with the same vertex in common).

SIC5 – all points in the space belong either to a parcel or to a street (alternatively, there are no missing parcels).

SIC6 – the relations inherent in the Euler equality checking procedure.

SIC7 – there are no free-standing edges or vertices.

SIC8 – there are no dangling edges.

SIC9 – stored information (such as lengths, perimeters and areas) must be consistent with the results of mathematical formulae.

As topological information provides perhaps the most trustworthy foundation for checking, we develop this topic in some detail.

15.4 THE USE OF TOPOLOGY IN CREATING INTEGRITY CHECKING MECHANISMS

As discussed in Chapter 5, various measures have been devised for revealing properties of particular graphs based on the Euler equality:

$$V + F = E + S$$

in which V stands for the number of vertices, E the number of edges, F the number of faces, and S the number of disconnected sub-objects. In the toy cadastre example given in Figure 15.18, we have 25 vertices, 30 edges, 9 faces and 4 objects, giving $V + F = 25 + 9 = 34$, and $E + S = 30 + 4 = 34$.

However, when the result holds, as noted in section 5.3, some compensatory miscounts can exist: perhaps one more segment and one more vertex than is correct. When this formula does not hold, though, then we know there is a problem. A second difficulty is that when the Euler–Poincaré equality is not met, it is very difficult to know what to do first. If much topological information is stored in the database, the correction will be easy; but alas, the more tuples we have, the more inconsistencies should occur. For instance, sometimes it is very difficult to know whether there is a missing polygon in a tessellation or a hole in it.

15.4.1 The topology of tessellations

We now look at different spatial representations, beginning with tessellations. Recall that there are three distinct cases:

1. The complete tessellation of a space or subspace, for instance most countries, in which any point must belong to a face and all faces are contiguous.
2. The discontinuous tessellation made of disconnected pieces, that is, a set of non-intersecting polygons, for example lakes or city blocks.
3. A mixture of both, or, more exactly, a discontinuous set of tessellations.

As an example, using Figure 15.18, we see that:

Vertices + Parcels = Segments + City blocks

In a cadastre database, we should use this formula as a check when inserting a new parcel: that is, we should insert not only a parcel identifier but also its geometric and topological information. So, let us examine the frequent case of parcel subdivision (Figure 15.21). Suppose we have a parcel, say Number 24, and that this parcel should be divided into two different parcels (say 24A and 24B). For that purpose, the surveyor has determined a segment, say Number 56, splitting parcel 24 as shown in Figure 15.21b. In some cases the extremities of this segment will not fall exactly on another segment, producing an undershoot or overshoot. So

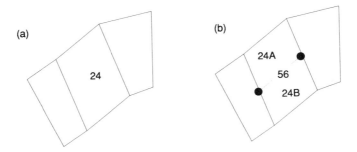

Figure 15.21 Consistency checking after parcel updates. (a) Before parcel division. (b) After parcel division.

the exact intersection must be determined, cutting each neighbouring segment into two parts. The checking procedure will be as follows:

1. Determine the exact intersections, creating two new vertices and causing the replacement of two segments by five new ones, including the segment used for subdivision.
2. Determine the boundary of the two new parcels.
3. Change the boundary of the neighbouring parcels by incorporating one more vertex and one more segment to each.

So the result of the Euler equality checking process is:

1. The total number of city blocks is unchanged.
2. The total number of parcels is increased by one.
3. The total number of vertices is increased by two.
4. The total number of edges is increased by three.

15.4.2 The topology of networks

The case of graphs can be treated similarly. Suppose we have a water supply system described by the following relations:

PIPE (Pipe-ID, From_node-ID, To_node-ID)
NODE (Node-ID, X, Y)

It is easy to find the number of edges (pipes) and vertices (nodes), but the problem is how to detect the number of faces and disconnected pieces. For computing this number, a special algorithm must be run. For some kinds of utility networks, we know that we must have only one subgraph ($S = 1$). Should we have several disconnected pieces, it means that an error occurs and we have to detect it. In this case, the Euler checking procedure is not easy to perform, but some other topological tests can be readily performed, such as:

1. Testing whether each node mentioned in the PIPE relation has coordinates.
2. Testing whether each to-node mentioned in a PIPE tuple has one or several corresponding from-nodes, except when it is a node located at the end of the graph (called a sink node).
3. Testing if a pipe is free-standing.

One of the pecularities of urban utility networks is the problem of crossings. For instance, when two electrical wires cross, sometimes they must be independent and sometimes they must be connected. This example implies that the reconstruction of network topology from geometry can be difficult to implement.

15.4.3 The topology of digital terrain models

The conventional model of contour lines is given in the following relations in which we suppose that we have interpolation between points:

TERRAIN (Terrain-ID, Z_level)
LEVEL (Z_level, Piece-ID, Closure_flag)
PIECE (Piece-ID, Point-ID, Order)
CONTROL_POINT (Point-ID, X, Y)

Among integrity constraints, let us mention:

SIC1 – points must be mentioned only once.
SIC2 – contour lines do not intersect.
SIC3 – a contour line does not have a piece missing.

For a grid model, we have the following relations:

TERRAIN (Terrain-ID, Mesh-ID)
MESH (Mesh-ID, NW_POINT-ID, NE_POINT-ID, SW_POINT-ID, SE_POINT-ID)
MESHED_POINT (Point-ID, X, Y, Z)

For integrity constraints, we can have:

SIC1 – no points mentioned in the mesh have a null ID, and they must be different.
SIC2 – no point has null coordinates.
SIC3 – any point identifier mentioned in the MESHED_POINT relation must appear at the minimum once and the maximum in four MESH tuples.
SIC4 – in the MESH relation, the corner points coordinates are associated in a simple way.

Possibly, for accelerating some procedures, it could be nice to add the neighbouring meshes as attributes to the MESH relation, giving:

MESH (Mesh, NW_point, NE_point, SW_point, SE_point, North_mesh, East_mesh, South_mesh, West_mesh)

But the consequences will be to give more integrity constraints. For instance, for a mesh and its corresponding North_mesh, the SW_point-ID and the SE_point-ID of the latter will coincide with the NW_point-ID and the NE_point-ID of the former.

Triangulated irregular networks modelled via the following relations:

TERRAIN (Terrain, Triangle-ID)
TRIANGLE (Triangle-ID, Edge1-ID, Edge2-ID, Edge3-ID)
EDGE (Edge-ID, Vertex1-ID, Vertex2-ID, Triangle1-ID,
    Triangle2-ID)
VERTEX (Vertex-ID, X, Y, Z)
BOUNDARY (Edge-ID)

must have spatial integrity constraints like the following:

SIC1 – every point of the space is covered by a triangle.
SIC2 – an edge is generally mentioned twice except at the boundary of
    the tessellation.
SIC3 – for the EDGE relation, an edge must have two end-points (no
    null values) although a triangle identifier can be null (at the
    boundary).
SIC4 – no vertex is inside a triangle.
SIC5 – an edge must have a different vertex ID.
SIC6 – no free standing edges or vertices.

15.5 AN EXAMPLE OF CONSISTENCY CHECKING FOR A TERRAIN MODEL

Integrity constraints can be established for regular tessellation representations as well as for vector models. Indeed, the ease of making spatial consistency checks may be one factor in a list of criteria for deciding upon a particular form of representation.

15.5.1 Triangulated irregular network representation

As a demonstration of these concepts and practical tools let us examine a toy terrain database described by triangulated irregular networks:

R1 (Triangle-ID, Edge1-ID, Edge2-ID, Edge3-ID)
R2 (Edge-ID, Point1-ID, Point2-ID)
R3 (Point-ID, X, Y, Z)

Depicted in Figure 15.22, this triangulation has the following true information:

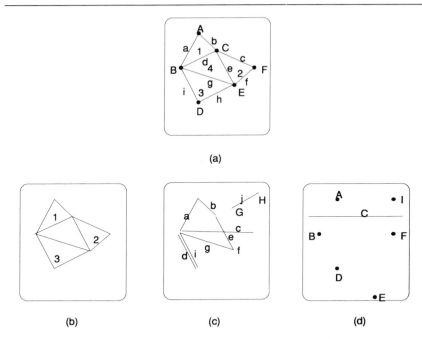

(a)

(b)   (c)   (d)

Figure 15.22 Some inconsistencies for a triangulated irregular net-work. (a) Example of the triangulation truth. (b) Triangles as stored in the database. (c) Edges as stored in the database. (d) Vertices as stored.

R1	Triangle-ID	Edge1-ID	Edge2-ID	Edge3-ID
	1	a	b	d
	2	e	c	f
	3	g	h	i
	4	d	e	g

R2	Edge-ID	Point1-ID	Point2-ID
	a	A	B
	b	A	C
	c	C	F
	d	B	C
	e	C	E
	f	E	F
	g	B	E
	h	D	E
	i	D	B

R3	Point-ID	X	Y	Z
	A	2	5	10
	B	1	3	3
	C	3	4	12
	D	2	1	3
	E	4	2	8
	F	5	3	15

A variant of R2 adds the triangle neighbours to the point information; this can be useful for checking the convex hull of the triangulated network:

R2′ (Edge-ID, Point1-ID, Point2-ID, Triangle1-ID, Triangle2-ID)

The true data for this relation are:

R2′	Edge-ID	Point1-ID	Point2-ID	Triangle1-ID	Triangle2-ID
	a	A	B	1	Null
	b	A	C	1	Null
	c	C	F	2	Null
	d	B	C	1	4
	e	C	E	2	4
	f	E	F	Null	2
	g	B	E	4	3
	h	D	E	Null	3
	i	D	B	3	Null

Data stored in the database are as follows, and shown in Figure 15.22:

R1	Triangle-ID	Edge1-ID	Edge2-ID	Edge3-ID
	1	a	Null	d
	2	e	c	b
	3	g	h	i

A datum is missing for one edge; Triangle 4 is missing altogether; and Edge3 for the Triangle 2 is recorded as b rather than f.

R2	Edge-ID	Point1-ID	Point2-ID	
	a	A	B	
	b	A	C	
	c	B	F	* Point1 different
	d	B	D	* Point2 different; see i
	e	C	E	
	f	E	E	* Same end-points to edge
	g	B	E	
	i	D	B	
	j	G	H	* Dangling edge; h is missing

R3	Point-ID	X	Y	Z	
	A	4	3	10	* The vertex is in another triangle
	B	1	3	3	
	C	Null	4	Null	* Two coordinates are missing
	D	2	1	3	
	E	0	2	8	* One coordinate out of range
	F	5	3	15	
	I	5	5	18	* A free-standing point

The Euler equality is balanced – three triangles and seven vertices equate with nine edges and one terrain; but there are numerous errors that have to be detected otherwise. The attribute data can be used more profitably than the counts of tuples. While Triangle-3 would not fail a test looking for three edges, Triangle-3 does not appear a correct number of times in the extended table, R2', as data for the neighbours for edges.

This condition is revealed in the extended R2 table, along with further possible inconsistencies:

R2'	Edge-ID	Point1-ID	Point2-ID	Triangle1-ID	Triangle2-ID	
	a	A	B	1	Null	
	b	A	C	Null	Null	* Double null
	c	B	F	2	Null	* Point1
	d	B	D	1	4	* Point
	e	C	E	2	2	* Same triangle
	f	E	E	Null	2	* Same points
	g	B	E	4	2	* Triangle 2
	i	D	B	3	Null	
	j	G	H	Null	7	* Dangling

15.5.2 Regular cell representation

Similarly we can have spatial integrity constraints concerning raster and quadtree representations. We have terrain stored as a devil's staircase (Figure 11.12) at the raster level:

STAIRCASE (Terrain-ID, X, Y, Elevation)

So, we may have rules:

SIC1 – X and Y must be within a special range: $X_{min} \leqslant X \leqslant X_{max}$ and $Y_{min} \leqslant Y \leqslant Y_{max}$.

SIC2 – every grid cell must be mentioned, that is, holes are not allowed

For instance, if X is noted, then $X + 1$ must exist, unless $X = X_{max}$. In the case of run-length encoding we can check whether two run-length regroupings do not overlap.

For a terrain represented by a devil's staircase:

PR (Terrain-ID, P, Side_length, Elevation)

the spatial integrity constraints simply correspond to the conformance levels (section 13.6):

SIC1 – the quadrants must be well positioned.

SIC2 – there must be no overlaps.

The third conformance level, necessary for obtaining compact objects, is not strictly speaking an integrity constraint. When the second level condition is met, the database is consistent, although it does not have compact objects. The concept of the non-compact object for spaghetti modelling is exemplified by a polyline in which three intermediate points are aligned.

15.6 CONCLUSION ABOUT SPATIAL INDEXING AND INTEGRITY

Although people are conscious of the importance of dealing with a consistent spatial database, little appears to be undertaken in practical settings to check the global consistency of (huge) spatial databases. Generally speaking, if anything is done at all, error checking is undertaken by special procedures, but there is nothing at the level of **declaration**, that is explicit statements, within the database system itself. Moreover, because in some cases spatial information system practitioners are content to work with databases that are, say, 95 per cent consistent, consistency checking can be done in other ways, perhaps through sampling.

As a practical matter, some commercial spatial information systems (Figure 11.12) software products offer spatial integrity checking, although the procedures may not be identified, and some may rely on visual rather than numerical methods. Some systems offer topological checking during data capture and editing, as well as attribute data consistency checks.

It is possible to write some integrity rules in the SQL using the **ASSERT** clause, checking that certain values are Null or not Null as the case may be, or by comparing equality of values of point coordinates. Writing some spatial integrity constraints such as the closure of a polygon requires that the **ASSERT** clause be extended by using parameters in a function or by counting possibilities (although this is not generally implemented).

Particularly important in the design of large spatial databases, such as currently contemplated for global environmental monitoring, integrity checking can be installed at two levels:

1. At each spatial data update or modification, especially by applying constraints linked to a single database item (tuple or object).
2. At the level of the entire database.

However, due to the complexity of some checking procedures implying the use of large computational geometry algorithms, spatial constraint declaration most likely will involve the calling of procedures external to the database system.

Integrity checking procedures will depend not only on the spatial data model used for geometric representation, but also on the database model.

In the relational databases checking will ordinarily require use of more than one tuple because the spatial object is disseminated across several tuples. In contrast, for the object oriented model discussed in Chapter 17, an integrity checking procedure can be bundled with the particular object type or instance of it. Some checking procedures will need to work at a global level, for example, for a tessellation or a network, whereas some relate to particular instances of non-complex objects, like an individual polyline.

At present it seems that we have insufficient knowledge of the nature of spatial consistency. It would appear that several problems deserve attention. First of all, we need to know how to get complete lists of spatial integrity constraints (topological, geometric and so on) for a given purpose. Secondly, we need to consider the possibility of determining the minimum set of spatial constraints in order to avoid redundancy in checking. Next it appears desirable to know both how to harmonize automatically the geometric representations and spatial constraint declaration, and how to produce ways of checking regularly, by sampling, at the database creation and at each update. Fifthly, there will be a need to determine how to alert data users of the type and the location of inconsistencies.

The problem of consistency is different from the problem of accuracy. In our opinion, for many domains of use of spatial information systems, consistency checking is the first step towards a high quality database through validating spatial information with sound connections and relationships. In contrast, accuracy, being more concerned with the numerical precision of coordinates (northings, eastings, and elevation) is a secondary element of data quality.

Returning to the topic of spatial indexing, we mention by way of summary that our logical models of the spatial reality are insufficient for capturing the semantics associated with continuous space. In this chapter we have presented several indexing tools that might foster the point-in-polygon and region queries, the operations that constitute the basic requests in a spatial database system. Of the several alternatives, space-filling curves, quadtrees, R-trees, no one approach is emerging as the best. Clearly, different index structures may be better for particular purposes, for example, binary searches for range queries, or inverted lists for Boolean queries.

Even though the techniques of spatial indexing presented are relatively efficient for the georelational model, they are not sufficient for spatial databases in which zillions of chunks of spatial knowledge must be organized. Similarly, as we consider in the next chapter, the indexing methods are not enough for spatially referencing huge collections of overlapping maps with different themes, especially because the earth is approximately spherical rather than planar.

In presenting here the two difficult topics of spatial indexing and integrity constraint checking, we can see that spatial information systems are in their infancy. It is quite likely that the world's largest databases are, and will be, contained within spatial information systems. As databases grow, automation of good procedures to provide rapid access to high quality data will be most important.

15.7 BIBLIOGRAPHY

For further background on indexing and integrity constraints for relational database management systems, the reader can refer to Elmasri and Navathe.

Abel, David J. and David M. Mark. 1990. A comparative analysis of some two-dimensional orderings. *International Journal of Geographical Information Systems* 4(1): 21–31.

Bentley, J. L. and J. H. Friedman. 1979. Data structure for range searching. *ACM Computing Surveys* 11(4).

Codd, E. F. 1990. *The Relational Model for Database Management*. Version 2. Reading, Massachusetts, USA: Addison-Wesley.

Codd, E. F. 1970. A relational model for large shared data banks. *Communications of the ACM* 13(6): 377–387.

Date, Chris J. 1988. *An Introduction to Database Systems*, 4th edn. Reading, Massachusetts, USA: Addison-Wesley.

Elmasri, Ramez and Shamkant B. Navathe. 1989. *Fundamentals of Database Systems*. New York, New York, USA: Benjamin Cummings.

Faloutsos, Christos, Timoleon K. Sellis and Nicholas Roussopoulos. 1987. Analysis of object-oriented spatial access methods. *Proceedings of the Conference of the Association for Computing Machinery Special Interest Group, Management of Data*, San Francisco, California, USA, pp. 426–439.

Faloutsos, Christos and S. Roseman. 1989. *Fractals for Secondary Key Retrieval*. Technical Report CS-TR-2242, Institute for Advanced Computer Studies, University of Maryland, College Park, Maryland, USA.

Faloutsos, Christos and Y. Rong. 1989. *Spatial Access Methods Using Fractals: Algorithms and Performance Evaluation*. Technical Report CS-TR-2214, Institute for Advanced Computer Studies, University of Maryland, College Park, Maryland, USA.

Goodchild, Michael, F. and S. Gopal (editors). 1989. *Accuracy of Spatial Databases*. London, UK: Taylor and Francis.

Günther, Oliver. 1990. Spatial database techniques for remote sensing. *Proceedings of the Fourth International Symposium on Spatial Data Handling*, Zurich, Switzerland, pp. 961–970.

Günther, Oliver. 1989. The design of the cell tree: an object-oriented index structure for geometric databases. *Proceedings of the Institute of Electrical and Electronics Engineers Fifth International Conference on Data Engineering, Los Angeles, California, USA*, pp. 598–605.

Guttman, A. 1984. R-trees: a dynamic index structure for spatial searching. *Proceedings of the Conference of Association for Computing Machinery, Special Interest Group, Management of Data, Boston, Massachusetts, USA*.

Howe, David R. 1983. *Data Analysis for Data Base Design*. London, UK: Edward Arnold.

Laurini, Robert. 1985. Graphics databases built on Peano space-filling curves. *Proceedings of the Eurographics Conference, Nice*. Amsterdam: North-Holland, pp. 327–338.

Laurini, Robert and Françoise Milleret-Raffort. 1989. *L'ingenierie des Connaissances Spatiales*. Paris, France: Hermès.

Laurini, Robert and Françoise Milleret-Raffort. 1991. Using integrity constraints for checking consistency of spatial databases. *GIS/LIS 1991 Proceedings, Atlanta, Georgia, USA*, paper 634–642.

Ling, Tok W. 1987. Integrity constraint checking in deductive databases using the PROLOG not-predicate. *Data and Knowledge Engineering* 2: 145–168.

Marx, Robert W. and Alan J. Saalfeld. 1988. Programs for assuring map quality at the Bureau of the Census. *Fourth Annual Research Conference, US Bureau of the Census, Arlington, Virginia*. Washington DC, USA: US Government Printing Office.

Morton, G. M. 1966. *A Computer-oriented Geodetic Database and a New Technique in File Sequencing*. Ontario, Canada: IBM.

Pizano, Arturo, Allen Klinger and Alfonso Cardenas. 1989. Specifications of spatial integrity constraints in pictorial databases. *Computer* (December) 59–71.

Sellis, Timoleon, Nicholos Roussopoulos and Christos Faloutsos. 1987. The R$^+$-tree: dynamic index for multidimensional objects. *Proceedings of the Thirteenth Very Large Data Bases Conference, Brighton, UK*.

Smith, Terence R. and Peng Gaó. 1990. Experimental performance evaluations on spatial access methods. *Proceedings of the Fourth International Symposium on Spatial Data Handling, Zurich, Switzerland*, pp. 991–1002.

Van Oosterom, Peter and Eric Claasen. 1990. Orientation insensitive indexing methods for geometric objects. *Proceedings of the Fourth International Symposium on Spatial Data Handling, Zurich, Switzerland*, pp. 1016–1029.

White, Marvin. 1986. *N-trees: Large Ordered Indexes for Multi-dimensional Space*. Report of the Statistical Research Division, US Bureau of the Census, Washington, DC, USA.

Wilson, Peter R. 1985. Euler formulas and geometric modelling. *Institute of Electrical and Electronics Engineers Transactions on Computer Graphics and Applications* 5(8): 24–36.

16
Hypermedia

Multimedia spatial information systems and hypermaps

Julius Caesar is interested in acquiring all information about any piece of the Roman Empire, in the form of maps, texts, photographs and voice information. Marcus Antonius proposed that this information should be organized into hypertext, while Cleopatra preferred hypermaps. She wanted to browse in cobwebs and in pyramids, but finally opted for navigation for solving queries.

In this and the next chapter we move closer to data organization for a higher semantic level. In this chapter, we deal with data of different media, for example photographs and sound, as well as text and graphic materials that, ideally, we would like in the same digital spatial information system environment. The organization of such diverse data through the hyperdocument approach is then presented as the specific topic of hypermaps. This chapter employs the relational and other conceptual approaches described in Chapters 13 and 14, and is a precursor to the discussion of intelligent spatial databases in Chapter 17.

16.1 HYPERDOCUMENTS

Hypertexts and multimedia hyperdocuments are an increasingly common type of documentation. Whereas most, but not all, conventional documents, essentially textual print media, have a logical and a physical layout structure organized sequentially and hierarchically, hyperdocuments are a modern version of non-linearly organized materials. That is, they are electronic documents with a direct access to information of diverse form by means of window presentation and mouse clicking on important words or other displayed information.

One forerunner of this modern form of hyperdocument is the printed atlas. The atlas is not necessarily read in a sequential fashion from beginning to end like a novel; it contains a set of maps at different scales; it may have insets providing detail of congested areas, and it often has ancillary information like a key and scale bar. An atlas may have textual details about the source of data, perhaps even explanations about a graphic technique used, and it may have map sheet numbers at page edges, as pointers, to move easily to adjacent places in space from the one map sheet currently viewed.

There are two principal facets to the domain of spatial data beyond the realm of textual information in print media:

1. The non-print media.
2. The non-sequential organization and access.

The non-print media, often referred to as **multimedia** in the jargon of the computer and electronics industries, comprise a variety of analogue and digital forms of data that come together via common channels of communication. Traditional audiovisual and text forms are assembled electronically, but the net can be cast wider to include seismic signals, sound, and even data produced via other human sensations, tactile or olfactory. We use the term **content portion** to refer to the individual unit for data, of whatever form. The non-sequential organization, or **hyper** structure refers to a form of communication beyond or over the linear style that is associated with most books.

16.1.1 Multimedia spatial data

In libraries the non-print media collection encompasses picture material in the form of photographs, filmstrips, 35 mm slides, videotapes or videodisks. Some printed materials may include or consist entirely of monotone or colour photographs. Many scientific fields have long used sketches, maps, graphs, charts, diagrams or other non-textual materials. For about twenty-five years many data have been remotely acquired in electronic form via field instruments recording such phenomena as river turbulence or soil conditions, and in digital form from spacecraft-borne sensors. Dealing with varied information is not a new dimension to many fields; but being able to integrate analogue and digital forms, and to work with print and non-print media, continue to challenge the electronic engineers and software designers.

In the context of spatial information systems there are many uses of non-print media, and some interesting applications blending the world of pictures, sounds and signals with the conventional text- and map-oriented databases, software and hardware. Among the possible domains of use of multimedia data in spatial information systems are:

Urban and regional planning
Environmental planning
Hazard prevention and management
Fire fighting
Estate management and sales
Tourist industry
Archeological studies
Geology and geomorphology studies
Teaching geography

Of the increasing numbers of uses combining images and maps, we can mention the integration of photologger data for highway conditions with the ARC/INFO geographic information system by the State of Wisconsin Department of Transportation in the USA (Fletcher, 1989); and the compilation of movies, two hundred maps, and sixteen hundred still pictures making up the school oriented videodisk, GTV, created by the National Geographic Society of the United States for teaching history and geography. A few years ago geographers, computer specialists, and others assisted the British Broadcasting Corporation in the production of a pair of videodisks containing text, tables of numeral data, maps and images of different parts of Britain. Called the Domesday project, the effort commemorated the original survey of 1086 for William the Conqueror (Rhind *et al.*, 1988). Companies in Barcelona, Spain, Paris, France and elsewhere are marketing software products that allow the integrated use of text, graphic and image data.

But we have more than a combined use of multimedia data; we have also multimedia documents for which a **geographic access** is most important. By geographic access, we mean a coordinate based access in which by referencing a place or a region on a map, we can retrieve all information dealing with the identified point or zone. Referring to Figure 16.1, we need, for example, to demarcate a region by any arbitrary boundary defined by a random computer mouse clicking on a map image, as a way to retrieve not only numerical or alphanumerical data, but also documents, images, seismic measures and so on, 'grounded' at some point or in some region.

At the moment we often have huge collections of paper documents serving many purposes. These consist of maps, architectural drawings, engineering blueprints, land use zoning maps, sketches and the like for which an access based on coordinates is quite desirable. A typical query can be to retrieve all documents describing the totality or a portion of a territory defined by mouse-clicking on a point or on the boundaries of a

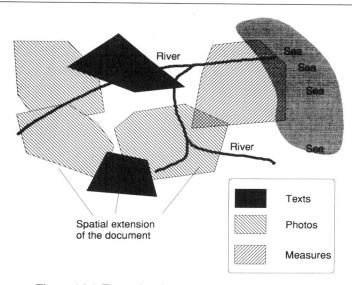

Figure 16.1 Example of coordinate-based documents.

zone. Consequently, we need to store, to archive, to select and to retrieve materials such as maps with different scales and topics, sketches with different approximation degrees, paper reports and documents with various origins and structures, photographs and three-dimensional landscape images with different geometric perspectives, aerial photographs and satellite images with unknown exact positions, seismic signals, temperature measures and all sorts of electromagnetic signals.

16.1.2 The hypermap concept

Traditional texts, linear or hierarchical in form, have a double structure: logical, comprising chapters, sections, paragraphs and so on, and layout, consisting of volumes, pages, blocks and so on. In contrast, hypertexts are organized by **semantic units** called nodes and associations between nodes referred to as links.

The semantic units may be simple elements like a single word or proper name, or may be more complex like a table of numbers, or a map, or photograph, or a paragraph of words describing an intellectual concept, or an entire document or file of data. Most users of hypertext favour using **nodes** which express a single concept or idea. A hypertext system invites the creators to modulate their own thoughts into units in a

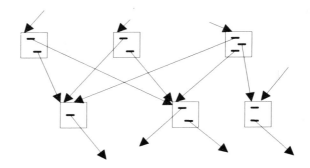

Figure 16.2 A cobweb of semantic units.

way that allows an individual idea to be referenced elsewhere. A node may have several successors and predecessors; readers can choose their own path of reading.

Generally, a hypertext possesses a great deal of short semantic units and every unit is connected to others by means of reference links. **Links** can be of several types, such as:

1. To connect a node reference to the node itself.
2. To connect an annotation or source citation to a document portion.
3. To provide relationships between two objects within the same document.
4. To connect two successive document portions.

There is in this network of nodes a large quantity of information in separate fragments which can be linked in different ways, and for which only a few items may be used at any one time. Linked by semantic associations, the set of information is inherently a loose, unbound structure, something like the pages of a thematic atlas, with unlimited possibilities of connection, but with some tighter bindings, like an inset map for a smaller scale map. Knowledge of the substantive domain can reveal patterns of connection that make more sense than others, so that we create particular graph instances, called **webs** representative of some order.

With this form of linkages a hypertext has a network structure. When the number of references among the semantic units is high, this structure becomes a **cobweb** of links. Figure 16.2 depicts a cobweb of hyperdocuments in which nodes appear as squares, a reference location as a dash and links as unbroken lines.

In an electronic environment the connecting is done at two levels. Icons, windows or other forms on a screen represent entities in a

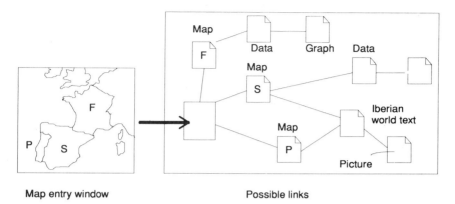

Map entry window   **Possible links**

Figure 16.3 An example of an hypermedia network.

database, and can be linked by graphical means on the screen, while in the database there are connections by pointers or some other device. For navigating from one node to another, we have to click a reference word or picture which constitutes a bridge to another semantic unit. So, by a man–machine interface, especially based on windows and a mouse, it is possible to move from one semantic unit to another. This provides a way to 'read' electronically a dossier about some topics, regrouping several small documents tightly together, and it provides a means for learners to make associative linkages between apparently unconnected, discrete pieces of information. This hyperdocument concept is far from the conventional printed book with a fixed hierarchical structure.

The same structure can be built not only for text material (hypertext), but also for multimedia contents such as images, graphics, sounds and signals. In this case, we deal with hyperdocuments and hypermedia. Typically, browsing a hyperdocument is done via a path of several mouse clicks as for dictionaries in which the reader is going from one word to another. We can, by this approach, intertwine the main text and footnotes, figures, tables and images as suggested by Figure 16.3.

To illustrate this situation we present first a crude relational model:

WEB (Document_node-ID, (Word_locator (To_document_node-ID, Type_of_link)*)*, (From document_node-ID, Type_of_link)*)

This model can be useful for tracking navigation paths through the database and for recording uses of particular content portions. The Type_of_link element refers to the nature of the path from one node to another. The Word_locator is the word, graph unit, or pixel, or other element, in the first document, akin to the HYPERCARD 'button' or the keyword in HYPERTIES. Entry to the hyperdocument may be included via:

HOME_NODE (User-ID, Document_node-ID)

The main realizations of the hyper concepts are:

1. A hypertext is a network of content portions which are always textual.
2. A hyperdocument is a network of content portions which can always be displayed on a screen or presented via a loudspeaker.
3. In hypermedia, the content portions are like the hyperdocument but can also include digitized speech, audio recordings, movies, film clips and presumably tastes, odours and tactile sensations.

Let us use the term **hypermap** to refer to multimedia hyperdocuments with a geographic coordinate based access via mouse clicking or its equivalent. It is the objective of this chapter to define exactly the principles of hypermaps, their applications, and give some highlights on their physical structures. First, though, we discuss the representation of multimedia data and the organization of maps and related documents into libraries.

16.2 MULTIMEDIA IMAGE DATA

Images, that is, aerial and landscape photographs, infra-red images, film based pictures, satellite data and other kinds of remotely sensed digital data, are perhaps some of the most important kinds of multimedia information. For them the most complex matters to deal with are:

1. Image representation, essentially in order to reduce storage space occupancy.
2. The representation of dynamic images like movies.
3. Modelling and retrieving picture objects.
4. The creation of operators for efficient image manipulation.

Since much data capture and storage of image information requires specialized electronic hardware, analogue or digital form, our review will not be comprehensive as it concentrates on the data modelling and data organization aspects.

16.2.1 Image modelling

We approach the topic of image representation via relational modelling,

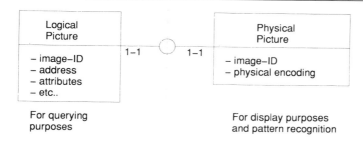

For querying   For display purposes
purposes   and pattern recognition

Figure 16.4 Relationships between logical and physical pictures.

thinking of logical pictures and elements. A general relational logical model is:

IMAGE (Image-ID, Format, Resolution, Capture_date,
    (X, Y, Colour)*)

in which we have a matrix of picture elements identified by row and column identifiers, Y and X, and their content represented by a colour, or a grey scale value, and ancillary information as to data capture conditions. A more compact form of the relation, ignoring the supplementary data is:

ABBREV_IMAGE (Image-ID, (X, Y, Colour)*)

A common image architecture splits the picture into two parts (Figure 16.4). The first, generally feasible for integration into a relational databases, refers to the image identifiers and some attributes. The second corresponds to physical storage, possibly on an analogue videodisk or an optical erasable disk. While the first of these is used for query purposes, the other is the basis for pattern recognition processing or picture display.

16.2.2 Physical encoding

The physical picture may be encoded in several forms, of which the common are: raster or bitmap, run length, quadtree and pyramid. The **bitmap**, the easiest physical encoding technique, stores each picture element or pixel value in consecutive order, usually row by row, beginning in the top-left corner (Figure 16.5a). Since all elements are mentioned without any regrouping, no coordinates are necessary:

BITMAP (Image-ID, X_format, Y_format, (Colour)*)

in which the format refers to the horizontal and vertical sizes, in number

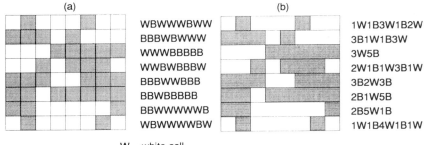

W = white cell
B = black cell
3 = number of cells in a run in a row

Figure 16.5 Bitmap and run-length encoding. (a) Bitmap. (b) Run-length.

of pixels. Customarily, arrays are 512 × 512 for a satellite image, or 1,980 × 1,980 for an air photograph. To solve the query as to the content (colour) of a pixel we search by a direct addressing system such as via arrays at the bit level.

Run-length encoding recognizes that there may be stretches of like content in either horizontal or vertical dimensions (Figure 16.5b), so it regroups runs of pixels with the same colour, generally operating row-wise:

RUNLENGTH (Image-ID, (Y, (Min_X, Max_X, Colour)*)*)

This representation is valuable if we deal with only a few categories (colours) and have homogeneous blocks. A run-length encoding is fine for the Great Lakes, all bodies of water, or the desert of Australia, but is not so good for checkerboard patterns.

Instead of grouping along a line or column, two-dimensional blocks, of which the quadtree is the most well-known, may be used. The pyramid model can be used to convey different levels of resolution, perhaps for having alternative representations of all details and general patterns. Usually applied to bitmap encoding, this structure increases the occupancy by only one-third, so that the extra costs here may be offset by the benefit of having different resolutions available. At times it may be necessary to have overlapping pyramids, and sometimes there are difficulties regarding continuity at boundaries of the pyramid blocks.

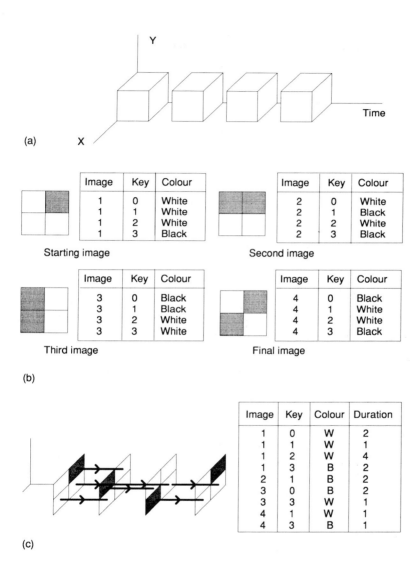

Figure 16.6 Example of image sequences encoded by linear quad-trees. (a) Sequence of cubes with a 3D Peano curve with breakpoint (temporal octtrees). (b) Example of succession of images. (c) Example of image sequence encoded with a linear quadtree.

16.2.3 Dynamic image models

Movies, dynamic image models, or animations are represented as three dimensonal images with a time element:

FILM (Film-ID, (Time, Image-ID, (X, Y, Colour)*)*)

Encoding may use methods previously discussed, but the quadtree and hypercube are especially valuable if there are only a few colours (Figure 16.6). The intent is not to display the entire image at each time slice, usually 24 or 30 images per second, but to show the changing pieces (Figure 16.6c). For quadtrees the relational representation is:

FILM-PR (Film-ID, Image-ID, (Peano_key, Side_length, Colour, Duration)*)

While this kind of representation is effective for movie displays because each image is reconstituted from its predecessors, it is not so valuable for sole image display. However, each whole image can, of course, be encoded at given time intervals. There are, though, substantial electronic limitations in achieving good signal synchronization.

16.2.4 Picture object modelling for retrieval

For many purposes it is necessary to store and retrieve picture objects, the individual components within a physical picture, for example a person or house. On the one hand, objects must be recognized; on the other, thay have to be retrieved. We pass over the object recognition as being beyond the scope of this book to look briefly at modelling once objects are identified:

IMAGEDATA (Image-ID, (Object-ID, Type, Locator)*)

where locator refers to one of several alternatives: by an enclosing rectangle or other regular figure, by a boundary representation, or by a Peano key based area (Figure 16.7). At times, objects may be grouped into object categories, and then accessed by key words, like searching for all wetlands.

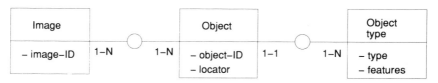

Figure 16.7 Relationships between images and objects.

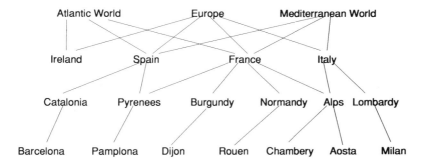

Figure 16.8 Example of a thesaurus.

Possibly a **thesaurus** can be of value for retrieval purposes. A thesaurus links keywords via a graph structure in which two kinds of association exist:

1. The synonym relationship for equivalent keywords like forest and wood.
2. The generalization relationship which extends or reduces the meaning of a word.

For example, as in Figure 16.8, France can be generalized by Europe, the Atlantic World and so on, and can be broken down into regions like Britanny. The same label, for instance, Mississippi, from a different part of the world, can be given to several geographic objects. The same geographic entities can have several names, perhaps in different languages, for instance, London, Londres, Londra, or by having a change in name during time, as for Saint Petersburg, Petrograd and Leningrad. Name relationships are, of course, not only part of the domain of picture data; they may also be desirable for other databases for spatial data.

Currently, a limitation to more widespread use of image data is the large amount of physical storage required. Compression techniques like run-length encoding may be helpful, but other methods are being sought. Images may be approximately reconstructed after compression by statistical or other means, or they may be represented exactly by techniques without loss of information, such as run-length encoding. In general for spatial information systems purposes, whether the data are images or maps, bitmap encoding is better if there are many different values in a picture; run-length encoding is better if there are a moderate number, say under twenty, and quadtrees for the fewest, say just a handful of values on a grey scale or colour image. Quadtrees and run-length encoding are awkward if some spatial operations like rotation are

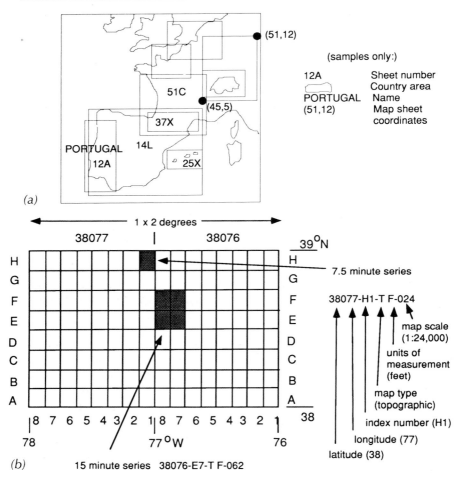

Figure 16.9 Map retrieval referencing. (a) Possible methods. (b) Actual case of some US Geological Survey maps. (Taken from Maryland, Delaware, and District of Columbia Catalog of topographic and other published maps, US Geological Survey, Reston, Virginia, USA, December 1987.)

needed, requiring rebuilding of the database. Some research laboratories are currently experimenting with fractal geometry for data compression for images.

Otherwise, there have been some developments in a database management system context. Recalling our example of the use of the long data type for coordinates in the GEOVIEW design, we mention now that a bulk storage data type can be used for image data. Indeed, the growing interest in and viability of multimedia databases has led to the creation of the BLOB, a binary large object data type. It is special in being able to contain unformatted objects with no inherent size limit. In contrast to the 64 kilobytes restriction mentioned in the GEOVIEW design, the BLOB apparently can contain large chunks of data, like motion video sequences measured in hundreds of megabytes, or page size colour photographs consuming 2–10 megabytes of storage space.

16.3 ORGANIZATION OF COLLECTIONS OF MAPS AND IMAGES

Several structures based either on names or locators are possible for the computer organization of huge quantities of maps and images. That is to say, firstly, that we can access documents by feature names or place names, retrieving a document if a scan of the object contents for that document reveals the place or feature. Or, secondly, retrieval may make use of some positional referencing data such as longitude/latitude coordinates for the corners of a map (Figure 16.9).

In a well organized map library, all maps have an identifier and a locator, and for a well developed digital spatial information system each cartographic object will have one or more names and one or several locators. The geographic information system ARC/INFO has a concept of a map library, allowing any chunks of space, regular or irregular in shape, to be demarcated as a map sheet. Each map sheet has real world entities encoded spatially and by a user-provided name. Ideally, there will also be metadata about the different origin, processing trail, borrowing and returns of the digital equivalent of the map sheets in a paper map library.

Maps and other documents with data needed for a spatial information system may have several properties that affect their organization and use. They may be:

Single sheets with only one map
Single sheets with one or more insets or outsets
Single sheets or pages that are part of a series (for example, the tourist maps of the Ordnance Survey of Britain, or a collection (atlas), or a road travel *Triptik*)
Joined exactly at edges or having overlap
Of different scales or orientation
Of different shape and size
Of different map graticule or projection

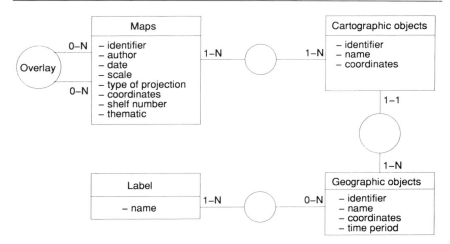

Figure 16.10 Structure of a map library.

Of different perception or viewing point
Part of other kinds of documens or single pages

Consequently, both the logical and physical organization must be considered in an information system allowing access to objects as well as documents.

A conceptual model for a simple map library is given as Figure 16.10. Original sketches, maps and photographs, having content and locators, may be classified and then accessed by type of document or thematic content, or may be a basis for more 'pseudo' maps created via spatial information system tools. Any one map may have several cartographic objects, which may be contained within one map or several. (We all know that our own neighbourhood always happens to fall at the corner intersection of four different pages in a city street atlas.) Real world geographic entities can encompass several map objects, as demonstrated via the conceptual model.

The artificial tiling segmentation associated with most analogue map series may be advantageous in a physical sense of facilitating the creation of manageable paper documents (except perhaps if you are an airline pilot trying to deal with many rolled paper maps), but does have limitations in the splitting of logical entities by map sheet edges. We have already noted earlier in this book of the need for a double encoding to preserve the topological integrity of the geographic objects while maintaining the physical unit of the map sheet or the tile.

Extending our context beyond paper maps, suppose a municipal

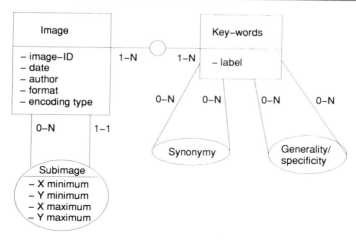

Figure 16.11 Example of an image database structure.

library needs to organize an image database, incorporating several kinds of photographs with different colours and formats. Each photograph is described by a set of keywords and is located in both space and time. We also recognize that there may be a need to group objects into thematic sub-images, requiring identification by box coordinates of the particular objects in the main image. The conceptual model (Figure 16.11) includes both this aspect and the use of a keyword extension. Not only are synonyms desirable but also the degree of specificity associated with them.

If it is necessary to locate objects explicitly, then a different model is needed (Figure 16.12). Assuming the source documents are satellite images, then a representative scenario includes the identification of one or more regions on the image, and provision of a locator. Particular phenomena, for example cities or rivers, may then be recognized as occurring within the region. Collections of aerial photographs are modelled in terms of crude and geometrically corrected images (Figure 16.13).

Currently, most map and photograph collections are not established in the ways presented above by the examples of relational models. Not only are most collections not inventoried and stored in a database, but where they are, access is largely via words, although there is some software that allows graphic referencing by geographical windows. As one example, the CARTONET system designed at the University of Edinburgh (Morris, 1987) has graphic geographic access; the reader can use a screen cursor to identify a geographic region on a world or continental map background,

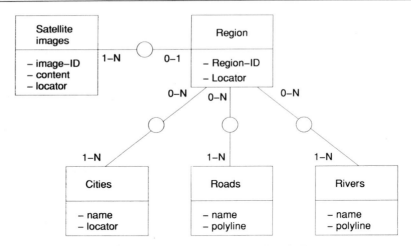

Figure 16.12 Example of a picture description.

or smaller regions by graphic zooming. The database of records of information about map series, individual sheets or photographs contain coordinates for the geographical limits, usually expressed as latitude–longitude coordinates. A simplified model for the CARTONET database is:

MAPDOC (Map_series-ID, (Map-ID)*)
MAP_SERIES (Map_series-ID, Title, Date, Publisher, Scale)
MAP_SHEET (Map-ID, X_1Y_1, X_2Y_2, X_3Y_3, X_4Y_4, Scale)

where the XY (longitude, latitude) coordinate pairs indicate the four corners of specific map sheets because not all maps are square and access may need to use more than the diagonal range deduced from only two corner points.

Indexing and abstracting services like Petroleum Abstracts of Oklahoma and GeoRef, an online bibliographic database service based on a world-

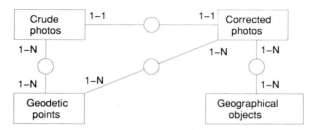

Figure 16.13 Typical structure for an aerial photograph database.

wide earth science file maintained by the American Geological Institute, use word-based vocabulary for geographic concepts, not map-based representations, for locating bibliographic records about printed documents of interest (Hill, 1990). In such systems the spatial access is usually done via keywords, possibly with a hierarchical thesaurus, as in the case of Petroleum Abstracts' Geographic Thesaurus. Sometimes, as is the case of GeoRef, retrieval may be based on text data geographic coordinates, matching a range established by a rectangular extent against coordinate values stored as character data in bibliographic records.

In the case of global hypermaps, it is interesting that we need a spatial indexing mechanism that takes into account the fact that the earth is spherical. For this reason it seems appropriate to take advantage of the tessellation system proposed by Dutton (1990). Another possibility is to use latitude and longitude expressed as spherical coordinates. In this case, portions of the earth can be delimited by, for example, particular parallels and meridians expressed as Longitude_minimum, Longitude_maximum, Latitude_minimum, and Latitude_maximum, with the possibility of building a set of R- or R^+-trees.

Among the many issues involved in the further development of spatially based access to map libraries or printed material containing geographical locators or spatial images we mention here just a few:

1. The boundaries of geographical regions may be indefinite.
2. The same places may have different names.
3. Names and even positions change over time.
4. There is a preference by users for word based searches.
5. There may be overlap of different delimitations for a given concept or spatial unit.
6. It may be desirable for integration of bibliographic databases with spatial data in geographical information systems.
7. The spatial indexing device may be crude or refined.

In summary, careful planning is required to get a good match between target area or object and source area or object. In the future, map libraries will perhaps be seen as an organization of archives of multimedia documents. In similar vein, many engineering companies today are faced with storing their technical drawings in libraries. The current world of computerized maps and drawings is different, but old paper items pose interesting challenges in how best to create a computer based archive.

16.4 HYPERMAPS

The special idea of hypermaps is to extend the hyperdocument concepts by integrating geographic referencing. Should geographic references remain at a literal level, such as by place names, they can be modelled by hypertext links. For many documents, though, this kind of reference is not enough to cover correctly the space involved; a coordinate based referencing method through a cartographic system is desirable. We now examine the implications of coordinate based documents according to two features: spatial referencing and spatial queries.

16.4.1 Spatial referencing of hyperdocuments

Spatial referencing of hyperdocuments presents two aspects:

1. Spatial referencing of document nodes.
2. Spatial referencing of maps and other cartographic documents.

A node, supposedly representing a single or a few semantic concepts, may have one or many spatial references. For example, a particular document node can describe a geodetic point, a region or a linear feature. Some other nodes can compare points and regions; thus, we can have multi-point document nodes, multi-line document nodes, multi-area document nodes, and several other combinations.

Let us take the conventional way of dividing spaces into zero-dimensional elementary objects (for point), one-dimensional (for line), two-dimensional (for area) and three-dimensional (for volume) entities, as in Figure 16.14. So if documents can be associated to any elementary object, here referred to as zones, elementary objects can also be associated to documents. A sort of many-to-many relationship can then be defined. In order not to have problems with elementary objects, let us suppose that spatial referencing is made through elementary pieces of space, via a tessellation. Indeed, any kind of spatial indexing can be used,

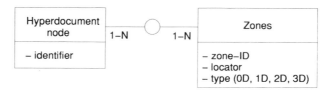

Figure 16.14 Relationships between nodes and zero-, one, two-, and three-dimensional spatial zones.

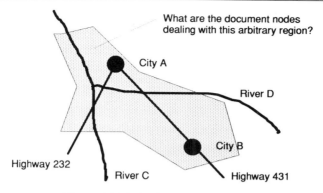

Figure 16.15 An hypermap query.

as seen in the discussion of Chapter 15 which is in fact quite relevant to the concept of hypermaps. For limited territorial extent, a tessellation of squares may be satisfactory; the quaternary triangular mesh provides a possibility for global hypermap referencing.

16.4.2 Spatial queries for retrieving hypermap nodes

To retrieve a spatial document, the basic starting point is a map for which we query by delimiting a region by means of a mouse. Four types of spatial query are of relevance: point, buffer zone, segment, and region query. So, navigating in a hypermap must combine these types of query and conventional hyperdocument scanning via reference links (Figure 16.15).

When making a hypermap query, several situations can occur (Figure 16.16). We can see that documents D2 and D3 are fully relevant, document D1 is irrelevant, and documents D4 and D5 are partially

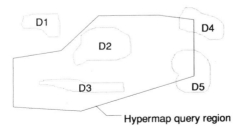

Figure 16.16 Relevant, partially relevant, and irrelevant documents.

relevant. Sometimes, it is difficult to have the exact demarcation zone for a document node, in which case a membership degree can be given to several embedded zones. In the example given in Figure 16.16, notice that document D2 has 100 per cent fit, but D4 is only about 20 per cent.

However, the main challenge is how to organize both locational relationships and document relationships. We have two aspects to consider:

1. The document-to-map relationships.
2. The map-to-map relationships.

We have two alternatives for each: either using Peano relations or using map pyramids and R-trees.

16.4.3 Encoding hypermap spatial references by Peano relations

Here, we deal with documentary objects with spatial relationships in connection with elementary spatial objects. We have shown in Chapter 14 that Peano relations are a good way to organize this kind of information because they replace the computational geometry accesses by tuple algebra algorithms. With this approach, areas have to be modelled by quadtrees:

DOCUMENT (Document_node-ID, (Peano_key, Side_length)*)

This approach is interesting for the document-to-space and the map-to-map relationships. When the membership degree of a quadrant is not 100 per cent, we can transform the previous Peano relationship by adding this membership degree as an attribute:

DOC (Node-ID, (Peano_key, Side_length, Membership_degree)*)

16.4.4 R-trees and map pyramids

The basic assumption is that a document space is delimited by a rectangle. When the space is not rectangular, it must be delimited by a set of rectangles in order to achieve some resolution level. The R-tree organization, based on rectangles encompassing several other rectangles, fits in well with the organization of sets of maps. An example for map organization is presented as Figure 16.17.

The pyramid structure seems to be effective for storing and organizing maps and similar documents, as recently proposed for a map library for fire fighting. The map pyramid structure is a special use of the image

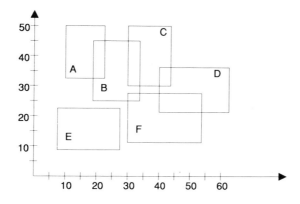

Figure 16.17 A located map set.

pyramid structure and R-trees already discussed in section 7.6. Node documents can be grounded to several documents with different scales. In addition, geographic documents that possess inset maps can also be integrated into this structure.

Starting from the ideas of pyramids for images and applying this structure to maps (Figure 16.18) we obtain:

1. The base of the pyramid is formed by maps with the smallest scales.
2. Each scale corresponds to a pyramid level.
3. The maps in the upper level involve the maps at the lower level.
4. Neighbouring maps at the same level are connected.
5. Maps may overlap or be askew.

The example showing three levels by the scales a, b and c, and a perfect matching to higher levels from the lower, can in practice become much more complex as a result of map overlaps, a failure to coincide with orthogonal reference system axes, or unusual shapes.

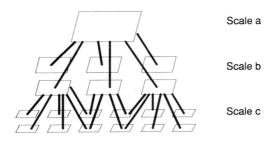

Figure 16.18 Map scales and pyramid levels.

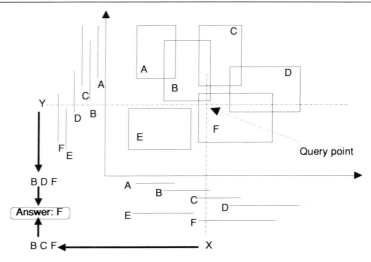

Figure 16.19 Solving the point query for the R-tree structure.

16.4.5 Navigation in hypermaps

Navigating in a hypermap has two aspects: thematic navigation, as is usually done in hypertext and in hypermedia, and spatial navigation which is particular to hypermaps. By **thematic navigation**, we mean the way to navigate from a text to something else. In order to handle several

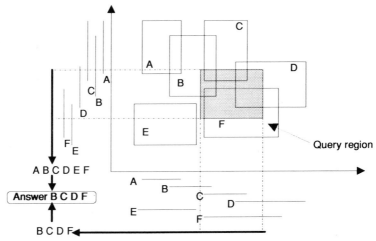

Figure 16.20 Solving the region query for the R-tree structure.

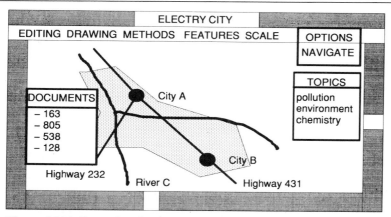

Figure 16.21 Example of a human-machine interface for hypermap navigation.

kinds of user, several modes of navigation can be implemented. Thus we can define a novice mode when the user is a newcomer, and an expert mode when the user has a good background in the domain.

For solving the **spatial navigation**, let us first examine the point-in-polygon query. To retrieve whether a point P_0 belongs to one or some rectangles, we compare its x-coordinates to the rectangle x-coordinates, and then also for the y-coordinates. As a result, we obtain two lists of possible rectangles. The answer is given by the intersection of both lists as depicted in Figure 16.19. The region query is answered similarly (Figure 16.20).

Via Figure 16.21, an example of a human–machine interface emphasizing both the spatial and the thematic aspects of hypermap navigation, we provide a context as to how this environment might be. The map is the window to the world, the entry point to the hypermap system, the access to a multimedia resource. Imagine a tourist visiting New York City, browsing through a map showing the main attractions by pictorial icons. A click or touch on a symbol for the Statue of Liberty provides a connection not only to a picture of that gift from the people of France, but also a connection to a different part of the world, possibly to retrieve some textual information describing the context of that donation.

16.5 SUMMARY

We have defined a hypermap as a combination of cartographic documents and hyperdocuments for which a coordinate based access system is needed. Then, we have shown how to organize a hypermap, emphasizing the aspects of document-to-map relationships and map-to-map relationships. The double consideration of thematic and spatial linkages, so vital and traditional a part of geographic enquiry, is only now being adequately dealt with in a digital environment. Spatial indexing techniques are most important in this regard. Indeed, a hypermap can be seen as a mixture of hyperdocuments and spatial indexing. The utilization of image data is more a hardware question than a database logical structuring issue; the navigation is a cognitive matter.

In the academic community there is a strong belief that hypermedia based criss-crossing of semantic landscapes is a very beneficial form of learning. Geographers have long practised and preached that several journeys across a physical landscape were necessary for a full, multi-viewed understanding of the earth as a home of man. Informaticians are providing tools for associative browsing and particular queryings of the semantic content of spatial information systems that utilize graphic and picture data substantially and even primarily, not just text and numbers. Hypertext and hyperdocuments are not necessary if all we want to know is a street address or a latitude/longitude coordinate. However, if we want true learning and understanding in the spatial domain then there is definitely a place for a hypermap. We think that in the near future hypermaps will be widely used in spatial information systems, or more exactly, that the spatial information systems of the future will be based on a hypermap concept.

16.6 BIBLIOGRAPHY

Extensive surveys of the hypertext concepts are provided by Conklin and by Shneiderman and Kearsley; while Nielsen offers a more comprehensive review of hypermedia in general.

Bush, Vanevar. 1986. As we may think [originally in] *Atlantic Monthly*, July, 1945, reprinted in CDROM.: the new papyrus, Microsoft Press, pp. 101–108.

Conklin, E. Jeffrey. 1987. Hypertext: an introduction and survey. *Computer*; September, 2(9): 17–41.

Dutton, Geoffrey. 1990. Locational properties of quaternary triangular meshes. *Proceedings of the 4th International Symposium on Spatial Data Handling; Zurich, Switzerland*, pp. 901–910.

Fletcher, David and Tom Ries. 1989. Integrating network data into a transportation oriented geographic information sytem [paper presented at]. *The Annual Conference of the Association of American Geographers*; March 21, Baltimore, Maryland.

Healey, Richard G. and Barbara A. Morris. 1987. CARTO-NET: a relational database approach to automated map cataloguing. *Cartographic Journal* 24(1).

Hill, Linda. 1990. Access to geographic concepts in online bibliographic files: what if we had a graphic interface? *Proceedings of the University of Pittsburg 1990 American Society for Information Science Student Conference*; March 14, University of Pittsburgh, Pittsburgh, Pennsylvania, pp. 55–68.

Laurini, Robert. 1989. Spatial knowledge, engineering and reasoning in geomatic applications. *Ekistics* (September/October) 56: 353–361.

Laurini, Robert and François Milleret-Raffort 1990. Principles of geomatic hypermaps. *Proceedings of the 4th International Symposium on Spatial Data Handling*; July 23–27, Zurich, Switzerland: pp. 642–651.

Laurini, Robert and R. Hori. 1989. Spatial information system for fire-combat. *Proceedings of the 13th Urban Data Management Symposium*; May 29–June 2, Lisbon, Portugal: pp. 79–90.

Morris, Barbara A. 1987. CARTO-NET: A Cartographic Information Retrieval System. British Library Research Paper, No. 33. London: The British Library.

Nielsen, Jakob. 1990. The art of navigating through hypertext. *Communications of the Association for Computing Machinery* (March) 33(3): 296–310.

Nielsen, Jakob. 1990. *Hypertext and Hypermedia*. London, UK: Academic Press.

Pinon, Jean Marie. 1989. Multimedia communication system. A tutorial presented at *The 13th Urban Data Management Symposium*, Lisbon, Portugal; May 29–June 2.

Rhind, David, Peter Armstrong and Stan Openshaw. 1988. The Domesday machine: a nationwide geographical information system. *The Geographical Journal* 154: 56–68.

Shneiderman, Ben and Greg Kearsley. 1988. *Hypertext: an Introduction to a New Way of Organizing and Accessing Information*. Reading, Massachusetts: Addison-Wesley Publishing.

Smith, Karen E. and Stanley B. Zdonik. 1987. Intermedia: a case study of the differences between relational and object-oriented database systems. *Proceedings of the Object-Oriented Programming System, Languages, and Applications 1987 Conference*; pp. 1–16.

Thompson, Derek. 1990. G.I.S. – a view from the other (dark?) side: the perspective of an instructor of introductory courses at university level [paper presented at]. *The GIS Education and Training Conference*; March 20–21, Leicester University. Leicester, UK.

Thompson, Derek and Charles Murphy. 1975. *Atlas of Maryland*. College Park, Maryland: University of Maryland.

Utting, Kenneth and Nicole Yankelovich. 1989. Context and orientation in hypermedia networks. *The Association for Computing Machinery Transactions on Information Systems* 7(1): 58–84.

Wallin, Erik. 1990. The map as hypertext: on knowledge support systems for the territorial concern. *Proceedings of the First European Geographical Information Systems Conference*, Amsterdam, Netherlands; April 10–13. Amsterdam, Netherlands: pp. 1125–1134.

17
Spatial Knowledge

Intelligent spatial information systems

After having made their spatial databases with the relational approach, John Johnson in the USA, Ivan Ivanovitch in the USSR, Juan Juárez in Spain, Jan Janssens in the Netherlands and Johann Johannsohn in Germany discover that all information regarding a single plot of land is disseminated into an unknown number of tables. So the international expert Ohannes Ohannessian proposes that they should regroup all this information into a single dossier. And Giannis Giannopoulos examines all consequences of this interesting idea.

Mohamed El Raisuli tries to simulate town planner reasoning by means of an expert system. Jack Colombo tries to deduce other information from raw data in a database, Moise Rosenblatt is trying to understand images and visual descriptions. Fran Redford is looking for edelweiss in the Alps. Gerry Mander thinks he can get a computer to do all boundary drawing for him. Nicos Grecopoulos is inspired by geometric reasoning and its application to geomatics. Finally, Geo Matix dreams about 'intelligent' spatial information systems.

With so many varied geomatic needs for which data manipulation is complex, we require reasoning more than numerical computing or retrieving data. For this, new kinds of computer software seem to be desirable. So, in this chapter, we will examine the cross-fertilization between the fields of artificial intelligence and spatial information systems, drawing on the former's attention to not only raw data but also higher level knowledge. Developing some ideas already presented in Chapter 12, the chapter first presents details about object-oriented databases, leading to a review of the representation of spatial knowledge. Finally, with the aid of examples, we describe the process of reasoning to the solution of spatial problems.

```
INTELLIGENT      > OBJECT ORIENTATION
SPATIAL          > HYPERMAPS
INFORMATION      > REASONING FACILITIES
SYSTEM           > FRIENDLY
                   INTERFACES
```

Figure 17.1 Components for an intelligent spatial information system.

17.1 TOWARDS INTELLIGENT SPATIAL INFORMATION SYSTEMS

Parsaye *et al.* recently wrote (1989) that intelligent database systems will encompass several new technologies including: object-oriented programming, expert systems and deductive facilities, hypermedia and interfaces. It now appears to be feasible to design new spatial information systems integrating such technologies (Laurini and Milleret-Raffort, 1990). Indeed, object orientation can very easily be conveyed to this domain without major problems except in dealing with some geometric representations. Expert system coupling and deductive facilities are important for spatial reasoning; they are perhaps the main steps towards intelligent spatial information systems. The utilization of hypermedia technologies already appears in the sense of hypermaps.

In other words, future 'intelligent' spatial information systems (Figure 17.1) must include:

1. Object orientation, the ability not only to store and manipulate data about the real world but also to model it more adequately, encapsulating its behaviour as well as its form, so as better to capture the semantics.
2. Hypermedia and hypermaps, that is the storage of any kind of multimedia spatial knowledge, and an environment for navigating in a person-compatible way through the contents of a base of information, however stored.
3. Facilities for logical deduction and geometric computation, that is reasoning facilities, giving the user more powerful tools for solving problems efficiently and effectively.
4. Excellent person-compatible interfaces to allow efficient interaction with the computer.

We have the feeling that much work must be done in order to specify intelligent spatial databases precisely. Indeed, this is a challenge for the current decade. For now, though, let us propose a framework for better understanding needs, trends and possibilities.

17.2 FROM RECORD-ORIENTED TO OBJECT-ORIENTED DATABASES

The object-oriented approach, a relatively new method in computing, is an attempt to improve modelling of the real world. Whereas previous modelling approaches were more record oriented, essentially too close to the computers, this new paradigm is a framework for generating models closer to real world features. The ideal would seem to be to provide an isomorphy, that is a **direct correspondence**, between real world entities and their computer representation. This chapter gives an overview of the object-oriented approach and presents some examples in spatial information systems.

17.2.1 Rationale and objectives

One main drawback of the relational data model is that information is scattered. In fact, the implementation of the normalization requirements tends to allocate data to small relations, so disseminating all information concerning a particular entity. For example, information pertaining to a person, a dossier or a building may be scattered into several separate but connectable tables. Consequently queries are cumbersome. Therefore, using the Structured Query Language, it is impossible to pose a query to retrieve *all* information regarding some specific person except if we can mention in the **FROM** clause the names of all relations in which we have the requisite information, together with joins in the **WHERE** clauses.

To avoid this condition, the object orientation tends to group all data concerning an object (Figure 17.2). This feature, known as **encapsulation**, combining the data and operations appropriate for those data, provides a simple basis for retrieving all information regarding a single object. In order to do so, a different model of reality is needed, one that is more oriented to the phenomenon itself than to the physically oriented record, say a row out of a relational table.

Considering a portion of the real world, comprising several entities, the object-oriented model will represent it with a one-to-one correspondence, whereas the conventional record-oriented data model implies a sort of one-to-many correspondence (Figure 17.3). Moreover, in the relational model it is possible to access any and all information pertaining to an object, located in any tuple, even if this is not desired. Protection is usually installed at the relation level or view level instead of at the object level. For instance, it is not possible to declare secret all information regarding Mrs Smith. However, in the object-oriented approach, because of its compact structure the protection can be posed at the object level, thus providing an easy way to secure all information concerning Mrs Smith.

While conventional database management systems have proved beneficial as data transaction record processing for the corporate business and institutional worlds, the domains of geometric and multimedia spatial

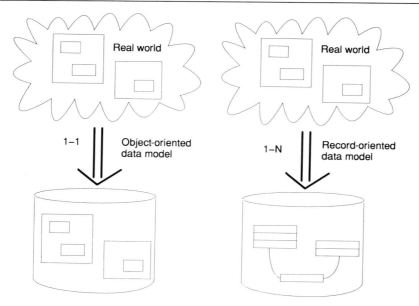

Figure 17.2 Differences between record and object orientation. (After Dittrich and Dayal, 1986.)

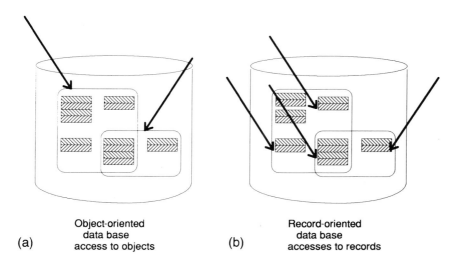

Figure 17.3 Different accesses. (a) Object-oriented. (b) Record-oriented. (After Dittrich and Dayal, 1986.)

information systems require more flexible and expressive approaches than even the relational model can provide. So object-oriented databases, developed as a result of the grafting of several roots including artificial intelligence techniques, particularly the concept of frames; object oriented programming languages, for example, *SMALLTALK*; abstract data types; operating systems, especially the feature of the protection of information; system design, and hardware features, offer a promise for the future.

In our view an object orientation refers to a computer system support for the independent existence of an entity. That is, a feature, simple or complex, is reached not via attributes but directly: it has an existence separate from descriptive thematic data. While this concept is not directly a goal of a relational database system, it is conceivable to think of a row in a table as an abstract entity reachable directly provided it has a unique identifier. So there are other characteristics differentiating object- and record-oriented database models.

These differences will become apparent when we discuss the main concepts of the model: classes, abstract data types, instances, methods and inheritance. Some of the characteristics of the object-oriented approach have already been discussed under the general topic of semantic data models (section 12.4). It was there, too, that we introduced the concept of the hypergraph based data modelling, an approach which underlies both the discussion that followed and the study by Salgé and Sclafer (cited in Chapter 12). These constructs go beyond the more elementary, albeit powerful, entity-relation modelling to richer semantic content, to include notions of generalization and specialization, associations, functions, events and behaviours.

17.2.2 Classes, subclasses and instances

Classification, in a scientific sense, is the assignment of individual occurrences of some phenomena to categories defined on the basis of selected attributes or functions. Those categories will be more or less homogeneous with regard to the defined properties of the phenomena. Thus, cities may be grouped into large or small, clean or dirty, or be located on a river estuary or inland on a river. Notwithstanding the many mechanical procedures for undertaking such classification, there are only a few basic ideas inherent in the process of building taxonomies. These are:

1. Classification may be singular, or use more than one attribute of real world entities.

2. Attributes may be static characteristics or dynamic properties like functions or behaviour.
3. The categorization may proceed from the bottom up, that is by grouping individual occurrences to make larger sets, or may subdivide existing groups into subgroups.
4. The resultant classification may be hierarchical or allow for overlap, that is an instance may lie in more than one group.
5. There is usually an index of the overall homogeneity of the class relative to others, or there can be an indication of the degree of differences among categories.
6. There can be a measure of the degree to which an individual belongs to one or more categories.

In spatial contexts, especially in the field of geography, there is often the imposition of spatial requirements to produce compact regions, or regions based on functional ties like trade areas, port hinterlands and the like (as suggested earlier in discussions of polygon entities), containment hierarchies, and overlaps of spatial units. In the case of a spatially qualified classification, then, we need locational information: spatial classification may be based on measures of contiguity or proximity, or linkages between different chunks of earth space, as in the example of districting.

Creating unambiguous comprehensive classifications is not easy for other than trivial problems. The hierarchical model, while simple and clear, is often superimposed on a more complex reality. Examples of some of the difficulties have appeared earlier as Figures 3.4 and 5.23, and in the case of political districting described in section 8.5.1. The hierarchical taxonomy, often referred to as nomothetic, can be very comprehensive, but its categories are disjoint, and mutually exclusive. In contrast, **polythetic categorizations** may allow for fuzzy boundaries and overlapping memberships.

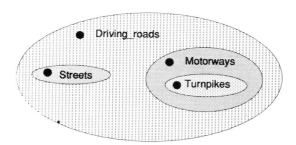

Figure 17.4 Example of classes and subclasses.

In an object-oriented database context, classification is a mapping of objects or instances to a common type. Objects can be regrouped into classes. By **classes** we mean a collection of objects with the same behaviour, that is responsiveness to some computing operations, or properties. We distinguish between the real world functions like travel for commuters, or erosion of landscapes, which could be directly modelled, and functions or behaviours defined only for computer encoded data, like drawing a polygon. Instances are particular occurrences of objects for a given class. Within classes, it is possible to define **subclasses**: an example is the splitting of the class Road into two subclasses: Street and Motorway (Figure 17.4). The resultant elements, using a conventional written formalism, are:

CLASS Motorway : SUBCLASS_OF Road
CLASS Street   : SUBCLASS_OF Road
CLASS Turnpike : SUBCLASS_OF Motorway

We also say that Motorway is a **superclass** of Turnpike, and Road is a superclass of Motorway. As discussed in Chapter 13, in the sense class → subclass we speak about specialization of a class, and in the reverse sense, class → superclass, we will use the term generalization.

17.2.3 Attributes and data types

Recalling from our discussions of databases in Chapter 9, an attribute for an entity has a data type such as integer, Boolean, float, string, money, bitmap and so on, associated with it; and each object is identified by a special attribute, the identifier (ID). As an example, the attributes of a road can be its number, its width, its origin and its destination. For the turnpike, we have its toll. A more formal description including data types is:

CLASS Turnpike   : SUBCLASS_OF Motorway
      Turnpike-ID : integer
      Name   : string
      Toll_fee   : money

A class is a sort of **prototypical object** defined with attributes. Instances of an object have all those class attributes in common. Attribute values can be defined at either the class or the instance level. For example, residents in a city can have a value of City_name in common, but will have their own personal names. Sometimes, as a practical matter, a default mechanism can be used in a database in order to simplify

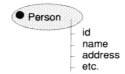

id
name
address
etc.

Figure 17.5 Example of attributes.

assigning values for attributes.

Attributes are not only basic data types, in the sense of physical form, but also issue from other classes. For example (Figure 17.5), we know:

CLASS City
        City-ID   : string
        Lord_Mayor : person

given that:

CLASS Person
        Person-ID   : string
        etc.

In some circumstances, combinations of attributes are necessary. In this situation, we can define a set or a list of attributes by SET_OF or LIST_OF:

CLASS City
        City-ID   : string
        Buildings   : LIST_OF Building

Sometimes, we have a list of other instances of the same class:

CLASS City
        City-ID   : string
        Twin_cities : LIST_OF City

Such devices serve to link several objects while saving on data repetition. For example, lists of names can be referred to as needed from appropriate objects.

In object-oriented databases and in some extensions of the relational database model the concept of user-defined data types is encountered. Sometimes referred to as **abstract data types**, these will be defined for a particular need, possibly constrained by the limitations of the particular database system, reflecting the underlying programming language used. For spatial information systems the more common data types that might be encountered are polygon, rectangle, region, line, point or image.

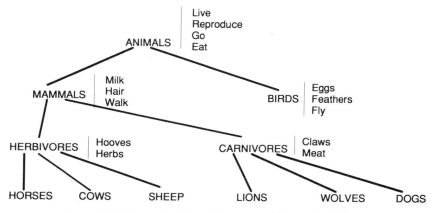

Figure 17.6 Example of an inheritance hierarchical diagram.

17.2.4 Inheritance

In classification hierarchies, an object in a subclass inherits all attributes of the corresponding higher level class. As an example, let us consider some animal categories, beginning with the subdivision into mammals and birds (Figure 17.6). Animals are characterized by being breathers of oxygen (shown as Live in the diagram) and having an ability to reproduce. Mammals are here characterized by hair and milk; birds by feathers and eggs. Mammals can be split into herbivores and carnivores. Herbivores are distinguished by having hooves and eating grass or other plants, whereas carnivores are characterized by having claws and by eating flesh. Among carnivores, some subclasses can be distinguished, such as dogs, wolves and lions; for herbivores, we can have horses, cows, sheep and so on.

Now, we can **instantiate**, that is identify instances for each class or subclass. For example, the White Horse belongs to the subclass Horses and so it is characterized by the stored attributes for all horses. But since Horses is a subclass of Herbivores, the horses inherit the attributes of having hooves and eating herbs. In other words, each object is automatically characterized by its superclass attributes and by the properties of progressively higher classes.

In particular cases, an object, say A, can be a subcategory of two different categories, say B and C. It will inherit attributes from both. Sometimes the same attributes can be defined differently in both supercategories, a condition known as **inheritance conflict**. For solving those conflicts, some special conditions must hold or a priority rule must

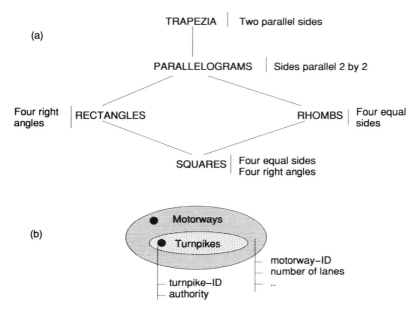

Figure 17.7 Examples of inheritance hierarchies. (a) Attributes in multiple inheritance hierarchies. (b) Attributes of classes when inheritance exists.

be invoked. For example, an ostrich is a bird which does not fly; a mule is an animal which cannot reproduce itself; a whale is a swimming mammal. In other words, some objects have attributes different from the class to which they belong; special attention must be paid in order to handle correctly such objects. In the object-oriented database jargon this condition is called overloading. For a spatial example, we have the case of the United States federal district, the District of Columbia (D.C.), having a representative in Congress, but that person is not a senator. That is, the D.C. fits, like the states, into a category of political units, but is an exception in the sense of being represented by other than a senator.

In some situations, we need to work with classes as subclasses of several superclasses. That is, a strict hierarchical classification does not exist; instead there is a state of **multiple inheritance**. Consider Figure 17.7a showing that the class Square is derived from Rectangles and Rhombs. In this theoretical example, no difficulty can occur. Nor is there any in the example of the turnpikes, for which the turnpike objects take on the characteristics of the motorways like number of lanes.

In some practical contexts, though, as depicted in Figure 17.8, a

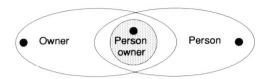

Figure 17.8 Example of a multiple inheritance.

Person-owner class is defined from the class Owner and from the class Person. In such a situation inheritance conflicts can occur. Indeed, supposing that both superclasses (Owner and Person) have attributes with the same name, but with different definitions or data types, there is a difficulty in choosing:

        CLASS Person-owner : SUBCLASS_OF Person, SUBCLASS_OF
                              Owner
              PO-ID   : integer
              Attributes   : . . .

    CLASS Person
          Person-ID   : integer
          Name   : string
          Date   : date

    CLASS Owner
          Owner-ID   : integer
          Parcels   : LIST_OF Parcel
          Date   : string

In this example, a request for data for Date.Person_owner, leads to an ambiguity about the type since the two invoked possibilities, Date.Person and Date.Owner, have different data types.

17.2.5 Links between classes and instances

As at the semantic level, so objects may be linked in different ways in a database, and classes of objects may be linked, such as that between owner class and parcel class shown in Figure 17.9. As noted in Chapter

Figure 17.9 Links between classes.

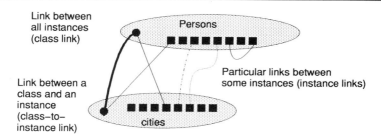

Figure 17.10 Differences between class links and instance links.

12, links may represent different kinds of association between objects or classes. They may be indicative of derived attributes (the most common kind), or possibly real world ownership, functional dependencies, an assemblage of components of complex entities, as in all graph edges which make up a network. In this object level modelling, though, we do not customarily mention cardinalities; we begin with the conventional assumption that links correspond to many-to-many relationships.

It can be interesting to create special links between instances when we do not want to generalize these links to all class instances. In the example given in Figure 17.10, a link is drawn between the two classes, meaning, in one sense, inhabitants. There are also two special links between just specific instances, such as the previous Mayor of the city and the richest person living in the city.

17.2.6 Methods

Objects are not only characterized by attributes but also by methods. We distinguish between the attributes (called **declarative knowledge** in some fields and which are the descriptive properties of the real phenomena as encoded) and the methods (the **procedural knowledge**, or some information as to what to do with those objects as encoded). The term **method** refers to an operation on the data, a procedure which can be applied to a class of objects. Thus we identify, and emphasize, the **combination** of data and operations that characterize the object-oriented model in contrast to the record-oriented model.

We can define general methods such as **CREATE, DELETE, RETRIEVE** or **MODIFY** an object. But some particular methods can also be defined for objects, including several operations very desirable for geometric objects, like **PICK, MEASURE, LOCATE** and **DRAW**. As is the situation with attributes, methods can be inherited (Figure 17.11). In this circumstance

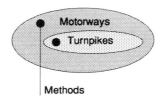

Methods

Figure 17.11 Inheritance of methods.

the operation must be valid not only for a class but also for all corresponding subclasses and instances. If necessary, only the method name can be defined at the upper level and the algorithm or particulars for the action can be given at the lower level. For instance, if we say 'go' for an animal object in the database, a bird will have the behaviour of flying, a mammal (except for a whale) will have the behaviour of walking, a snail will crawl and a kangaroo will jump.

For spatial information systems, methods may include those dealing directly with individual pieces of data, like **DELETE**; those modifying attributes, like **RECODE**; those invoking computations, as in **FIND AVERAGE DISTANCE**; and those for plotting, such as **SHADE**. For many of these, for example, the tasks for which many particular drawing procedures are general, the methods will ordinarily be defined at a higher level, say for a country, and then called by inheritance when needed for lower level objects like city traffic zones. Methods or daemons might be used for checking integrity constraints.

17.3 UTILIZATION FOR GEOMATICS

In this section, before discussing general issues about object-oriented spatial information systems, we will give some examples of the potential utility of the object-orientation approach in practical contexts in geomatics.

URBAN PLANNING EXAMPLE

For an example in urban planning (Figure 17.12), let us create a point class; defined by an ID and the coordinates. Bearing in mind that each attribute has only one particular data type, we have:

{CLASS Point
        Point-ID   : integer
        X   : float

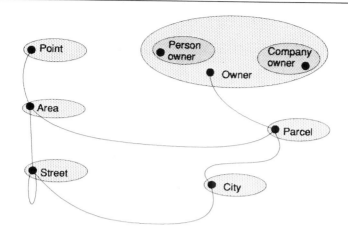

Figure 17.12 Example of an object-oriented model for a city.

> Y   : float
> Z   : float}

From this class it is possible to define a subclass Urban_point. Urban_points inherit all point attributes but also have another attribute Accuracy:

> {CLASS Urban_point : SUBCLASS_OF Point
>   Accuracy   : integer}

A mixtiline class can then be defined as a list of urban points:

> {CLASS Mixtiline
>   Mixtiline-ID   : integer
>   Points   : LIST_OF Urban_point}

Now, we can define an object Parcel with some other attributes such as Co_owner, for which we have a set of pairs (Owner and Share); we have also its geometry as a mixtiline:

> {CLASS Parcel
>   Parcel-ID   : integer
>   Co_owner   : LIST_OF {owner, share : percentage}
>   Purchase_date   : date
>   Geometry   : Mixtiline}

where owner is defined by:

> {CLASS Owner
>   Owner-ID   : integer

```
Owner_address  : string
Parcels        : LIST_OF Parcel}
```

But we can have several kinds of owner, for example persons or companies. We can then define two subclasses of Owner, each with different features:

```
{CLASS Person_owner  : SUBCLASS_OF Owner
       Owner_name     : string
       Date_of_birth  : date}
```

and:

```
{CLASS Company_owner  : SUBCLASS_OF Owner
       Company_name    : string
       Date_of_creation : date
       CEO              : string
       Capital          : integer}
```

A city also consists of streets. At the geometric level, it can be interesting to store street axes as mixtilines and the border of streets as two sets of mixtilines; at the topological level, crossroads can be invoked, giving a recursive definition of streets:

```
{CLASS Street
       Street-ID      : integer
       Street_name    : string
       Axis           : Mixtiline
       Right_border   : LIST_OF Mixtiline
       Left_border    : LIST_OF Mixtiline
       Cross_roads    : LIST_OF Street}
```

Now, we can describe a city as a set of several objects as previously defined:

```
{CLASS city
       City_name    : string
       Lord_mayor   : string
       Parcels      : LIST_OF Parcel
       Owners       : LIST_OF Owner
       Buildings    : LIST_OF Building
       Streets      : LIST_OF Street}
```

A very nice aspect to the object-orientation approach is to allow the definition of methods within objects. In our example, the methods might be to compute the parcel area or perimeter, or to draw parcels, streets or

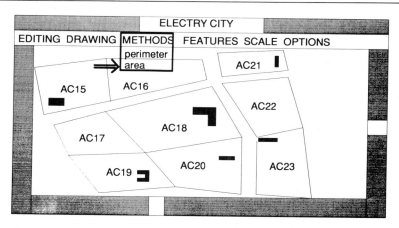

Figure 17.13 Example of an interface for accessing the object-oriented structure of an urban space.

other objects. Figure 17.13 not only shows this component, but also illustrates a graphic interface for manipulating and accessing the object oriented structure of an urban space.

An example for urban multimedia documentation, Figure 17.14, illustrates the use of the object-oriented approach for storing maps and images. The images may be either aerial photographs or satellite images, which we regard as two subclasses, each of them with different characteristics. Within those images, pictorial objects can be discovered, based on homogeneous textures. Similarly, some cartographic objects can be recognized for maps. Lastly, cartographic and pictorial objects can be linked to geographic objects, representing real urban objects in the city.

17.4 OBJECT ORIENTED DATABASES AND SPATIAL INFORMATION SYSTEMS

Some commercial spatial information systems have some of the features of the object orientation discussed above. However, in the world of software products, there are many ways to implement and deliver object-oriented features. The TIGRIS of Intergraph Corporation, the SYSTEM 9 of Prime Corporation, and the SMALLWORLD system of Smallworld Systems Ltd offer different design strategies. However, some prototypes are emerging for other products; and several non-spatially oriented, object-oriented systems are available commercially, including at least the

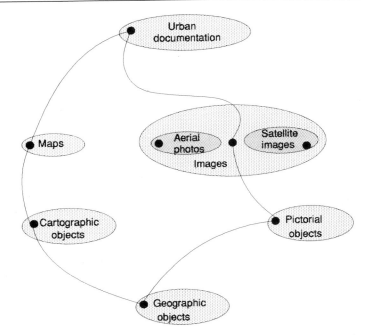

Figure 17.14 Example of object-oriented multimedia documentation.

following: GEMSTONE (Brelt, 1989), ORION (Kim *et al.*, 1989), O$_2$ (Deux, 1990), and IRIS (Fishman *et al.*, 1989).

In evaluating commercial offerings, or just the various models absent a particular implementation as a product, it is important to bear in mind several dimensions that invoke the idea of an object in spatial information systems, at this stage of development characterized by confusing uses of terminology. In a sense, all databases deal with objects, or units of data, but these units vary from the tables, records and fields of tabular types, to the relations and tuples of relational models, to the combination of the data and methods in the encapsulation unit of the object-oriented databases. Moreover, the computer working level unit does not have to have a direct correspondence to the conceptual spatial unit.

The matters to be aware of are:

1. The difference between an entity orientation and a field orientation for conceptualizing spatial problems.
2. The difference between an entity or object orientation and a thematic map layer system architecture.

3. The use of object-oriented techniques for graphic display purposes in contrast to analytical or query-processing purposes.
4. The difference between object-oriented and relational type databases.
5. The distinction between the semantic conceptual level discussions of entities or objects and the level of data encoding of objects.

For the first of these, as we have stressed the distinction between entity and field views enough already, suffice it to say that approaches and particular techniques are available for encompassing both in a practical sense. For example, lattices of points contrast with the discretization of tessellations or networks. This topic, primarily a question of spatial data representation, can be addressed in object-oriented databases by the definition of data types like polygon or image.

The second dimension, again a matter of practice as opposed to concept, points to the general architecture of systems for handling spatial data, comparing those which are implemented with a thematic map data layer model, as exemplified by ARC/INFO and several other systems, with those which have some features of object orientation, like SYSTEM 9 or APIC. Of course, the layered system deals with objects: the points, lines or polygons, or annotations or nets or triangulated cells, and these elementary level units are, on the one hand, those to which computing operations apply, and, on the other, the units with which users must work. Complex objects, defined as those new units produced from mixtures of others, are primarily virtual, although, it is possible, through polygon-overlay processing techniques, to create new units for permanent use. The non-layered system will define objects – simple, complex or compound – at the time of database creation, using data-structuring techniques to avoid spatial unit overlay processing.

Thirdly, some spatial information systems that are not implemented via object-oriented databases provide object-oriented capabilities for purposes of graphic display. That is to say, map design, a principal output tool for spatial information systems, may be effected by a map compilation toolbox allowing designation of sets of basic elements as graphic objects, with facilities for rescaling, repositioning and realignment. Such objects may be individual lines, map titles, key boxes, scale bars, sets of polygons, and ancillary information.

The fourth dimension, perhaps of most interest at this place in this book, contrasts the object-oriented databases with the more conventional relational databases, reflecting the principal contrast today for spatial information systems design. Adding to the brief comments earlier as to differences between record-oriented and object-oriented approaches, we point to the capability of the latter to recognize abstract data types, giving the users and programmers considerable flexibility; the additional scope provided by working with a network of classes of objects, as opposed merely to designating a membership of an entity set via an attribute in a relational table; the computational comprehensiveness implied by the encapsulation of procedures with objects, as well as the general ability to handle complex spatial objects and operations.

As a practical matter, there is more of a continuum of differences rather than a polarity between record- and object-oriented databases. Indeed, as noted elsewhere in this book, some of the shortcomings of the relational variety of record-oriented systems have been addressed, somewhat successfully, by using a bulk data type for variable length sets of data (for example, GEOVIEW and SYSTEM 9), and a higher level query language to overcome the awkwardnesses of the SQL. Some features of the object-oriented world, the abstract data types and classification, may now be found in extensions to the relational model, as represented by the POSTGRES model. Otherwise, hybrid systems exist, combining tabular data processing technology with newer object-oriented programming languages and data structures, as is the case with the TIGRIS and SMALLWORLD systems.

Finally, for the fifth dimension mentioned above, let us recall the earlier discussion in section 12.4 about semantic data models. Historically, the general ideas of entity-relation modelling were followed by the early semantic data models, leading to the object-oriented model. While entity-relation modelling is today generally recognized as the primary tool for data modelling, its initial restriction to entities, attributes, and relations is too limited for many purposes. Additional concepts like aggregation and generalization appear in the other varieties of semantic data modelling, and, in object oriented semantic modelling, all (conceptual) entities are modelled as (computer) objects, bearing in mind the sharing of attributes and operations among classes and/or instances. While semantic modelling has in the past emphasized the structural aspects of objects, especially class relationships, object-oriented modelling also treats the functions of those objects.

Even though developments have been occurring at the level of the computer operations associated with the encoded elements, it is nonetheless apparent that attention must be given at the conceptual level to the dynamic nature of real world objects. It is here, in the scientific fields, that further development is needed, if only to articulate phenomenologically the subject matter of a particular domain of study.

17.5 ARTIFICIAL INTELLIGENCE AND EXPERT SYSTEMS

In addition to the ideas from object-oriented databases and programming languages, the fields of artificial intelligence and expert systems also provide building blocks for the development of intelligent spatial information systems. In this preview, we first present the general nature of artificial intelligence and then some particulars as to definitions and components.

The ambition of the field of artificial intelligence, in a practical computer sense, is to simulate human reasoning in situations like voice recognition, automatic language translation, vision understanding, and reasoning by experts in particular fields of study. We can think of urban and regional planning activities, in a certain aspect, as a complex situation needing not only the manipulation of data but also any kind of information. So we can envisage the exploitation of artificial intelligence techniques in geomatics leading to a so-called intelligent spatial information system.

Expertise expression is perhaps the main success of the field of artificial intelligence via the development of tools called **expert systems**. These tools, computer programs designed to aid reasoning for particular purposes, are built on logics and comprise rules and facts governed by an inference engine, a set of procedures for undertaking some kind of reasoning.

Perhaps scores of expert systems have been built for various tasks ranging from cartographic production name placement to land area development zoning to finding areas of exploitable mineral resources. Some examples are cited by Fisher (1989) although not all involve spatial reasoning, an important part of intelligent spatial information systems.

17.5.1 Facts and rules

Facts can be defined as single events, such as the vehicle collision that occurred at the intersection of First Street and Broad Avenue, or single features, for example the value of an attribute of an entity. Other examples are: it is raining, Peter is three years old, the Eiffel Tower is 300 metres high, Moscow is the capital of the USSR.

Rules can be many kinds of statement that establish a regulation, a process, a method, a standard, a principle, a code of conduct, a law, a procedure, an ordnance. Generally speaking, they are written with IF–THEN statements such as:

IF a set of conditions occurs THEN a set of actions follows

or with a mathematical expression:

IF C_1 and C_2 and . . . and C_N THEN A_1 and A_2 and . . . and A_M

for which the semantic context can be stated as:

IF a context is realized THEN do so and so

For example:

IF it rains THEN take an umbrella

In this trivial example we have two specific facts: 'it rains', and 'take an umbrella', linked by a rule when the first fact, the condition, occurs. Now let us examine other examples more germane to spatial information systems.

In strategic urban planning, some examples of rules might be:

IF a zone is a marshland THEN prohibit construction

or

IF there is unemployment
THEN encourage business enterprise creation
AND create industrial areas

or

IF a parcel is close to an airport THEN limit building height

or

IF a parcel is near to a firestation THEN prohibit hospital construction

In urban renewal:

IF a building has a good architectural design
AND IF it is more than 100 years old
AND IF the building condition is mediocre
AND IF the owner does agree THEN suggest restoration

or

IF a building falls into ruins
AND IF nobody dwells in it THEN demolish it

or

IF a building has a poor architectural design
AND IF inner rooms are deteriorating
AND IF money is raised THEN suggest rehabilitation

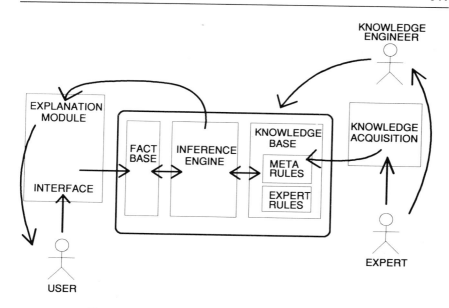

Figure 17.15 General structure of an expert system.

17.5.2 General structure of an expert system

In an expert system, knowledge is regrouped, often as scores of different pieces, in order to constitute a sort of 'program'. Expert 'rules' model behavioural or functional rules and 'facts' describe single values such as basic information or events. So, an expert system is an integration of a set of rules and a set of facts together with an inference engine (Figure 17.15). Notice the three parts of:

1. The core of the expert system itself.
2. A module for knowledge acquisition.
3. A module interfacing the core with the user.

Note too that three categories of persons are usually involved in the design and the use of the expert system: the user, the subject matter expert and a knowledge engineer. **Knowledge engineering** is a process of codifying human knowledge. The role of the knowledge engineer is to hold discussions with the expert in order to extract her or his knowledge. This phase of knowledge acquisition from experts is crucial. The knowledge engineer's art is to pose the right questions in order to understand what knowledge the experts are using and what their ways of reasoning are; and then to structure and encode that knowledge and logic.

When a new nugget of knowledge is integrated into the knowledge base, some verifications are performed in order to check whether this chunk is consistent with the knowledge already included. Indeed, a new piece should not contradict the previous knowledge otherwise some corrections must be performed. Often, for the user's benefit, the expert system is completed with an explanation module in order to help users to understand the result of their acquisition of information from that knowledge base. The core of the expert system is the inference engine linked with the base of facts and the base of knowledge comprised of metarules and expert rules (to be defined shortly).

17.5.3 Inference engine

The role of the inference engine is to deduce, starting from input facts, some other facts, either intermediate or final output, using the encoded rules. There are several methods of reasoning that can be used.

First we have **deduction** (*modus ponens*):

$$(P; P \rightarrow R) \Rightarrow R$$

meaning that if P is true, and that a rule $P \rightarrow R$ is also true, then we derive by deduction that R is true.

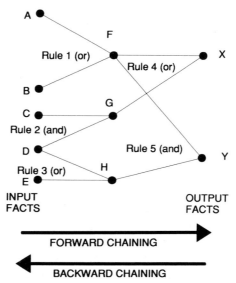

Figure 17.16 Diagram illustrating a set of facts linked with rules.

Secondly, there is **abduction** (*modus tollens*), also called reasonable explanation:

$$(R; P \rightarrow R) \Rightarrow P$$

meaning that if R is true, and we have a rule stating that $P \rightarrow R$, then we obtain by abduction that P is true.

When two facts are always concomitant, by **induction** we can affirm that there is a rule expressing a relationship between them:

$$(P; R) \Rightarrow (P \rightarrow R)$$

whose signification is, knowing that when P is true, R is also true, then we can derive (induce) a rule $P \rightarrow R$.

Fourthly, we have reasoning by **transitivity**:

$$(P \rightarrow Q; Q \rightarrow R) \Rightarrow (P \rightarrow R)$$

That is, if we have two different rules, the first implying Q and the second starting from Q, then by transitivity, a new rule can be produced, covering both P and R.

While expert systems may be based on these four types of logical reasoning, the more commonly used are sets of deductions, also called forward chaining, or sets of abductions, called backward chaining, by means of transitivity. **Forward chaining** is the way to test the consequences of some starting context. It can be mainly used for 'what-if?' reasoning. On the other hand, **backward chaining** is interesting for diagnosis, that is to discover the reasons generating the observed situation. For a toy example (Figure 17.16), the input facts are: A, B, C, D and E. Output facts are X and Y and the rules are:

Rule 1:   A or B → F
Rule 2:   C and D → G
Rule 3:   D or E → H
Rule 4:   F or G → X
Rule 5:   F and H → Y

In this example, which also illustrates the conventional use of a graph structure for showing rules, we use Boolean conditions with OR and AND operations. However, in the majority of expert systems, only AND operators, known as Horn clauses, are used.

For this same example, in the case of forward chaining we have information about input facts, and we want to deduce information concerning the output facts. Suppose we know that B and D are true, A and C are wrong, and E unknown. By applying rule 1, we deduce that F is true, and by rule 4 that X is true. Moreover, rule 3 implies that H is also true, and since F is true, then Y is also true. Suppose now that A is true, B, D and E are false, and C unknown, we can then deduce that F is true, G is unknown, and H is false. So X is true and Y is false.

Consider now that we impose values on the output facts, and that we are interested in knowing what the values of the input facts could be. In other words, we are looking for input values implying the observed output (diagnosis). As an example, suppose we look for a configuration so that X is true and Y is false. We can obtain (by abduction) immediately that:

F must be true
H must be false
G is indifferent

In this case, one of the solutions is:

A must be true
B and C must be indifferent
D and E must be false

In practical applications, sometimes a mixing of forward chaining and backward chaining is necessary. To elaborate a little on the rules, the previous example illustrates the case in which the rules are organized with a direct acyclic graph. Sometimes, though, loops between rules exist, causing the navigation of the expert system to be much more complex, and possibly producing some inconsistencies.

17.5.4 Metarules

Besides rules, an expert system can possess metarules. A metarule is a rule concerning knowledge, for instance, a procedure for selecting other rules. Often, strategies for solving problems are included in metarules. For example, IF such a condition occurs, THEN APPLY Rule 234 and Rule 456. When designing an expert system it is important to delineate metarules because their usage can accelerate logical inferences.

17.6 SPATIAL KNOWLEDGE REPRESENTATION

By spatial knowledge we mean chunks of knowledge in which the spatial component is important, especially via locators like coordinates, or place names. For instance 'in the United Kingdom, people drive on the left' or 'above 2,000 metres of elevation, trees give way to grass and then no vegetation', or 'along commercial strips used car dealers are found next to car repair shops' are pieces of spatial knowledge.

We now examine some specificities of spatial knowledge, covering the topics of spatial facts, spatial relations, spatial rules, fuzzy spatial knowledge, and spatial knowledge derived from logic rules and numeric formulae.

17.6.1 Spatial facts

Spatial facts are factual statements that include some spatial properties such as locations and shapes, dealt with initially in Chapter 3. Examples of such facts are: 'I am twenty miles from the Empire State Building', 'Sicily is at the toe of Italy', 'my handkerchief is in my pocket', 'the United States borders Canada', 'from here, I can see the Mont Blanc'.

For present purposes we consider there are three ways of expressing spatial facts:

1. By narrative description, as in, 'I am situated at 3545 King Street, second storey'.
2. By measurements such as, 'I am situated at 24°23′34″N and 56° 14′48″W, and 'my elevation is 345 metres under the sea level'.
3. Pictures and other images, that is visual description.

Visual description, an important subset of spatial knowledge, is dealt with in more detail in section 17.6.8.

17.6.2 Spatial relations

There are several ways in which separate spatial units may be related. Recalling some prior discussions in Chapter 4, for our purposes here we recognize (Figure 17.17):

1. Spatial relations oriented to an observer, for example object A is behind, and object B is in front of a point of reference.
2. Spatial relations linked to axes: object D is north of C.
3. Intrinsic relations between objects, such as object F is between objects E and G.

17.6.3 Spatial metarules

Generally speaking, using the areal spatial units as examples, spatial rules can apply to several zones and a zone can have several rules. Figure 17.18 depicts an example of relationships between zones and spatial rules: zone 1 has rule B; zone 2 has rules A and C; zone 3 has rules C and D, and zone 4 has rule B. And so one task is, given an (x, y) point, to ascertain the applicable spatial rules. This task first requires a point-in-polygon query to be solved; and then metarules can be defined:

IF (x, y) belongs to zone 1 THEN rule B is applicable
IF (x, y) belongs to zone 2 THEN rules A and C are applicable
IF (x, y) belongs to zone 3 THEN rules C and D are applicable
IF (x, y) belongs to zone 4 THEN rule B is applicable

Conversely, a **rule index** can be defined. With the Peano relational model, we can use:

PR-index (Peano_key, Side_length (rule)*)

In either case, rules may be valuable for undertaking integrity checking.

17.6.4 Fuzzy spatial knowledge

Spatial knowledge is often spatially fuzzy. By this term we mean that shape and location are not totally known. For describing fuzzy knowledge, we need some **membership degree** (Zimmermann *et al.*, 1984; Zimmermann, 1985), a statement of the probability of a piece of data meeting the criteria specified in a statement (Figure 17.19). Suppose we are asked to find the number of young people in the Left Bank of Paris. We can say that some city blocks belong 100 per cent to the Left Bank, another one only at 95 per cent, some in the middle at 50 per cent and so on. Similarly, for dealing with young people: at twenty years of age a person is definitely young (100 per cent); at thirty years, perhaps 60 per cent; at thirty-five years, possibly 5 per cent, and so on. Finally, by mixing these individual membership degrees, an answer can be found. In this case, the previous Peano relation becomes:

PR-fuzzy (Peano_key, (Object, Membership_degree)*)

17.6.5 Spatial knowledge from logical deduction

As an example, the following transitivity rule holds:

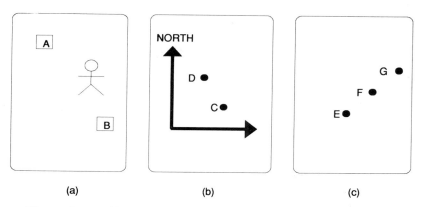

Figure 17.17 Different types of spatial relations. (a) Linked to an observer. (b) Linked to axes. (c) Intrinsic.

North (X, Y) and North (Y, Z) → North (X, Z)

Suppose, also, that we have the following facts:

North (A, B)   and   North (B, C)

Then we can deduce that:

North (A, C)

by using the transitivity rule, which states that North (X, Y) and North (Y, Z) → North (X, Z). However, if we have:

Neighbour (D, E)   and   Neighbour (E, F)

we cannot deduce that D is a neighbour to F.

Let us take another example:

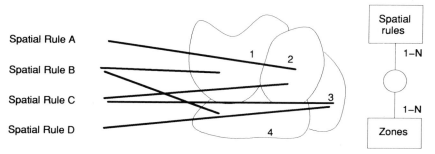

Figure 17.18 Correspondence between spatial rules and zones.

Between (J, I, K)
Position (J, 3, 2)
Position (K, 4, 3)

where the numbers are (x, y) coordinates, we can deduce:

North (K, J)

The role of deduction can be seen in another example, from cartography. An important problem in map design is the task of name placement. Indeed, the positioning of the names of cities, rivers, states, and so on must be done according to several rules such as avoiding having more than one name at the same place, and avoiding putting names on the top of objects, especially point features. The process of map compilation and design can possibly be enhanced by expert system techniques, provided that the rules can be well codified (Cook and Jones, 1990).

17.6.6 Spatial knowledge derived from numerical formulae

Suppose we have a set of triangles defined only by vertex coordinates and the query is to retrieve all triangles with a 30° angle. To answer this query we need to store some geometric or trigonometric rules, as follows.

Some data are stored in the database, where A, B, C are vertex identifiers:

R1 (Triangle-ID, A, B, C)
R2 (Point-ID, X, Y)

or not stored in the database, but are possible to access through rules,

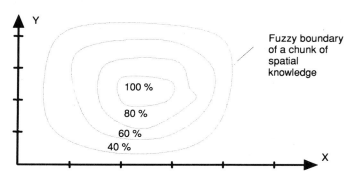

Figure 17.19 Fuzzy boundary of a chunk of spatial knowledge.

where a, b and c are segment lengths, and alpha, beta and gamma are angles, as illustrated in Figure 17.20:

R3 (Triangle-ID, a, b, c)
R4 (Triangle-ID, alpha, beta, gamma)

There are rules to compute segment lengths:

Rule 1: $a^2 = (X_B - X_C)^2 + (Y_B - Y_C)^2$

Rule 2: $b^2 = (X_C - X_A)^2 + (Y_C - Y_A)^2$

Rule 3: $c^2 = (X_A - X_B)^2 + (Y_A - Y_B)^2$

and rules to compute angles:

Rule 4: alpha $= \arccos\left[(b^2 + c^2 - a^2)/2bc\right]$

Rule 5: beta $= \arccos\left[(c^2 + a^2 - b^2)/2ca\right]$

Rule 6: gamma $= \arccos\left[(a^2 + b^2 - c^2)/2ab\right]$

These rules are sufficient if we need only the absolute value of the angles. Should we require the angle orientation, we have to use other trigonometric rules to derive sines. One example of this need is the application of the refraction law to determining the path of a route across different transportation media, as presented later in section 17.7.2.

17.6.7 Examples of spatial process representation

In this section we will present the xGEM2 system for spatial knowledge engineering, a method developed in Australia (Davis *et al.*, 1990). Other methods can be seen in Peuquet (1984) or Webster (1990). The xGEM2 system is based on production rules such as:

IF premises THEN conclusion

Or more precisely, formed with quadruplets:

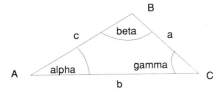

Figure 17.20 Example of a triangle and derivation of certain other features.

IF (parameter   relation   expression   spatial-expression)
THEN (parameter   relation   expression   spatial-expression)

The first term (parameter) corresponds to the name of an entity relevant to the problem, such as geology or vegetation cover. The second term (relation) gives the type of relation between the parameter and the expression, which is the third element, defined below, and the fourth one (spatial-expression) defines the current region of interest.

These terms are defined by the following constructs, depicted in conventional BNF notation in which ::= means a definition, possibly recursive, | means another possibility, and * indicates zero or more times:

relation	::=	'is' | 'is not' | 'is one of' | 'is not one of' | 'is between' | 'is not between' | 'is greater than' | 'is less than' | 'is greater than or equal to' | 'is less than or equal to'
expression	::=	parameter | value | list | arithmetic-expression
spatial-expression	::=	(spatial-selector (spatial-description)*) | NULL
spatial-selector	::=	for all | for any
spatial-description	::=	explicit-regionlist | implicit-regionlist | NULL
implicit-regionlist	::=	compass-direction explicit-regionlist | distance-expression explicit-regionlist | adjacency-relation explicit-regionlist
compass-direction	::=	'north' | 'south' | 'east' | 'west'
distance-expression	::=	'within' distance | 'further' distance
adjacency-relation	::=	'adjacent to'
explicit-regionlist	::=	region-name explicit-regionlist | NULL
distance	::=	real number
region-name	::=	'Sydney', 'Melbourne', etc.

The following examples of this xGEM2 formalism illustrate statements incorporating spatial information and consequently demonstrates that spatial inferencing is possible:

Item A	IF	fuel load is low for all regions to the north and within 3 km
	THEN	fire danger is not high
Item B	IF	fuel load is low for any region within 3 km
	THEN	fire danger is low for all within 5 km
Item C	IF	fire danger is known for any Region1, Region2, Region3
	THEN	fire danger is known for all Region1, Region2, Region3

Item D IF   fuel load is low for any region to the north and
                within 1 km
        THEN   fire danger is high for all regions adjacent to Region3

17.6.8 Visual knowledge encoding

As noted earlier, the extraction of spatial knowledge from visual descriptions is a very important endeavour. In simple terms, starting from image pixels, the aim is to detect pictorial objects and to check whether they can be recognized as spatial objects via pattern recognition techniques (Figure 17.21). Also, the field of robotics is full of spatial knowledge processing and visual image procesing procedures (Davis, 1986). In geomatics there are four main areas that need image understanding techniques:

1. The analysis of satellite images.
2. The analysis of aerial photographs.
3. The analysis of other photography, like landscape pictures.
4. The analysis of scanned maps.

While it is not the aim of this book to describe the way to address these different needs, in order to give just a taste, we mention the following steps of a process for recognizing geographic objects (Figure 17.22):

1. Image segmentation especially by texture and morphometric analysis in order to delineate pictorial objects.
2. Combining some pictorial objects by pattern recognition, and then confronting them with ideal spatial objects as stored in a library, so that pictorial and geographic information may be assimilated.
3. The result of this process, a set of some recognized objects and some residual pixels, the unrecognized pixels.

The task can be formulated in the following manner. Starting from a relation such as:

IMAGE (Pixel-ID, Red_level, Green_level, Blue_level)

Figure 17.21 Image understanding: from pictorial to spatial objects.

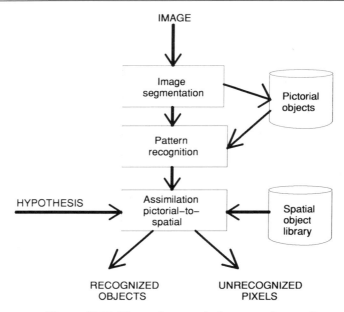

Figure 17.22 The main steps in image understanding.

we produce:

R1 (Object-ID, Locator)
R2 (Object-ID, Shape)
R3 (Object1-ID , Object2-ID, Relation_type)

where Relation_type specifies which, if any, kind of association there is between two specified objects, Object1 and Object2. The objects may be references to classes via identifier specification, or nothing at all, meaning any kind of object. Associations, in a spatial sense, may be such as given below, in which objects can be the same or different types:

R3	Object1	Object2	Relation_type	Type1/Type2
	345	678	Vicinity of	Zone/Zone
	278	804	Contained within	Zone/Zone
	301	345	North of	City/City
	451	901	Outside	Building/Road
	341	935	More than 1,000 m	Coastline/House
	136	730	Near	Farm/Farm
	276	778	Crossing	Road/Road

Notice that some hypotheses may be assumed in order to confront all the objects, and that there also is need for an hypothesizer/verifier procedure.

17.6.9 Examples in spatial knowledge engineering

In this section, we present two small examples in spatial knowledge engineering. The first concerns the discovery of which flowers might be found in different parts of the European Alps. The second is a decision to site a new hospital.

ALPINE FLOWER EXAMPLE

We have the following data, with a map representation as Figure 17.23.

ITEM A   The Gagea Fistulosa can be found on humid soils and near the chalets. It blooms in June.

ITEM B   The Epilobium Alsinifolium can be found near the mountain streams on decalcified soils. It blooms in July and August.

ITEM C   The Rumex Scutatus can be found in screes and in dry streams. It blooms in June and July.

ITEM D   The Ranunculus Aconitifolius can be found in humid and rich meadows. It blooms in May and June.

ITEM E   The Orchis Ustulata can be found in humid and meagre meadows. It blooms in May and June.

We can encode this knowledge as the following facts:

IF soil = humid   THEN flower = gagea
IF near_chalet   THEN flower = gagea

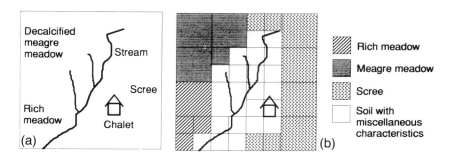

Figure 17.23 Example of soil characteristics and flower location. (a) Geographical data. (b) Quadtree representation.

IF near_stream   THEN flower = epilobium
IF soil = decalcified   THEN flower = epilobium
IF soil = scree   THEN flower = rumex
IF dry_stream   THEN flower = rumex
IF soil = rich_meadow   THEN flower = ranunculus
IF soil = meagre_meadow THEN flower = orchis

We also have a lot of places, defined as polygons with sets of points for the boundaries:

Zone (1) = polygon (1, {point1})
Zone (2) = polygon (2, {point2})
Zone (3) = polygon (3, {point3})

where the symbols { } denote a set. For these zones we have the type of soils:

IF zone1   THEN soil = meagre_meadow
IF zone2   THEN soil = rich_meadow
IF zone3   THEN soil = scree

We have also some other locations defined by distances:

Near_chalet (C) = polygon (ID-C, {p} : dist (p, C) < 100 metres)
Near_stream (S) = polygon (ID-S, {p} : dist (p, S) < 100 metres)

where C and S refer respectively to the chalet and stream.; and ID-C and ID-S are respectively the identifiers of the polygon, the inner points of which satisfy the criterion. We use a geometric rule transforming a zone into a linear quadtree:

IF polygon (N, {point})
    THEN quadtree (N, {Peano_key, Side_length})

where N is, as above, the identifier number for the polygon. By applying all these rules, we finally obtain:

IF quadtree (A, {Peano_key, Side_length})
    THEN flower = gagea
IF quadtree (B, {Peano_key, Side_length})
    THEN flower = gagea
IF quadtree (C, {Peano_key, Side_length})
    THEN flower = epibolium
IF quadtree (C, {Peano_key, Side_length})
    THEN flower = epibolium
IF quadtree (D, {Peano_key, Side_length})
    THEN flower = epibolium

IF quadtree (E, {Peano_key, Side_length})
THEN flower = rumex
IF quadtree (F, {Peano_key, Side_length})
THEN flower = ranunculus
IF quadtree (G, {Peano_key, Side_length})
THEN flower = orchis

Since in the same location, several flowers can be found, we can reverse the results by creating a sort of spatial index (Chapter 15):

IF Spatial_index ({Peano_key, Side_length})
THEN flower = {flowers}

Now suppose that we have a spatial query (point-in-polygon, region or whatsoever). We transform the zone into a quadtree and the result is given in picking a good Peano key in this index, that is the spatial or Peano join. Note that we can use these rules conversely, that is starting from existing flowers in the soil, to find the soil features. Also the rules may be extended to include the temporal aspects, based on the data provided for the month of blooming.

NEW HOSPITAL LOCATION EXAMPLE

In this second example we treat the category of location problems known as location–allocation, as discussed initially in section 2.4. For the particular case of deciding upon a satisfactory location for a new hospital (Figure 17.24) we have information such as:

ITEM 1   The potential demand for hospitals, as measured by the

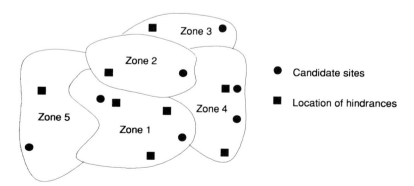

Figure 17.24 Information necessary for a new hospital location.

number of people accumulated to zones, and assigned to the centroids of the zones:

R1 (Zone-ID, Population, X_zone, Y_zone)

ITEM 2   The location of some places that are candidates for the site of the hospital. The alternative locations are characterized by land availability, land purchase and development costs, neighbourhood character and neighbourhood existing amenities:

R2 (Site-ID, X_site, Y_site, (Features)*)

ITEM 3   The accessibility from zones to sites, measured as both kilometric and temporal distance:

R3 (Site-ID, Zone-ID, Km_distance)
R4 (Site-ID, Zone-ID, Time_distance)

ITEM 4   However, there are some places in which a hospital is not welcome or does not fit well, or is prohibited by local building or zoning regulations, perhaps to avoid being close to a fire station or factory because of the noise associated with those kinds of land use. Moreover, these hindrances may be evaluated by degree or level of inutility:

R5 (Hindrance-ID, X_hinder, Y_hinder, Hindrance_type, Hindrance_degree)

Beginning with these pieces of knowledge, imagine performing a spatial reasoning type of solution similar to that done conventionally and based on operations research techniques. Perhaps the most important aspect to this reasoning approach is the selection of several weighting functions, the influence of which will confer different desirability values on the different candidate sites.

The reasoning steps consist of:

1. An initial weight function must be selected for creating an accessibility potential index built upon the individual distances for the population zones:

A1 (Site-ID, Accessibility_potential_index)

2. For each site, starting from hindrances, we can compute a special statistic recognizing the hindrance degree and the inverse of the distance between the site and the hindrance locations:

A2 (Site-ID, Hindrance_potential_index)

3. In parallel, we can confer a measure of utility on the individual candidate sites based on their features, and the indicators of accessibility and hindrances:

   A3 (Site-ID, Utility_weight)

4. We now have the most important candidate as that with the highest utility index. Alternative solutions can be created by using different weightings or objective criteria, such as minimizing the average travel distance.

17.7 SPATIAL REASONING IN SPATIAL INFORMATION SYSTEMS

In our everyday life we have to perform some spatial reasoning. Walking, moving, eating, driving are some of them. The situation in spatial reasoning is typified by the 'piano-mover' task in which a complex object has to be moved through narrow corridors.

In spatial information systems reasoning is important for many geomatic tasks; there is much to do beyond data retrieval, manipulation and mapping. More sophisticated computing tools are needed for tasks such as:

Site selection for commercial or industrial establishments
Urban area housing permit and re-zoning decisions
Geomarketing needs like demarcating best sales territories
Path-finding
Emergency vehicle dispatching
Delivery truck routing and scheduling
Hydrology research like modelling chemical discharges
Understanding climatological processes
Hazard prevention
Pollution fighting
Disease distribution studies
Inferring the geometric shape of strata from borings
Autonomous land vehicle driving
Name placement on maps
Layout design or arrangement of premises
Proximity to services design
Crime prediction
The 'where am I?' problem

For these decision making activities, geomatic reasoning capabilities are necessary. But what exactly is reasoning? What is the difference with problem-solving methodologies? What are the specificities for this kind of reasoning? The aim of this section is to provide some initial enlightenment on a difficult topic. For instance, consider some specifications for the 'where am I?' problem. Imagine that a parachutist, with a spatial knowledge base in his pocket, seeing three oak trees and a river, infers from those pieces of information that he is near Newton City.

First of all, we explain what the learning possibilities are. Then, the two main types of reasoning, the logico-deductive and spatial, will be examined. We conclude by briefly addressing the implications to intelligent spatial information systems software design.

17.7.1 Learning possibilities

Learning seems to be the key element of reasoning. Indeed, we can reason from our experiences, or, more exactly, from the knowledge originating from our experience. Suppose we wish to go from the Eiffel Tower in Paris, France to the Parthenon in Athens, Greece. In order to find the path, it would be nice to have *a priori* knowledge such as:

1. If we want to walk, we should take seas and mountains into account.
2. If we intend to sail, we should first go to the nearest harbour.
3. We should avoid Albania because crossing it is very complex.
4. If we drive, we should take Italian motorways because they are very efficient.
5. If we drive, we can use a ferryboat from Brindisi to Patras.
6. If we fly, we need to go to the airport that has direct flights.

Indeed, if we do not have this knowledge acquired from empirical evidence from books or friends or previous travels from France to Greece, any algorithm will consider all roads in all countries to be crossed and all streets in all cities to be passed through. In other words, any conventional procedure will examine all streets issuing from the Eiffel Tower place and examine all possibilities, checking whether the Greek Parthenon is on the Paris Champs-Elysées, and so on.

Turning to computer systems, two kinds of knowledge can be introduced: prior empirical evidence or knowledge issued from previous task completions. Algorithms designed in a good learning system must have the capability of incorporating any kind of knowledge in order to modify their behaviour accordingly. Let us mention that very few systems in artificial intelligence incorporate efficient learning capabilities.

17.7.2 Logico-deductive and spatial reasoning

Two kinds of reasoning must be considered, logico-deductive and spatial reasoning. Logico-deductive reasoning uses true information such as in expert systems: there is a route from A to B; there is a route from B to C; there is a route from C to D. From these facts we can deduce, by transitivity and forward chaining, that there is a route from A to D. But this route is not necessarily the best one according to several constraints or some performance factors like travel time, distance or costs. For spatial information systems, we do not think that the utilization of logico-deductive reasoning is sufficient to solve real problems. However, while in some special circumstances it can be useful, solutions to real geomatic problems need overall spatial reasoning capabilities.

Spatial reasoning (Woodwark, 1989) may encompass not only the use of computational geometry but also the use of some other mathematical theories or practices in fields like operations research or complex computations, in order to find a solution among an infinity of possibilities. In the field of geomatics, two possibilities of spatial reasoning are emerging: topological and pure geometrical reasoning. Pure geometrical reasoning, though, often implies the division of the space in order to scan every possibility. For that, space-filling curves can be useful (Chapters 4 and 13).

As an example, let us examine the selection of paths between two points A and B for various circumstances and contexts. Recalling from section 2.4.1 that paths can be of many kinds, including: within networks, for hierarchized graphs, in the sense of a travelling salesman, or across terrains. Here the problem is how to solve by reasoning more complex situations like multimodal paths and paths within a hostile environment (Mitchell, 1988).

First of all, for a **multimodal path** (Figure 17.25), it is often the case that a path must be divided into many components with different characteristics or involving various means of travel (for instance, running and swimming). In Figure 17.26 we demonstrate a situation in which someone has to choose where to dive in order to reach a goal as rapidly as possible, say an island or even a drowning person. In this case, a problem known as the path within differently weighted regions, it can be shown that the optimum route is similar to the path of light as expressed in the law of refraction, otherwise known as the Fermat principle or the Snell (or Descartes) law, as demonstrated in the case of shipments across land and sea media.

Another possibility is to use different modes of transport physically restricted to particular channels and expressed as the choice of a path

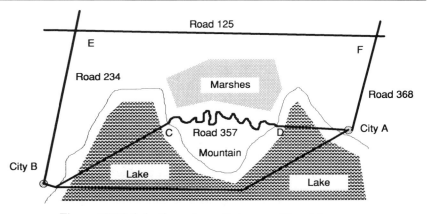

Figure 17.25 Selection of a path in a multimodal context.

within a topological graph. Another situation is to try to find paths in a hostile environment, that is from a starting point to an objective with the minimum concealment time *vis-à-vis* an enemy or an observer (Figure 17.27).

17.7.3 Example of districting

Among the many facets of spatial reasoning, districting is an important real world task. We use the term zoning for the division of a continuous space of land into a tessellation, and the term **districting** to refer to any regrouping of basic zones into another tessellation. Generally speaking, districting is undertaken for a special purpose, according to certain rules and criteria to be optimized.

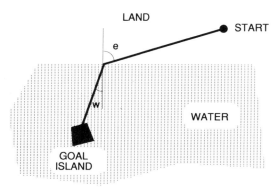

Figure 17.26 A path within differently weighted regions.

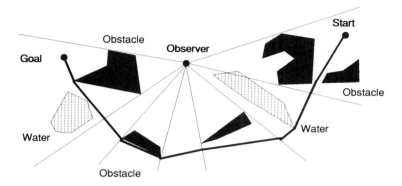

Figure 17.27 Finding a path through an hostile environment (after Mitchell, 1988).

Drawing boundaries is an element common to the situations of a company splitting some territory into sectors for management or sales purposes, of a country modifying its electoral precincts, or a municipality creating new school regions. Among the different instances, we examine only the case of electoral districting. Boundaries are redrawn every so often in order to match more closely population distributions. In any event, as discussed earlier in section 8.5, redistricting encompasses several spatial properties.

We set out now the nature of a spatial reasoning approach to the task of districting an hypothetical state with three hundred elementary zones into ten electoral districts in order to elect ten representatives. For each zone we have data for the number of registered voters, the number of votes cast in prior elections for each party, and the geometric description (Figure 17.28a). We utilize a process beginning with some seed zones, adding neighbours to form the districts and stopping the process when some conditions are met. Assuming that the protection of incumbents is important (to reduce their competition one against another), then we will initiate the process from the zones of incumbent representatives, or a spatially random selection if there is no incumbent.

Starting from this basis, a spatial reasoning mechanism can be invoked to create different sets of districts. These alternative plans will be evaluated against three kinds of criterion. First of all, we employ a population fairness criterion. For this element we state that the population of all districts should be approximately the same, possibly within a range of 50,000 plus or minus 5 per cent. Secondly, we require district shapes not to be suspiciously strange. By shape fairness we mean that the districts shall not have holes or isolated pieces, and no long

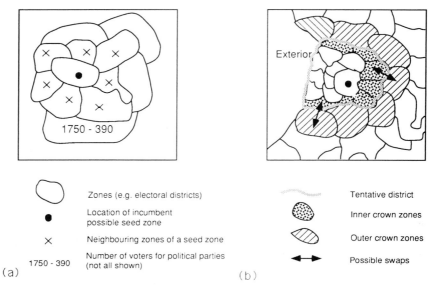

(a)

Zones (e.g. electoral districts)

● Location of incumbent possible seed zone

× Neighbouring zones of a seed zone

1750 - 390 Number of voters for political parties (not all shown)

(b)

Tentative district

Inner crown zones

Outer crown zones

Possible swaps

Figure 17.28 An example for illustrating the districting reasoning procedure. (a) Map emphasizing a seed zone and its neighbours. (b) Map presenting the inner crown zones and the outer crown zones of a possible district.

tentacles. Thirdly, we require political fairness, meaning that no political party is drastically over- or under-represented. We define the following:

Z is the number of zones
D is the number of districts
P is the number of political parties
VOTERS is the total number of voters in the state
VOTERS (z, p) is the total number of voters in the zone z for the party p
VOTERS (d, p) is the total number of voters in the district d for the party p

The **population fairness** criterion, for one example, can be based on the standard deviation of the number of voters per district:

$$CPOP = \Sigma \, [\bar{D} \, VOTERS - \Sigma \, VOTERS \, (d, p)]^2$$

For the criterion of **shape fairness** we can optimize any index which favours the shape of a circle. Among several, we use a simple measure comparing the ratio of the squared perimeter to area, which is at a minimum for a circle. Calling the perimeter and area for a district d respectively $L(d)$ and $A(d)$, we try to minimize the following statistic:

$$CSHAPE = \Sigma \; [L^2(d)/A(d)]$$

Regarding the **political fairness** criterion, one possibility is to aim for a result in which for each political party the ratio of elected representatives to the total number will be approximately the same as the proportion of voters by party preference. So we define $REPR(d, p)$ as an integer variable, having the value of 1 when the result is that the district gets a representative of the party p, and 0 in the other case. The political fairness criterion may be measured by the difference (or ratio) between the percentage of people voting for each party and the percentage of elected representatives for each party.

We start with two data structures, the first for zones with all necessary data for geometry, voters and political party preference, and a list of neighbouring zones in order to facilitate aggregation. The other data will be the political party and location of the incumbent representative. So we have:

STARTl (Incumbent-ID, (Segment-ID)*, (Party, Votes)*,
  (Neighbour_zone-ID)*)
START2 (Incumbent-ID, Zone-ID, Party)

The result of the districting process will consist of similar relations for each alternative plan produced, except that for a district we must also record the list of member zones:

RESULT1 (Plan-ID, District-ID, (Segment-ID)*, (Party, Votes)*,
  Zone-ID)*)
RESULT2 (Plan-ID, Incumbent-ID, Party, District-ID)

In order to facilitate reasoning we need also to keep a record of the list of zones which can easily be exchanged with other zones in other districts:

T1 (Plan-ID, District-ID, (Inner_crown_zone-ID)*,
 (Outer_crown_zone-ID)*)

That is, with reference to Figure 17.28b, we recognize an **inner crown**, a set of zones bordering a currently active grouping as legitimate candidates to be added. There also exists a set of zones bordering a completed district, called here the **outer crown**. Such zones are candidates for swapping with one or more inner crown zones as adjustments are made to a district to make its population total closer to the desired number. While the map shows the possibility of swapping neighbours in the inner and outer crowns, the process can be set up to allow swapping of non-adjacent zones.

With regard to the rules utilized in the redistricting process, we describe only a few in order to convey a flavour of what is possible, without over-simplifying a complex spatial problem. We can distinguish several kinds of rule:

1. Starting rules in order to assign seed zones.
2. Stopping rules, in essence invoked by the comparison of results with the criteria.
3. Rules for integrating a new zone into a district, and for exchanging zones between districts.

The starting rules will assign seed zones taking incumbent legislator location into account. If the number of incumbents is different from the number of districts, in one case a random assignment can be used, and, in the other, incumbents can be eliminated randomly.

Rule 1 IF   the number of incumbents = the number of zones
        THEN   assign Incumbent_location_zones to Seed_zones
Rule 2 IF   the number of incumbents exceeds the number of zones
        THEN   eliminate randomly some incumbents (*sic*!) and apply rule 1
Rule 3 IF   the number of incumbents is greater than the number of zones
        THEN   select one or more extra seed zones randomly and apply rule 1.

At this step, each zone has a seed, so the process of aggregation can be invoked. The idea is to visit the list of the outer crown zones in order to select a zone to be included in the district being created.

Rule 4 IF   the district number of voters exceeds the target number
        THEN   choose randomly a zone from the outer crown list
Rule 5 IF   the zone assigned to the district is effected
        THEN   compute the number of voters for each party, modify the boundaries, and modify the inner crown and outer crown

Whenever a district has too many people, similar rules can be invoked in order to eliminate a zone from the inner crown (rules 6 and 7). We could possibly exchange zones, especially when two districts are neighbours, one with minimum population, and one with maximum population allowed:

> Rule 8 IF two districts are neighbours AND a district is under-populated and the other one over-populated
>
> THEN exchange a boundary zone sufficient to rectify the population imbalance

At this stage, the criteria of population fairness is respected, and all zones are assigned to a specific district. Now we need to examine districts for their shape, either for the existence of holes or for the existence of elongated shapes. The latter can be quantified as an excessive shape index value; the former can be ascertained by identifying one or more zones surrounded by zones all of which are in a different district.

> Rule 9 IF a district has tentacles
>
> THEN look for the farthest zone and attach it to its neighbouring district with fewer voters
>
> Rule 10 IF a zone is surrounded by a different district
>
> THEN exchange it with one or more zones to remove the enclave

For optimizing the political fairness criterion, we must determine globally the state of under- or over-representation for one or more political parties. So, we can write a rule similar to rule 8:

> Rule 11 IF two districts are neighbours AND a district is under-represented AND the other over-represented
>
> THEN exchange boundary zones to rectify the political imbalance

Thus, when a solution is reached, it corresponds to a new plan, which can then be compared with alternatives in order to make a selection. However, in the practical world of districting, some criteria can be very difficult to codify using a given artificial intelligence programming language, especially PROLOG, for implementing computational geometry and operations research algorithms.

17.8 SUMMARY

Looking back to Figure 2.17, which set out major orientations of spatial information systems, we have, in these last two chapters, presented a framework for the current developments of real products that deal with information browsing and use in decision-making contexts. Notwithstanding the need to explore further how people think spatially, we are now moving close to the creation of decision support systems that incorporate some spatial reasoning capabilities involving identifiable objects and explicit rules for somewhat well-structured problems.

It is premature to talk about practical experiences regarding intelligent spatial information systems for at this time, for very good reasons, there are very few object oriented spatial databases. In this regard, though, we can mention that users or prospective users of object-oriented tools are faced with two choices in building their databases. On the one hand, they can create from scratch, dealing painstakingly and systematically with each and every instance and object in detail. On the other hand, building from existing sets of data, possibly scattered across many unconnected computer files, is also difficult, especially if some measure of consistency is to be conveyed to the new database. Even so, this second option seems to be the more likely because many organizations already hold much geographical information in a computerized form.

The evolution from the line printer, pixel oriented, single theme mapping of the late 1960s to the current excitement about the development of comprehensive information systems incorporating reasoning capabilities has been short in time but great in effort, involving people in many fields and disciplines. At this point in the evolution of sophisticated computer-based tools we can but say that the cross-fertilization of work in database systems, computer hardware, semantic data modelling, programming languages and scientific knowledge acquisition is leading to the creation of software systems and databases that are object oriented, include hypermap ideas, have person-compatible interfaces, and provide spatial reasoning facilities.

We look towards the realization of an object-oriented, intelligent, dynamic, multimedia spatial database concept, combining the power of hypermaps with the efficiency of database management system concepts. We look for high level tools for knowledge discovery, data analysis and display, multimedia management, data integrity, and support for a great range of spatially oriented decisions. We think a user should be able to operate directly on objects of his own choosing and combination, moving freely across different semantic resolutions and spatial scales.

On the technical side, encoding spatial knowledge is difficult because logic programming and computational geometry must be combined. Another difficulty is that the use of the spatial facts needed to describe spatial objects implies the employment of intensional rather than extensional representation. On the knowledge side, the formal codifying of spatial knowledge in fields dealing with spatial concepts, or diverse phenomena, is in a sense only just beginning, notwithstanding centuries of contributions by many productive people. Perhaps soon we shall be able to say that hypermaps, multimedia and automated spatial reasoning will have revolutionized people's views of the spaces and worlds in which they live.

17.9 BIBLIOGRAPHY

An extensive conceptual overview of intelligent databases, including hypermedia and object oriented varieties, is that of Parsaye *et al*. More technical coverages of object oriented systems are provided by Kim and Lochovsky, and Zdonik and Maier.

Abel, David J. 1989. SIRO-DBMS: a database tool-kit for geographic information systems. *International Journal of Geographical Information Systems* 3(2): 103–111.

Abiteboul, Serge and R. Hull. 1987. IFO: a formal semantic database model. *ACM Transactions on Database Systems* 12: 525–533.

Anthony, R. and P. J. Emmermann. 1985. Spatial reasoning and knowledge representation. In Bruce K. Opitz (ed.). *Geographic Information Systems in Government, Proceedings of a Symposium, Springfield, Virginia, USA*. Hampton, Virginia, USA: Deepak Publishing, pp. 795–813.

Barr, Avron, E. Felgenbaum and P. Cohen. 1982. *The Handbook of Artificial Intelligence*. Reading, Massachusetts, USA: Addison-Wesley.

Bennis, Karima *et al*. 1990. GéoTROPICS: database support alternatives for geographic applications. *Proceedings of the Fourth International Symposium on Spatial Data Handling, Zurich*, Switzerland, pp. 599–610.

Bretl, R. *et al*. 1989. The GemStone data management system. In W. Kim and F. H. Lochovsky (eds). *op. cit.*, pp. 283–308.

Cook, Anthony C. and Christopher B. Jones. 1990. A PROLOG interface to a cartographic database for name placement. *Proceedings of the Fourth International Symposium on Spatial Data Handling, Zurich*, pp. 701–710.

Davis, Ernest. 1986. *Representing and Acquiring Geographic Knowledge, San Mateo*, California, USA: Morgan Kaufmann.

Davis, John R., P. Whigham and I. W. Grant. 1988. Representing and applying knowledge about spatial processes in environment management. *Artificial Intelligence Applications*, vol. 2. Reprinted in Donna J. Peuquet and Duane F. Marble (eds). *Introductory Readings in GIS*. London, UK: Taylor and Francis.

Deux, O. 1990. The story of O_2. *IEEE Transactions on Knowledge and Data Engineering* 2(1): 91–108.

Dittrich, K. and Dayal U. (eds). 1986. *Proceedings of the International Workshop on Object-oriented Database Systems, Asilomar, California*. New York, New York, USA: IEEE Computer Society Press.

Elmakhchouni, Mohamed and Robert Laurini. 1989. An expert system for the visual simulation of land use built forms. *Sistemi Urbani* 11(1): 47–58.

Fisher, Peter R. F. 1989. Expert system applications in geography. *Area* 21(3): 279–287.

Fisher, R. B. 1989. *From Surfaces to Objects. Computer Vision and Three Dimensional Scene Analysis*. New York: Wiley.

Fishman, D. H. *et al.* 1989. Overview of the IRIS DBMS. In W. Kim and F. H. Lochovsky (eds), *op. cit.*, pp. 219–250.

Gahegan, Mark N. and Stuart A. Roberts. 1988. An intelligent, object-oriented geographical information system. *International Journal of Geographical Information Systems* 2(2): 101–110.

Harel, David. 1988. On visual formalisms. *Communications of the ACM* 31(5): 514–530.

Herring, John R. 1987. TIGRIS: topologically integrated geographic information system. *Proceedings of the Auto Carto 8 Conference, Baltimore*, Maryland, USA, pp. 282–291.

Herring, John R. 1990. TIGRIS: a data model for an object-oriented geographic information system. Paper presented at the GIS Design Models and Functionality Conference, Leicester University, Leicester, UK.

Hopkins, Lewis D. and Douglas M. Johnston. 1990. Locating spatially complex activities with symbolic reasoning: an object-oriented approach. *Proceedings of the Fourth International Symposium on Spatial Data Handling, Zurich*, Switzerland, pp. 762–771.

Hull, Richard and Roger King. 1987. Semantic database modeling: survey, applications, and research issues. *ACM Computing Surveys* 19(3): 201–260.

Kemp, Zarine. 1990. An object-oriented model for spatial data. *Proceedings of the Fourth International Symposium on Spatial Data Handling, Zurich*, Switzerland, pp. 659–668.

Kierne, Daniel and Kenneth J. Dueker. 1990. Modeling cadastral spatial relationships using Smalltalk-80. *Urban and Regional Information Systems Association Journal* 2(1): 26–37.

Kim, W. and F. H. Lochovsky (eds). 1989. *Object-oriented Concepts, Databases and Applications*. New York: ACM.

Kim, W. *et al.* 1989. Features of the ORION object-oriented system. In W. Kim and F. H. Lochovsky (eds), *op. cit.*, pp. 250–282.

Laurini, Robert. 1988. Expert systems and image generation for urban planning: the case of French land use plans. In Maria Giaoutzi and Peter Nijkamp (eds), *Informatics and Regional Development*: Avebury, UK: pp. 279–291.

Laurini, Robert. 1989. Introduction to expert systems for town planning. *Sistemi Urbani* 11(1): 3–5.

Laurini, Robert. 1990. My city is object-oriented. Paper presented at the Israeli Conference on Urban Data Management, Jerusalem, Israel.

Laurini, Robert and Françoise Milleret-Raffort. 1990. Towards intelligent GIS. *Proceedings of the Urban and Regional Information Spatial Analysis/Network for Education and Training Seminar, Patras, Greece*, pp. 127–156.

Manola, Frank, Jack Orenstein and Umeshwar Dayal. 1987. Geographic information processing in the PROBE database system. *Proceedings of the Auto Carto 8 Conference*, Baltimore, Maryland, USA, pp. 316–326.

Meynowitz, Norman K. 1986. Intermedia: the architecture and construction of an object-oriented hypermedia system and applications framework. *Proceedings of the 1986 Conference on Object-oriented Programming Systems, Languages, and Applications*, pp. 186–201.

Mitchell, J. S. B. 1988. An algorithmic approach to some problems in terrain navigation. *Artificial Intelligence* 37: 171–201.

Naqvi, Shamin and S. Tsur. 1989. *A Logical Language for Data and Knowledge Bases*. New York, New York, USA: Computer Sciences Press.

Orenstein, Jack A. 1986. Spatial query processing in an object-oriented database system. *Proceedings of the ACM, Special Interest Group, Management of Data*, Washington, DC, USA, pp. 326–336.

Orenstein, Jack A. 1989. An object-oriented approach to spatial data processing. *Proceedings of the Fourth International Symposium on Spatial Data Handling*, Zurich, Switzerland, pp. 669–678.

Parsaye, Kamran *et al.* 1989. *Intelligent Databases, Object-oriented, Deductive and Hypermedia Technologies*. New York, New York, USA: Wiley.

Peuquet, Donna J. 1984. Data structures for a knowledge-based geographic information system. *Proceedings of the Fourth International Symposium on Spatial Data Handling*, Zurich, Switzerland, pp. 372–391.

Rosenfeld, Azriel and Avinash C. Kak. 1982. *Digital Image Processing*, 2 volumes. New York, New York, USA: Academic Press.

Smith, Terence, Donna Peuquet, Sudhakar Menon and Pankaj, Agorwal. 1987. KBGIS-II: A knowledge-based geographical information system. *International Journal of Geographical Information Systems* 1(2): 149–172.

Teorey, Toby J., Dongoing Yang and James P. Fry. 1986. A logical design methodology for relational databases using the extended entity-relationship model. *ACM Computing Surveys* 10(2): 197–222.

Thompson, Derek *et al.* 1983. *The Maryland Reapportionment Information System*. University of Maryland, College Park, Maryland, USA.

Webster, Chris. 1990. Rule-based spatial search. *International Journal of Geographical Information Systems* 4(3): 241–259.

Woodwark, John (ed.). 1989. *Geometric Reasoning*. Oxford, UK: Clarendon Press.

Worboys, Michael F., Hilary M. Hearnshaw and David J. Maguire. 1990. Object-oriented data modelling for spatial databases. *International Journal of Geographical Information Systems* 4(4): 369–383.

Worboys, Michael F.. Hilary M. Hearnshaw and David J. Maguire. 1990. Object-oriented data and query modelling for geographical information systems. *Proceedings of the Fourth International Symposium on Spatial Data Handling*, Zurich, Switzerland, pp. 679–688.

Zdonik, Stanley B. and David Maier (eds). 1989. *Readings in Object-oriented Data Base Systems*. San Mateo, California, USA: Morgan Kaufmann.

Zimmerman, H. J. 1985. *Fuzzy Sets Theory and Its Application*, International Series in Management Science Operations Research. Dordrecht, The Netherlands: Kluwer Academic.

Zimmerman, H. J. *et al.* (eds). 1984. *Fuzzy Sets and Decision Analysis*. Amsterdam, The Netherlands: North-Holland.

Afterword

Over time, we have successively passed from automated cartography, (computerization of mapping and drawings) to structuring spatial data. This book is a dated survey of what is possible now in the domain of organizing and manipulating spatial information. We think that the next steps will be to invent a sort of second generation set of tools for the treatment of geographic information. As explained, the toolbox will include more intelligent capabilities and will pass from structuring spatial data to assisting the decision-making process based on territorial information. However, here the adjective 'intelligent' has a narrow meaning and spatial information systems of the future will surely integrate other features.

First of all, we identify the third dimension as most important. By this we mean that our actual spatial information systems, reflecting their origin in a map model, are currently very much two-dimensionally oriented. This bias is also fostered by the use of two-dimensional computer screens. Actual spatial information system products offer only some limited perspective view possibilities. However, in several domains, as in landscape design, geological studies, and automated land (or sky) vehicle control, we really need to deal with the third dimension. For instance, in designing a new ski resort, we need to display skiing areas, ski tracks, lifts and so on, including their three-dimensional features.

Secondly, the impact of the time dimension is an important component for developing more comprehensive systems. Two kinds of approach would be possible: static cartography and dynamic cartography. By the latter, we mean the combination of computer graphics animation with mapping. At the moment there are some software products that allow the display of temporal evolution of geographic variables, the growth of some zoning schemes and so on, but nothing really in the spirit of animation. Continuing with the example of the ski resort, it would be interesting to visualize the behaviour of skiers in the resort, to see where they go, how and for what reason they gather together, and the locations of any congested areas.

Thirdly, we expect to see developments in the near future in spatial simulation. By computer simulation, we mean the necessity of combining tools to allow modelling of different types of behaviour linked to different scenarios of development. For comparing those solutions, several criteria must hold. Again using the skiing example, different variables will be simulated, studied and compared, for instance defining the scheduling of skichairs and determining the number of people they hold in order to study the length of time for waiting in queues; proposals for the placement of new ski tracks and new lifts, and so on.

Yet the most important feature is the possibility of providing reasoning capabilities. Existing software systems are modest in this regard. The tools offered by artificial intelligence specialists are too much alpha-numerically oriented, whereas spatial knowledge, as appears in mental maps, vision data, landscape design, urban and regional planning, spatial analyses and geographical studies, is much more topologically and geometrically oriented. In order to have real spatial reasoning tools, cross-fertilization of computing and spatially-oriented disciplines with cognitive science and linguistics is also necessary.

Taking our skiing example for the last time, starting from mountain topography and using language and some expert rules, it would be interesting to help the resort's landscape designer by providing some reasoning capabilities with a sort of geometry based inference engine, for instance regularity of the ski track gradients, rules for the placement of lifts and so forth.

We think that a hypermap style is one future scenario for spatial information systems. Recalling that we presented some elements for cartographic hyperdocuments, the feeling is that we need to develop this kind of tool in order to know more about the complexity of combining geographical, thematic and multimedia navigation. For the skiing example, suppose during the simulated animation, we can click on a selected skilift in order to get a window, either presenting all these static and dynamic characteristics, or displaying the technical drawings showing a particular part or a textual contract about it.

So we now come to the point that spatial information systems are more that just computer based technical instruments. They are resources for helping people to learn and to make decisions. Indeed, we think that the understanding of technical matters is very much ahead of our under-standing of spatial data mental processing by people. So the spatial information systems community has a lot more work to do in order to understand the world and master it for the benefit of everyone. We hope that the readers, whatever their discipline, now have a more realistic view of what is possible regarding spatial information, and have some hints allowing them to plan better and dream for the future.

Index